Current Topics in Membranes, Volume 62

The Eye's Aqueous Humor

Current Topics in Membranes, Volume 62

Series Editors

Dale J. Benos
Department of Physiology and Biophysics
University of Alabama
Birmingham, Alabama

Sidney A. Simon
Department of Neurobiology
Duke University Medical Centre
Durham, North Carolina

Current Topics in Membranes, Volume 62

The Eye's Aqueous Humor

Edited by

Mortimer M. Civan
Departments of Physiology and Medicine
University of Pennsylvania
Philadelphia, Pennsylvania

ELSEVIER

AMSTERDAM • BOSTON • HEIDELBERG • LONDON
NEW YORK • OXFORD • PARIS • SAN DIEGO
SAN FRANCISCO • SINGAPORE • SYDNEY • TOKYO
Academic Press is an imprint of Elsevier

Academic Press is an imprint of Elsevier
525 B Street, Suite 1900, San Diego, California 92101-4495, USA
84 Theobald's Road, London WC1X 8RR, UK

This book is printed on acid-free paper. ∞

For information on all Academic Press publications
visit our website at books.elsevier.com

ISBN: 978-0-12-373894-3
ISSN: 1063-5823

Printed and bound in the United Kingdom
Transferred to Digital Printing, 2011

To Judith: my wife, friend and inspiration

Contents

Contributors

Numbers in parentheses indicate the pages on which the authors' contributions begin.

Nicholas W. Baetz (47) Department of Cell Biology and Anatomy, The University of Arizona, Tucson, Arizona 85711

Teresa Borrás (323) Department of Ophthalmology, University of North Carolina School of Medicine, Chapel Hill, North Carolina

Peter R. Brink (71) Department of Physiology and Biophysics, State University of New York, Stony Brook, New York 11794

Carl B. Camras (231) Department of Ophthalmology, University of Nebraska Medical Center, Omaha, Nebraska, 68198-5840

Mortimer M. Civan (1, 97) Departments of Physiology and Medicine, University of Pennsylvania, Philadelphia, Pennsylvania 19101-6085

Abbot F. Clark (427) Glaucoma Research, Alcon Research, Ltd., Fort Worth, Texas, USA; Department of Cell Biology and Genetics University North Texas Health Science Center Fort Worth, Texas, USA

Miguel Coca-Prados (123) Department of Ophthalmology and Visual Science, Yale University School of Medicine, New Haven, Connecticut 06510

Thomas F. Freddo (161) School of Optometry, University of Waterloo, Ontario N2L 3G1, Canada

Sikha Ghosh (123) Department of Ophthalmology and Visual Science, Yale University School of Medicine, New Haven, Connecticut 06510

Mark Johnson (161) Departments of Biomedical Engineering and Ophthalmology, Northwestern University, Illinois 60208

Jeffrey W. Kiel (273) Department of Ophthalmology, University of Texas Health Science Center at San Antonio, San Antonio, Texas 78229

Wennan Lu (301) Department of Physiology, University of Pennsylvania, Philadelphia, Pennsylvania 19104

Anthony D. C. Macknight (97) Department of Physiology, University of Otago Medical School, Dunedin, New Zealand

Richard T. Mathias (71) Department of Physiology and Biophysics, State University of New York, Stony Brook, New York 11794

Claire H. Mitchell (301) Department of Physiology, University of Pennsylvania, Philadelphia, Pennsylvania 19104

Iok-Hou Pang (427) Glaucoma Research, Alcon Research, Ltd., Fort Worth, Texas, USA

Herbert A. Reitsamer (273) Department of Ophthalmology, Paracelsus Medical University, Salzburg, Austria

W. Daniel Stamer (47) Department of Pharmacology, The University of Arizona, Tucson, Arizona 85711; Department of Ophthalmology and Vision Science, The University of Arizona, Tucson, Arizona 85711

Ernst R. Tamm (379) Institute of Human Anatomy and Embryology, University of Regensburg, 93053 Regensburg, Germany

Carol B. Toris (193, 231) Department of Ophthalmology, University of Nebraska Medical Center, Omaha, Nebraska, 68198-5840

Thomas W. White (71) Department of Physiology and Biophysics, State University of New York, Stony Brook, New York 11794

Andrea J. Yool (47) Department of Cell Biology and Anatomy, The University of Arizona, Tucson, Arizona 85711; Discipline of Physiology, University of Adelaide, SA 5005, Australia; Department of Pharmacology, The University of Arizona, Tucson, Arizona 85711

Preface

The current volume updates the book first published by Academic Press in 1998. The first edition was well received. The book is concerned both with basic physiology and with its implications in addressing glaucoma.

Glaucoma is a major cause of irreversible blindness throughout the world and is usually associated with elevated intraocular pressure. The only currently validated intervention to delay the onset and slow the rate of progression of glaucomatous is to lower intraocular pressure. Thus, the mechanisms and regulation of maintaining normal intraocular pressure are of interest both physiologically and clinically.

The intraocular pressure is directly dependent on the rate of inflow of aqueous humor into the posterior chamber of the eye and the resistance to outflow of that fluid from the anterior chamber of the eye. Part of the blood plasma delivered to the ciliary epithelium of the eye is transferred (secreted) into the aqueous humor. Many of the basic mechanisms involved in secretion have been identified, but their integration and regulation are less well understood. How the aqueous humor leaves the eye is even less well understood, since the precise functional pathways through the trabecular and uveoscleral pathways have not been documented. Despite these uncertainties, the outflow process is of particular importance because glaucoma is thought to arise from a poorly understood increase in resistance to outflow. Given the substantial gaps in our knowledge and the clear clinical relevance of the work, there is value in periodically assessing recent advances and relating these advances to an integrated view of the regulation of intraocular pressure and the implications for addressing glaucoma. I believe that this integrative purpose is ill-served by a large, multivolume work, which serves primarily as a repository of advances in compartmentalized knowledge. Such an approach, while useful in its own right, tends to encourage investigators to continue thinking within the box.

I hope that the new edition of my initial book will be read as a whole. It incorporates a number of substantive changes. First, the perspective has been considerably broadened, introducing entirely new chapters dealing with inflow and outflow of aqueous humor and with glaucomatous blindness, while consolidating focus on specific transport mechanisms to three, rather than the original five chapters. Recent insights concerning inflow are provided in new chapters on the circulatory regulation and topography of inflow and on the potential coupling of inflow and outflow. Outflow and glaucoma are considered in greater breadth and depth, including new chapters on the pathogenesis of retinal ganglion cell death, on functional genomics, on clues to the molecular bases of glaucoma, and on innovative strategies for controlling intraocular pressure and for neuroprotection. Given the importance of whole-animal studies and the conundra frequently arising from interpreting results obtained with different species, the original single chapter has been expanded to two, with separate consideration of nonhuman whole-animal models and of clinical studies.

In broadening the perspective, the number of authors contributing chapters has also increased. This necessarily leads to some overlap in subject material. I regard this overlap as positive in providing both emphasis of new, important concepts and in expressing a spectrum of views on those new concepts that are as yet incompletely accepted.

In addition to presenting new concepts, the second edition expands discussion of measurement techniques in isolated tissues, in nonhuman animals and in humans. These techniques include electron-probe X-ray microanalysis of *in vitro* tissues, measurements of the circulation, inflow and outflow of nonhuman and human subjects, and the techniques of functional genomics.

I express my appreciation to the contributors, both to the first edition and to this second edition of the book. I am also grateful to the reviewers who offered constructive suggestions of the individual chapters: Drs. Nicholas A. Delamere, Tejvir S. Khurana, Jeffrey W. Kiel, Michael H. Koval, Rajkumar V. Patil, W. Daniel Stamer, Richard A. Stone, and Chi-ho To. Each of the authors has published significant contributions in journals. It is my hope that this book has succeeded in placing these contributions in a broader perspective, providing insight into seminal developments and future possibilities of addressing aqueous humor dynamics and glaucoma.

Previous Volumes in Series

Current Topics in Membranes and Transport

Volume 23 Genes and Membranes: Transport Proteins and Receptors* (1985)
Edited by Edward A. Adelberg and Carolyn W. Slayman

Volume 24 Membrane Protein Biosynthesis and Turnover (1985)
Edited by Philip A. Knauf and John S. Cook

Volume 25 Regulation of Calcium Transport across Muscle Membranes (1985)
Edited by Adil E. Shamoo

Volume 26 Na^+–H^+ Exchange, Intracellular pH, and Cell Function* (1986)
Edited by Peter S. Aronson and Walter F. Boron

Volume 27 The Role of Membranes in Cell Growth and Differentiation (1986)
Edited by Lazaro J. Mandel and Dale J. Benos

Volume 28 Potassium Transport: Physiology and Pathophysiology* (1987)
Edited by Gerhard Giebisch

Volume 29 Membrane Structure and Function (1987)
Edited by Richard D. Klausner, Christoph Kempf, and Jos van Renswoude

Volume 30 Cell Volume Control: Fundamental and Comparative Aspects in Animal Cells (1987)
Edited by R. Gilles, Arnost Kleinzeller, and L. Bolis

Volume 31 Molecular Neurobiology: Endocrine Approaches (1987)
Edited by Jerome F. Strauss, III, and Donald W. Pfaff

**Part of the series from the Yale Department of Cellular and Molecular Physiology.*

Volume 32 Membrane Fusion in Fertilization, Cellular Transport, and Viral Infection (1988)
Edited by Nejat Düzgünes and Felix Bronner

Volume 33 Molecular Biology of Ionic Channels* (1988)
Edited by William S. Agnew, Toni Claudio, and Frederick J. Sigworth

Volume 34 Cellular and Molecular Biology of Sodium Transport* (1989)
Edited by Stanley G. Schultz

Volume 35 Mechanisms of Leukocyte Activation (1990)
Edited by Sergio Grinstein and Ori D. Rotstein

Volume 36 Protein–Membrane Interactions* (1990)
Edited by Toni Claudio

Volume 37 Channels and Noise in Epithelial Tissues (1990)
Edited by Sandy I. Helman and Willy Van Driessche

Current Topics in Membranes

Volume 38 Ordering the Membrane Cytoskeleton Trilayer* (1991)
Edited by Mark S. Mooseker and Jon S. Morrow

Volume 39 Developmental Biology of Membrane Transport Systems (1991)
Edited by Dale J. Benos

Volume 40 Cell Lipids (1994)
Edited by Dick Hoekstra

Volume 41 Cell Biology and Membrane Transport Processes* (1994)
Edited by Michael Caplan

Volume 42 Chloride Channels (1994)
Edited by William B. Guggino

Volume 43 Membrane Protein–Cytoskeleton Interactions (1996)
Edited by W. James Nelson

Volume 44 Lipid Polymorphism and Membrane Properties (1997)
Edited by Richard Epand

Volume 45 The Eye's Aqueous Humor: From Secretion to Glaucoma (1998)
Edited by Mortimer M. Civan

Volume 46 Potassium Ion Channels: Molecular Structure, Function, and Diseases (1999)
Edited by Yoshihisa Kurachi, Lily Yeh Jan, and Michel Lazdunski

Volume 47 AmilorideSensitive Sodium Channels: Physiology and Functional Diversity (1999)
Edited by Dale J. Benos

Volume 48 Membrane Permeability: 100 Years since Ernest Overton (1999)
Edited by David W. Deamer, Arnost Kleinzeller, and Douglas M. Fambrough

Volume 49 Gap Junctions: Molecular Basis of Cell Communication in Health and Disease
Edited by Camillo Peracchia

Volume 50 Gastrointestinal Transport: Molecular Physiology
Edited by Kim E. Barrett and Mark Donowitz

Volume 51 Aquaporins
Edited by Stefan Hohmann, Søren Nielsen and Peter Agre

Volume 52 Peptide–Lipid Interactions
Edited by Sidney A. Simon and Thomas J. McIntosh

Volume 53 CalciumActivated Chloride Channels
Edited by Catherine Mary Fuller

Volume 54 Extracellular Nucleotides and Nucleosides: Release, Receptors, and Physiological and Pathophysiological Effects
Edited by Erik M. Schwiebert

Volume 55 Chemokines, Chemokine Receptors, and Disease
Edited by Lisa M. Schwiebert

Volume 56 Basement Membrances: Cell and Molecular Biology
Edited by Nicholas A. Kefalides and Jacques P. Borel

Volume 57 The Nociceptive Membrane
Edited by Uhtaek Oh

Volume 58 Mechanosensitive Ion Channels, Part A
Edited by Owen P. Hamill

Volume 59 Mechanosensitive Ion Channels, Part B
Edited by Owen P. Hamill

Volume 60 Computational Modelling of Membrane Bilayers
Edited by Scott E. Feller

Volume 61 Free Radical Effects on Membranes
Edited by Sadis Matalon

CHAPTER 1

Formation of the Aqueous Humor: Transport Components and Their Integration

Mortimer M. Civan
Departments of Physiology and Medicine, University of Pennsylvania School of Medicine, Philadelphia, Pennsylvania 19104

Current Topics in Membranes, Volume 62 1063-5823/08 $35.00

DOI: 10.1016/S1063-5823(08)00401-8

I. OVERVIEW

In large part, this volume focuses on the aqueous humor, its inflow from the blood and its outflow from the eye into the venous circulation. This chapter addresses the first step in establishing that flow, the secretion of the aqueous humor by the ciliary epithelium. The major aims are to present the underlying transport components and regulatory elements of that secretion. The chapter will also introduce relatively recent changes in our thinking concerning the regulatory role of the circulation, functional topography and species variation in forming the aqueous humor. The latter issues will be addressed in depth in subsequent chapters

II. INTRODUCTION

A. Function of Aqueous Humor

One major function of aqueous humor inflow is to maintain inflation of the globe, stabilizing its optical properties. For this purpose, it might be expected that the intraocular pressure (IOP) of the eye would be relatively constant about the observed median of 16–17 mm Hg (Brubaker, 1998). Early reports of a circadian rhythm of IOP proved inconsistent (Liu, 1998; Asejczyk-Widlicka and Pierscionek, 2007). Furthermore, the variations in IOP of a few mm Hg observed during the day in individuals do not detectably alter image quality, presumably because of unidentified compensating mechanisms (Asejczyk-Widlicka and Pierscionek, 2007). A second major function of aqueous humor is to deliver oxygen and nutrients and to remove metabolic waste products from the avascular anterior segment consisting of the lens, cornea, and trabecular meshwork. Other functions ascribed to aqueous humor inflow have been less clearly defined (Krupin and Civan, 1996), and include the delivery of antioxidants, such as ascorbate, and participation in local immune responses. The ciliary epithelium concentrates ascorbate in the aqueous humor 40-fold over the plasma concentration (Krupin and Civan, 1996). In so doing, the intracellular ascorbate concentration of the ciliary epithelium likely increases to millimolar levels (Helbig *et al.*, 1989b) through a Na^+-ascorbate cotransporter (Socci and Delamere, 1988; Helbig *et al.*, 1989b). This is comparable to the levels of ascorbate in the cerebrospinal fluid and brain cells (Rice, 2000). Recently, evidence has been reported that ascorbate may be a regulator of ion channel activity, and not simply a scavenger of reactive oxygen species (ROS) (Nelson *et al.*, 2007). Ascorbate concentrations in the extracellular fluids of rat brain cycle during the day and can be correlated with total motor activity

(Fillenz and O'Neill, 1986). However, this ascorbate cycling in the brain is diurnal in being reversed by inverting the light–dark cycle, and cannot therefore be causally related to the circadian rhythm of aqueous humor inflow.

B. Inflow and Outflow Pathways

The aqueous humor is secreted by the ciliary epithelium into the posterior chamber bounded by the vitreous humor and lens posteriorly, and the iris and pupil anteriorly. The bulk of the fluid flows through the pupil into the anterior chamber, and finally exits at the angle formed by the iris and cornea. Most of the primate aqueous humor has long been considered to leave the anterior chamber through a "conventional" trabecular pathway (Bill and Phillips, 1971), consisting of the trabecular meshwork, juxtacanalicular tissue, Schlemm's canal, collector channels, and venous outflow in series. More recent work has raised the possibility that a substantial fraction of the aqueous humor may exit through a complex, parallel uveoscleral outflow system. These outflow pathways are considered in depth in Chapters 6 (Freddo and Johnson, 2008), 7 (Toris, 2008), and 8 (Toris and Camras, 2008).

In contrast to IOP, the rate of inflow of aqueous humor undergoes an unequivocal and striking circadian rhythm. From 8 am to 12 pm, inflow in the normal young human reaches ~3 μl/min, but falls by some 60% to ~1.3 μl/min from 12 to 6 am (Brubaker, 1998). Although the basis for this circadian rhythm is unclear (Toris and Camras, 2008), the magnitude of the decline is greater than that achievable by currently available drugs.

The rate of aqueous humor secretion can be altered by second messengers and drugs, as discussed below. Furthermore, the phenomenon of circadian cycling suggests that inflow is physiologically regulated. However, that regulation seems insensitive to IOP since inflow does not change in glaucomatous patients (Brubaker, 1998). The importance of understanding aqueous humor secretion lies not in clarifying the pathogenesis of glaucoma, but in facilitating development of strategies for lowering IOP. Lowering the IOP is the only intervention as yet documented to delay the onset and reduce the rate of progression of glaucomatous blindness (Collaborative Normal-Tension Glaucoma Study Group, 1998a,b; The AGIS investigators, 2000; Kass *et al.*, 2002; Leske *et al.*, 2003; Higginbotham *et al.*, 2004). Recent interest has actually focused more on increasing outflow facility (reducing outflow resistance) than on reducing inflow in order to lower IOP, largely because of two theoretical considerations (Gabelt and Kaufman, 2005; Toris and Camras, 2008). First, concern has been expressed about reducing flow to the avascular anterior segment. However, the baseline flow rate is reasonably rapid, resulting in the total replacement of the ciliary epithelial intracellular fluid in ~4 min. This calculation is based on the known area of the rabbit ciliary epithelium (5.72 cm^2) [Table I, p. 120 of Cole (1966)] and rabbit inflow [2.72 $\pm$ 0.12 μl/min, averaged from data of Table 3 of Toris (2008)], and taking the total height of nonpigmented ciliary epithelial (NPE) and pigmented ciliary epithelial (PE) cells to be ~20 μm. Furthermore, as noted above, the physiological circadian reduction in flow during nighttime is actually greater than that achievable with currently available drugs. Second, increasing outflow facility to lower IOP has been thought to be a possibly more physiological strategy since glaucoma is associated with reduced outflow facility and never with increased inflow. However, recent results from studies of the uveoscleral component of total outflow (Gabelt and Kaufman, 2005; Toris and Camras, 2008) raise the possibility that lowering inflow may prove to be the more physiological way to address glaucomatous ocular hypertension. Patients with ocular hypertension display normal inflow rates, but their uveoscleral outflow is reduced by a third (Toris *et al.*, 2002). In order to match outflow to inflow, patients elevate IOP in order to increase outflow through the more pressure-sensitive trabecular outflow pathway (Bill, 1966; Toris and Pederson, 1985). The outflow facility of these patients is also reduced by a third (Toris *et al.*, 2002), but it is unclear whether the fall in outflow facility is a cause or a result of the ocular hypertension. It is also unclear whether drugs that increase outflow facility act at the same outflow site affected in glaucoma. Arguably, it may be more physiological to reduce inflow to match the fall in uveoscleral outflow, rather than stimulate outflow through a pathway possibly different from the physiological routes and different from the site of glaucomatous obstruction.

C. Mode of Aqueous Humor Formation

As recently as 35 years ago, some publications still postulated that the aqueous humor was primarily an ultrafiltrate of the blood (Green and Pederson, 1972). Subsequent data have rendered that view untenable (Krupin and Civan, 1996). From measurements of capillary hydrostatic pressure and stromal oncotic pressure, Bill (1973) concluded that ultrafiltration across the ciliary epithelium would lead to absorption, and not secretion, of aqueous humor. Furthermore, metabolic poisons and selective transport inhibitors such as cardiotonic steroids (Cole, 1960, 1977; Shahidullah *et al.*, 2003) inhibit aqueous humor inflow by 60–80%. In addition, alterations of <25% in systemic arterial pressure about the physiological value have little effect on the rate of aqueous humor formation (Bill, 1973; Reitsamer and Kiel, 2008). The higher concentrations of many amino acids (Reddy *et al.*, 1961) and ascorbate in the aqueous humor than in the plasma also indicate that the secretion is transcellular, crossing plasma membranes, and is not simply a largely protein-free, paracellular ultrafiltrate.

Although likely of minor direct importance in forming aqueous humor, the arterial pressure is critical for delivering the solutes and water required for transcellular secretion. Progressive reductions by >25% in baseline perfusion pressure or ciliary blood flow lead to progressive falls in aqueous humor secretion (Reitsamer and Kiel, 2003, 2008). The important role of the circulation may also be indicated by the substantially lower net ion (Do and Civan, 2004) and water transfer (Candia *et al.*, 2005, 2007) produced *in vitro* by iris-ciliary bodies isolated from multiple species. In the absence of capillary perfusion, collapse of ciliary processes and a marked increase in unstirred fluid layers would be expected to reduce *in vitro* secretion. When unstirred layers were minimized by removing the underlying stroma, the isolated rabbit ciliary epithelium was reported to produce a 30- to 50-fold higher rate of net Cl^- secretion (Crook *et al.*, 2000; Table I). Furthermore, the arterially perfused bovine eye forms aqueous humor at 2.7 ± 0.5 μl/min (Shahidullah *et al.*, 2005), which can be estimated to be approximately threefold higher than that expected from the net Cl^- flux across the isolated bovine ciliary epithelium (Do and To, 2000).

TABLE I

Cl^- Fluxes Across the Ciliary Body or Ciliary Epithelial (CE) Bilayer Under Short-Circuited Condition

Investigators	Year	Species	J_{sa}	J_{as}	Net flux
(Holland and Gipson, 1970)	1970	Cat	12.28	9.39	2.89[a]
(Saito and Watanabe, 1979)	1979	Toad	7.67	4.12	2.60[a]
(Kishida *et al.*, 1982)	1982	Rabbit	15.69	13.44	2.25[a]
(Pesin and Candia, 1982)	1982	Rabbit	10.9	9.2	1.7
(Do and To, 2000)	2000	Bovine	4.74	3.71	1.03[a]
(Crook *et al.*, 2000)	2000	Rabbit CE bilayer	180.3	72.3	108.0[a]

Flux expressed as $\mu Eq/h/cm^2$. J_{sa}, stromal-to-aqueous flux; J_{as}, aqueous-to-stromal flux. Reprinted (Do and Civan, 2004) with the permission of Springer.

[a]Statistically significant net Cl^- secretion.

III. STRUCTURE OF CILIARY EPITHELIUM

The ciliary epithelium that covers the ciliary body consists of a major *pars plicata* anteriorly and a minor *pars plana* posteriorly in the human. The *pars plicata* is composed of ~70 villiform processes extending anteriorly to the pupil. Connective tissue, vessels, and nerve endings comprise the stroma of each process. The *pars plana* is flat and extends posteriorly to the *ora serrata*, the demarcation with the neuroretina (Pei and Smelser, 1968).

As a result of the embryological invagination of the optic vesicle to form the optic cup, the microanatomy of the ciliary epithelium is unique (Krupin and Civan, 1996). Unlike other epithelia, the two cell layers adjoin each other at their apical surfaces (Fig. 1). The basolateral surfaces of the outer PE cells abut the stroma and those of the inner NPE cells face the aqueous humor. Gap junctions provide low-resistance pathways interconnecting the intracellular fluids of cells within and between the two cell layers (Raviola and Raviola, 1978). The gap junctions of the ciliary epithelium are considered in depth by Mathias *et al.* (2008; Chapter 3) in this volume.

IV. UNIDIRECTIONAL SECRETION OF AQUEOUS HUMOR

A. Basic Strategy of the Ciliary Epithelium

1. Relationship of Solute and Water Secretion

Secretion has long been thought to be based upon a primary transfer of net solute from stroma to aqueous humor, thereby establishing an osmotic gradient. Water has been considered to follow secondarily by local osmosis. The discovery of aquaporin (AQP) water channels has unequivocally demonstrated that water movement can indeed be dissociated from solute movement in response to local osmotic gradients (King *et al.*, 2004). AQPs are considered briefly in a later section and in depth in Chapter 2 (Stamer *et al.*, 2008).

Whether all transmembrane water movement proceeds through local osmosis has recently been questioned (Loo *et al.*, 2002; Fischbarg *et al.*, 2006). A series of publications has reported evidence suggesting that water may also be transferred across biological membranes in fixed stoichiometry to ions and nonelectrolytes simultaneously cotransported (Loo *et al.*, 2002). Whether water is ever cotransported at a fixed stoichiometry, and if so, whether it is quantitatively significant, has been controversial (Lapointe, 2007; Zeuthen and Zeuthen, 2007). In addition, electroosmosis has long been considered a possible contributor to transepithelial water movement (McLaughlin and

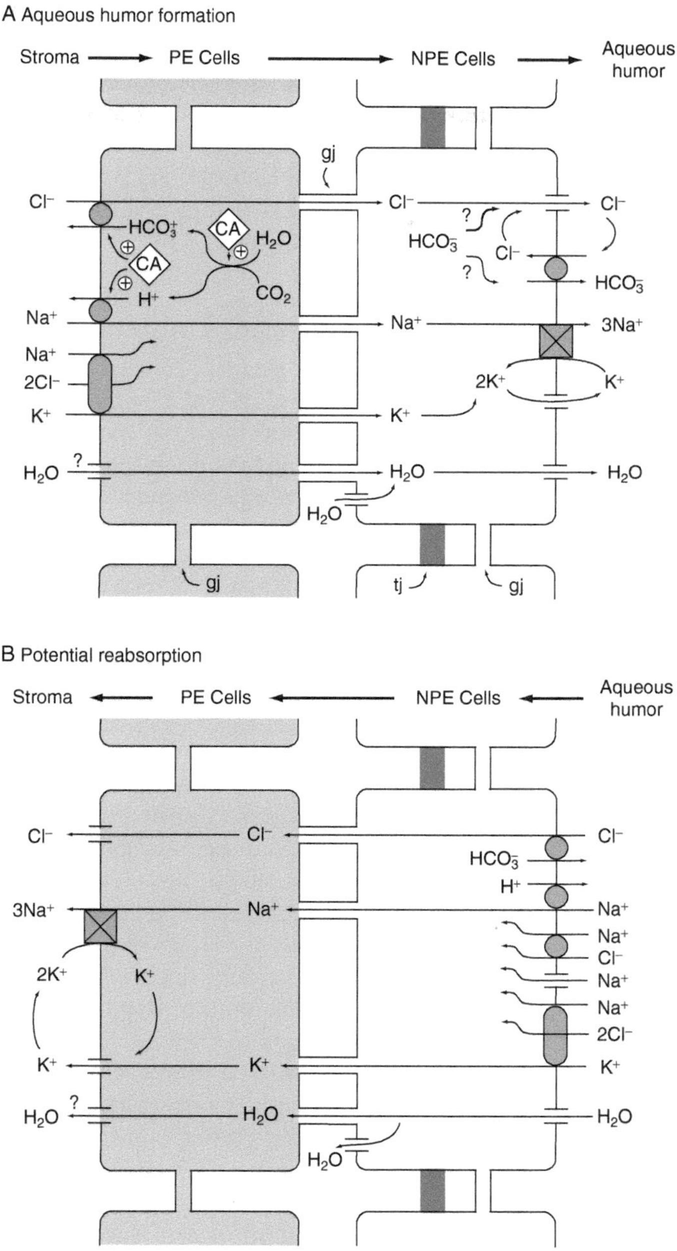

FIGURE 1 Transport components underlying unidirectional secretion (A) and possible unidirectional reabsorption (B) across the ciliary epithelium. Tight junctions (tj) between the NPE cells provide a barrier between the stromal and aqueous compartments. Gap junctions (gj) subserve intercellular communication between adjoining PE cells, NPE cells, and PE–NPE cell couplets. Carbonic anhydrase (CA) directly stimulates the Na^+/H^+ and Cl^-/HCO_3^- antiports.

Mathias, 1985), a possibility readdressed in a recent series of studies (Fischbarg *et al.*, 2006). The analyses of Lapointe (2007) and Mathias and Wang (2005) raise doubt whether it is necessary to invoke either water cotransport or electroosmosis, respectively, to account for transepithelial water movement. The current prevailing view is that transepithelial water flow generally proceeds by local osmosis. ATP is expended in order to transfer solute across epithelia in order to establish an osmotic gradient for secondary, uncoupled secretion of water.

2. Centrality of NaCl Secretion

As noted above (Sections II.A and C), the composition of the aqueous humor differs from that of the plasma. Nevertheless, both plasma and aqueous humor are largely solutions of NaCl, with Na^+ and Cl^- concentrations of ~150 and ~130 mM, respectively, in the human aqueous humor (Krupin and Civan, 1996). Thus, the formation of the aqueous humor can be viewed essentially as a primary, energy-dependent transfer of NaCl, and a secondary transfer of water, across the ciliary epithelium. Consistent with this view, blocking Na^+ or Cl^- transepithelial transport reduces the rate of aqueous humor formation (Shahidullah *et al.*, 2003). The minor constituents of the stromal extracellular fluid and aqueous humor, especially HCO_3^-, K^+, and Ca^{2+}, are known to modulate secretion, but those important effects are exerted indirectly on Na^+ and Cl^- transfer (Krupin and Civan, 1996; To *et al.*, 2001; Do and Civan, 2004).

3. Transcellular and Paracellular Components of Secretion

In principle, solutes and water can be transferred through both the transcellular pathway through the cells and the paracellular pathway between the epithelial cells. Taking the convoluted surface of the isolated, full-thickness ciliary epithelium into account, the transmural resistance of the rabbit preparation is ~1 KΩ·cm^2 (Krupin *et al.*, 1984). However, much of that resistance may reflect contributions of the stroma underlying the epithelium. After isolating small areas of the rabbit epithelial bilayer, Sears *et al.* (1991) found that the transmural resistance was reduced to 40 Ω·cm^2, even though the transepithelial potential was still −0.65 mV, a value comparable to that measured across the full-thickness preparation (Krupin and Civan, 1996). This purely transepithelial resistance corresponds to that of a leaky epithelium (Rose and Schultz, 1971; Frömter and Diamond, 1972), suggesting that the paracellular pathway provides a substantial transmural electrical shunt. The observation that transport inhibitors can reduce inflow in experimental preparations by as much as 60–80% indicates that the secretory pathway is largely transcellular. However, the isolated ciliary epithelium of multiple species displays a transepithelial potential difference of ~1 mV, with the aqueous humor negative to the stroma. Whether this small driving force produces a significant paracellular Na^+ contribution to total secretion is unknown.

Transcellular epithelial transfer of NaCl fundamentally depends on direct coupling of ATP utilization with Na^+ movement through Na^+, K^+-activated ATPase, but is also mediated by an ensemble of ion and water channels, cotransporters (symports) coupling flows in the same direction, and countertransporters (antiports) coupling solute flows in opposite directions. These transport components are introduced in the following section. A comprehensive discussion of the ion channels described in the following section is provided in the monograph by Hille (2001).

B. Transport Components Underlying Transcellular Secretion

The ciliary epithelium expresses a wide range of ion channels and transporters responsible for facilitated diffusion, cotransport, and countertransport (Jacob and Civan, 1996; Krupin and Civan, 1996). Many of these transport elements perform housekeeping tasks necessary for individual cell viability and function. This chapter focuses on the channels and transporters likely to be directly involved in transepithelial secretion of ions and water (Fig. 1A).

1. Uptake of Stromal NaCl

The first step in transepithelial secretion is the uptake of NaCl from the stromal extracellular fluid by the PE cells (Fig. 1A). The intracellular potentials of the PE and NPE cells are very similar to each other and highly negative (Green *et al.*, 1985), so that the electroneutral transporters indicated in Fig. 1A permit the PE cells to take up Cl^- against a strong electrochemical gradient. Measured with the same extracellular bathing solution containing 152 mM Cl^-, the intracellular potential of rabbit ciliary epithelium was found to be -67.0 ± 0.2 mV ($N = 110$) (Carré *et al.*, 1992), and the intracellular Cl^- concentration was estimated by electron-probe X-ray microanalysis to be 46 ± 5 mmol/kg intracellular water ($N = 99$) (Bowler *et al.*, 1996). This value is fourfold higher than the intracellular Cl^- concentration calculated for an equilibrium distribution at the measured membrane potential. At least two sets of electroneutral transporters support uptake of NaCl from the stroma (Wiederholt *et al.*, 1991) as described below.

a. Na^+-K^+-2Cl– Cotransporters (Symports). Following the report by Geck *et al.* (1980), the Na^+-K^+-$2Cl^-$ cotransporter has been identified as a major mechanism for uptake of NaCl by both secretory and absorptive epithelia. This symport has been immunolocalized at the basolateral surface of PE cells of young calves (Dunn *et al.*, 2001). Inhibition of the symport with furosemide or bumetanide has been found to reduce intracellular Cl^- activity in shark ciliary epithelium (Wiederholt and Zadunaisky, 1986), reduce Na^+ and Cl^- uptake by cultured bovine PE cells (Helbig *et al.*, 1989a), and shrink native bovine PE cells (Edelman *et al.*, 1994). Blocking the Na^+-K^+-$2Cl^-$ cotransporter with bumetanide also inhibits net Cl^- secretion across ciliary epithelium from the rabbit (Crook *et al.*, 2000) and cow (Do and To, 2000), and inhibits aqueous humor formation in isolated, arterially perfused bovine eyes (Shahidullah *et al.*, 2003). In all of these reports, the thermodynamic driving force evidently favored net uptake of Na^+, K^+, and Cl^- from the stromal surface into the PE cells. However, the Na^+-K^+-$2Cl^-$ cotransporter supports bidirectional movement of solute. Reversal of the thermodynamic driving force by reducing ionic concentrations in the bath has been reported to cause bumetanide-inhibitable cell shrinkage (Edelman *et al.*, 1994). The strong dependence of the net thermodynamic driving force on intracellular Cl^- concentration and its implications are considered in greater depth in Chapter 4 (Macknight and Civan, 2008).

b. Parallel Na^+/H^+ *and* Cl^-/HCO_3^- *Countertransporters (Antiports)* Measurement of radioactive tracer uptake by cultured bovine PE cells led to the suggestion that Na^+/H^+ and Cl^-/HCO_3^- exchange might also be important mechanisms underlying uptake of NaCl from the stroma *in vivo* (Helbig *et al.*, 1989a; Wiederholt *et al.*, 1991). These antiports were later identified as Na^+/H^+ exchanger NHE-1 and Cl^-/HCO_3^- exchanger AE2 by pharmacological and immunostaining approaches, respectively (Counillon *et al.*, 2000). As discussed in Chapter 4 (Macknight and Civan, 2008), electron-probe X-ray microanalyses have indicated that the antiports are important both on the stromal (Fig. 1A) and aqueous (Fig. 1B) surfaces of intact rabbit ciliary epithelium. Carbonic anhydrase II (CAII) stimulates the turnover of the antiports, both directly and indirectly (Fig. 1A). Intracellular CAII is now known to bind directly to NHE1 (Li *et al.*, 2002) and AE2 (Sterling *et al.*, 2001). CAII also increases the turnover rates of the antiports by catalyzing the production of H^+ and HCO_3^- from CO_2 and water (Meldrun and Roughton, 1933). The importance of CA in catalyzing the turnover rates of the antiports suggests that CA inhibitors act here to reduce inflow and lower IOP.

2. Passage of NaCl from PE to NPE Cells Through Gap Junctions

Gap junctions, considered in depth in Chapter 3 (Mathias *et al.*, 2008), are formed of two hemichannels (half gap junctions or connexons), one at each abutting surface of two adjoining cells. In turn, each connexon consists of six connexin (Cx) monomers that may be generated from a single connexin (homomeric) or may arise from different connexins (heteromeric). The full gap junction is formed by the linking of connexons of adjoining cells. The full junction may be composed either of identical connexons (homotypic) or of different connexons (heterotypic). Connexin-generated gap junctions exclude molecules greater than ~1 kDa mass, or 6 Å radius. The gap junctions may be a site for secretory regulation under certain physiological conditions and could provide a target for pharmacological inhibition.

A great range of techniques has demonstrated the presence of gap junctions linking cells within and between the PE and NPE cell layers, including structural (Reale, 1975; Raviola and Raviola, 1978), biochemical (Coca-Prados *et al.*, 1992; Wolosin *et al.*, 1997b; Sears *et al.*, 1998; Do and To, 2000; Coffey *et al.*, 2002; Do, 2002), and functional (Green *et al.*, 1985; Wiederholt and Zadunaisky, 1986; Carré *et al.*, 1992; Edelman *et al.*, 1994; Oh *et al.*, 1994; Bowler *et al.*, 1996; Stelling and Jacob, 1997) analyses. Each of the connexin gap junctions thus far identified is both homomeric and homotypic (Coffey *et al.*, 2002). The gap junctions known to link the PE and NPE cells are homomeric, homotypic structures formed from the connexins Cx40 and Cx43, and those known to link adjoining cells in the NPE cell layer arise from connexins Cx26 and Cx31 (Coffey *et al.*, 2002). The molecular basis for the gap junctions linking adjoining PE cells is, as yet, unknown, and might reflect unidentified connexins or the newly recognized, ubiquitous pannexins (Panchin *et al.*, 2000; Panchin, 2005; Barbe *et al.*, 2006; Li *et al.*, 2008). As discussed more fully in Chapter 4 (Macknight and Civan, 2008), the gap junctions linking the PE and NPE cells are more numerous (Raviola and Raviola, 1978) and possibly more robust to certain experimental stresses (McLaughlin *et al.*, 2004) than those linking cells within the PE and NPE cell layers. These observations have led to the view that the PE–NPE cell couplets form the fundamental functional unit of the ciliary epithelium (McLaughlin *et al.*, 2004). The supporting evidence, obtained by electron-probe X-ray microanalysis, is considered in Chapter 4 (Macknight and Civan, 2008).

The PE–NPE gap junctions are interrupted by the nonselective blockers octanol (Stelling and Jacob, 1997) and heptanol (Mitchell and Civan, 1997). Heptanol also inhibits short-circuit current across rabbit (Wolosin *et al.*, 1997a) and bovine (Do and To, 2000) ciliary epithelium and reduces net Cl^- transport across the bovine preparation (Do and To, 2000). Under baseline conditions, the gap junctions do not likely limit the rate of transcellular NaCl secretion since the elemental compositions of the PE and NPE cells are similar (Bowler *et al.*, 1996). Were the gap junctions to present a substantial barrier under baseline conditions, we would expect to find a higher concentration in the PE cells. However, recent evidence suggests that second-messenger cascades can downregulate solute passage through the PE–NPE gap junctions. Gap junctions are known to be regulated at translational, trafficking, and functional levels (Warn-Cramer and Lau, 2004). However, 3′,5′-cyclic adenosine monophosphate (cAMP) has been reported to activate Cx40 (van Rijen *et al.*, 2000) but both to increase (Somekawa *et al.*, 2005) and decrease (Lampe and Lau, 2000) communication through Cx43 gap junctions. Transmural measurements of bovine ciliary epithelium have suggested that the overall effect of cAMP is to block the PE–NPE gap junctions (Do *et al.*, 2004a), a conclusion confirmed by very recent dye-transfer and dual-cell patch clamping of bovine cell couplets (Do *et al.*, 2008). The multiple roles of cAMP in regulating aqueous humor inflow are further considered in the following sections.

3. Extrusion of NaCl from NPE Cells to Aqueous Humor

a. Na^+, K^+-Activated ATPase. The formation of the aqueous humor ultimately rests upon activity of ciliary epithelial Na^+, K^+-activated ATPase (Cole, 1960, 1977). Hydrolysis of ATP to ADP is coupled to the extrusion of three intracellular Na^+ in exchange for two extracellular K^+. Thus, ATP utilization provides energy both for secreting Na^+ and for establishing the ionic asymmetries and membrane potential needed for secretion of other ions and of nonelectrolytes.

Although required for secretion, Na^+, K^+-activated ATPase is actually expressed at both surfaces of the ciliary epithelium (Fig. 1A and B). Data obtained by molecular probes (Ghosh *et al.*, 1990, 1991), immunocytochemistry (Mori *et al.*, 1991), and transepithelial electrical measurements (Krupin *et al.*, 1984) have localized the ATPase to the basolateral membranes of both the PE and NPE cells. In principle, Na^+ might be actively transported in opposite directions by the ciliary epithelium toward the stroma and toward the aqueous humor. Nevertheless, net secretion clearly proceeds from stroma to aqueous humor, and that secretion is strongly inhibited by blocking Na^+, K^+-activated ATPase of the arterially perfused bovine eye with ouabain (Shahidullah *et al.*, 2003). The dominant role of the ATPase of the NPE over that of the PE cells may reflect at least three factors. First, the number of pumps, assayed by tritiated-ouabain binding, is much greater at the aqueous than at the stromal surface of rabbit ciliary epithelium (Usukura *et al.*, 1988). Second, Na^+, K^+-activated ATPase may be modulated by different regulators in the NPE and PE cells. This possibility is supported by the observation that DARPP-32 (dopamine- and cAMP-regulated phosphoprotein of M_r 32 kDa), a component of phosphorylation-mediated modulation of ATPase activity in some cells (Therien and Blostein, 2000), is localized immunohistochemically only to the NPE and not to the PE cells of the rat, cat, rhesus monkey, and human (Stone *et al.*, 1986). Third, the NPE and PE cell layers express different isoforms of Na^+, K^+-activated ATPase (Martin-Vasallo *et al.*, 1989; Ghosh *et al.*, 1990, 1991; Coca-Prados *et al.*, 1995b; Wetzel and Sweadner, 2001), although the isozyme topography appears to be species dependent (Wetzel and Sweadner, 2001). These isozymes display different ionic binding affinities and selectivities and different turnover rates (Blanco and Mercer, 1998; Crambert *et al.*, 2000).

The Na^+, K^+-activated ATPase activity of other cells has long been known to be regulated by cAMP-activated kinase (protein kinase A, PKA) (Aperia *et al.*, 1991; Therien and Blostein, 2000). For example, ATPase activity of the rat-collecting duct was found to be inhibited by a number of agonists that increase cAMP, such as dopamine, vasopressin, and forskolin (Satoh *et al.*, 1993). In part, PKA acts directly by phosphorylating the ATPase at Ser943, thereby reducing its activity. Furthermore, PKA-mediated phosphorylation of DARPP-32 inhibits protein phosphatase 1, locking ATPase in a phosphorylated, downregulated state. PKA can affect ATPase in more complex ways, as well (Therien and Blostein, 2000), by altering the number of plasma-membrane pumps, by altering Na^+ and K^+ concentrations, by interacting with protein kinase C (PKC), and by activating intermediate proteins. For example, PKA appears to inhibit Na^+, K^+-activated ATPase activity of rat cortical collecting duct by stimulating the cytochrome P450-monooxygenase pathway of arachidonic acid metabolism (Satoh *et al.*, 1993). Whether PKA increases or decreases ATPase activity is species and tissue specific, and depends upon Ca^{2+} concentration and ROS (Therien and Blostein, 2000). Given these complexities, it is scarcely surprising that reports of the effects of cAMP on NPE-cell Na^+, K^+-activated ATPase have been in incomplete agreement. Administration of db-cAMP, a membrane-permeant form of cAMP, was found to reduce ouabain-sensitive phosphate release from rabbit ciliary epithelium (Delamere and King, 1992). However, the β-adrenergic agonist isoproterenol, which increases intracellular cAMP, was reported to increase ouabain-sensitive Rb^+ uptake by a line of cultured human NPE cells; the β-adrenergic antagonist propranolol prevented that stimulation (Liu *et al.*, 2001).

The effects of PKC, dopamine, and endothelin-1 on NPE-cell ATPase have also been complex. For example, activating PKC has stimulated ouabain-sensitive Rb^+ uptake by a cultured line of rabbit NPE cells (Mito and Delamere, 1993; Delamere *et al.*, 1997). In contrast, PKC activation was reported to inhibit cytohistochemically measured K^+-dependent *p*-nitrophenyl phosphatase in rabbit ciliary epithelium (Nakano *et al.*, 1992).

Divergent results have also been obtained by stimulating NPE-cell dopamine (DA) receptors. An agonist of DA_1 was found to reduce ouabain-sensitive Rb^+ uptake by a rabbit NPE cell line (Nakai *et al.*, 1999), but dopamine did not affect ouabain-sensitive, bumetanide-insensitive $^{86}Rb^+$ uptake by cultured fetal human NPE monolayers (Riese *et al.*, 1998). The different results could have reflected differences in cell preparation and experimental conditions. However, the divergence could also reflect the complexity of hormone action. Dopamine is thought to affect Na^+, K^+-activated ATPase activity of other cells through both DA_1- and DA_2-receptor-stimulated, PKC-dependent mechanisms and DA_1-stimulated, PKA-associated pathways (Therien and Blostein, 2000).

Endothelin-1 also exerts complex effects on the NPE cells. The hormone produced a direct inhibition of enzyme activity, but also increased mRNA for its synthesis in transformed human NPE cells (Krishnamoorthy *et al.*, 2003).

The second-messenger nitric oxide (NO) also reduces ouabain-sensitive Na^+, K^+-activated ATPase activity of native porcine NPE cells (Shahidullah and Delamere, 2006). The inhibition is observed whether NO is delivered by donor molecules or generated by nitric oxide synthase (NOS). In contrast, NOS-generated NO has recently been reported to stimulate Na^+, K^+-activated ATPase activity of rabbit cardiac myocytes, measured as whole-cell, electrogenic Na^+-K^+ pump current (White *et al.*, 2008).

In summary, multiple hormones and second-messenger cascades modulate Na^+,K^+-activated ATPase activity of the ciliary epithelium, but their actions can be direct or indirect, and depend on isoform specificity and interactions with parallel signaling cascades. A further complexity arises from increasing evidence that Na^+,K^+-activated ATPase itself plays a key role in signaling cascades, which is independent of its effects on intracellular Na^+ and K^+ concentration (Xie and Askari, 2002). This newly appreciated role includes effects on gene regulation and cell growth, mediated through protein–protein interactions.

b. Cl^- Channels. Extrusion of Na^+ through Na^+, K^+-activated ATPase is accompanied by release of Cl^- into the aqueous humor through anion channels of the NPE cells. Several observations suggest that this release is a rate-limiting factor in aqueous humor formation. Of the three steps comprising aqueous humor formation, stromal uptake of NaCl is not rate limiting under baseline conditions since the PE-cell Cl^- concentration is fourfold higher than that expected at electrochemical equilibrium. As noted in Section IV.B.1, this relatively high intracellular Cl^- concentration is established by the electroneutral symports and antiports of the PE cells. The second step, transfer of NaCl, from the PE to NPE cells, is also not likely rate limiting since the Cl^- contents (McLaughlin *et al.*, 2007), Cl^- concentrations (Bowler *et al.*, 1996), and intracellular potentials (Green *et al.*, 1985) of the two cell layers are closely similar. By exclusion, the aqueous surface of the ciliary epithelium is likely to be the major site of regulation. As discussed in Section IV.B.3.a, Na^+, K^+-activated ATPase at this surface can certainly be modified, but its continuous activity, necessary for maintenance of transmembrane ionic asymmetries, is readily detected under baseline conditions (Krupin *et al.*, 1984). In contrast, Cl^--channel activity of native bovine NPE cells is low under baseline conditions, and can be enhanced by a number of perturbations (Section VI).

The molecular identity of Cl^- channels at the aqueous surface has not yet been established. More than one channel is likely expressed since hypotonic swelling of native bovine NPE cells was found to activate Cl^- channels with unitary conductances of 7.3 and 18.8 pS (Zhang and Jacob, 1997). Several lines of evidence have suggested that ClC-3 (Coca-Prados *et al.*, 1996; Civan, 2003) or pICln (Anguíta *et al.*, 1995; Coca-Prados *et al.*, 1995a) might play substantial roles in NPE-cell Cl^--channel activity.

ClC-3 has been implicated by the observations that: (1) NPE cells express ClC-3 transcripts and protein product (Coca-Prados *et al.*, 1996; Sanchez-Torres *et al.*, unpublished observation); (2) activation of PKC lowers NPE-cell Cl^--channel activity (Civan *et al.*, 1994; Coca-Prados *et al.*, 1995a, 1996; Shi *et al.*, 2003; Do *et al.*, 2005), a signature property of Cl^- currents associated with ClC-3 (Kawasaki *et al.*, 1994); (3) antisense oligonucleotides knockdown ClC-3 message and protein product in NPE cells, and also reduce volume-activated Cl^- currents (Wang *et al.*, 2000); and (4) blocking antibody directed against ClC-3 (Wang *et al.*, 2003) reduces swelling-activated Cl^- currents of both transformed rabbit NPE cells (Vessey *et al.*, 2004) and native bovine NPE cells (Do *et al.*, 2005). These results link ClC-3 to Cl^- channels, but its precise role is unclear, both in the NPE and other cells. Whether ClC-3 is necessary for expression of swelling-activated Cl^- channels in any cell has been controversial (Hermoso *et al.*, 2002; Jentsch *et al.*, 2002). At issue has been whether swelling-activated Cl^- channels in other cells of ClC-3-null mice are different from those of the wild-type mice (Stobrawa *et al.*, 2001; Gong *et al.*, 2004; Yamamoto-Mizuma *et al.*, 2004; Wang *et al.*, 2005). Another issue has been whether ClC-3 is a Cl^- channel, like ClC-1, ClC-2, ClC-Ka, and ClC-Kb, or whether ClC-3 functions as a Cl^-/H^+ antiport exchanger, like ClC-4, ClC-5, and the bacterial homologue ClC-ec1 (Jentsch, 2007; Zifarelli and Pusch, 2007). One possible interpretation is that ClC-3 may form part of a protein complex constituting the swelling-activated Cl^- channel. Another possibility is that ClC-3 plays roles in the posttranslational processing, trafficking, and/or regulation of other swelling-activated Cl^- channels. The latter possibility is consistent with the observation that PKC activation initially inhibited swelling-activation of NPE-cell Cl^- channels, but did not affect steady-state activation (Do *et al.*, 2005). Among other interpretations that result may reflect a role of ClC-3 in the trafficking or regulation of different Cl^- channels capable of mediating swelling-activated Cl^- channels.

Substantial experimental work has also raised the possibility that pICln (Paulmichl *et al.*, 1992) might underlie or regulate swelling-activated NPE-cell Cl^--channels. pICln is not only found in, but its human form was first cloned from, the NPE cells (Anguíta *et al.*, 1995; Coca-Prados *et al.*, 1995a). Furthermore, an antisense oligonucleotide directed against pICln downregulated both protein and swelling-activated Cl^- currents in native bovine NPE cells (Chen *et al.*, 1999). Nevertheless, as for ClC-3, the potential role of pICln in expressing swelling-activated NPE-cell Cl^- currents has been, and remains, controversial (Clapham, 1998; Strange, 1998; Fürst *et al.*, 2006). At issue have been the questions whether pICln is physiologically present in the plasma membrane, whether it functions as a channel, and if so, whether its selectivity conforms to a Cl^- channel. The question has even been raised that the role of this ubiquitous, abundant, and conserved protein may not be directly related to swelling-activation of Cl^- currents in other cells (Strange, 1998). In the case of the NPE cells (Sanchez-Torres *et al.*, 1999), pICln was immunolocalized to the cytoplasm and perinuclear region and was not translocated to the plasma membrane by hypotonic challenge. These results have suggested that the functional effects of antisense knockdown of pICln (Chen *et al.*, 1999) may be mediated indirectly, possibly through restructuring of the cytoskeleton.

c. K^+ Channels. K^+ channels subserve at least three main functions. In addition to providing a pathway for release of K^+ down its electrochemical gradient to the aqueous humor (Fig. 1A), these channels are needed to maintain the intracellular potential more negative than the Cl^- equilibrium (Nernst) potential. The more negative the intracellular potential, the greater is the thermodynamic force driving Cl^- secretion. The third function of the K^+ channels is to provide a conduit for K^+ to act as a catalyst, enhancing physiological turnover of other transporters. At the basolateral surface of the NPE cells (Fig. 1A), release of intracellular K^+ ensures a high enough extracellular K^+ concentration to support rapid cycling of the Na^+, K^+-exchange pump. At the stromal surface, K^+ channels (Fig. 1B) ensure that the K^+ concentration is high enough to help drive NaCl into the PE cell through the Na^+-K^+-$2Cl^-$ symport. In either case, the K^+ channels act to accelerate cycling either of the symport and/or of Na^+, K^+-activated ATPase. This function is particularly well illustrated by the loss-of-function mutation of the luminal ROMK2 K^+ channel that interferes with symport uptake of Na^+-K^+-$2Cl^-$ by the thick ascending limb of the renal loop of Henle, producing one form of Bartter's syndrome with urinary loss of salt and volume depletion (Hebert, 2003).

Both the NPE and PE cells express multiple K^+ channels, including inward rectifiers, delayed outward rectifiers, and Ca^{2+}-activated outward rectifiers (Jacob and Civan, 1996; Bhattacharyya *et al.*, 2002). Inward rectifiers pass more current into the cell than out of it in response to voltage steps of the same magnitude, and the opposite is true for outward rectifiers. Both delayed (Lang *et al.*, 1998) and Ca^{2+}-activated outward rectifiers (Vázquez *et al.*, 2001; Fernández-Fernández *et al.*, 2002) have been thought to provide exit pathways for K^+ in parallel with Cl^- channels in mediating swelling-activated release of KCl from other cells. The physiological delivery of fluid from the PE cell layer to the NPE cells may sustain the activity of these K^+ channels and thus be particularly relevant to K^+ secretion into the aqueous humor.

4. Transfer of Water from Stroma to Aqueous Humor

The pathways for water secretion across the ciliary epithelium are incompletely understood (Fig. 1A). The specialized AQP water channels (Agre and Kozono, 2003; King *et al.*, 2004) are thought to play a major role (Nielsen *et al.*, 1993; Hasegawa *et al.*, 1994; Stamer *et al.*, 1994; Frigeri *et al.*, 1995; Hamann *et al.*, 1998; Zhang *et al.*, 2002; Yamaguchi *et al.*, 2006). AQP1 has been localized to the apical and basolateral membranes (Stamer *et al.*, 1994; Hamann *et al.*, 1998; Yamaguchi *et al.*, 2006) and AQP4 to the basolateral surfaces of the NPE cells (Hamann *et al.*, 1998; Yamaguchi *et al.*, 2006). Agreement is incomplete whether AQP4 is (Hamann *et al.*, 1998) or is not (Yamaguchi *et al.*, 2006) also expressed in the NPE-cell apical membranes.

In contrast, no AQP has yet been found in the PE cells. Possibly, water is taken from the stroma through unidentified AQPs. Alternatively, water might permeate other transporters, such as sodium-glucose symports (Loike *et al.*, 1996). Another possibility is that water might diffuse across the plasma membranes of these cells. In the absence of high contents of sphingomyelin and cholesterol, plasma membranes can display relatively high water permeability (Finkelstein, 1976). The lipid composition of the PE plasma membranes is unknown.

Irrespective of the precise permeating pathway, water is thought to follow uptake of solute from the stroma into the PE cells, cross the gap junctions into the NPE cells, and be released by local osmosis through AQP1 and AQP4 channels into the aqueous humor (Fig. 1A). This hypothesis is consistent with the observation that double knockout of AQP1 and AQP4 reduced IOP in mice (Zhang *et al.*, 2002).

It is increasingly recognized that AQPs not only provide a conduit for water, and in some cases glycerol or gases, but may interact with other transporters in the plasma membranes with which they are clustered. In particular, proteins incorporating PDZ domains can interact with AQP1 and AQP2 (Cowan *et al.*, 2000) and with AQP9 (Cowan *et al.*, 2000; Pietrement *et al.*, 2008). The full significance of this clustering is not yet clear. The eye's AQPs and their regulation are considered more fully in Chapter 2 (Stamer *et al.*, 2008).

V. POTENTIAL UNIDIRECTIONAL REABSORPTION OF AQUEOUS HUMOR

A. Transport Components Underlying Potential Transcellular Reabsorption Across the Ciliary Epithelium

In addition to mechanisms supporting transcellular transfer of solute and water (Fig. 1A), a number of transporters have been identified that can underlie translocation of fluid in the opposite direction (Fig. 1B). At the aqueous surface, NaCl may be reabsorbed by Na^+/H^+ and Cl^-/HCO_3^- antiports, Na^+-K^+-$2Cl^-$ and Na^+-Cl^- symports, and amiloride-sensitive Na^+-channels (Crook *et al.*, 1992; Von Brauchitsch and Crook, 1993; Crook and Polansky, 1994; Dong and Delamere, 1994; Civan *et al.*, 1996; Crook and Riese, 1996; Riese *et al.*, 1998) functionally identified in cultured NPE cells. The AQP1 and AQP4 channels at the basolateral membranes of the NPE cells (Hamann *et al.*, 1998; Yamaguchi *et al.*, 2006) can subserve water movement back into the cells from the aqueous humor. The fluid reabsorbed can be transferred back to the PE cells through the gap junctions linking the two cell layers.

Once the reabsorbed fluid reaches the PE cells, mechanisms are also in place for subsequent solute release into the stroma. Albeit less numerous in the PE cells (Usukura *et al.*, 1988), Na^+, K^+-activated ATPase is expressed at the stromal, as well as at the aqueous, surface (Krupin *et al.*, 1984; Ghosh *et al.*, 1990; Ghosh *et al.*, 1991). Thus, Na^+ can be extruded by the PE cells back into the stroma, in parallel with Cl^- channels. At least one population of these PE-cell channels comprises maxi-Cl^- channels that can be synergistically activated by ATP and tamoxifen (Mitchell *et al.*, 2000). The effect of ATP appears mediated by stimulating cAMP (Fleischhauer *et al.*, 2001) that acts directly on the channels (Do *et al.*, 2004a). As illustrated by Figs. 2 and 3, the cAMP increases open-channel probability at physiological membrane potentials. This effect is larger when the PE cells have higher concentrations of intracellular Cl^-, which would enhance their ability to cope with increased rates of reabsorptive Cl^- delivery from the NPE cells. The maxi-Cl^- channels are also activated by swelling (Zhang and Jacob, 1997), which might result from delivery of reabsorbed aqueous humor transferred via the NPE cells.

As discussed in greater depth in Chapter 4 (Macknight and Civan, 2008), electron microprobe analysis suggests that the relative importance of the potential reabsorptive pathway varies across different regions of the rabbit ciliary epithelium. The physiological importance of regional transcellular reabsorption has not yet been defined. However, Na^+ reabsorbed at the aqueous surface is now known to be a major determinant of the PE-cell Na^+ content in the anterior region of the intact rabbit ciliary epithelium (McLaughlin *et al.*, 2007).

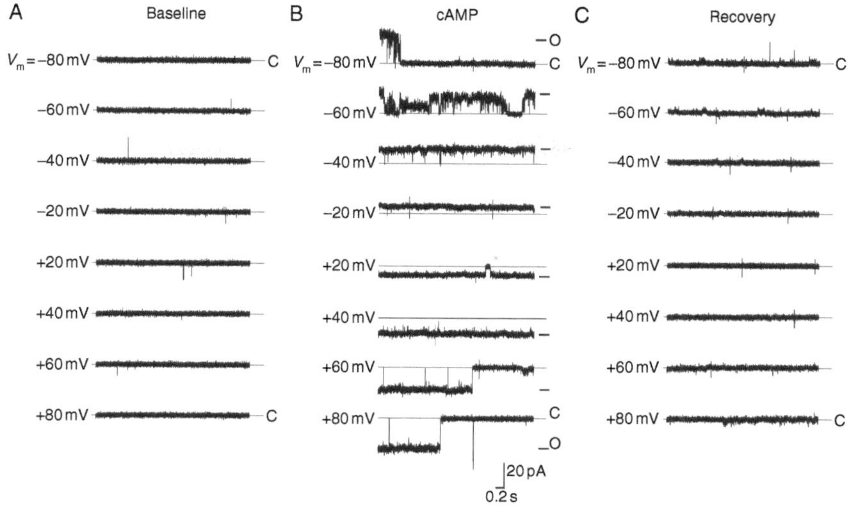

FIGURE 2 Activation of maxi-Cl^- channels by cAMP (500 μM) in an excised inside-out patch from native bovine PE cells (Do *et al.*, 2004a). The holding potential (V_h) was 0 mV, and patches were clamped at membrane potentials (V_m) from −80 to +80 mV in steps of 20 mV. The channel was usually open when V_m was within the range ±40 mV; channels inactivated outside this voltage range. Dotted and solid lines symbolize closed (c) and open (o) states of the channel, respectively. Upward current deflections indicate inward currents and vice versa. Channel activity was not observed before adding or after removing cAMP. (A) Before adding cAMP. (B) During exposure to cAMP. (C) Following removal of cAMP. Reprinted with the permission of the American Physiological Society.

B. *Reabsorption via Iris Root*

Passage across the iris root provides direct communication for diffusion of proteins from the posterior to the anterior chamber in rabbits (Freddo *et al.*, 1990), monkeys (Barsotti *et al.*, 1992), and humans (Bert *et al.*, 2006). However, as noted above (Section II.C), net flow of aqueous flow across the iris root must be in the direction of reabsorption (Fig. 1B) in response to the net hydrostatic and oncotic driving force (Bill, 1973). The quantitative significance of reabsorption through this pathway is unknown.

VI. REGULATION OF NET AQUEOUS HUMOR SECRETION

Many hormones and second messengers modify the transport components subserving net ciliary epithelial secretion. How these modifiers are integrated in regulating aqueous humor formation is unknown. In addition to the many modifiers noted elsewhere (Do and Civan, 2004), bestrophin-2 (Best2) has

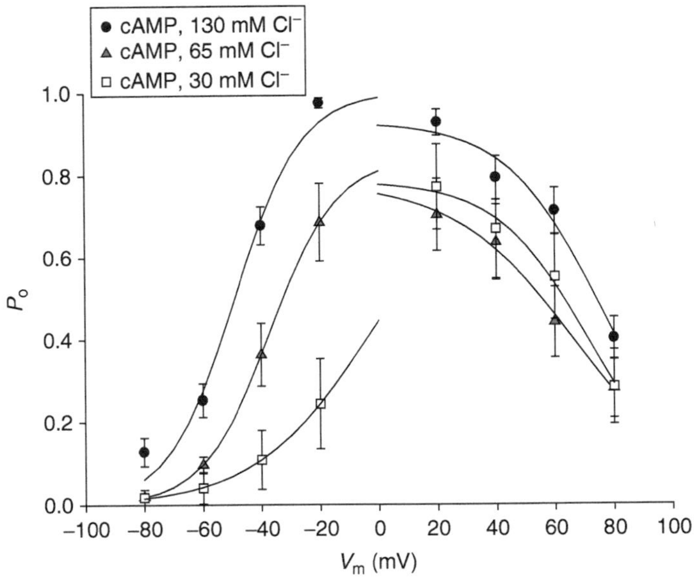

FIGURE 3 V_m-dependence of open probability (P_o) for maxi-Cl$^-$ channels in the presence of 500 μM cAMP (Do *et al.*, 2004a). Averages were calculated from patches that displayed open events at all applied voltages. The channel displayed V_m-dependent inactivation, especially when V_m was either greater than +40 mV or smaller than −40 mV. The topmost curve represented the baseline conditions in which Cl$^-$ concentrations in the micropipette and bath were 130 mM. Reducing the cytoplasmic Cl$^-$ concentration from 130 mm to either 65 or 30 mm reduced P_o at all potentials. The extracellular NaCl concentration was constant at 130 mM, whereas the cytoplasmic Cl$^-$ concentration was varied. Curves were fitted to two Boltzmann equations. Reprinted with the permission of the American Physiological Society.

recently been reported to accelerate inflow into the mouse eye (Bakall *et al.*, 2008). Best2 is associated with Cl$^-$ currents, but its potential physiological role is unclear, in part because it also appears to facilitate outflow of aqueous humor from the eye (Zhang *et al.*, 2008). In the absence of a comprehensive hypothesis, four regulatory pathways, which have received particular attention, are considered here.

A. Swelling-Activation of Cl$^-$ Channels

Over periods of minutes, swelling-activation of Cl$^-$ channels may be the dominant mechanism for ensuring that release of NaCl and water into the aqueous humor by the NPE cells match stromal fluid delivery through the PE cells. For example, whole-cell Cl$^-$ currents of isolated NPE cells can be increased 40-fold by swelling-activation (e.g., Do *et al.*, 2005). The orientation of these channels in the intact epithelium does support transepithelial secretion of Cl^-. Bathing both surfaces of the isolated bovine ciliary epithelium with hypotonic solution triggers a large increase in short-circuit current (Fig. 4) that can be inhibited by Cl^--channel blockers or by leaching Cl^-

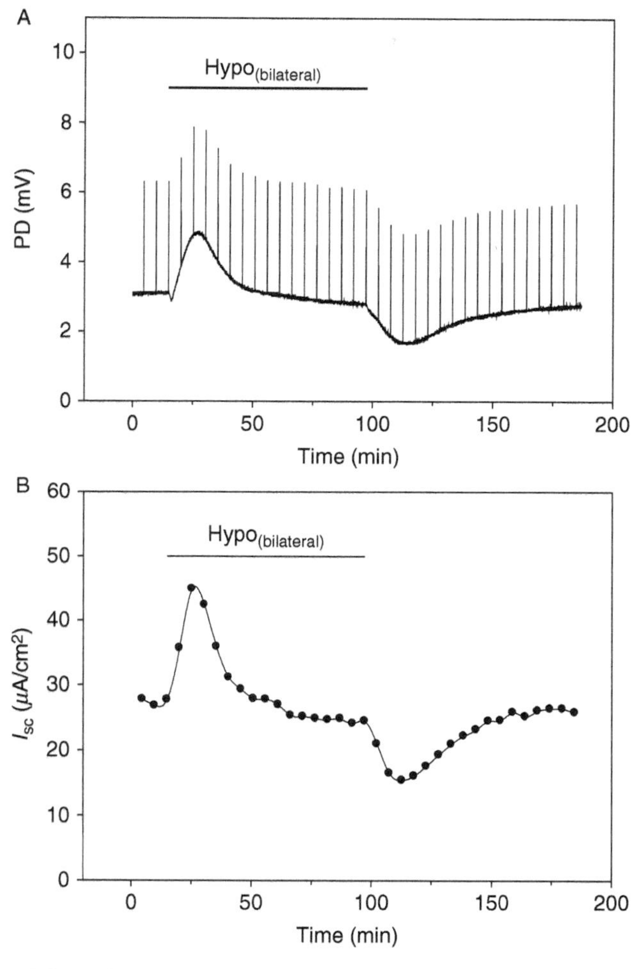

FIGURE 4 Effects of bilateral hypotonicity on electrical parameters in native bovine ciliary epithelium (Do *et al.*, 2006). (A) Measurement of transepithelial PD. Constant-current pulses (3 s) of 10 A were applied to the preparation every 5 min, and the deflections (△PD) were recorded as an index of R. I_{sc} was calculated from the measured PD and R. The aqueous surface was negative to the stromal surface. (B) The calculated I_{sc} from the preparation of (A). Reprinted with the permission of the Association for Research in Vision and Ophthalmology.

from the tissue (Do *et al.*, 2006). The time course of the swelling-activated transepithelial current is closely similar to that of the regulatory volume decrease of the isolated bovine NPE cells (Fig. 5). Cl^--channel activity is enhanced by cell swelling, and thereafter returns to baseline values once release of K^+, Cl^-, and water restores the initial cell volume. The importance of aqueous-surface Cl^- channels is supported by reports that blocking their activity with NPPB inhibits both net Cl^- transfer by the isolated bovine ciliary epithelium (Do and To, 2000) and aqueous humor inflow by the isolated, arterially perfused bovine eye (Shahidullah *et al.*, 2003). Increased transfer of fluid from the PE cells is expected to swell the NPE cells transiently, thereby activating the Cl^- channels at the aqueous surface (Section IV.B.3.b) and stimulating secretion across much of the ciliary epithelium. In those regions of the ciliary epithelium that may possibly reabsorb aqueous humor, delivery of fluid from the NPE cells is expected to trigger swelling-activation of PE-cell Cl^- channels (Section V.B), thereby reducing net secretion.

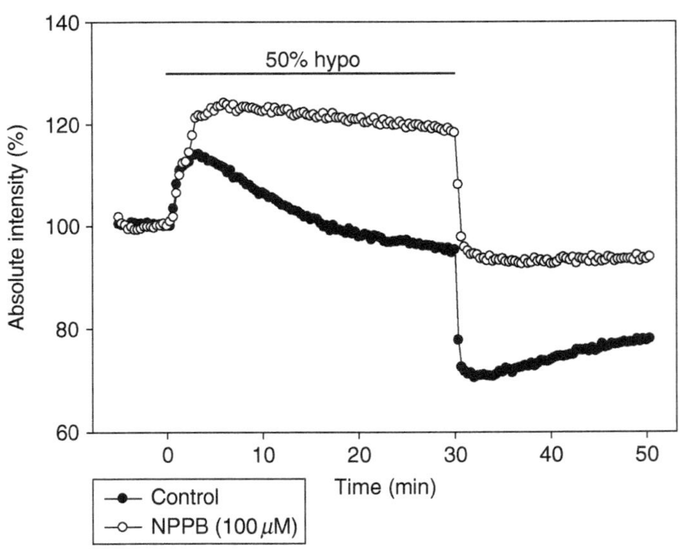

FIGURE 5 Responses of total calcein fluorescence to anisosmotic changes in volume (Do *et al.*, 2006). Fluorescence, normalized to the baseline value in isotonic solution, increases with cell swelling. The regulatory volume decrease (RVD) after hypotonic challenge was markedly inhibited by the Cl^--channel blocker NPPB. Reprinted with the permission of the Association for Research in Vision and Ophthalmology.

B. Cyclic Adenosine Monophosphate

Antagonists of β-adrenergic receptors lower IOP, and topical nonselective β-adrenergic antagonists have long been mainstays of glaucoma therapy (Toris and Camras, 2008). Agonists to all three β-receptors, β_1, β_2, and β_3, stimulate adenylyl cyclase to produce cAMP, an effect mediated by the heterotrimeric G protein G_s (Hoffman *et al.*, 1996). The β-blockers both reduce cAMP production and lower IOP by reducing inflow of aqueous humor. A causal relationship between these two actions has been widely presumed. However, many observations, summarized elsewhere (Yorio, 1985; McLaughlin *et al.*, 2001a; Do and Civan, 2004), have seemed at odds with the idea that the inflow reduction by β-antagonists is necessarily mediated by a fall in intracellular concentration of cAMP. Particularly puzzling have been the reports that increasing cAMP by directly stimulating adenylyl cyclase with forskolin actually lowers inflow (Caprioli *et al.*, 1984; Lee *et al.*, 1984), and that the β-agonist isoproterenol, also expected to increase cAMP, lowers IOP in water-loaded rabbits (Vareilles *et al.*, 1977). In addition, as discussed in Chapter 4 (Macknight and Civan, 2008), application of cAMP also does not reverse the effects of the β-blocker timolol on the intracellular elemental composition of intact rabbit ciliary epithelium (McLaughlin *et al.*, 2001a).

In part, the unexpected observations concerning the effects of cAMP and β-adrenergic agents may reflect the multiple actions of the second messenger on sites within the ciliary epithelium (Do and Civan, 2004; Table II). Several known effects of cAMP are indeed expected to stimulate aqueous humor formation, including (Fig. 1A) activation of the Na^+-K^+-$2Cl^-$ PE-cell symports (Crook *et al.*, 2000) and of some the NPE-cell Cl^- channels (Chen *et al.*, 1994, Edelman *et al.*, 1995). In addition, the β-adrenergic agonist isoproterenol has been observed to increase Na^+, K^+-activated ATPase activity in cultured human NPE cells (Liu *et al.*, 2001). In contrast, direct application of cAMP can reduce net secretion (Fig. 1B) by inhibiting the Na^+, K^+-pump (Delamere and King, 1992), by blocking PE–NPE gap junctions (Do *et al.*, 2008), and by activating maxi-Cl^- channels of the PE cells (Fleischhauer *et al.*, 2001; Do *et al.*, 2004b). Given these opposing actions of cAMP on ciliary epithelial secretion, the consistently ocular-hypotensive effect of β-blockers raises the possibility of compartmentation of cAMP. This possibility has been substantiated in Calu-3 cells. Huang *et al.* (2001) found that 1 μM adenosine increased local cAMP concentration enough to activate CFTR Cl^- channels with little increase in the total cAMP content. Taken together with additional results, these authors concluded that clustering of receptors, G proteins, adenylyl cyclase, and PKA permitted local activation of the target, CFTR.

The immediately foregoing considerations suggest that part of the apparent inconsistencies in the results obtained with β-agonists, β-antagonists, and cAMP may reflect drug-triggered effects on cAMP production in the local

TABLE II

Effects of cAMP on Transport Components of the Ciliary Epithelium

Transporter target	Effect	Predicted action on net secretion	References
• Na^+-K^+-$2Cl^-$ of PE	↑net Cl^- uptake from stroma by PE	↑	(Delamere and King, 1992)
• PE–NPE gap junctions	↓transfer to NPE	↓	(Do *et al.*, 2008)
• NPE Cl^- channels	↑Cl^- release to aqueous	↑	(Chen *et al.*, 1994; Edelman *et al.*, 1995)
• NPE Na^+, K^+-ATPase	↓Na^+ pump activity	↓	(Delamere and King, 1992)
• PE maxi-Cl^-	↑Cl^- release from PE to stroma	↓	(Do *et al.*, 2004b)

PE, pigmented ciliary epithelial; NPE, nonpigmented ciliary epithelial.

microenvironment of the adrenergic receptors. In addition, cAMP does not mediate all of the actions of β-adrenergic agonists (Torphy, 1994). A number of reports have recently documented that β-adrenergic receptors can couple to G_i proteins, and not exclusively to G_s proteins (Denson *et al.*, 2005). For example, Denson *et al.* (2005) found that the β-agonist isoproterenol activates BK potassium channels by coupling to G_i, activating cytosolic phospholipase A_2 (c-PLA_2), and stimulating production of arachidonic acid. Isoproterenol's action was blocked by the β-antagonist propranolol. It is entirely possible that the isoproterenol-triggered activation of NPE-cell BK channels is also mediated by arachidonic acid. Stimulation of BK channels of rabbit native NPE cells by isoproterenol is not mediated by cAMP, but does depend on G-protein coupling (Bhattacharyya *et al.*, 2002). Furthermore, arachidonic acid has long been known to activate NPE-cell K^+ channels (Civan *et al.*, 1994).

In summary, β-blockers effectively lower ciliary epithelial secretion, IOP, and cAMP formation. However, discordant results obtained by applying β-agonists, β-antagonists, and cAMP have raised the possibility that changes in total cellular cAMP concentration do not necessarily mediate the drug-triggered changes in aqueous humor dynamics. Recent studies have now led to at least two possible explanations. First, cAMP exerts many, sometimes opposing, effects on ciliary epithelial secretion (Table II). Administration of large concentrations of membrane-permeant forms of cAMP is likely to affect all of these transport targets. In contrast, drugs, hormones, and biologically active peptides that bind to receptors at specific membrane areas may elevate cAMP in circumscribed microenvironments, targeting a narrow range of membrane transporters. Second, although *β*-agonists have been widely presumed to act solely through G_s-mediated production of cAMP, at least one alternative pathway has been demonstrated. The agonists and antagonists can also trigger G_i-mediated activation of phospholipase A_2, enhancing arachidonic acid formation.

C. Carbonic Anhydrase

Inhibition of CA provided the first successful approach for lowering IOP by reducing the rate of aqueous humor inflow (reviewed by Brubaker, 1998). The first successful clinical trials were reported more than half-a-century ago and the inhibitor acetazolamide has been long known to reduce accessibility of plasma HCO_3^- to the aqueous humor (Maren, 1976). Nevertheless, understanding of the probable mechanism of action of CA inhibitors has developed much more recently (Helbig *et al.*, 1989a; Wiederholt *et al.*, 1991). As discussed in Secton III. B.1.b, CA directly stimulates (Sterling *et al.*, 2001; Li *et al.*, 2002) the NHE1 Na^+/H^+ and AE2 Cl^-/HCO_3^- antiports (Fig. 1A; Counillon *et al.*, 2000). Thus, CA inhibitors, such as acetazolamide and dorzolamide, likely block the first step in aqueous humor formation by inhibiting NaCl uptake from the stroma.

This hypothesis has been supported by measurements of IOP in living mice during topical inhibition of the symports (Avila *et al.*, 2002a). Measurements were conducted with an electrophysiological approach (the servo-null micropipette system) that permits continuous monitoring of IOP in the small mouse eye (Avila *et al.*, 2001a). Topical application of each of three selective inhibitors of Na^+/H^+ antiports (Figs. 6A, and 7A, B, and D) reduced IOP. The promptness of the IOP response likely reflects enhanced delivery of drug from the tear film into the aqueous humor around the tip of the exploring micropipette (Wang *et al.*, 2007). Bumetanide alone had no significant effect during the period of recording (Fig. 6B). However, bumetanide further lowered IOP if applied after either the selective Na^+/H^+-exchange inhibitors (Figs. 7A, B, and D) or after blocking CA with dorzolamide (Fig. 7C). These data are consistent with the notion that the Na^+/H^+ and Cl^-/HCO_3^- antiports play a major role in secretion, and that CA inhibitors act on these exchangers to slow aqueous humor formation.

D. A_3 Adenosine Receptors

Among other potential regulators of aqueous humor dynamics, A_3-subtype adenosine receptors (A_3ARs) are of particular interest since knockout of these receptors reduces the IOP of living mice (Avila *et al.*, 2002b). Furthermore,

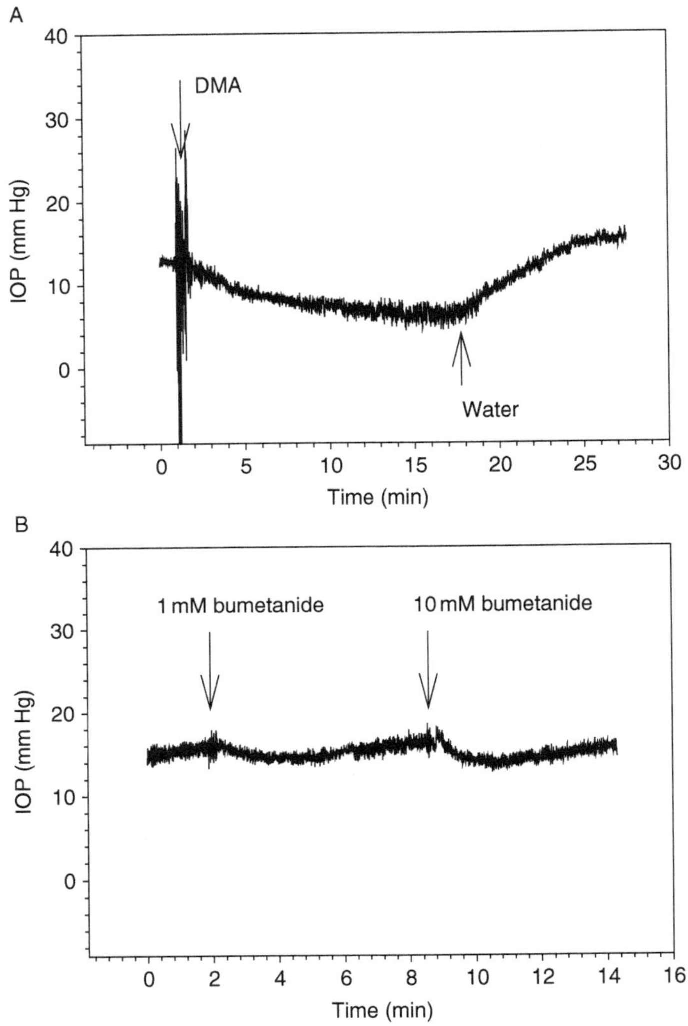

FIGURE 6 Responses of mouse IOP to inhibition of Na^+/H^+ antiports with dimethylamiloride (DMA) or to inhibition of Na^+-K^+-$2Cl^-$ symports with bumetanide (Avila *et al.*, 2002a). (A) DMA (1 mM) lowered IOP. Water was added at the conclusion of the experiment to verify the patency of the micropipette by hypotonically raising IOP. (B) Neither 1 nor 10 mM bumetanide itself changed mouse IOP. Reprinted with the permission of the Association for Research in Vision and Ophthalmology.

these receptors are greatly overexpressed in NPE cells from patients with the pseudo-exfoliation syndrome, which is a major cause of open-angle glaucoma (Schlötzer-Schrehardt *et al.*, 2005).

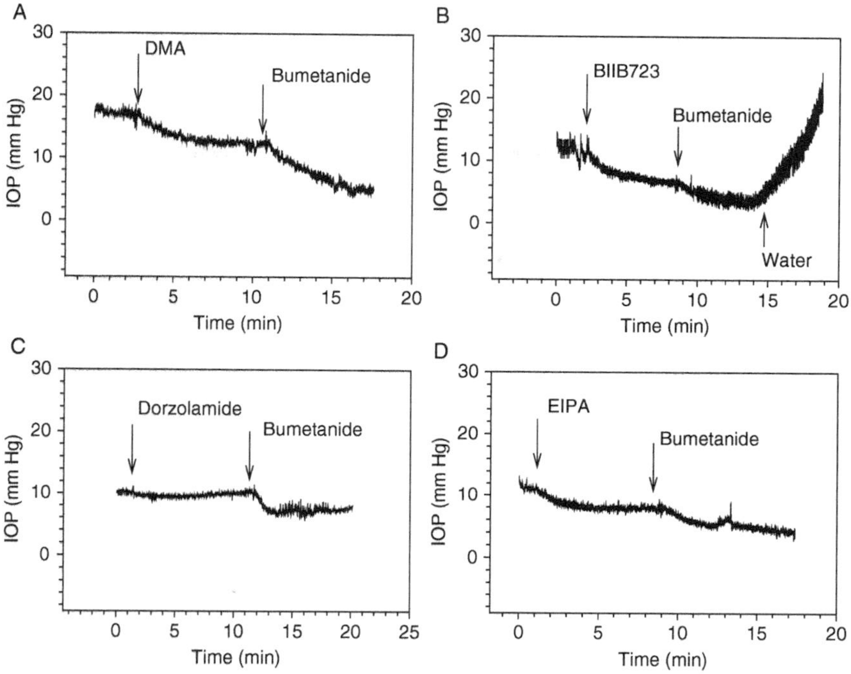

FIGURE 7 Responses to topical addition of direct or indirect inhibitors of Na^+/H^+ antiports, followed by bumetanide (Avila *et al.*, 2002a). (A) 1 mM DMA followed by 1 mM bumetanide, (B) 1 mM BIIB723 followed by 1 mM bumetanide, (C) 55.4 mM dorzolamide followed by 1 mM bumetanide, and (D) 1 mM EIPA followed by 1 mM bumetanide. Bumetanide significantly reduced IOP after prior inhibition of the Na^+/H^+ antiports. Reprinted with the permission of the Association for Research in Vision and Ophthalmology.

Adenosine was first noted to stimulate transepithelial Cl^- transport in studies of frog cornea nearly 30 years ago (Reinach *et al.*, 1979; Spinowitz and Zadunaisky, 1979). The adenosine was subsequently thought to act by increasing Cl^- permeability across the apical membrane of the corneal epithelium (Patarca *et al.*, 1983). At the concentration applied (200 μM), the action of adenosine could have been mediated by any of the currently recognized adenosine receptors (A_1, A_{2A}, A_{2B}, and A_3) (Fredholm *et al.*, 1994). Adenosine has subsequently been found to activate Cl^- channels of isolated mammalian preparations, native bovine and cultured human NPE cells, and intact rabbit ciliary epithelium (Carré *et al.*, 1997). Whole-cell patch-clamp recording and volumetric measurements have established that the adenosine-triggered activation of Cl^- channels is mediated by A_3ARs (Mitchell *et al.*, 1999; Carré *et al.*, 2000). This activation is inhibited by selective A_3AR antagonists (Mitchell *et al.*, 1999; Carré *et al.*, 2000). Message for A_3ARs is expressed by cultured human NPE cells and the ciliary processes of rabbit (Mitchell *et al.*, 1999). The similarity of the macroscopic current characteristics of the A_3AR- and swelling-activated Cl^- currents suggests that both currents permeate the same channels (Carré *et al.*, 2000).

Adenosine can be physiologically delivered to the aqueous surface by ATP release and ectoenzymatic metabolism of ATP by the NPE cells themselves (Mitchell *et al.*, 1998; Fig. 8). ATP can also be released to the stromal surface by the PE cells. Binding of ATP to P2Y2 receptors (Shahidullah and Wilson, 1997) initiates a cascade leading to direct stimulation of maxi-Cl^- channels (Fleischhauer *et al.*, 2001; Do *et al.*, 2004b). Tamoxifen synergistically enhances the ATP-triggered activation of Cl^- channels, likely by binding to a plasma-membrane estrogen receptor (Mitchell *et al.*, 2000), but the mode of interaction with the ATP-induced signaling cascade is unknown.

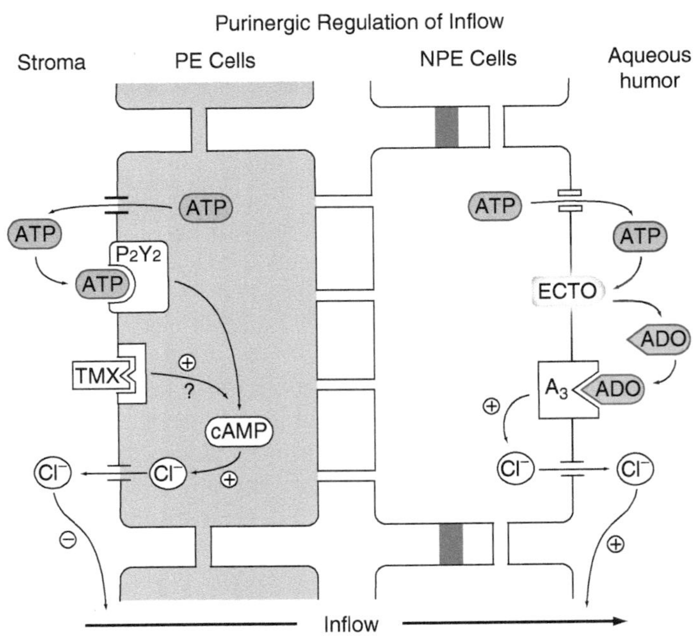

FIGURE 8 Purinergic regulation of ciliary epithelial secretion. Following its autocrine release by the NPE cells, ATP is metabolized by ectoenzymes to adenosine, stimulating A_3 adenosine receptors to activate Cl^- channels and enhance inflow. At the stromal surface, ATP released from the PE cells directly stimulates ATP receptors to initiate a cascade leading to activation of maxi-Cl^- channels, thereby reducing net inflow. Tamoxifen synergistically enhances the effect of ATP.

Release of ATP at both surfaces of the ciliary epithelium leads to a potential push–pull mechanism of purinergic regulation, with adenosine-activated NPE-cell Cl^- channels enhancing and ATP-activated PE-cell Cl^- channels diminishing the rate of net aqueous humor formation. Which effect predominates would depend on gating of the conduits for ATP release, local ectoenzyme activity, the membrane concentration of the Cl^- channels, and the influence of other regulators of the Cl^--channel activities at the opposite surfaces.

The role of adenosine in regulating IOP has been examined in the living mouse. A_3-null mice display lowered baseline IOP (Fig. 9; Avila *et al.*, 2002b). In wild-type mice, topical adenosine elicits a large increase in IOP (Fig. 10B; Avila *et al.*, 2002b; Yang *et al.*, 2005), as do the selective A_3AR agonists Cl-IB-MECA (Avila *et al.*, 2001b) and IB-MECA (Avila *et al.*, 2001b; Yang *et al.*, 2005). As expected, the selective A_3AR antagonists MRS 1191 and MRS 1097 (Avila *et al.*, 2001b, 2002b) and MRS 1292 (Yang *et al.*, 2005) exert an opposite effect, lowering IOP. In contrast, the effects of the agonist adenosine and the antagonist MRS 1191 are very much reduced in the knockout mouse (Avila *et al.*, 2002b). Parenteral administration of adenosine to normal humans has been reported to produce a small decrease in IOP (Polska *et al.*, 2003), possibly mediated by systemic effects.

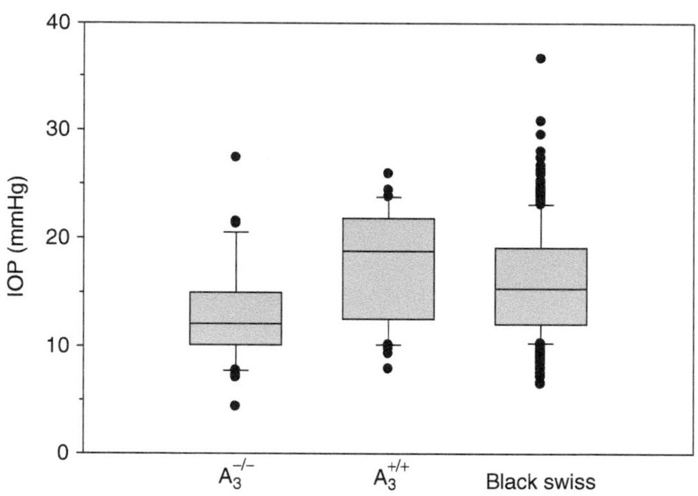

FIGURE 9 Baseline IOP in $A_3AR^{-/-}$ ($n = 44$) and $A_3AR^{+/+}$ control ($n = 42$) mice (Avila *et al.*, 2002b) and in black Swiss outbred mice ($n = 292$) measured in earlier studies (Fig. 1 from Avila *et al.*, 2002b). Central horizontal lines, medians; lower and upper lines, all data points between the 25th and 75th percentiles; whiskers, range of results between the 10th and 90th percentiles. Circles are individual data lying beyond this range. The IOP in the $A_3AR^{-/-}$ mice was significantly lower than that in the two control groups. Reprinted with the permission of the Association for Research in Vision and Ophthalmology.

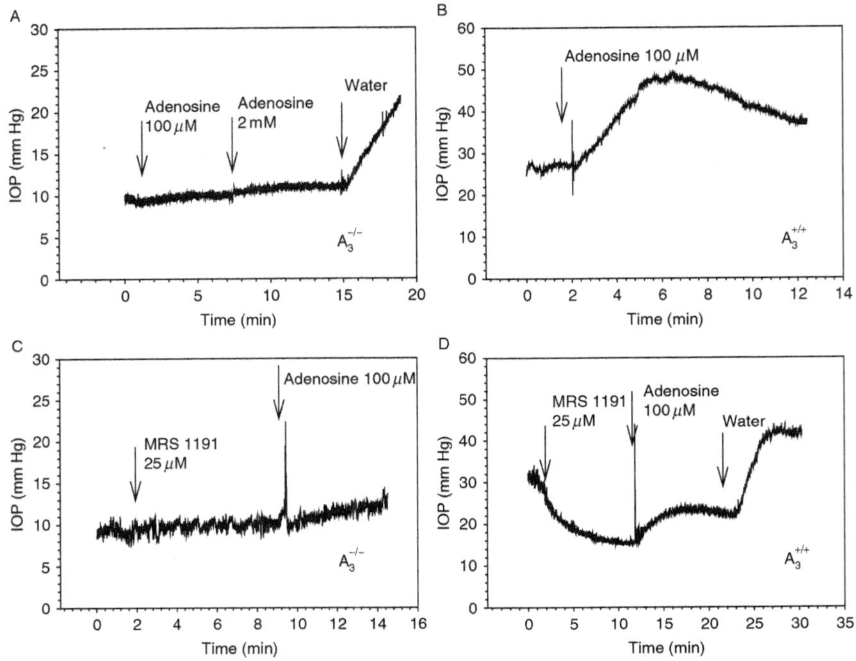

FIGURE 10 Effects of the nonselective AR agonist adenosine and the A_3-selective antagonist MRS 1191 on IOP in $A_3AR^{-/-}$ and $A_3AR^{+/+}$ mice (Avila *et al.*, 2002b). Each trace was obtained from continuous measurement of a single mouse. (A) Adenosine had little effect on IOP in $A_3AR^{-/-}$ mice at a droplet concentration of 100 μM or 2 mM, whereas intraperitoneal water elevated IOP, as noted in wild-type mice. (B) In contrast, the lower adenosine concentration markedly elevated IOP in control $A_3AR^{+/+}$ mice. (C) Application of 25 μM MRS 1191 did not alter baseline IOP in $A_3AR^{-/-}$ mice and did not inhibit the subsequent slight response to 100 μM adenosine. (D) The same droplet concentration of MRS 1191 markedly lowered baseline IOP in control $A_3AR^{+/+}$ mice and strongly inhibited the subsequent response to 100 μM adenosine. Intraperitoneal water produced the expected increase in IOP. Reprinted with the permission of the Association for Research in Vision and Ophthalmology.

Topical administration of two selective A_3AR antagonists has been found to reduce IOP of nonhuman primates (Okamura *et al.*, 2004), as expected from the *in vitro* and living-mouse studies.

VII. SUMMARY OF CURRENT VIEWS, RECENT ADVANCES, AND FUTURE DIRECTIONS

A. Fundamental Basis of Ciliary Epithelial Secretion

Aqueous humor is formed by the transfer of solute from the stroma to the posterior chamber of the eye (Fig. 1A). Although gap junctions subserve intercommunication between cells both in the PE and NPE layers, the fundamental secretory unit is the PE–NPE couplet, a point that will be developed more fully in Chapter 4 (Macknight and Civan, 2008). The current prevailing view is that water flows from stroma to aqueous humor by local osmosis in response to the osmotic gradient established by the solute transfer.

B. Species Variation

The ionic compositions of the aqueous humor and of the plasma are largely conserved among mammals. One of the largest differences reported has been in the HCO_3^- concentration of the anterior aqueous humor, which is some 28 mM in the rabbit and 22 mM in the human (Krupin and Civan, 1996). The anion gap, defined as the [Na^+ concentration+K^+ concentration–Cl^- concentration] is commonly taken as an approximate index of the HCO_3^- concentration. Calculated from the data of Gerometta *et al.* (2005), the anion gap in the aqueous humor of the anterior chamber in several species is 19 mM (sheep), 28 mM (pig and cow), and 40 mM (rabbit). This ranking does not correlate with the calculated values of the anion gap in the plasma of these species. The corresponding anion gap calculated from Table 12-1 of Krupin and Civan (1996) is 45 mM for the rabbit (an overestimate of the measured bicarbonate concentration of 28 mM) and 25 mM for the human (close to the measured value of 22 mM). One interpretation of these measurements would be that there may be a spectrum of bicarbonate concentrations in the aqueous humor, with the sheep at the low end and the rabbit at the high end of the scale. The human bicarbonate concentration is likely close to that of the pig and cow.

With the exception of these relatively minor differences, the formation of the aqueous humor largely consists in secreting an isosmotic NaCl solution. It seems reasonable to presume that this secretion is conducted by much the same transporters in different mammalian species. Indeed, bumetanide, Cl^--channel blockers, and CA inhibitors inhibit transport across the ciliary epithelia of rabbit and cow, and pig as well (Wu *et al.*, 2004; Kong *et al.*, 2006). However, there is increasing awareness of functional differences among the several mammalian preparations currently used for experimental study. For example, removing bicarbonate from the bathing solutions qualitatively depolarizes the transepithelial potential (PD) across the isolated ciliary epithelium of several species. However, there are quantitative differences. Bicarbonate removal only partially lowers the PD across the ox preparation (Do and To, 2000), abolishes the PD across the pig preparation (Kong *et al.*, 2006), and reverses the PD across the rabbit preparation (Kishida *et al.*, 1981; Krupin *et al.*, 1984). The nonselective Cl^--channel blocker NPPB is commonly used to block Cl^- channels in ciliary epithelial cells of other preparations, but is ineffective in changing PD across the pig ciliary epithelium (Kong *et al.*, 2006). In contrast, another nonselective Cl^--channel blocker, niflumic acid, also used in studying Cl^- channels from cells of other species, nearly completely abolishes the PD across the pig preparation (Kong *et al.*, 2006). Whether this pharmacological profile reflects fundamental biophysical differences in the porcine channel, or perhaps simply accessibility to the blocking sites, is unknown. In view of these observed differences, further study of species variance would be welcome.

C. Circulation

The ciliary plasma flow can be roughly estimated to be ~73 μl/min in humans and ~50 μl/min in monkeys (Reitsamer and Kiel, 2008). Thus, the maximal diurnal flow of aqueous humor (~3 μl/min) constitutes only some 4–6% of the plasma flow delivered. As the plasma flow rate falls, the percentage extraction of water from that plasma increases in order to sustain the same rate of aqueous humor secretion. Once the flow rate is reduced by more than 25%, further lowering of plasma flow produces progressive reductions in the rate of aqueous humor formation (Reitsamer and Kiel, 2003, 2008). This phenomenon is analogous to the relationship between the renal plasma flow and glomerular filtration rate (Fig. 33-6D; Giebisch and Windhager, 2005). As in the kidney, the progressive extraction of water necessarily increases the protein concentration of the capillary plasma. This increase in protein concentration elevates the plasma oncotic pressure, restraining further release of water (and with it, solute) from the capillary lumen to the stroma of the ciliary processes. The recent information concerning the dependence of aqueous inflow on circulatory dynamics and its potential significance are considered in Chapter 9 of this volume.

D. Topography

Regional differences in the expression of Na^+, K^+-activated ATPase, other proteins and biologically active peptides, summarized by McLaughlin *et al.* (2001b), led to the suggestion that net ion transport might actually be reversed across some area of the ciliary epithelium (Ghosh *et al.*, 1991). Electron-probe X-ray microanalyses of intact rabbit ciliary epithelium have provided support for this possibility (McLaughlin *et al.*, 2001b, 2004, 2007). As discussed in Chapter 4 (Macknight and Civan, 2008), this functional topography might provide the basis for a novel approach to reducing net inflow and IOP.

E. Regulation

Among many known modifiers of net secretion, swelling-activation of Cl^- channels at the two surfaces of the ciliary epithelium may provide the major minute-to-minute regulation of net secretion. Swelling-activated Cl^- channels at the aqueous surface are predominant since swelling the entire intact bovine epithelium enhances baseline net Cl^- current directed toward the aqueous surface (Do *et al.*, 2006).

The second-messenger cAMP is an important regulator of multiple transporters subserving aqueous humor formation. However, agonists and antagonists of β-adrenergic receptors probably alter inflow by changing G_s-mediated cAMP concentration in microenvironments of these targets, rather than by altering the total cytosolic concentration. In addition, these β-adrenergic drugs appear to act through at least one additional signaling cascade, triggering G_i-mediated changes in arachidonic acid.

CA is likely important in regulating aqueous humor formation by stimulating Na^+/H^+ and Cl^-/HCO_3^- exchange activity at the stromal surface of the epithelium, the likely target of CA inhibitors.

Agonists of A_3ARs stimulate NPE-cell Cl^- channels *in vitro* and elevate IOP in the living mouse. Antagonists exert opposite actions. In view of the increasingly evident species variations, the development of A_3 antagonists that are effective across species enhances the potential human relevance of their ocular hypotensive effects (Yang *et al.*, 2005; Wang *et al.*, 2008).

References

Agre, P., and Kozono, D. (2003). Aquaporin water channels: Molecular mechanisms for human diseases. *FEBS Lett.* **555,** 72–78.

Anguíta, J., Chalfant, M. L., Civan, M. M., and Coca-Prados, M. (1995). Molecular cloning of the human volume-sensitive chloride conductance regulatory protein, pICln, from ocular ciliary epithelium. *Biochem. Biophys. Res. Commun.* **208,** 89–95.

Aperia, A., Fryckstedt, J., Svensson, L., Hemmings, H. C., Jr., Nairn, A. C., and Greengard, P. (1991). Phosphorylated Mr 32,000 dopamine- and cAMP-regulated phosphoprotein inhibits Na+,K(+)-ATPase activity in renal tubule cells. *Proc. Natl. Acad. Sci. USA* **88,** 2798–2801.

Asejczyk-Widlicka, M., and Pierscionek, B. K. (2007). Fluctuations in intraocular pressure and the potential effect on aberrations of the eye. *Br. J. Ophthalmol.* **91,** 1054–1058.

Avila, M. Y., Carré, D. A., Stone, R. A., and Civan, M. M. (2001a). Reliable measurement of mouse intraocular pressure by a servo-null micropipette system. *Invest. Ophthalmol. Vis. Sci.* **42,** 1841–1846.

Avila, M. Y., Stone, R. A., and Civan, M. M. (2001b). A(1)-, A(2A)- and A(3)-subtype adenosine receptors modulate intraocular pressure in the mouse. *Br. J. Pharmacol.* **134,** 241–245.

Avila, M. Y., Seidler, R. W., Stone, R. A., and Civan, M. M. (2002a). Inhibitors of NHE-1 Na^+/H^+ exchange reduce mouse intraocular pressure. *Invest. Ophthalmol. Vis. Sci.* **43,** 1897–1902.

Avila, M. Y., Stone, R. A., and Civan, M. M. (2002b). Knockout of A(3) adenosine receptors reduces mouse intraocular pressure. *Invest. Ophthalmol. Vis. Sci.* **43,** 3021–3026.

Bakall, B., McLaughlin, P., Stanton, J. B., Zhang, Y., Hartzell, H. C., Marmorstein, L. Y., and Marmorstein, A. D. (2008). Bestrophin-2 is involved in the generation of intraocular pressure. *Invest. Ophthalmol. Vis. Sci.* **49,** 1563–1570.

Barbe, M. T., Monyer, H., and Bruzzone, R. (2006). Cell-cell communication beyond connexins: The pannexin channels. *Physiology (Bethesda)* **21,** 103–114.

Barsotti, M. F., Bartels, S. P., Freddo, T. F., and Kamm, R. D. (1992). The source of protein in the aqueous humor of the normal monkey eye. *Invest. Ophthalmol. Vis. Sci.* **33,** 581–595.

Bert, R. J., Caruthers, S. D., Jara, H., Krejza, J., Melhem, E. R., Kolodny, N. H., Patz, S., and Freddo, T. F. (2006). Demonstration of an anterior diffusional pathway for solutes in the normal human eye with high spatial resolution contrast-enhanced dynamic MR imaging. *Invest. Ophthalmol. Vis. Sci.* **47,** 5153–5162.

Bhattacharyya, B. J., Lee, E., Krupin, D., Hockberger, P., and Krupin, T. (2002). (−)-Isoproterenol modulation of maxi-K(+) channel in nonpigmented ciliary epithelial cells through a G-protein gated pathway. *Curr. Eye Res.* **24,** 173–181.

Bill, A. (1966). Conventional and uveo-scleral drainage of aqueous humour in the cynomolgus monkey (Macaca irus) at normal and high intraocular pressures. *Exp. Eye Res.* **5,** 45–54.

Bill, A. (1973). The role of ciliary blood flow and ultrafiltration in aqueous humor formation. *Exp. Eye Res.* **16,** 287–298.

Bill, A., and Phillips, C. I. (1971). Uveoscleral drainage of aqueous humour in human eyes. *Exp. Eye Res.* **12,** 275–281.

Blanco, G., and Mercer, R. W. (1998). Isozymes of the Na-K-ATPase: Heterogeneity in structure, diversity in function. *Am. J. Physiol.* **275,** F633–F650.

Bowler, J. M., Peart, D., Purves, R. D., Carré, D. A., Macknight, A. D., and Civan, M. M. (1996). Electron probe X-ray microanalysis of rabbit ciliary epithelium. *Exp. Eye Res.* **62,** 131–139.

Brubaker, R. F. (1998). Clinical measurement of aqueous dynamics: Implications for addressing glaucoma. *In* "Eye's Aqueous Humor: From Secretion to Glaucoma" (M. M. Civan, ed.), pp. 234–284. Academic Press, San Diego.

Candia, O. A., To, C. H., Gerometta, R. M., and Zamudio, A. C. (2005). Spontaneous fluid transport across isolated rabbit and bovine ciliary body preparations. *Invest. Ophthalmol. Vis. Sci.* **46,** 939–947.

Candia, O. A., To, C. H., and Law, C. S. (2007). Fluid transport across the isolated porcine ciliary epithelium. *Invest. Ophthalmol. Vis. Sci.* **48,** 321–327.

Caprioli, J., Sears, M., Bausher, L., Gregory, D., and Mead, A. (1984). Forskolin lowers intraocular pressure by reducing aqueous inflow. *Invest. Ophthalmol. Vis. Sci.* **25,** 268–277.

Carré, D. A., Tang, C. S., Krupin, T., and Civan, M. M. (1992). Effect of bicarbonate on intracellular potential of rabbit ciliary epithelium. *Curr. Eye Res.* **11,** 609–624.

Carré, D. A., Mitchell, C. H., Peterson-Yantorno, K., Coca-Prados, M., and Civan, M. M. (1997). Adenosine stimulates Cl^- channels of nonpigmented ciliary epithelial cells. *Am. J. Physiol.* **273,** C1354–C1361.

Carré, D. A., Mitchell, C. H., Peterson-Yantorno, K., Coca-Prados, M., and Civan, M. M. (2000). Similarity of A(3)-adenosine and swelling-activated Cl(−) channels in nonpigmented ciliary epithelial cells. *Am. J. Physiol. Cell Physiol.* **279,** C440–C451.

Chen, S., Inoue, R., Inomata, H., and Ito, Y. (1994). Role of cyclic AMP-induced Cl conductance in aqueous humour formation by the dog ciliary epithelium. *Br. J. Pharmacol.* **112,** 1137–1145.

Chen, L., Wang, L., and Jacob, T. J. (1999). Association of intrinsic pICln with volume-activated Cl- current and volume regulation in a native epithelial cell. *Am. J. Physiol.* **276,** C182–C192.

Civan, M. M. (2003). The fall and rise of active chloride transport: Implications for regulation of intraocular pressure. *J. Exp. Zoolog. A Comp. Exp. Biol.* **300,** 5–13.

Civan, M. M., Coca-Prados, M., and Peterson-Yantorno, K. (1994). Pathways signaling the regulatory volume decrease of cultured nonpigmented ciliary epithelial cells. *Invest. Ophthalmol. Vis. Sci.* **35,** 2876–2886.

Civan, M. M., Coca-Prados, M., and Peterson-Yantorno, K. (1996). Regulatory volume increase of human non-pigmented ciliary epithelial cells. *Exp. Eye Res.* **62,** 627–640.

Clapham, D. E. (1998). The list of potential volume-sensitive chloride currents continues to swell (and shrink). *J. Gen. Physiol.* **111,** 623–624.

Coca-Prados, M., Ghosh, S., Gilula, N. B., and Kumar, N. M. (1992). Expression and cellular distribution of the alpha 1 gap junction gene product in the ocular pigmented ciliary epithelium. *Curr. Eye Res.* **11,** 113–122.

Coca-Prados, M., Anguíta, J., Chalfant, M. L., and Civan, M. M. (1995a). PKC-sensitive Cl-channels associated with ciliary epithelial homologue of pICln. *Am. J. Physiol.* **268,** C572–C579.

Coca-Prados, M., Fernandez-Cabezudo, M. J., Sanchez-Torres, J., Crabb, J. W., and Ghosh, S. (1995b). Cell-specific expression of the human Na^+,K($^+$)-ATPase beta 2 subunit isoform in the nonpigmented ciliary epithelium. *Invest. Ophthalmol. Vis. Sci.* **36,** 2717–2728.

Coca-Prados, M., Sánchez-Torres, J., Peterson-Yantorno, K., and Civan, M. M. (1996). Association of ClC-3 channel with Cl^- transport by human nonpigmented ciliary epithelial cells. *J. Membr. Biol.* **150,** 197–208.

Coffey, K. L., Krushinsky, A., Green, C. R., and Donaldson, P. J. (2002). Molecular profiling and cellular localization of connexin isoforms in the rat ciliary epithelium. *Exp. Eye Res.* **75,** 9–21.

Cole, D. F. (1960). Effects of some metabolic inhibitors upon the formation of the aqueous humour in rabbits. *Br. J. Ophthalmol.* **44,** 739–750.

Cole, D. F. (1966). Aqueous humor formation. *Doc. Ophthalmol.* **21,** 116–238.

Cole, D. F. (1977). Secretion of the aqueous humour. *Exp. Eye Res.* **25**(Suppl.), 161–176.

Collaborative Normal-Tension Glaucoma Study Group (1998a). Comparison of glaucomatous progression between untreated patients with normal-tension glaucoma and patients with therapeutically reduced intraocular pressures. *Am. J. Ophthalmol.* **126,** 487–497.

Collaborative Normal-Tension Glaucoma Study Group (1998b). The effectiveness of intraocular pressure reduction in the treatment of normal-tension glaucoma. *Am. J. Ophthalmol.* **126,** 498–505.

Counillon, L., Touret, N., Bidet, M., Peterson-Yantorno, K., Coca-Prados, M., Stuart-Tilley, S., Wilhelm, S., Alper, S. L., and Civan, M. M. (2000). Na^+/H^+ and Cl^-/HCO_3^- antiporters of bovine pigmented ciliary epithelial cells. *Pflügers Arch.* **440,** 667–678.

Cowan, C. A., Yokoyama, N., Bianchi, L. M., Henkemeyer, M., and Fritzsch, B. (2000). EphB2 guides axons at the midline and is necessary for normal vestibular function. *Neuron* **26,** 417–430.

Crambert, G., Hasler, U., Beggah, A. T., Yu, C., Modyanov, N. N., Horisberger, J. D., Lelievre, L., and Geering, K. (2000). Transport and pharmacological properties of nine different human Na, K-ATPase isozymes. *J. Biol. Chem.* **275,** 1976–1986.

Crook, R. B., and Polansky, J. R. (1994). Stimulation of Na^+,K^+,Cl^- cotransport by forskolin-activated adenylyl cyclase in fetal human nonpigmented epithelial cells. *Invest. Ophthalmol. Vis. Sci.* **35,** 3374–3383.

Crook, R. B., and Riese, K. (1996). Beta-adrenergic stimulation of Na,K^+, Cl^- cotransport in fetal nonpigmented ciliary epithelial cells. *Invest. Ophthalmol. Vis. Sci.* **37,** 1047–1057.

Crook, R. B., von Brauchitsch, D. K., and Polansky, J. R. (1992). Potassium transport in nonpigmented epithelial cells of ocular ciliary body: Inhibition of a Na^+,K^+,Cl^- cotransporter by protein kinase C. *J. Cell. Physiol.* **153,** 214–220.

Crook, R. B., Takahashi, K., Mead, A., Dunn, J. J., and Sears, M. L. (2000). The role of NaKCl cotransport in blood-to-aqueous chloride fluxes across rabbit ciliary epithelium. *Invest. Ophthalmol. Vis. Sci.* **41,** 2574–2583.

Delamere, N. A., and King, K. L. (1992). The influence of cyclic AMP upon Na,K-ATPase activity in rabbit ciliary epithelium. *Invest. Ophthalmol. Vis. Sci.* **33,** 430–435.

Delamere, N. A., Parkerson, J., and Hou, Y. (1997). Indomethacin alters the Na,K-ATPase response to protein kinase C activation in cultured rabbit nonpigmented ciliary epithelium. *Invest. Ophthalmol. Vis. Sci.* **38,** 866–875.

Denson, D. D., Li, J., Wang, X., and Eaton, D. C. (2005). Activation of BK channels in GH3 cells by a c-PLA2-dependent G-protein signaling pathway. *J. Neurophysiol.* **93,** 3146–3156.

Do, C. W. (2002). Characterization of chloride and bicarbonate transport across the isolated bovine ciliary body/epithelium (CBE). *In* "Department of Optometry and Radiography," p. 168. The Hong Kong Polytechnic University, Hong Kong.

Do, C.-W., and Civan, M. M. (2004). Basis of chloride transport in ciliary epithelium. *J. Membr. Biol.* **200,** 1–13.

Do, C. W., and To, C. H. (2000). Chloride secretion by bovine ciliary epithelium: A model of aqueous humor formation. *Invest. Ophthalmol. Vis. Sci.* **41,** 1853–1860.

Do, C. W., Kong, C. W., and To, C. H. (2004a). Cyclic AMP inhibits transepithelial chloride secretion across bovine ciliary body/epithelium. *Invest. Ophthalmol. Vis. Sci.* **45,** 3638–3643.

Do, C. W., Peterson-Yantorno, K., Mitchell, C. H., and Civan, M. M. (2004b). cAMP-activated maxi-Cl^- channels in native bovine pigmented ciliary epithelial cells. *Am. J. Physiol. Cell Physiol.* **287,** C1003–C1011.

Do, C. W., Lu, W., Mitchell, C. H., and Civan, M. M. (2005). Inhibition of swelling-activated Cl- currents by functional anti-ClC-3 antibody in native bovine non-pigmented ciliary epithelial cells. *Invest. Ophthalmol. Vis. Sci.* **46,** 948–955.

Do, C. W., Peterson-Yantorno, K., and Civan, M. M. (2006). Swelling-activated Cl^- channels support Cl^- secretion by bovine ciliary epithelium. *Invest. Ophthalmol. Vis. Sci.* **47,** 2576–2582.

Do, C. W., Wang, Z., Valiunas, V., Clark, A. F., Wax, M. B., Chatterton, J., and Civan, M. M. (2008). Regulation of Gap-Junction Coupling in Bovine Ciliary Epithelium. E-Abstract #2103, Association for Research in Vision and Ophthalmology (ARVO), Fort Lauderdale.

Dong, J., and Delamere, N. A. (1994). Protein kinase C inhibits Na(+)-K(+)-2Cl- cotransporter activity in cultured rabbit nonpigmented ciliary epithelium. *Am. J. Physiol.* **267,** C1553–C1560.

Dunn, J. J., Lytle, C., and Crook, R. B. (2001). Immunolocalization of the Na-K-Cl cotransporter in bovine ciliary epithelium. *Invest. Ophthalmol. Vis. Sci.* **42,** 343–353.

Edelman, J. L., Sachs, G., and Adorante, J. S. (1994). Ion transport asymmetry and functional coupling in bovine pigmented and nonpigmented ciliary epithelial cells. *Am. J. Physiol.* **266,** C1210–C1221.

Edelman, J. L., Loo, D. D., and Sachs, G. (1995). Characterization of potassium and chloride channels in the basolateral membrane of bovine nonpigmented ciliary epithelial cells. *Invest. Ophthalmol. Vis. Sci.* **36,** 2706–2716.

Fernández-Fernández, J. M., Nobles, M., Currid, A., Vázquez, E., and Valverde, M. A. (2002). Maxi K+ channel mediates regulatory volume decrease response in a human bronchial epithelial cell line. *Am. J. Physiol. Cell Physiol.* **283,** C1705–C1714.

Fillenz, M., and O'Neill, R. D. (1986). Effects of light reversal on the circadian pattern of motor activity and voltammetric signals recorded in rat forebrain. *J. Physiol.* **374,** 91–101.

Finkelstein, A. (1976). Water and nonelectrolyte permeability of lipid bilayer membranes. *J. Gen. Physiol.* **68,** 127–135.

Fischbarg, J., Diecke, F. P., Iserovich, P., and Rubashkin, A. (2006). The role of the tight junction in paracellular fluid transport across corneal endothelium. Electro-osmosis as a driving force. *J. Membr. Biol.* **210,** 117–130.

Fleischhauer, J. C., Mitchell, C. H., Peterson-Yantorno, K., Coca-Prados, M., and Civan, M. M. (2001). PGE_2, Ca^{2+}, and cAMP mediate ATP activation of Cl^- channels in pigmented ciliary epithelial cells. *Am. J. Physiol. Cell Physiol.* **281,** C1614–C1623.

Freddo, T. F., and Johnson, M. (2008). Aqueous humor outflow resistance. *In* "The Eye's Aqueous Humor" (M. M. Civan, ed.). (2nd) Elsevier, San Diego.

Freddo, T. F., Bartels, S. P., Barsotti, M. F., and Kamm, R. D. (1990). The source of proteins in the aqueous humor of the normal rabbit. *Invest. Ophthalmol. Vis. Sci.* **31,** 125–137.

Fredholm, B. B., Abbracchio, M. P., Burnstock, G., Daly, J. W., Harden, T. K., Jacobson, P., Leff, P., and Williams, M. (1994). Nomenclature and classification of purinoceptors. *Pharmacol. Rev.* **46,** 143–156.

Frigeri, A., Gropper, M. A., Turck, C. W., and Verkman, A. S. (1995). Immunolocalization of the mercurial-insensitive water channel and glycerol intrinsic protein in epithelial cell plasma membranes. *Proc. Natl. Acad. Sci. USA* **92,** 4328–4331.

Frömter, E., and Diamond, J. (1972). Route of passive ion permeation in epithelia. *Nat. New Biol.* **235,** 9–13.

Fürst, J., Bottà, G., Saino, S., Dopinto, S., Gandini, R., Dossena, S., Vezzoli, V., Rodighiero, S., Bazzini, M. L., Garavaglia, M. L., Meyer, G., Jakab, M., Ritter, M., Wappl-Kornherr, E., and Paulmichl, M. (2006). The ICln interactome. *Acta Physiol. (Oxf)* **187,** 43–49.

Gabelt, B. T., and Kaufman, P. L. (2005). Changes in aqueous humor dynamics with age and glaucoma. *Prog. Retin. Eye Res.* **24,** 612–637.

Geck, P., Pietrzyk, C., Burckhardt, B. C., Pfeiffer, B., and Heinz, E. (1980). Electrically silent cotransport on Na^+, K^+ and Cl^- in Ehrlich cells. *Biochim. Biophys. Acta* **600,** 432–447.

Gerometta, R. M., Malgor, L. A., Vilalta, E., Leiva, J., and Candia, O. A. (2005). Cl- concentrations of bovine, porcine and ovine aqueous humor are higher than in plasma. *Exp. Eye Res.* **80,** 307–312.

Ghosh, S., Freitag, A. C., Martin-Vasallo, P., and Coca-Prados, M. (1990). Cellular distribution and differential gene expression of the three alpha subunit isoforms of the Na,K-ATPase in the ocular ciliary epithelium. *J. Biol. Chem.* **265,** 2935–2940.

Ghosh, S., Hernando, N., Martin-Alonso, J. M., Martin-Vasallo, P., and Coca-Prados, M. (1991). Expression of multiple Na^+,K($^+$)-ATPase genes reveals a gradient of isoforms along the nonpigmented ciliary epithelium: Functional implications in aqueous humor secretion. *J. Cell Physiol.* **149,** 184–194.

Giebisch, G., and Windhager, E. (2005). Glomerular filtration and renal blood flow. *In* "Medical Physiology" (W. F. Boron and E. L. Boulpaep, eds.), pp. 757–773. Elsevier, Philadelphia.

Gong, W., Xu, H., Shimizu, T., Morishima, S., Tanabe, S., Tachibe, T., Uchida, S., Sasaki, S., and Okada, Y. (2004). ClC-3-independent, PKC-dependent activity of volume-sensitive Cl channel in mouse ventricular cardiomyocytes. *Cell. Physiol. Biochem.* **14,** 213–224.

Green, K., and Pederson, J. E. (1972). Contribution of secretion and filtration to aqueous humor formation. *Am. J. Physiol.* **222,** 1218–1226.

Green, K., Bountra, C., Georgiou, P., and House, C. R. (1985). An electrophysiologic study of rabbit ciliary epithelium. *Invest. Ophthalmol. Vis. Sci.* **26,** 371–381.

Hamann, S., Zeuthen, T., La Cour, M., Nagelhus, E. A., Ottersen, O. P., Agre, P., and Nielsen, S. (1998). Aquaporins in complex tissues: Distribution of aquaporins 1–5 in human and rat eye. *Am. J. Physiol.* **274,** C1332–C1345.

Hasegawa, H., Lian, S. C., Finkbeiner, W. E., and Verkman, A. S. (1994). Extrarenal tissue distribution of CHIP28 water channels by *in situ* hybridization and antibody staining. *Am. J. Physiol.* **266,** C893–C903.

Hebert, S. C. (2003). Bartter syndrome. *Curr. Opin. Nephrol. Hypertens* **12,** 527–532.

Helbig, H., Korbmacher, C., Erb, C., Nawrath, M., Knuuttila, K. G., Wistrand, P., and Wiederholt, M. (1989a). Coupling of 22Na and 36Cl uptake in cultured pigmented ciliary epithelial cells: A proposed role for the isoenzymes of carbonic anhydrase. *Curr. Eye Res.* **8,** 1111–1119.

Helbig, H., Korbmacher, C., Wohlfarth, J., Berweck, S., Kuhner, D., and Wiederholt, M. (1989b). Electrogenic Na+-ascorbate cotransport in cultured bovine pigmented ciliary epithelial cells. *Am. J. Physiol.* **256,** C44–C49.

Hermoso, M., Satterwhite, C. M., Andrade, Y. N., Hidalgo, J., Wilson, S. M., Horowitz, B., and Hume, J. R. (2002). ClC-3 is a fundamental molecular component of volume-sensitive outwardly rectifying Cl^- channels and volume regulation in HeLa cells and Xenopus laevis oocytes. *J. Biol. Chem.* **277,** 40066–40074.

Higginbotham, E. J., Gordon, M. O., Beiser, J. A., Drake, M. V., Bennett, G. R., Wilson, M. R., and Kass, M. A. (2004). The ocular hypertension treatment study: Topical medication delays or prevents primary open-angle glaucoma in African American individuals. *Arch. Ophthalmol.* **122,** 813–820.

Hille, B. (2001). "Ion Channels of Excitable Membranes" Sinauer Associates, Inc, Sunderland, MA.

Hoffman, B. B., Lefkowitz, R. L., and Taylor, P. (1996). Neurotransmission: The autonomic and somatic motor nervous system. *In* "Goodman and Gilman's The Pharmacological Basis of Therapeutics" (J. G. Hardman, L. E. Limbird, P. B. Molinoff, R. W. Ruddon, and A. G. Gilman, eds.), (9th) pp. 105–139. McGraw-Hill, New York.

Holland, M. G., and Gipson, C. C. (1970). Chloride ion transport in the isolated ciliary body. *Invest. Ophthalmol.* **9,** 20–29.

Huang, P., Lazarowski, E. R., Tarran, R., Milgram, S. L., Boucher, R. C., and Stutts, M. J. (2001). Compartmentalized autocrine signaling to cystic fibrosis transmembrane conductance regulator at the apical membrane of airway epithelial cells. *Proc. Natl. Acad. Sci. USA* **98,** 14120–14125.

Jacob, T. J., and Civan, M. M. (1996). Role of ion channels in aqueous humor formation. *Am. J. Physiol.* **271,** C703–C720.

Jentsch, T. J. (2007). Chloride and the endosomal-lysosomal pathway: Emerging roles of CLC chloride transporters. *J. Physiol.* **578,** 633–640.

Jentsch, T. J., Stein, V., Weinreich, F., and Zdebik, A. A. (2002). Molecular structure and physiological function of chloride channels. *Physiol. Rev.* **82,** 503–568.

Kass, M. A., Heuer, D. K., Higginbotham, E. J., Johnson, C. A., Keltner, J. L., Miller, J. P., Parrish, R. K., II, Wilson, M. R., and Gordon, M. O. (2002). The ocular hypertension treatment study: A randomized trial determines that topical ocular hypotensive medication delays or prevents the onset of primary open-angle glaucoma. *Arch. Ophthalmol.* **120,** 701–713discussion 829–730.

Kawasaki, M., Uchida, S., Monkawa, T., Miyawaki, A., Mikoshiba, K., Marumo, F., and Sasaki, S. (1994). Cloning and expression of a protein kinase C-regulated chloride channel abundantly expressed in rat brain neuronal cells. *Neuron* **12,** 597–604.

King, L. S., Kozono, D., and Agre, P. (2004). From structure to disease: The evolving tale of aquaporin biology. *Nat. Rev. Mol. Cell Biol.* **5,** 687–698.

Kishida, K., Sasabe, T., Manabe, R., and Otori, T. (1981). Electric characteristics of the isolated rabbit ciliary body. *Jpn. J. Ophthalmol.* **25,** 407–416.

Kishida, K., Sasabe, T., Iizuka, S., Manabe, R., and Otori, T. (1982). Sodium and chloride transport across the isolated rabbit ciliary body. *Curr. Eye Res.* **2,** 149–152.

Kong, C. W., Li, K. K., and To, C. H. (2006). Chloride secretion by porcine ciliary epithelium: New insight into species similarities and differences in aqueous humor formation. *Invest. Ophthalmol. Vis. Sci.* **47,** 5428–5436.

Krishnamoorthy, R. R., Prasanna, G., Dauphin, R., Hulet, C., Agarwal, N., and Yorio, T. (2003). Regulation of Na,K-ATPase expression by endothelin-1 in transformed human ciliary non-pigmented epithelial (HNPE) cells. *J. Ocul. Pharmacol. Ther.* **19,** 465–481.

Krupin, T., and Civan, M. M. (1996). The physiologic basis of aqueous humor formation. *In* "The Glaucomas" (R. Ritch, M. B. Shields, and T. Krupin, eds.), pp. 251–280. Mosby, St. Louis.

Krupin, T., Reinach, P. S., Candia, O. A., and Podos, S. M. (1984). Transepithelial electrical measurements on the isolated rabbit iris-ciliary body. *Exp. Eye Res.* **38,** 115–123.

Lampe, P. D., and Lau, A. F. (2000). Regulation of gap junctions by phosphorylation of connexins. *Arch. Biochem. Biophys.* **384,** 205–215.

Lang, F., Busch, G. L., Ritter, M., Volkl, H., Waldegger, S., Gulbins, E., and Haussinger, D. (1998). Functional significance of cell volume regulatory mechanisms. *Physiol. Rev.* **78,** 247–306.

Lapointe, J. Y. (2007). Response to Zeuthen and Zeuthen's comment to the editor: Enough local hypertonicity is enough. *Biophys. J.* **93,** 1417–1419.

Lee, P. Y., Podos, S. M., Mittag, T., and Severin, C. (1984). Effect of topically applied forskolin on aqueous humor dynamics in cynomolgus monkey. *Invest. Ophthalmol. Vis. Sci.* **25,** 1206–1209.

Leske, M. C., Heijl, A., Hussein, M., Bengtsson, B., Hyman, L., and Komaroff, E. (2003). Factors for glaucoma progression and the effect of treatment: The early manifest glaucoma trial. *Arch. Ophthalmol.* **121,** 48–56.

Li, A., Lu, W., Leung, G. C.-T., Peterson-Yantorno, K., Mitchell, C. H., and Civan, M. M. (2008). Potential Role of Pannexin Hemichannels in ATP Release from Native Bovine Ciliary Epithelial Cells. E-Abstract #3161, Association for Research in Vision and Ophthalmology (ARVO), Fort Lauderdale.

Li, X., Alvarez, B., Casey, J. R., Reithmeier, R. A., and Fliegel, L. (2002). Carbonic anhydrase II binds to and enhances activity of the Na^+/H^+ exchanger. *J. Biol. Chem.* **277,** 36085–36091.

Liu, J. H. (1998). Circadian rhythm of intraocular pressure. *J. Glaucoma* **7,** 141–147.

Liu, R., Hintermann, E., Erb, C., Tanner, H., Flammer, J., Eberle, A. N., and Haefliger, I. O. (2001). Modulation of Na/K-ATPase activity by isoproterenol and propranolol in human non-pigmented ciliary epithelial cells. *Klin. Monatsbl. Augenheilkd.* **218,** 363–365.

Loike, J. D., Hickman, S., Kuang, K., Xu, M., Cao, L., Vera, J. C., Silverstein, S. C., and Fischbarg, J. (1996). Sodium-glucose cotransporters display sodium- and phlorizin-dependent water permeability. *Am. J. Physiol.* **271,** C1774–C1779.

Loo, D. D., Wright, E. M., and Zeuthen, T. (2002). Water pumps. *J. Physiol.* **542,** 53–60.

Macknight, A. D. C., and Civan, M. M. (2008). Regional dependence of inflow. *In* "The Eye's Aqueous Humor" (M. M. Civan, ed.). (2nd) Elsevier, San Diego.

Maren, T. H. (1976). The rates of movement of Na^+, Cl^-, and $HCO3^-$ from plasma to posterior chamber: Effect of acetazolamide and relation to the treatment of glaucoma. *Invest. Ophthalmol.* **15,** 356–364.

Martin-Vasallo, P., Ghosh, S., and Coca-Prados, M. (1989). Expression of Na,K-ATPase alpha subunit isoforms in the human ciliary body and cultured ciliary epithelial cells. *J. Cell. Physiol.* **141,** 243–252.

Mathias, R. T., and Wang, H. (2005). Local osmosis and isotonic transport. *J. Membr. Biol.* **208,** 39–53.

Mathias, R. T., White, T. W., and Brink, P. R. (2008). The role of gap junction channels in the ciliary body secretory epithelium. *In* "The Eye's Aqueous Humor" (M. M. Civan, ed.). (2nd) Elsevier, San Diego.

McLaughlin, C. W., Peart, D., Purves, R. D., Carré, D. A., Peterson-Yantorno, K., Mitchell, A. D., Macknight, A. D., and Civan, M. M. (2001a). Timolol may inhibit aqueous humor secretion by cAMP-independent action on ciliary epithelial cells. *Am. J. Physiol. Cell Physiol.* **281,** C865–C875.

McLaughlin, C. W., Zellhuber-McMillan, S., Peart, D., Purves, R. D., Macknight, A. D., and Civan, M. M. (2001b). Regional differences in ciliary epithelial cell transport properties. *J. Membr. Biol.* **182,** 213–222.

McLaughlin, C. W., Zellhuber-McMillan, S., Macknight, A. D., and Civan, M. M. (2004). Electron microprobe analysis of ouabain-exposed ciliary epithelium: PE-NPE cell couplets form the functional units. *Am. J. Physiol. Cell Physiol.* **286,** C1376–C1389.

McLaughlin, C. W., Zellhuber-McMillan, S., Macknight, A. D., and Civan, M. M. (2007). Electron microprobe analysis of rabbit ciliary epithelium indicates enhanced secretion posteriorly and enhanced absorption anteriorly. *Am. J. Physiol. Cell Physiol.* **293,** C1455–C1466.

McLaughlin, S., and Mathias, R. T. (1985). Electro-osmosis and the reabsorption of fluid in renal proximal tubules. *J. Gen. Physiol.* **85,** 699–728.

Meldrun, N. U., and Roughton, R. J. W. (1933). Carbonic anhydrase. Its preparation and properties. *J. Physiol.* **80,** 113–142.

Mitchell, C. H., and Civan, M. M. (1997). Effects of uncoupling gap junctions between pairs of bovine NPE-PE ciliary epithelial cells of the eye. *FASEB J.* **11,** A301.

Mitchell, C. H., Carré, D. A., McGlinn, A. M., Stone, R. A., and Civan, M. M. (1998). A release mechanism for stored ATP in ocular ciliary epithelial cells. *Proc. Natl. Acad. Sci. USA* **95,** 7174–7178.

Mitchell, C. H., Peterson-Yantorno, K., Carré, D. A., McGlinn, A. M., Coca-Prados, M., Stone, R. A., and Civan, M. M. (1999). A_3 adenosine receptors regulate Cl^- channels of nonpigmented ciliary epithelial cells. *Am. J. Physiol.* **276,** C659–C666.

Mitchell, C. H., Peterson-Yantorno, K., Coca-Prados, M., and Civan, M. M. (2000). Tamoxifen and ATP synergistically activate Cl^- release by cultured bovine pigmented ciliary epithelial cells. *J. Physiol.* **525**(Pt. 1), 183–193.

Mito, T., and Delamere, N. A. (1993). Alteration of active Na-K transport on protein kinase C activation in cultured ciliary epithelium. *Invest. Ophthalmol. Vis. Sci.* **34,** 539–546.

Mori, N., Yamada, E., and Sears, M. L. (1991). Immunocytochemical localization of Na/K-ATPase in the isolated ciliary epithelial bilayer of the rabbit. *Arch. Histol. Cytol.* **54,** 259–265.

Nakai, Y., Dean, W. L., Hou, Y., and Delamere, N. A. (1999). Genistein inhibits the regulation of active sodium-potassium transport by dopaminergic agonists in nonpigmented ciliary epithelium. *Invest. Ophthalmol. Vis. Sci.* **40,** 1460–1466.

Nakano, T., Fujimoto, K., Honda, Y., and Ogawa, K. (1992). Cytochemistry of protein kinase C and Na-K-ATPase in rabbit ciliary processes treated with phorbol ester. *Invest. Ophthalmol. Vis. Sci.* **33,** 3455–3462.

Nelson, M. T., Joksovic, P. M., Su, P., Kang, H. W., Van Deusen, A., Baumgart, J. P., David, T. P., Snutch, T. P., Barrett, P. Q., Lee, J. H., Zorumski, C. F., Perez-Reyes, E., *et al.* (2007). Molecular mechanisms of subtype-specific inhibition of neuronal T-type calcium channels by ascorbate. *J. Neurosci.* **27,** 12577–12583.

Nielsen, S., Smith, B. L., Christensen, E. I., and Agre, P. (1993). Distribution of the aquaporin CHIP in secretory and resorptive epithelia and capillary endothelia. *Proc. Natl. Acad. Sci. USA* **90,** 7275–7279.

Oh, J., Krupin, T., Tang, L. Q., Sveen, J., and Lahlum, R. A. (1994). Dye coupling of rabbit ciliary epithelial cells *in vitro*. *Invest. Ophthalmol. Vis. Sci.* **35,** 2509–2514.

Okamura, T., Kurogi, Y., Hashimoto, K., Sato, S., Nishikawa, H., Kiryu, K., and Nagao, Y. (2004). Structure-activity relationships of adenosine A_3 receptor ligands: New potential therapy for the treatment of glaucoma. *Bioorg. Med. Chem. Lett.* **14,** 3775–3779.

Panchin, Y. V. (2005). Evolution of gap junction proteins—the pannexin alternative. *J. Exp. Biol.* **208,** 1415–1419.

Panchin, Y., Kelmanson, I., Matz, M., Lukyanov, K., Usman, N., and Lukyanov, S. (2000). A ubiquitous family of putative gap junction molecules. *Curr. Biol.* **10,** R473–R474.

Patarca, R., Candia, O. A., and Reinach, P. S. (1983). Mode of inhibition of active chloride transport in the frog cornea by furosemide. *Am. J. Physiol.* **245,** F660–F669.

Paulmichl, M., Li, Y., Wickman, K., Ackerman, M., Peralta, E., and Clapham, D. (1992). New mammalian chloride channel identified by expression cloning. *Nature* **356,** 238–241.

Pei, Y. F., and Smelser, G. K. (1968). Some fine structural features of the ora serrata region in primate eyes. *Invest. Ophthalmol.* **7,** 672–688.

Pesin, S. R., and Candia, O. A. (1982). Na^+ and Cl^- fluxes, and effects of pharmacological agents on the short-circuit current of the isolated rabbit iris-ciliary body. *Curr. Eye Res.* **2,** 815–827.

Pietrement, C., Da Silva, N., Silberstein, C., James, M., Marsolais, M., Van Hoek, A., Brown, N., Pastor-Soler, N., Ameen, N., Laprade, R., Ramesh, V., and Breton, S. (2008). Role of NHERF1, cystic fibrosis transmembrane conductance regulator, and cAMP in the regulation of aquaporin 9. *J. Biol. Chem.* **283,** 2986–2996.

Polska, E., Ehrlich, P., Luksch, A., Fuchsjager-Mayrl, G., and Schmetterer, L. (2003). Effects of adenosine on intraocular pressure, optic nerve head blood flow, and choroidal blood flow in healthy humans. *Invest. Ophthalmol. Vis. Sci.* **44,** 3110–3114.

Raviola, G., and Raviola, E. (1978). Intercellular junctions in the ciliary epithelium. *Invest. Ophthalmol. Vis. Sci.* **17,** 958–981.

Reale, E. (1975). Freeze-fracture analysis of junctional complexes in human ciliary epithelia. *Albrecht Von Graefes Arch. Klin. Exp. Ophthalmol.* **195,** 1–16.

Reddy, D. V., Rosenberg, C., and Kinsey, V. E. (1961). Steady state distribution of free amino acids in the aqueous humours, vitreous body and plasma of the rabbit. *Exp. Eye Res.* **1,** 175–191.

Reinach, P. S., Schoen, H. F., and Candia, O. A. (1979). Metabolic requirements for anaerobic active Cl and Na transport in the bullfrog cornea. *Am. J. Physiol.* **236,** C268–C276.

Reitsamer, H. A., and Kiel, J. W. (2003). Relationship between ciliary blood flow and aqueous production in rabbits. *Invest. Ophthalmol. Vis. Sci.* **44,** 3967–3971.

Reitsamer, H. A., and Kiel, J. W. (2008). Effects of circulatory events on aqueous humor inflow and intraocular pressure. *In* "The Eye's Aqueous Humor" (M. M. Civan, ed.). (2nd) Elsevier, San Diego.

Rice, M. E. (2000). Ascorbate regulation and its neuroprotective role in the brain. *Trends Neurosci.* **23,** 209–216.

Riese, K., Beyer, A. T., Lui, G. M., and Crook, R. B. (1998). Dopamine D1 stimulation of Na^+, K^+, Cl^- cotransport in human NPE cells: Effects of multiple hormones. *Invest. Ophthalmol. Vis. Sci.* **39,** 1444–1452.

Rose, R. C., and Schultz, S. G. (1971). Studies on the electrical potential profile across rabbit ileum. Effects of sugars and amino acids on transmural and transmucosal electrical potential differences. *J. Gen. Physiol.* **57,** 639–663.

Saito, Y., and Watanabe, T. (1979). Relationship between short-circuit current and unidirectional fluxes of Na and Cl across the ciliary epithelium of the toad: Demonstration of active Cl transport. *Exp. Eye Res.* **28,** 71–79.

Sanchez-Torres, J., Huang, W., Civan, M. M., and Coca-Prados, M. (1999). Effects of hypotonic swelling on the cellular distribution and expression of pI(Cln) in human nonpigmented ciliary epithelial cells. *Curr. Eye Res.* **18,** 408–416.

Satoh, T., Cohen, H. T., and Katz, A. I. (1993). Intracellular signaling in the regulation of renal Na-K-ATPase. II. Role of eicosanoids. *J. Clin. Invest.* **91,** 409–415.

Schlötzer-Schrehardt, U., Zenkel, M., Decking, U., Haubs, D., Kruse, F. E., Junemann, A., Coca-Prados, M., and Naumann, G. O. (2005). Selective upregulation of the A3 adenosine receptor in eyes with pseudoexfoliation syndrome and glaucoma. *Invest. Ophthalmol. Vis. Sci.* **46,** 2023–2034.

Sears, J., Nakano, T., and Sears, M. (1998). Adrenergic-mediated connexin43 phosphorylation in the ocular ciliary epithelium. *Curr. Eye Res.* **17,** 104–107.

Sears, M. L., Yamada, E., Cummins, D., Mori, N., Mead, A., and Murakami, M. (1991). The isolated ciliary bilayer is useful for studies of aqueous humor formation. *Trans. Am. Ophthalmol. Soc.* **89,** 131–152discussion 152–134.

Shahidullah, M., and Delamere, N. A. (2006). NO donors inhibit Na,K-ATPase activity by a protein kinase G-dependent mechanism in the nonpigmented ciliary epithelium of the porcine eye. *Br. J. Pharmacol.* **148,** 871–880.

Shahidullah, M., and Wilson, W. S. (1997). Mobilisation of intracellular calcium by P2Y2 receptors in cultured, non-transformed bovine ciliary epithelial cells. *Curr. Eye Res.* **16,** 1006–1016.

Shahidullah, M., Wilson, W. S., Yap, M., and To, C. H. (2003). Effects of ion transport and channel-blocking drugs on aqueous humor formation in isolated bovine eye. *Invest. Ophthalmol. Vis. Sci.* **44,** 1185–1191.

Shahidullah, M., Yap, M., and To, C. H. (2005). Cyclic GMP, sodium nitroprusside and sodium azide reduce aqueous humour formation in the isolated arterially perfused pig eye. *Br. J. Pharmacol.* **145,** 84–92.

Shi, C., Szczesniak, A., Mao, L., Jollimore, C., Coca-Prados, M., Hung, O., and Kelly, M. E. (2003). A3 adenosine and CB1 receptors activate a PKC-sensitive Cl(−) current in human nonpigmented ciliary epithelial cells via a Gbetagamma-coupled MAPK signaling pathway. *Br. J. Pharmacol.* **139,** 475–486.

Socci, R. R., and Delamere, N. A. (1988). Characteristics of ascorbate transport in the rabbit iris-ciliary body. *Exp. Eye Res.* **46,** 853–861.

Somekawa, S., Fukuhara, S., Nakaoka, Y., Fujita, H., Saito, Y., and Mochizuki, N. (2005). Enhanced functional gap junction neoformation by protein kinase A-dependent and Epac-dependent signals downstream of cAMP in cardiac myocytes. *Circ. Res.* **97,** 655–662.

Spinowitz, B. S., and Zadunaisky, J. A. (1979). Action of adenosine on chloride active transport of isolated frog cornea. *Am. J. Physiol.* **237,** F121–F127.

Stamer, W. D., Snyder, R. W., Smith, B. L., Agre, P., and Regan, J. W. (1994). Localization of aquaporin CHIP in the human eye: Implications in the pathogenesis of glaucoma and other disorders of ocular fluid balance. *Invest. Ophthalmol. Vis. Sci.* **35,** 3867–3872.

Stamer, W. D., Baetz, N. W., and Yool, A. J. (2008). Ocular Aquaporins and Aqueous Humor Dynamics. *In* "The Eye's Aqueous Humor" (M. M. Civan, ed.). (2nd) Elsevier, San Diego.

Stelling, J. W., and Jacob, T. J. (1997). Functional coupling in bovine ciliary epithelial cells is modulated by carbachol. *Am. J. Physiol.* **273,** C1876–C1881.

Sterling, D., Reithmeier, R. A., and Casey, J. R. (2001). A transport metabolon. Functional interaction of carbonic anhydrase II and chloride/bicarbonate exchangers. *J. Biol. Chem.* **276,** 47886–47894.

Stobrawa, S. M., Breiderhoff, T., Takamori, S., Engel, D., Schweizer, M., Zdebik, A. A., Bösl, K., Ruether, K., Jahn, H., Draguhn, A., Jahn, R., and Jentsch, T. J. (2001). Disruption of ClC-3, a chloride channel expressed on synaptic vesicles, leads to a loss of the hippocampus. *Neuron* **29,** 185–196.

Stone, R. A., Laties, A. M., Hemmings, H. C., Jr., Ouimet, C. C., and Greengard, P. (1986). DARPP-32 in the ciliary epithelium of the eye: A neurotransmitter-regulated phosphoprotein of brain localizes to secretory cells. *J. Histochem. Cytochem.* **34,** 1465–1468.

Strange, K. (1998). Molecular identity of the outwardly rectifying, swelling-activated anion channel: Time to reevaluate pICln. *J. Gen. Physiol.* **111,** 617–622.

The AGIS investigators (2000). The advanced glaucoma intervention study (AGIS): 7. The relationship between control of intraocular pressure and visual field deterioration. *Am. J. Ophthalmol.* **130,** 429–440.

Therien, A. G., and Blostein, R. (2000). Mechanisms of sodium pump regulation. *Am. J. Physiol. Cell Physiol.* **279,** C541–C566.

To, C. H., Do, C. W., Zamudio, A. C., and Candia, O. A. (2001). Model of ionic transport for bovine ciliary epithelium: Effects of acetazolamide and HCO_3^-. *Am. J. Physiol. Cell Physiol.* **280,** C1521–C1530.

Toris, C. B. (2008). Aqueous humor dynamics I: Measurement methods and animal studies. *In* "The Eye's Aqueous Humor" (M. M. Civan, ed.). (2nd) Elsevier, San Diego.

Toris, C. B., and Camras, C. B. (2008). Aqueous humor dynamics II: Clinical studies. *In* "The Eye's Aqueous Humor" (M. M. Civan, ed.). (2nd) Elsevier, San Diego.

Toris, C. B., and Pederson, J. E. (1985). Effect of intraocular pressure on uveoscleral outflow following cyclodialysis in the monkey eye. *Invest. Ophthalmol. Vis. Sci.* **26,** 1745–1749.

Toris, C. B., Koepsell, S. A., Yablonski, M. E., and Camras, C. B. (2002). Aqueous humor dynamics in ocular hypertensive patients. *J. Glaucoma* **11,** 253–258.

Torphy, T. J. (1994). Beta-adrenoceptors, cAMP and airway smooth muscle relaxation: Challenges to the dogma. *Trends Pharmacol. Sci.* **15,** 370–374.

Usukura, J., Fain, G. L., and Bok, D. (1988). [3H]ouabain localization of Na-K ATPase in the epithelium of rabbit ciliary body pars plicata. *Invest. Ophthalmol. Vis. Sci.* **29,** 606–614.

van Rijen, H. V., van Veen, T. A., Hermans, M. M., and Jongsma, H. J. (2000). Human connexin40 gap junction channels are modulated by cAMP. *Cardiovasc. Res.* **45,** 941–951.

Vareilles, P., Silverstone, D., Plazonnet, B., Le Douarec, J. C., Sears, M. L., and Stone, C. A. (1977). Comparison of the effects of timolol and other adrenergic agents on intraocular pressure in the rabbit. *Invest. Ophthalmol. Vis. Sci.* **16,** 987–996.

Vázquez, E., Nobles, M., and Valverde, M. A. (2001). Defective regulatory volume decrease in human cystic fibrosis tracheal cells because of altered regulation of intermediate conductance $Ca2^+$-dependent potassium channels. *Proc. Natl. Acad. Sci. USA* **98,** 5329–5334.

Vessey, J. P., Shi, C., Jollimore, C. A., Stevens, K. T., Coca-Prados, M., Barnes, S., and Kelly, M. E. (2004). Hyposmotic activation of ICl,swell in rabbit nonpigmented ciliary epithelial cells involves increased ClC-3 trafficking to the plasma membrane. *Biochem. Cell Biol.* **82,** 708–718.

Von Brauchitsch, D. K., and Crook, R. B. (1993). Protein kinase C regulation of a Na^+, K^+, Cl^- cotransporter in fetal human pigmented ciliary epithelial cells. *Exp. Eye Res.* **57,** 699–708.

Wang, G. X., Hatton, W. J., Wang, G. L., Zhong, J., Yamboliev, I., Duan, D., and Hume, J. R. (2003). Functional effects of novel anti-ClC-3 antibodies on native volume-sensitive osmolyte and anion channels in cardiac and smooth muscle cells. *Am. J. Physiol. Heart Circ. Physiol.* **285,** H1453–H1463.

Wang, J., Xu, H., Morishima, S., Tanabe, S., Jishage, K., Uchida, S., Sasaki, S., Okada, Y., and Shimizu, T. (2005). Single-channel properties of volume-sensitive Cl^- channel in ClC-3-deficient cardiomyocytes. *Jpn. J. Physiol.* **55,** 379–383.

Wang, L., Chen, L., and Jacob, T. J. (2000). The role of ClC-3 in volume-activated chloride currents and volume regulation in bovine epithelial cells demonstrated by antisense inhibition. *J. Physiol.* **524**(Pt. 1), 63–75.

Wang, Z., Do, C. W., Avila, M. Y., Stone, R. A., Jacobson, K. A., and Civan, M. M. (2007). Barrier qualities of the mouse eye to topically applied drugs. *Exp. Eye Res.* **85,** 105–112.

Wang, Z., Do, C. W., Avila, M. Y., Peterson-Yantorno, K., Stone, R. A., Gao, Z. G., Leong, K. A., Jacobson, K. A., and Civan, M. M. (2008). A Novel Cross-Species Antagonist to A_3 Adenosine Receptors Lowers Mouse Intraocular Pressure Measured Non-Invasively. E-Abstract #356, Association for Research in Vision and Ophthalmology (ARVO), Fort Lauderdale.

Warn-Cramer, B. J., and Lau, A. F. (2004). Regulation of gap junctions by tyrosine protein kinases. *Biochim. Biophys. Acta* **1662,** 81–95.

Wetzel, R. K., and Sweadner, K. J. (2001). Immunocytochemical localization of NaK-ATPase isoforms in the rat and mouse ocular ciliary epithelium. *Invest. Ophthalmol. Vis. Sci.* **42,** 763–769.

White, C. N., Hamilton, E. J., Garcia, A., Wang, D., Chia, K. K., Figtree, G. A., and Rasmussen, H. H. (2008). Opposing effects of coupled and uncoupled NOS activity on the Na+-K+ pump in cardiac myocytes. *Am. J. Physiol. Cell Physiol.* **294,** C572–C578.

Wiederholt, M., and Zadunaisky, J. A. (1986). Membrane potentials and intracellular chloride activity in the ciliary body of the shark. *Pflügers Arch.* **407**(Suppl. 2), S112–S115.

Wiederholt, M., Helbig, H., and Korbmacher, C. (1991). Ion transport across the ciliary epithelium: Lessons from cultured cells and proposed role of the carbonic anhydrase. *In* "Carbonic Anhydrase" (F. Botré, G. Gross, and B. T. Storey, eds.), pp. 232–244. VCH, New York.

Wolosin, J. M., Candia, O. A., Peterson-Yantorno, K., Civan, M. M., and Shi, X. P. (1997a). Effect of heptanol on the short circuit currents of cornea and ciliary body demonstrates rate limiting role of heterocellular gap junctions in active ciliary body transport. *Exp. Eye Res.* **64,** 945–952.

Wolosin, J. M., Schütte, M., and Chen, S. (1997b). Connexin distribution in the rabbit and rat ciliary body. A case for heterotypic epithelial gap junctions. *Invest. Ophthalmol. Vis. Sci.* **38,** 341–348.

Wu, R., Yao, K., Flammer, J., and Haefliger, I. O. (2004). Role of anions in nitric oxide-induced short-circuit current increase in isolated porcine ciliary processes. *Invest. Ophthalmol. Vis. Sci.* **45,** 3213–3222.

Xie, Z., and Askari, A. (2002). $Na(^{+})/K(^{+})$-ATPase as a signal transducer. *Eur. J. Biochem.* **269,** 2434–2439.

Yamaguchi, Y., Watanabe, T., Hirakata, A., and Hida, T. (2006). Localization and ontogeny of aquaporin-1 and -4 expression in iris and ciliary epithelial cells in rats. *Cell Tissue Res.* **325,** 101–109.

Yamamoto-Mizuma, S., Wang, G. X., Liu, L. L., Schegg, K., Hatton, W. J., Duan, D., Horowitz, F. S., Lamb, F. S., and Hume, J. R. (2004). Altered properties of volume-sensitive osmolyte and anion channels (VSOACs) and membrane protein expression in cardiac and smooth muscle myocytes from $Clcn3^{-/-}$ mice. *J. Physiol.* **557,** 439–456.

Yang, H., Avila, M. Y., Peterson-Yantorno, K., Coca-Prados, M., Stone, R. A., Jacobson, K. A., and Civan, M. M. (2005). The cross-species A_3 adenosine-receptor antagonist MRS 1292 inhibits adenosine-triggered human nonpigmented ciliary epithelial cell fluid release and reduces mouse intraocular pressure. *Curr. Eye Res.* **30,** 747–754.

Yorio, T. (1985). Cellular mechanisms in the actions of antiglaucoma drugs. *J. Ocul. Pharmacol.* **1,** 397–422.

Zeuthen, T., and Zeuthen, E. (2007). The mechanism of water transport in Na^{+}-coupled glucose transporters expressed in Xenopus oocytes. *Biophys. J.* **93,** 1413–1416discussion 1417–1419.

Zhang, D., Vetrivel, L., and Verkman, A. S. (2002). Aquaporin deletion in mice reduces intraocular pressure and aqueous fluid production. *J. Gen. Physiol.* **119,** 561–569.

Zhang, J. J., and Jacob, T. J. (1997). Three different Cl^- channels in the bovine ciliary epithelium activated by hypotonic stress. *J. Physiol.* **499**(Pt. 2), 379–389.
Zhang, Y., Bakall, B., McLaughlin, P., Marmorstein, L. Y., and Marmorstein, A. (2008). Bestrophin-2 and Generation of Intraocular Pressure. E-Abstract #5862, Association for Research in Vision and Ophthalmology (ARVO), Fort Lauderdale.
Zifarelli, G., and Pusch, M. (2007). CLC chloride channels and transporters: A biophysical and physiological perspective. *Rev. Physiol. Biochem. Pharmacol.* **158,** 23–76.

CHAPTER 2

Ocular Aquaporins and Aqueous Humor Dynamics

W. Daniel Stamer,[*,†] Nicholas W. Baetz,[§] and Andrea J. Yool[†,‡,§]

[*]Department of Ophthalmology and Vision Science, The University of Arizona, Tucson, Arizona 85711

[†]Department of Pharmacology, The University of Arizona, Tucson, Arizona 85711

[‡]Discipline of Physiology, University of Adelaide, SA 5005, Australia

[§]Department of Cell Biology and Anatomy, The University of Arizona, Tucson, Arizona 85711

I. OVERVIEW

Due to a requirement for transparent optical structures, vision depends upon the movement of water between and within ocular tissues and compartments. A class of integral membrane proteins, the aquaporins, functions to efficiently move water across biological membranes. Expressed by more ocular cell types than anywhere else in the body, aquaporins participate in the circulation of intraocular fluids. The purpose of this chapter is to review data that characterizes the role of aquaporins in aqueous humor dynamics.

Current Topics in Membranes, Volume 62

1063-5823/08 $35.00
DOI: 10.1016/S1063-5823(08)00402-X

II. INTRODUCTION

Vision is dependent upon the efficient movement of water between and within various structures of the eye. To facilitate the faithful transmission of light rays from the corneal surface to the retinal photoreceptors, the eye is pressurized, having three compartments that are filled with optically transparent fluids: Vitreous humor occupies ~80% of the interior volume of the eye and lies between the posterior face of the lens and the retina. Aqueous humor fills the other two compartments, the anterior and posterior chambers, located on either side of the iris. The circulation of aqueous humor from the posterior to the anterior chamber (and then out of the eye) enables the delivery of nutrients and removal of waste products from two specialized avascular tissues, the cornea and crystalline lens that function to focus light onto the retina. The clarity of these two organic lenses, and thus their ability to refract light, is exquisitely dependent upon water homeostasis within and the circulation of aqueous humor around their structures. For example, corneal clarity is reliant upon the maintenance of stromal water content by the cellular barriers that line either surface, while intraocular pressure is regulated within a narrow range by the balance of aqueous humor secretion and drainage. Not surprisingly, every tissue that produces, removes, or is in contact with aqueous humor contains specialized channels called aquaporins (AQPs) that facilitate the efficient and selective movement of water across ocular membranes.

The purpose of this chapter is to review data that characterize the role of aquaporins in the movement of water into and out of the eye (aqueous humor dynamics). First, we will provide an overview of aquaporin discovery and its molecular structure and function in cellular membranes. Second, we will summarize the specific distribution of aquaporin homologues in the eye. Third, we will discuss the specific role of aquaporin channels in aqueous humor dynamics. Finally, we will discuss the future direction of aquaporin research in aqueous humor dynamics and the potential of aquaporins as drug targets.

III. AQUAPORINS ARE ASSEMBLED AS FOUR HOMOMERIC SUBUNITS

Aquaporins provide molecular pathways for the movement of water and selected small solutes across cell membranes (King *et al.*, 2004). Aquaporins are found throughout the kingdoms of life, including prokaryotes and eukaryotes. In mammals, there are at least 12 classes of aquaporins (AQP0 to AQP11), which show tissue-specific patterns of expression. These channels are broadly classified as orthodox aquaporins, selective for water, and the aquaglyceroporins, such as AQP3, AQP7, and AQP9, that allow transmembrane movement of glycerol as well as water. Beyond the simple bimodal classification scheme, our understanding of permeability properties is being extended steadily to include roles for aquaporins in the transport of other compounds including ions, gases, and small organic compounds, as reviewed previously (Yool and Stamer, 2004). Much remains to be discovered about the full range of functional properties of this family of channels.

Crystal structural data now available for AQP1 have verified classic work in the field that first defined general principles of structure in the archetypal member of this family of proteins (Jung *et al.*, 1994; de Groot *et al.*, 2001; Ren *et al.*, 2001; Sui *et al.*, 2001). Aquaporins are tetrameric complexes of subunits (Fig. 1). Each subunit has six full transmembrane domains per subunit, intracellular N- and C-terminal domains, and water pores framed by loops B and E. The hourglass model of a subunit of AQP1, originally known as CHIP28, was envisioned as a narrow pore pathway within each subunit, with the hallmark asparagine-proline-alanine (NPA) motifs located near the center of the membrane interior at the junction of the folded B and E loops (Jung *et al.*, 1994). In the intrasubunit pores, the orthodox aquaporins show a high selectivity for water, excluding solutes, ions, and protons.

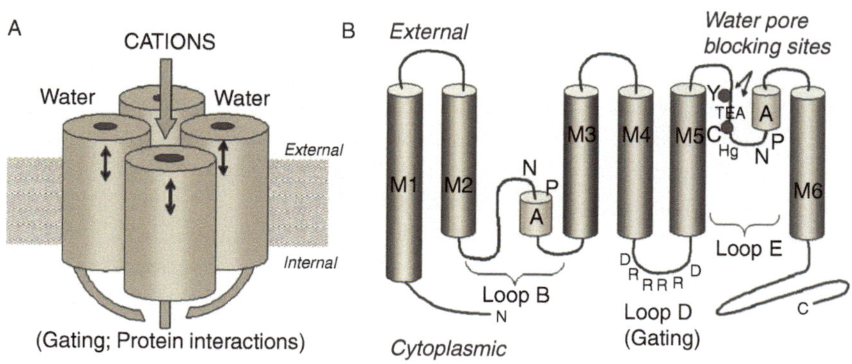

FIGURE 1 Schematic showing the tetrameric organization and subunit transmembrane topology of the aquaporin-1 (AQP1) protein. (A) Aquaporins are assembled as four homomeric subunits. The constitutive water-selective pores are located within each subunit, and for AQP1 the proposed ion channel is located in the center of the tetrameric complex (Yu *et al.*, 2006). (B) Diagram of main features of the transmembrane topology of a human AQP1 subunit, indicating loops A to E and six full transmembrane regions M1 to M6. Selected functional domains include the proposed gating region (loop D), with arginines (R159 and R160) suggested to serve in the cGMP-induced activation of the AQP1 ionic conductance, and the asparagine-proline-alanine (NPA) motifs in loops B and E that contribute to water-selective pore structures. Tyrosine (Y187) and cysteine (C189) in loop E have been shown to mediate block of water permeability by extracellular tetraethylammonium and mercuric compounds, respectively. The C-terminal domain contains regions that influence cGMP-induced activation, enable protein–protein interactions, and in addition to other intracellular protein domains of AQP1 might be sites of modulation. (See Color Insert.)

The central pore at the fourfold axis of symmetry in the tetramer may provide a parallel pathway for regulated movement of other molecules, such as CO_2 and ions, in specialized subsets of aquaporins (Yu *et al.*, 2006). For example, CO_2 permeation through AQP1 could serve a physiological role in membranes that have a low intrinsic permeability to the gas. Based on analysis of free energy barriers, the AQP1 central cavity is favored over the monomeric channel as a candidate pathway for CO_2 (Hub and de Groot, 2006). Molecular dynamics simulations suggest the central pore is a pathway for cations in AQP1 (Yu *et al.*, 2006). It is possible that the two AQP1 channel states exist as alternatives, with Na^+ permeating the hydrated central pore in the hypothetical "open" state, and CO_2 moving through the dehydrated pore in a "closed" state. These multifunctional properties add complexity to the potential roles of these channels in tissues such as the eye.

IV. OCULAR DISTRIBUTION OF AQUAPORINS

As is the kidney, the eye is a water-transporting organ. The eye rivals the kidney in terms of the number of aquaporin homologues that are expressed, and surpasses the kidney in terms of the number of different cell types that express aquaporin channels. To date, the selective expression of 6 different aquaporin homologues in 8 different cell types located in 10 different ocular tissues have been described (Table I). Thus, the distribution of AQP0, AQP1, AQP3, AQP4, AQP5, and AQP9 are for the most part nonoverlapping and found in epithelial cells, endothelial cells, fibroblasts, trabecular meshwork (TM) cells, lens fiber cells, neuronal, glial, and photoreceptors in the eye. These aquaporin-expressing cells populate the cornea, conjunctiva, lens, iris, TM, ciliary body, sclera, retina, choroid, and optic nerve. One aquaporin homologue, AQP0, is found primarily in the lens fiber cells of the crystalline lens, but has been recently detected in testis (Hermo *et al.*, 2004) and liver (Tietz *et al.*, 2005).

The clarity of the organic lenses of the eye is highly dependent upon water homeostasis, and thus upon aquaporin channel function. The crystalline lens expresses two aquaporins, AQP1 in the monolayer of epithelial cells, which covers the anterior surface, and AQP0 in the terminally differentiated lens fiber cells, which forms the bulk of the lens' mass. AQP0 constitutes almost half of the total protein at plasma membrane of lens fiber cells. AQP0 was the first aquaporin discovered (Gorin *et al.*, 1984); however, its role as a water channel was not completely understood until later because of its low capacity for water subject to physiological regulation (Mulders *et al.*, 1995; Chandy *et al.*, 1997). For example, changes in intracellular signals in the lens, Ca^{2+}-calmodulin and pH, regulate the water permeability of endogenously expressed AQP0 in lens fiber cells, but not that of AQP1 natively expressed in lens epithelial cells (Varadaraj *et al.*, 2005).

TABLE I

Summary of Aquaporin Expression in the Human Eye

Tissue	Cell type	Aquaporin	References
Cornea	Epithelia	AQP3, AQP5	(Raina *et al.*, 1995; Patil *et al.*, 1997b; Funaki *et al.*, 1998; Hamann *et al.*, 1998)
	Keratocytes	AQP1	(Hamann *et al.*, 1998)
	Endothelia	AQP1	(Echevarria *et al.*, 1993; Hasegawa *et al.*, 1993, 1994; Nielsen *et al.*, 1993; Stamer *et al.*, 1994; Patil *et al.*, 1997b; Hamann *et al.*, 1998)
Conjunctiva	Epithelia	AQP3, AQP5	(Frigeri *et al.*, 1995; Hamann *et al.*, 1998; Oen *et al.*, 2006)
Lens	Epithelia	AQP1	(Nielsen *et al.*, 1993; Hasegawa *et al.*, 1994; Stamer *et al.*, 1994; Patil *et al.*, 1997b; Hamann *et al.*, 1998)
	Fiber cells	AQP0	(Broekhuyse *et al.*, 1979; Fitzgerald *et al.*, 1983; Gorin *et al.*, 1984; Zampighi *et al.*, 1989)
Iris	Anterior and posterior epithelia	AQP1	(Nielsen *et al.*, 1993; Hasegawa *et al.*, 1994; Stamer *et al.*, 1994; Patil *et al.*, 1997b; Hamann *et al.*, 1998)
Conventional outflow tract	Trabecular meshwork	AQP1	(Stamer *et al.*, 1994, 1995, 2001)
	Schlemm's canal	AQP1	(Stamer *et al.*, 1994; Hamann *et al.*, 1998)
Ciliary body	Nonpigmented epithelia	AQP1, AQP4	(Hasegawa *et al.*, 1993, 1994; Nielsen *et al.*, 1993; Stamer *et al.*, 1994; Frigeri *et al.*, 1995; Patil *et al.*, 1997b; Hamann *et al.*, 1998)
Sclera	Fibroblasts	AQP1	(Hamann *et al.*, 1998)
Retina	Müller	AQP1, AQP4	(Frigeri *et al.*, 1995; Hamann *et al.*, 1998; Kim *et al.*, 1998; Nagelhus *et al.*, 1998)
	Retinal pigment epithelia	AQP1	(Stamer *et al.*, 2003)
	Amacrine	AQP1, AQP9	(Kim *et al.*, 1998, 2002; Kang *et al.*, 2005; Iandiev *et al.*, 2006)
	Photoreceptors	AQP1	(Nagelhus *et al.*, 1998; Iandiev *et al.*, 2006)
Optic nerve	Astrocytes	AQP4	(Nagelhus *et al.*, 1998)
Capillaries (sclera, ciliary body, choroid)	Endothelial	AQP1	(Hamann *et al.*, 1998)

In addition to a role as a regulated water channel, AQP0 is thought to have a structural function as a cell–cell adhesion protein (Mulders *et al.*, 1995). In the lens, microdomains located at the junctions between fiber cells form two-dimensional arrays of AQP0 proteins that are thought to provide cell–cell adhesion, and are surrounded by densely packed gap junction channels that mediate intercellular communication (Zampighi *et al.*, 2002; Buzhynskyy *et al.*, 2007). AQP0 arrays appear to be stabilized by physical associations with both gap junction proteins and the lens-specific intermediate filament proteins filensin and CP49 (Yu *et al.*, 2005; Lindsey Rose *et al.*, 2006).

Consistent with its role in maintaining lens clarity, mutations in AQP0 result in cataract (Shiels and Bassnett, 1996). Dominantly inherited cataracts were found in two families carrying different point mutations in the gene for AQP0, presenting different clinical features: the mutation E134G associates with a unilamellar cataract, whereas the mutation T138R correlates with multifocal opacities that increase with age (Francis *et al.*, 2000). Coexpression of mutant AQP0 with wild type in *Xenopus* oocytes decreases water permeability, and high levels of coexpression of the mutant impairs regulation of wild-type water fluxes by calcium. These findings suggest that the regulated water permeability of AQP0 could be an important component in lens homeostasis and development (Kalman *et al.*, 2006). Taken together, accumulating evidence suggests that AQP0 is more than a physical anchoring structure, but also serves a role in the movement of fluids within the lens, with details of its functional roles yet to be defined.

Maintenance of lens transparency depends not only on AQP0, but also on AQP1. The role of the high-capacity AQP1 channels in the lens epithelium is likely to be a more orthodox one—that is, to facilitate the efficient movement of water across its epithelial surface that will contribute to water circulation in the lens, and thus to lens health and transparency. For instance, osmotic water permeability was decreased almost 3-fold in epithelial cells of intact lenses from AQP1-deficient mice as compared to wild type, and the loss of lens transparency was accelerated more than 50-fold during osmotic stress (Ruiz-Ederra and Verkman, 2006).

The selective expression of AQP1, AQP3, AQP4, and AQP5 in distinct ocular epithelia compels an expectation that each aquaporin class has a distinct and specific role in complex regulation of water movements in the eye. At the anterior surface of the mammalian eye, both the corneal and conjunctival epithelia express two aquaporins, AQP3 and AQP5. In contrast, a single monolayer of endothelial cells at the posterior surface of the cornea expresses AQP1 channels and interfaces with aqueous humor in the anterior chamber. Here, AQP1 is thought to function in facilitating the efficient transport of water out of the corneal stroma and into the anterior chamber to help maintain clarity. Evidence of the role of aquaporin in maintaining corneal hydration, and thus clarity, was recently provided in aquaporin knockout mice. Corneal thickness was significantly decreased in AQP1 null mice and increased in AQP5 null mice (Thiagarajah and Verkman, 2002). While corneal transparency was not impaired under baseline conditions, the rate of corneal swelling was compromised in both AQP1 and AQP5 null mice when challenged upon exposure with hypotonic medium.

AQP1 is located in several ocular tissues where its function is unclear. For example, AQP1 is localized to the apical and basolateral membranes of pigmented posterior epithelial and anterior myoepithelial cells of the iris (Table I). The precise role of AQP1 in iris function is unknown, but may relate to changes in rapid water movement or cell volume that may occur upon contraction or relaxation during mydriasis or miosis, respectively. AQP1 is also found in fenestrated and nonfenestrated capillaries in ocular tissues that include the choroid, ciliary body, sclera, and iris. As in other capillary beds of the body, particularly fenestrated, their functional role is uncertain. Finally, AQP1 is highly expressed by resident fibroblasts of the sclera and corneal stroma (keratocytes). Unfortunately, the role of AQP1 in fibroblast function both in the eye and elsewhere still needs to be determined (Gallardo *et al.*, 2002; Maeda *et al.*, 2005).

Aquaporin channels are expressed by cells responsible for the production and removal of aqueous humor from the eye. AQP1 and AQP4 localize to both the apical and basolateral plasma membranes of nonpigmented epithelial cells of the ciliary body, but are completely absent from pigmented epithelial cells (Fig. 2). In the ciliary processes, aquaporins function to enable formation of aqueous humor with the efficient passage of water, following salt transport, from the ciliary stroma into the posterior chamber (discussed in detail in the following section). After flowing between the lens and iris into the anterior chamber, the majority of aqueous humor exits the eye via the conventional (~70%) and unconventional (~25%) routes (Bill and Phillips, 1971; Townsend and Brubaker, 1980; Toris *et al.*, 1999). A small portion of water (~5%) travels posteriorly through the vitreous, and exits across the retina. AQP1 is expressed by cells that populate the conventional and posterior outflow routes. In the conventional outflow pathway, cells that cover the trabecular lamellae and occupy the juxtacanalicular region of the TM express AQP1 (Fig. 2). Additionally, endothelial cells that form part of the blood–aqueous barrier, Schlemm's canal (SC) endothelia, express AQP1 channels. The role of AQP1 in regulating aqueous movement through the conventional route is yet uncertain (discussed in detail in the following section). A minor but significant amount of water exits the eye across a continuous monolayer of epithelial cells, the retinal pigment epithelium (RPE) that forms the blood–retina barrier and lies just posterior to the retina. In humans, AQP1 localizes to both the apical and basolateral membranes of RPE cells. The role of AQP1 in RPE function will also be discussed in the following section.

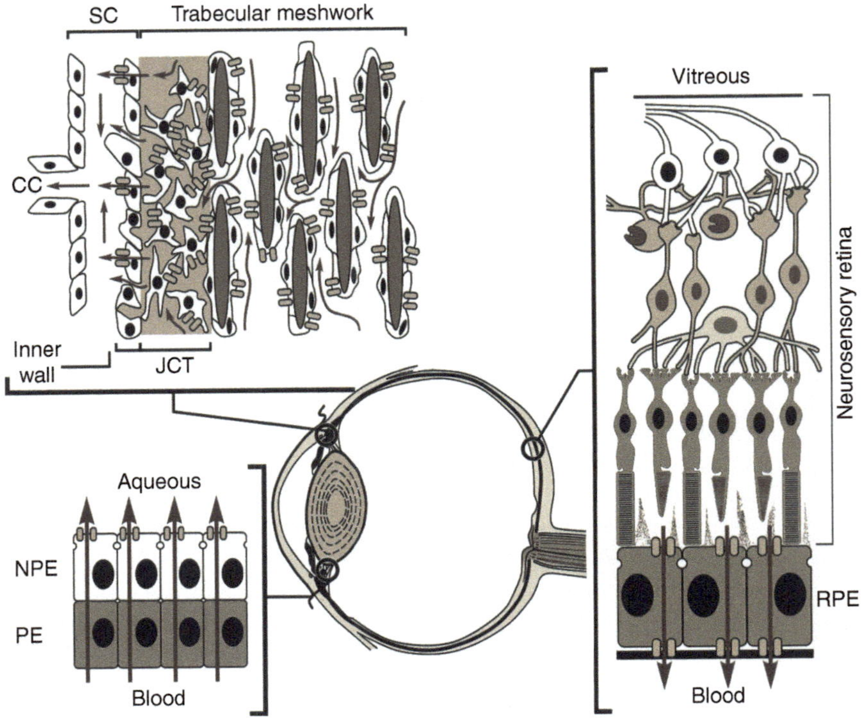

FIGURE 2 Distribution of aquaporin channels in ocular tissues that participate in aqueous humor dynamics. Shown is schematic of the human eye and three ocular tissues in cross section: the conventional outflow pathway (top left), the ciliary epithelial bilayer (bottom left), and retinal pigment epithelium (right). Indicted in green are putative locations of aquaporin channels that participate in aqueous humor dynamics in these three tissues. Red arrows indicate direction of flow across/through these ocular tissues. SC, Schlemm's canal; JCT, juxtacanalicular tissue; CC, collector channel; NPE, nonpigmented epithelium; PE, pigmented epithelium; RPE, retinal pigment epithelium. (See Color Insert.)

V. AQUAPORINS AND AQUEOUS HUMOR DYNAMICS

The rates of water movement across epithelial barriers in the renal tubular system and ciliary body are similar, among the highest measured in the body [0.6–1.2 cm^3/cm^2 (Brubaker, 1991)], and no doubt in part due to aquaporin expression (King *et al.*, 2004). In contrast to fluid resorption in renal tubules, secretion of aqueous humor by the ciliary epithelium occurs against both oncotic and hydrostatic gradients. To overcome these forces, water follows the active transport of salt across the two ciliary epithelia.

Aqueous humor formation is thought to involve a three-step process by the epithelial bilayer that lines the processes of the ciliary body (Civan and Macknight, 2004). First, paired sodium-proton and chloride-bicarbonate antiporters play a major role in transferring sodium and chloride from the ciliary body stroma into pigmented epithelial cells. Sodium and chloride easily pass by simple diffusion from pigmented cells into the nonpigmented cells through gap junctions before they are actively moved to the posterior chamber via a combination of Na^+-K^+ ATPases, Cl^- channels, and $Na^+K^+2Cl^-$ cotransporters. Because gap junctions between nonpigmented and pigmented cells allow the free movement of water and salt, and pigmented cells do not contain tight junctions; aquaporin channels (AQP1 and AQP4) appear to be needed only in nonpigmented cells. Interestingly, even though tight junctions between nonpigmented epithelial cells form the blood–aqueous barrier and eliminate paracellular passage of solute and water, aquaporins localize to plasma membranes on both apical and basolateral sides; suggesting that water is drawn into nonpigmented cells both from pigmented cells, through gap junctions, and from interstitial space on lateral sides, through aquaporins. Finally, water exits nonpigmented cells and enters the posterior chamber in part through AQP1 and AQP4 channels on the basal membranes.

The functional contribution of aquaporin channels to aqueous humor secretion *in vivo* was demonstrated in mice lacking AQP1, AQP4, or both (Zhang *et al.*, 2002). Despite probable compensatory mechanisms, intraocular pressure in the mice lacking aquaporins (AQP1, AQP4, or AQP1/AQP4) was significantly lower than their wild-type littermates (Fig. 3A). Depression of intraocular pressure varied between 1 and 2 mm Hg, depending upon the strain of mice and the missing aquaporin homologue(s). This decreased intraocular pressure in mice lacking aquaporins was found in part due to lower levels of aqueous humor production. In these animals, aqueous humor production was measured using *in vivo* confocal microscopy after introduction of fluorescein into the anterior chamber. Figure 3B shows that fluorescein had a longer half-life in the anterior chamber of mice lacking one or both of the aquaporins expressed by the nonpigmented ciliary epithelium. These data in living animals agree with data obtained with cultured cells showing that transport of fluid across monolayers of nonpigmented epithelial cells was inhibited upon treatment with mercuric chloride (a potent blocker of AQP1 channels) and antisense oligonucleotides specific for AQP1 RNA (Patil *et al.*, 2001).

In addition to effects of altered aquaporin expression on membrane permeability, regulation of AQP1 and AQP4 by second-messenger systems was also shown to impact water movement across cell membranes. For example, phosphorylation of AQP1 by cyclic adenosine monophosphate

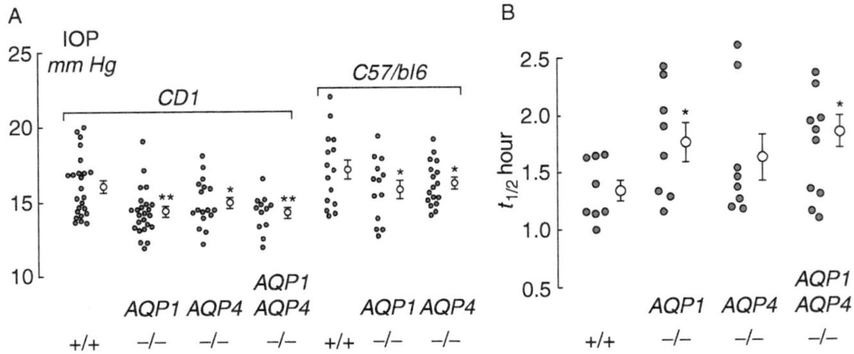

FIGURE 3 Intraocular pressure measurements and aqueous humor production in aquaporin (AQP) null mice. Panel A shows results of IOP measurements in two different strains of wild-type mice, AQP1 and AQP4 null mice and AQP1/AQP4 double null mice. Shown are data from individual eyes (filled circles) and mean ± SE (open circles). $*p < 0.05$, $**p < 0.002$ (ANOVA). Panel B shows measurements of individual eyes (filled circles) and mean±SE (open circles) for aqueous humor production in wild-type and AQP null mice. Data are expressed as half-times ($t_{1/2}$) for fluorescein disappearance. $*p < 0.05$ (ANOVA). Reprinted with permission from Zhang *et al.*(2002).

(cAMP)-dependent protein kinase A (PKA) was shown to increase fluid movement across cells heterologously expressing AQP1 by increasing AQP1 at the plasma membrane (Han and Patil, 2000). These data are consistent with known dependency of water movement across the ciliary epithelium upon intracellular levels of cAMP. However, the specific effects of AQP1 phosphorylation to cAMP-mediated changes in aqueous humor productions have not been demonstrated. With respect to AQP4, phorbol ester treatment resulted in AQP4 phosphorylation and a consequential decrease in membrane permeability of cells heterologously expressing AQP4 (Han *et al.*, 1998). Interestingly, phorbol ester treatment of rabbit eyes decreased intraocular pressure by 40%, although a role for aquaporin involvement is unknown (Mittag *et al.*, 1987b). Finally, atrial natriuretic peptide treatment of cells heterologously expressing AQP4 or AQP1 results in a decreased permeability to water (Patil *et al.*, 1997a). Atrial natriuretic peptide is also known to inhibit secretion of aqueous humor and lower IOP; however, the role of aquaporins again remains to be defined (Mittag *et al.*, 1987a; Crook and Chang, 1997; Fernandez-Durango *et al.*, 1999). In choroid plexus, a tissue that secretes cerebral spinal fluid and strongly expresses AQP1 channels, atrial natriuretic peptide similarly causes a decrease in fluid and salt transport; this process has been suggested to involve not only the water channel property of AQP1, but also the cGMP-activated cationic conductance mediated by AQP1 (discussed in the following section; Boassa *et al.*, 2006). It will be of interest in future studies to determine if the ion channel property has any contribution to the regulatory effects of atrial natriuretic peptide signaling in the eye.

While the role of aquaporins in aqueous humor inflow has been clearly demonstrated, the responsibility of aquaporins in outflow function is less certain. AQP1 is expressed abundantly in all regions of the conventional outflow tract, including the inner (uveal and corneoscleral) and outer (juxtacanalicular tissue, JCT) TM and the inner wall of SC. However, outflow facility measurements in AQP1 null mice were not significantly different from those of littermate controls (Zhang *et al.*, 2002). These results need to be interpreted with caution for several reasons. Since the conventional pathway regulates intraocular pressure by controlling the rate of aqueous humor drainage, there are likely multiple compensatory mechanisms to accommodate the loss of a single protein (AQP1 in this case). Next, effects of AQP1 deletion on outflow facility (hydrostatic-driven) may have been under the level of detection because hydrostatic-driven water permeability in other tissues is affected less by absence of AQP1 than is osmotic-driven water permeability (twofold versus tenfold) (Bai *et al.*, 1999). Additionally, appreciable differences in the anatomy and physiology of aqueous humor drainage exist between mice and humans. The TM is architecturally less complex in mice (composed of two to three layers of lamellae) than in humans (seven to eight layers). In mice, conventional outflow accounts for roughly half of total outflow, whereas in humans conventional outflow accounts for about three-quarters of the total outflow. Thus, clinically relevant analyses of the specific contribution of AQP1 to conventional outflow would benefit from use of animal models that are carefully matched with key human parameters or by use of human tissue such as perfused anterior segments in organ culture (Johnson and Tschumper, 1987). With the organ culture model, AQP1 protein can be manipulated using gene transfer, gene silencing, or pharmacological blockers and effects on outflow facility can be monitored over time.

Indications about the role of AQP1 in the conventional outflow tract were provided using primary cells that were isolated from human donor eyes (Stamer *et al.*, 2001). In these experiments, AQP1 expression was manipulated using adenovirus vectors that carried AQP1 cDNA oriented in the sense or antisense direction. Interestingly, AQP1 overexpression was found to increase resting intracellular volume by ~9%, and thus decrease paracellular permeability of trabecular cell monolayers. The inverse occurred upon knockdown of AQP1 protein (by ~70%); where resting TM cell volume decreased by ~8%. These data were among the first to implicate a role for AQP1 in cell volume regulation. Since this report, several laboratories have shown that aquaporins often exist in protein complexes that appear to sense or regulate cell volume (Krane *et al.*, 2001; Chan *et al.*, 2004; Kuang *et al.*, 2004; Liu *et al.*, 2006).

In the conventional outflow tract, changes in volume of cells in the juxtacanalicular region or inner wall of SC have been shown to influence outflow facility [and thus intraocular pressure (Gual *et al.*, 1997)]. Volume changes in the JCT affect the geometry of the conventional tissues and impact flow pathways for aqueous humor. For example, a 10% decrease in cell volume results in ~25% increase in outflow facility (Al-Aswad *et al.*, 1999). At the inner wall of SC, changes in cell volume would be expected to impact transcellular (vs paracellular) routes for fluid. Such routes have been referred to as "border pores" (Ethier *et al.*, 1998). Since the inner wall of SC is the only continuous cell barrier that aqueous humor encounters before entering the systemic circulation, changes in the number of AQP1 channels at the cell surface would likely affect the transcellular permeability of the barrier. The proportion of aqueous humor that utilizes transcellular versus paracellular routes presently is unknown, and thus the impact of aquaporin expression in SC cells on total outflow facility is uncertain (reviewed by Ethier, 2002; Johnson, 2006).

A role for AQP1 in the JCT cells and SC cells can be envisioned in light of their contribution to outflow resistance (i.e., regulation of fluid transport out of the eye), but the function of AQP1 channels in TM cells that reside on the lamellar beams—presumably providing no appreciable resistance to flow due to the wide opening between beams—is unknown. One possibility is that AQP1 channels may accommodate rapid volume changes that could occur in the conventional outflow tract when the outflow tissues are subjected to mechanical deformation. Trabecular cells reside in a unique environment that is under continuous mechanical stress, both repetitive and intermittent (Ethier, 2002). For instance, during accommodation, the TM is stretched and forces are transmitted throughout conventional outflow tissues via tendons that originate in the ciliary muscle and attach to the basement membrane below the SC inner wall. In addition, conventional outflow tissues are continually perturbed due to the ocular pulse, blinking, squinting, or eye rubbing (Coleman and Trokel, 1969). Such everyday activities can rapidly and transiently elevate intraocular pressure by up to an order of magnitude (from 10 to 100 mm Hg). As tissues deform, the resident cells can be forced to change volume, and aquaporin could be playing a key role in allowing TM cells to change volume in the meshwork. Interestingly, in skeletal muscle a similar role for AQP4 has been hypothesized where aquaporins are thought to facilitate the rapid transfer of water from blood to muscle during periods of intense activity, such as exercise (Frigeri *et al.*, 2004). If this hypothesis is true in the meshwork, the presence of AQP1 on uveal and corneoscleral meshwork cells emphasizes the dynamic biomechanical environment of the conventional outflow pathway.

A small but significant proportion of aqueous humor that is produced by the ciliary epithelia exits the eye posteriorly, across the retina and RPE. To facilitate this flux, there is a net apical to basolateral movement of solute across RPE cells

(Miller and Steinberg, 1977; Marmorstein, 2001). In addition to active transport of solute, two passive mechanisms, intraocular pressure and oncotic pressure from the choroid, contribute to water movement across the RPE and into the choroid. Because the paracellular route is restricted by the presence of highly complex tight junctions that are essential to maintain the blood–retinal barrier [resistance = $\sim$2000 Ω cm^2 (Joseph and Miller, 1991; Marmorstein, 2001)], water and solute must traverse cell membranes in a process likely to be facilitated by transporters and channels, including AQP1 channels.

The transport of solute across the RPE is dependent in part upon the concentration of potassium and sodium in the subretinal space (between the photoreceptors and RPE), which in turn is dependent upon ion conductances across photoreceptors during periods of light and darkness. During light onset for example, the subretinal potassium concentration decreases, causing changes in the activity of apically located potassium channels and transporters in the RPE that ultimately influence chloride transport (Gallemore *et al.*, 1998). Interestingly, the apically located Na^+-K^+ ATPase pump of the RPE does not contribute to vectorial transport of solute (in parallel with water movement) as it does in other epithelia, but instead is thought to regulate subretinal sodium concentration to support photoreceptor function. The transepithelial transport of chloride plays a major role in driving water movement across RPE, mediated by the Na^+-K^+-$2Cl^-$ cotransporter on apical membranes and chloride channels present in basolateral membranes (Joseph and Miller, 1992; Hughes and Segawa, 1993).

The high permeability of the RPE to water is enabled by AQP1. Localization of AQP1 to plasma membranes of RPE of human donor eyes and in RPE cells isolated from human donor eyes has been characterized for both fetal and adult (Stamer *et al.*, 2003). Figure 4 shows that modulation of AQP1 expression significantly impacts movement of water across fetal human RPE monolayers. Thus, the expression of AQP1 by human RPE facilitates water movement that is thought to be critical for sustaining retinal attachment and visual function. Not surprisingly, AQP1 channels are interesting as candidate therapeutic targets for visual disabilities associated with pathological states such as retinal edema.

Interestingly, there appears to be a species difference with respect to aquaporin expression by RPE. While AQP1 mRNA and protein are observed in human RPE, AQP1 protein was not detected in the RPE of rat eyes (Hamann *et al.*, 1998). The reason for this species difference is unclear, but may be related to differences in eye structure, the presence of other compensatory pathways for maintaining fluid balance, or the absence of selective pressure for longevity of the visual system in the aging rodent. Consideration of species differences is particularly important given that rats are used as a model organism for studies of transport properties in RPE (Eichhorn *et al.*, 1996; Maminishkis *et al.*, 2002).

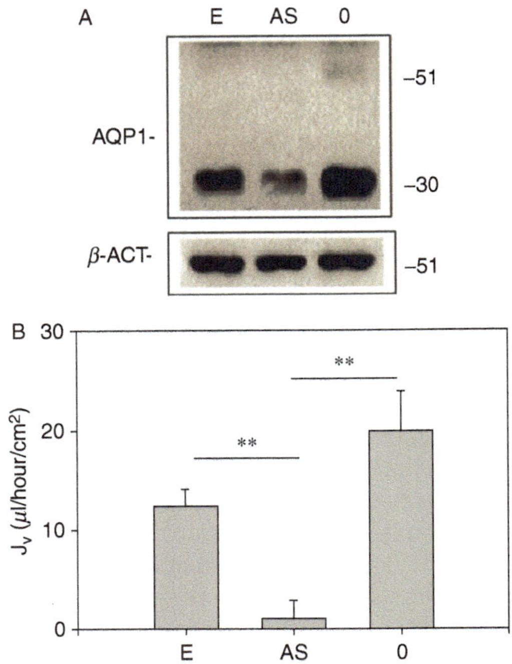

FIGURE 4 Expression and functional analyses of differentiated fetal human RPE monolayer in culture. Panel A shows the expression of native AQP1 in fetal RPE monolayers not infected (0) or after infection with control (empty, E) adenovirus or adenovirus-containing antisense AQP1 DNA (AS). Panel B shows amount of water movement across transduced monolayers in response to an osmotic gradient (~150 mOsm). Data are expressed as rate of water movement, J_v (μl/hour/cm^2), Asterisks indicate significant differences between AQP1-expressing monolayers versus control (**$p < 0.01$). Reprinted with permission from Stamer *et al.* (2003).

It is likely not a coincidence that all cells that form the blood–ocular barriers (blood–retina and blood–aqueous) and that limit paracellular transport with tight cell–cell junctions also express aquaporin channels. The RPE, nonpigmented ciliary epithelium and SC endothelium all express at least one aquaporin channel homologue. Such an expression pattern highlights the importance of the efficient water movement across barriers into and out of the eye.

VI. ION CHANNEL ACTIVITY OF AQP1

In addition to its constitutive function as a water channel, AQP1 contains a parallel pathway for cations that is regulated in part by the binding of intracellular cGMP (Fig. 1; Anthony *et al.*, 2000; Yool and Stamer, 2004). Water permeation occurs through individual pores located within single subunits of the aquaporin tetramer, and the central pore at the axis of fourfold symmetry has been suggested as a candidate ion pore. A conserved internal loop of AQP1 (loop D) has been modeled in molecular dynamic simulations as a flexible gatelike structure that could modify ion permeation at the putative central pore of the tetrameric AQP1 complex (Yu *et al.*, 2006).

AQP1 channels were shown to carry nonselective monovalent cationic currents after stimulation with PKA (Yool *et al.*, 1996), and cGMP but not cAMP (Anthony *et al.*, 2000). When reconstituted in lipid bilayers, AQP1 showed a cGMP-dependent cationic channel function; but only a very small proportion of the total population of water channels incorporated into the bilayer were available to be gated as ion channels (Saparov *et al.*, 2001), suggesting that other cellular components were missing in the reconstituted system. Further work has shown that native AQP1 channels in choroid plexus generate a robust cGMP-dependent cationic conductance that is lost after AQP1 knockdown by small interfering RNAs (Boassa *et al.*, 2006). This cationic conductance activated by atrial natriuretic receptor signaling (and associated cGMP generation) is blocked by Cd^{2+}, and appears to be physiologically relevant in governing fluid secretion (Boassa *et al.*, 2006). These data support a physiological role for AQP1 ion channel activity in tissues involved in fluid secretion and absorption. The dual ion and water channel function could in theory allow modification of local osmotic gradients, perhaps enabling adjustments in cell volume and morphology at a microscopic scale, or might serve in signal transduction by causing depolarization of the cGMP-stimulated cells.

In the eye, the importance of AQP1 as a water channel is obvious. An additional role for AQP1 in its mode as a gated cation channel remains to be assessed. Since not all tissues in which AQP1 is expressed would necessarily benefit from Na^+ entry and the depolarizing effects of the ion channel activity, it is likely that this additional function is under tissue-specific control. The presence of cGMP-sensitive cation channels in tissues of the eye that express AQP1 and are involved in aqueous humor dynamics is an intriguing observation, leading to the speculation that some component of the cation currents could be due to the activity of cGMP-gated AQP1 cation pores. A possible role for the dual water and ion channel function of AQP1 in the fine control of fluid secretion in ciliary epithelium and RPE is an interesting hypothesis that needs to be tested.

VII. AQUAPORIN AND ION CHANNEL INTERACTIONS

There is mounting evidence that aquaporins are incorporated into scaffolds at the plasma membrane with other proteins, suggesting that efficient fluid movement across tissues depends not on individual water channels but on complex associations with signaling and transport proteins (Cowan *et al.*, 2000). In many ocular tissues, chloride secretion provides a key component of the driving force for water movement; however, a parallel pathway for cation flow is required for electroneutral bulk flow. The coexpression of aquaporin water channels and the cystic fibrosis transmembrane conductance regulator (CFTR) channels for chloride enables effective salt and water transport in many types of tissues of the eye, such as corneal epithelia and endothelia, ciliary epithelium, and retinal pigmented epithelium (Levin and Verkman, 2006).

In the ciliary epithelium, sodium enters the pigmented layer from the stromal side along with chloride and transits through gap junctions to the NPE cells for secretion with the aqueous humor, primarily through Na^+-K^+-ATPase pumps, while chloride exits through various channels (Civan, 2003; Vessey *et al.*, 2004). A possible role for CFTR in chloride movement through the ciliary epithelium is supported by the presence of cAMP-activated chloride currents that result in movement of chloride between the pigmented and nonpigmented epithelium; however, there are conflicting results as to the presence of CFTR in the ciliary epithelium (Chu and Candia, 1985; Do *et al.*, 2004; Ni *et al.*, 2006). Fluid transport across the NPE cells relies on Na^+-K^+-ATPase pump activity and AQP1, as determined by sensitivity to block by mercuric chloride and by antisense knockdown (Patil *et al.*, 2001).

In the RPE, chloride is the primary driving force for water transport, moving through basal chloride channels including CFTR (Miller and Edelman, 1990; Hu *et al.*, 1996; Blaug *et al.*, 2003). Consistent with this idea, humans with cystic fibrosis or mice with mutations in CFTR exhibit decreased chloride transport across the RPE (Gallemore *et al.*, 1998; Wu *et al.*, 2006). Less well known are the means by which sodium, the likely counterion to chloride, is moved across the epithelium. Thus, while apical sodium entry is facilitated by the $Na^+K^+2Cl^-$ cotransporter, the basolateral membrane transport mechanism is unknown. AQP1's function as an ion channel on either membrane face might augment sodium flux down its electrochemical gradient. An intriguing possibility in both the NPE and RPE is that the cGMP-activated cation flux through AQP1 may modulate net water transport (Fig. 5). Because of differences in cellular distribution of Na^+-K^+-ATPase pumps between the NPE and RPE, and difference in the location of blood supply relative to transport direction across these barriers, it is conceivable that activated AQP1 ionic currents working by the same mechanism would have opposite effects in these two epithelia (i.e., in response to cGMP signaling, slowing the net secretion in the NPE and enhancing net secretion in RPE).

Signaling pathways involving cAMP and cGMP are known to influence salt and water transport in the ciliary epithelium and RPE. Interestingly, while the AQP1 ion conductance is activated by increased cGMP (Anthony

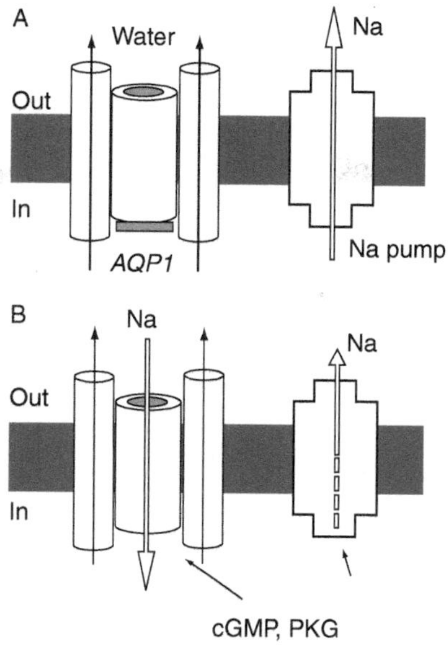

FIGURE 5 Schematic diagram of a hypothetical mechanism for controlling transmembrane salt and water flux by cGMP signaling. (A) Na active transport out of the cell by the Na^+-K^+-ATPase pump and water efflux through AQP1 water pores in the unstimulated state. (B) Downregulated Na^+-K^+-ATPase pump activity (Ellis *et al.*, 2000) and activation of the AQP1 ionic conductance after cGMP stimulation, resulting in a decrease in net secretion (an increase in net absorption) of fluid via possible local accumulation of Na^+ at the inner membrane.

et al., 2000; Boassa and Yool, 2002), it has been suggested to be antagonized by intracellular cAMP (Yool and Stamer, 2004). Water channel activity of AQP1 is increased by PKA, suggesting that a cAMP-responsive redistribution of AQP1 occurs by phosphorylation of AQP1 (Han and Patil, 2000). It is conceivable that independent regulation of the water and ion channel activity of AQP1 by intracellular signaling cascades would offer intricacy in the control of fluid transport. At present, there is no direct evidence for or against a role for AQP1 ion channels in inflow or outflow pathways, but there are lines of evidence indicating the presence of cGMP-sensitive ionic conductances (Carre *et al.*, 1996). For example, nitric oxide (NO) and cGMP cause a modest depolarization of the ciliary epithelial transmembrane potential (Fleischhauer *et al.*, 2001), activate cation conductances in rabbit ciliary epithelium (Carre *et al.*, 1996), and inhibit Na^+,K^+-ATPase via protein kinase G (PKG) but not PKA (Shahidullah and Delamere, 2006). Each of these instances is consistent with the known ability of NO donors to reduce aqueous humor secretion (Korenfeld and Becker, 1989; Shahidullah *et al.*, 2005).

Cyclic nucleotide-gated cation channels in the RPE have not been reported; however, chloride and potassium channels in basolateral membranes have been shown to be regulated by intracellular cAMP (Joseph and Miller, 1992; Hughes and Segawa, 1993). Studies have also shown that changes in cGMP levels increase with atrial natriuretic peptide treatment and induce changes in fluid and chloride transport across RPE (Mikami *et al.*, 1995). Further investigation is necessary to evaluate the mechanisms by which cGMP modulates fluid transport. The relationship between cGMP and AQP1 provides a potential way to control AQP1 ion channel function and fluid transport across ciliary and retinal epithelia.

In ocular epithelia, the role of cationic currents mediated by AQP1 channels in aqueous humor movement is an interesting possibility that remains to be tested (Anthony *et al.*, 2000; Yu *et al.*, 2006). In choroid plexus, the inhibition of Na,K-ATPase activity and the activation of AQP1 ion channels in response to cGMP stimulation lead to a decrease in net cerebral spinal fluid production; the inhibitory effect is reversed by application of an AQP1 ion channel blocker or by knockdown of AQP1 expression (Boassa *et al.*, 2006). These data prompt the hypothesis that the braking role of AQP1 ion channel activity on fluid export, in parallel with regulation of the Na^+ pump (Fig. 5), might be a conserved theme in the eye and brain ventricle. In ciliary epithelium, AQP1 ion channel activation would be expected to decrease the outflow of water across the membrane, decreasing aqueous humor production. In the RPE, the comparable mechanism of AQP1 ion channel activation will have an opposite effect, serving a complementary role in enhancing net fluid transfer into the RPE for subsequent removal into the blood.

VIII. FUTURE DIRECTIONS

While the dependency of aqueous humor secretion on aquaporin expression is clear in aquaporin null mice, such an *in vivo* model is not ideal for evaluating aquaporin expression in the conventional drainage tract or in the RPE. Thus, experiments are needed that test the role of AQP1 in the physiology and pathophysiology of these two tissues using model systems that more closely resemble the human case. For studying conventional drainage, the human anterior segment perfusion system or live nonhuman primates are commonly used. For studying retinal attachment, a live nonhuman primate model is likely best, unless a lower mammal that expresses AQP1 in the RPE and shows retina edema is identified. The species differences in these two outflow pathways for intraocular fluid is interesting, requires further study, and might offer new insights from comparative physiology into the diversity of strategies that allow management of intraocular pressure and maintenance of ocular homeostasis.

Due to the impact of aquaporin expression on aqueous humor production, aquaporins become logical targets for the development of therapeutics for ocular hypertension. However, several issues need to be considered regarding the specificity and level of action of a drug. These considerations include: (1) targeting the expression of an individual aquaporin homologue, (2) targeting a second-messenger system that modulates aquaporin function (water or ion permeability), or (3) blocking at a specific site on an aquaporin that selectively occludes a function (water or ion permeability).

Another consideration in development of therapeutics is the possible adverse effects on nontarget tissues, in addition to the desired effect at ciliary epithelium. Thus, topical administration of blockers to a specific aquaporin homologue such as AQP1 might adversely affect corneal and/or lens clarity, or the health of TM and iris. Results from AQP1 knockout studies indicate that this should not be a problem, although compensatory mechanisms in these animals cannot be excluded. In contrast to AQP1, blockers of AQP4 might have few or no ocular side effects due to a limited expression pattern that includes the nonpigmented epithelial cells (assuming that the retinal glial cells are not affected by topically applied agents).

Since intraocular pressure is a result of the balance of aqueous humor secretion and aqueous humor removal and aquaporins are expressed in tissues responsible for both, one might speculate at first glance that the blockade of aquaporins would have no net effect on overall fluid homeostasis. However, differences between outflow and inflow pathways with respect to the subcellular localization, forces driving flow (osmotic vs hydrostatic), density of aquaporins, and the potential differential influences of tissue-specific interacting proteins could create distinct functional roles for the same aquaporin homologue in these two tissues. Thus, aquaporin-selective blockers could generate asymmetrical effects. A beneficial decrease in inflow in principle could be achieved without substantially affecting outflow. Clearly, such issues need to be worked out in future studies, but the foundation of data accumulated thus far are promising for the development of new therapies involving aquaporins as novel anti-glaucoma targets.

References

Al-Aswad, L., Gong, H., Lee, D., O'Donnell, M., Brandt, J., Ryan, W., Schroeder, A., and Erikson, K. (1999). Effects of Na-K-2Cl cotransport regulators on outflow facility in calf and human eyes *in vitro*. *Inv. Oph. Vis. Sci.* **40,** 1695–1701.

Anthony, T., Brooks, H., Boassa, D., Leonov, S., Yanochko, G., Regan, J., and Yool, A. (2000). Cloned Aquaporin-1 is a cyclic GMP-gated ion channel. *Mol. Pha.* **57,** 576–588.

Bai, C., Fukuda, N., Song, Y., Ma, T., Matthay, M., and Verkman, A. (1999). Lung fluid transport in aquaporin-1 and aquaporin-4 knockout mice. *J. Clin. Inv.* **103,** 555–561.

Bill, A., and Philips, C. I. (1971). Uveoscleral drainage of aqueous humour in human eyes. *Exp. Eye. Res.* **12,** 275–281.

Blaug, S., Quinn, R., Quong, J., Jalickee, S., and Miller, S. S. (2003). Retinal pigment epithelial function: A role for CFTR. *Doc. Oph.* **106,** 43–50.

Boassa, D., and Yool, A. J. (2002). A fascinating tail: cGMP activation of aquaporin-1 ion channels. *Tre. Pharm. Sci.* **23,** 558–562.

Boassa, D., Stamer, W. D., and Yool, A. J. (2006). Ion channel function of aquaporin-1 natively expressed in choroid plexus. *J. Neur.* **26,** 7811–7819.

Broekhuyse, R. M., Kuhlmann, E. D., and Winkens, H. J. (1979). Lens membranes VII. MIP is an immunologically specific component of lens fiber membranes and is identical with 26K band protein. *Exp. Eye. Res.* **29,** 303–313.

Brubaker, R. (1991). Flow of Aqueous Humor in Humans. *Inv. Oph. Vis. Sci.* **32,** 3145–3166.

Buzhynskyy, N., Hite, R., Walz, T., and Scheuring, S. (2007). The supramolecular architecture of junctional microdomains in native lens membranes. *EMBO Rep.* **8,** 51–55.

Carre, D. A., Anguita, J., Coca-Prados, M., and Civan, M. M. (1996). Cell-attached patch clamping of the intact rabbit ciliary epithelium. *Cur. Eye. Res.* **15,** 193–201.

Chan, H., Butterworth, R. F., and Hazell, A. S. (2004). Primary cultures of rat astrocytes respond to thiamine deficiency-induced swelling by downregulating aquaporin-4 levels. *Neur. Let.* **366,** 231–234.

Chandy, G., Zampighi, G. A., Kreman, M., and Hall, J. E. (1997). Comparison of the water transporting properties of MIP and AQP1. *J. Mem. Bio.* **159,** 29–39.

Chu, T. C., and Candia, O. A. (1985). Effects of adrenergic agonists and cyclic AMP on the short-circuit current across the isolated rabbit iris-ciliary body. *Cur. Eye. Res.* **4,** 523–529.

Civan, M. M. (2003). The fall and rise of active chloride transport: Implications for regulation of intraocular pressure. *J. Exp. Zoo. Part A, Comp. Exp. Bio.* **300,** 5–13.

Civan, M. M., and Macknight, A. D. (2004). The ins and outs of aqueous humour secretion. *Exp. Eye. Res.* **78,** 625–631.

Coleman, D. J., and Trokel, S. (1969). Direct-recorded intraocular pressure variations in a human subject. *Arch. of Oph.* **82,** 637–640.

Cowan, C. A., Yokoyama, N., Bianchi, L. M., Henkemeyer, M., and Fritzsch, B. (2000). EphB2 guides axons at the midline and is necessary for normal vestibular function. [see comment] *Neu.* **26,** 417–430.

Crook, R. B., and Chang, A. T. (1997). Differential regulation of natriuretic peptide receptors on ciliary body epithelial cells. *Bio. J.* **324,** 49–55.

de Groot, B. L., Engel, A., and Grubmuller, H. (2001). A refined structure of human aquaporin-1. *FEBS Let.* **504,** 206–211.

Do, C. W., Kong, C. W., and To, C. H. (2004). cAMP inhibits transepithelial chloride secretion across bovine ciliary body/epithelium. *Inv. Oph. & Vis. Sci.* **45,** 3638–3643.

Echevarria, M., Kuang, K., Iserovich, P., Li, J., Preston, G. M., Agre, P., and Fischbarg, J. (1993). Cultured bovine corneal endothelial cells express CHIP28 water channels. *Am. J. Phy.* **265,** C1349–C1355.

Eichhorn, M., Schreckenberger, M., Tamm, E. R., and Lutjen-Drecoll, E. (1996). Carbonic anhydrase activity is increased in retinal pigmented epithelium and choriocapillaris of RCS rats. *Gra. Arc. for Clin. & Exp. Oph.* **234,** 258–263.

Ethier, C., Coloma, F., Sit, A., and Johnson, M. (1998). Two pore types in the inner wall endothelium of Schlemm's canal. *Inv. Oph. Vis. Sci.* **39,** 2041–2048.

Ethier, C. (2002). The inner wall of Schlemm's canal. *Exp. Eye. Res.* **74,** 161–172.

Fernandez-Durango, R., Moya, F. J., Ripodas, A., de Juan, J. A., Fernandez-Cruz, A., and Bernal, R. (1999). Type B and type C natriuretic peptide receptors modulate intraocular pressure in the rabbit eye. *Eur. J. Phar.* **364,** 107–113.

Fitzgerald, P. G., Bok, D., and Horwitz, J. (1983). Immunocytochemical localization of the main intrinsic polypeptide (MIP) in ultrathin frozen sections of rat lens. *J. Cell. Biol.* **97,** 1491–1499.

Fleischhauer, J. C., Beny, J. L., Flammer, J., and Haefliger, I. O. (2001). Cyclic AMP and anionic currents in porcine ciliary epithelium. *Klin. Mon. fur. Auge.* **218,** 370–372.

Francis, P., Chung, J. J., Yasui, M., Berry, V., Moore, A., Wyatt, M. K., Wistow, G., Bhattacharya, S. S., and Agre, P. (2000). Functional impairment of lens aquaporin in two families with dominantly inherited cataracts. *Hum. Mol. Gen.* **9,** 2329–2334.

Frigeri, A., Gropper, M., Umenishi, F., Kawashima, M., Brown, D., and Verkman, A. (1995). Localization of MIWC and GLIP water channel homologs in neuromuscular, epithelial and glandular tissues. *J. Cell Sci.* **108,** 2993–3002.

Frigeri, A., Nicchia, G. P., Balena, R., Nico, B., and Svelto, M. (2004). Aquaporins in skeletal muscle: Reassessment of the functional role of aquaporin-4. *FASEB J.* **18,** 905–907.

Funaki, H., Yamamoto, T., Koyama, Y., Kondo, D., Yaoita, E., Kawasaki, K., Kobayashi, H., Sawaguchi, S., Abe, H., and Kihara, I. (1998). Localization and expression of AQP5 in cornea, serous salivary glands, and pulmonary epithelial cells. *Am. J. Phy.* **275,** C1151–C1157.

Gallardo, P., Olea, N., and Sepulveda, F. V. (2002). Distribution of aquaporins in the colon of Octodon degus, a South American desert rodent. *Am. J. Phy. - Reg. Int. & Com. Phy.* **283,** R779–R788.

Gallemore, R., Hughes, B., and Miller, S. (1998). *In* "The Retinal Pigment Epithelium: Function and Disease" (Marmor, M., and Wolfensberger, T., eds.), pp. 103–198. Oxford University Press, Oxford, New York.

Gorin, M. B., Yancey, S. B., Cline, J., Revel, J. P., and Horwitz, J. (1984). The major intrinsic protein (MIP) of the bovine lens fiber membrane: Characterization and structure based on cDNA cloning. *Cell* **39,** 49–59.

Gual, A., Llobet, A., Gilabert, R., Borras, M., Pales, J., Bergamini, M., and Belmonte, C. (1997). Effects of time of storage, albumin and osmolality changes on outflow facility of bovine anterior segment *in vitro. Inv. Oph. Vis. Sci.* **38,** 2165–2171.

Hamann, S., Zeuthen, T., LaCour, M., Nagelhus, E., Ottersen, O., Agre, P., and Nielsen, A. (1998). Aquaporins in complex tissues: Distribution of aquaporins 1-5 in human and rat eye. *Am. J. Phy.* **274,** C1331–C1345.

Han, Z., and Patil, R. V. (2000). Protein kinase A-dependent phosphorylation of aquaporin-1. *Bio. & Bio. Res. Com.* **273,** 328–332.

Han, Z., Wax, M. B., and Patil, R. V. (1998). Regulation of aquaporin-4 water channels by phorbol ester-dependent protein phosphorylation. *J. Biol. Chem.* **273,** 6001–6004.

Hasegawa, H., Lian, S. C., Finkbeiner, W. E., and Verkman, A. S. (1994). Extrarenal tissue distribution of CHIP28 water channels by *in situ* hybridization and antibody staining. *Am. J. Phy.* **266,** C893–C903.

Hasegawa, H., Zhang, R., Dohrman, A., and Verkman, A. S. (1993). Tissue-specific expression of mRNA encoding rat kidney water channel CHIP28k by *in situ* hybridization. *Am. J. Phy.* **264,** C237–C245.

Hu, J. G., Gallemore, R. P., Bok, D., and Frambach, D. A. (1996). Chloride transport in cultured fetal human retinal pigment epithelium. *Exp. Eye. Res.* **62,** 443–448.

Hub, J. S., and de Groot, B. L. (2006). Does CO2 permeate through aquaporin-1. *Bio. J.* **91,** 842–848.

Hughes, B. A., and Segawa, Y. (1993). cAMP-activated chloride currents in amphibian retinal pigment epithelial cells. *J. Phy.* **466,** 749–766.

Iandiev, I., Biedermann, B., Reichenbach, A., Wiedemann, P., and Bringmann, A. (2006). Expression of aquaporin-9 immunoreactivity by catecholaminergic amacrine cells in the rat retina. *Neur. Let.* **398,** 264–267.

Johnson, D. H., and Tschumper, R. C. (1987). Human trabecular meshwork organ culture. A new method. *Inv. Oph. & Vis. Sci.* **28,** 945–953.

Johnson, M. (2006). 'What controls aqueous humour outflow resistance?'. *Exp. Eye. Res.* **82,** 545–557.

Joseph, D., and Miller, S. (1991). Apical and basal membrane ion transport mechanisms in bovine retinal pigment epithelium. *J. Phy.* **435,** 439–463.

Joseph, D. P., and Miller, S. S. (1992). Alpha-1-adrenergic modulation of K and Cl transport in bovine retinal pigment epithelium. *J. Gen. Phy.* **99,** 263–90.

Jung, J. S., Preston, G. M., Smith, B. L., Guggino, W. B., and Agre, P. (1994). Molecular structure of the water channel through aquaporin CHIP. The hourglass model. *J. Biol. Chem.* **269,** 14648–14654.

Kalman, K., Nemeth-Cahalan, K. L., Froger, A., and Hall, J. E. (2006). AQP0-LTR of the Cat Fr mouse alters water permeability and calcium regulation of wild type AQP0. *Bio. et. Bio. Act.* **1758,** 1094–1099.

Kang, T. H., Choi, Y. K., Kim, I. B., Oh, S. J., and Chun, M. H. (2005). Identification and characterization of an aquaporin 1 immunoreactive amacrine-type cell of the mouse retina. *J. Com. Neur.* **488,** 352–367.

Kim, I. B., Oh, S. J., Nielsen, S., and Chun, M. H. (1998). Immunocytochemical localization of aquaporin 1 in the rat retina. *Neur. Let.* **244,** 52–54.

Kim, I. B., Lee, E. J., Oh, S. J., Park, C. B., Pow, D. V., and Chun, M. H. (2002). Light and electron microscopic analysis of aquaporin 1-like-immunoreactive amacrine cells in the rat retina. *J. Com. Neur.* **452,** 178–191.

King, L. S., Kozono, D., and Agre, P. (2004). From structure to disease: The evolving tale of aquaporin biology. *Nat. Rev. Mol. Cell Biol.* **5,** 687–698.

Korenfeld, M. S., and Becker, B. (1989). Atrial natriuretic peptides. Effects on intraocular pressure, cGMP, and aqueous flow. *Inv. Oph. & Vis. Sci.* **30,** 2385–2392.

Krane, C. M., Melvin, J. E., Nguyen, H. V., Richardson, L., Towne, J. E., Doetschman, T., and Menon, A. G. (2001). Salivary acinar cells from aquaporin 5-deficient mice have decreased membrane water permeability and altered cell volume regulation. *J. Biol. Chem.* **276,** 23413–23420.

Kuang, K., Yiming, M., Wen, Q., Li, Y., Ma, L., Iserovich, P., Verkman, A. S., and Fischbarg, J. (2004). Fluid transport across cultured layers of corneal endothelium from aquaporin-1 null mice. *Exp. Eye. Res.* **78,** 791–798.

Levin, M. H., and Verkman, A. S. (2006). Aquaporins and CFTR in ocular epithelial fluid transport. *J. Mem. Biol.* **210,** 105–115.

Lindsey Rose, K. M., Gourdie, R. G., Prescott, A. R., Quinlan, R. A., Crouch, R. K., and Schey, K. L. (2006). The C terminus of lens aquaporin 0 interacts with the cytoskeletal proteins filensin and CP49. *Inv. Oph. & Vis. Sci.* **47,** 1562–1570.

Liu, X., Bandyopadhyay, B., Nakamoto, T., Singh, B., Liedtke, W., Melvin, J. E., and Ambudkar, I. (2006). A role for AQP5 in activation of TRPV4 by hypotonicity: Concerted involvement of AQP5 and TRPV4 in regulation of cell volume recovery. *J. Biol. Chem.* **281,** 15485–15495.

Maeda, S., Ito, H., Tanaka, K., Hayakawa, T., and Seki, M. (2005). Localization of aquaporin water channels in the airway of the musk shrew (Suncus murinus) and the rat. *J. Vet. Med. Sci.* **67,** 975–984.

Maminishkis, A., Jalickee, S., Blaug, S. A., Rymer, J., Yerxa, B. R., Peterson, W. M., and Miller, S. S. (2002). The P2Y(2) receptor agonist INS37217 stimulates RPE fluid transport *in vitro* and retinal reattachment in rat. *Inv. Oph. & Vis. Sci.* **43,** 3555–3566.

Marmorstein, A. (2001). The polarity of the retinal pigment epithelium. *Traffic* **2,** 867–872.

Mikami, Y., Hara, M., Yasukura, T., Uyama, M., Minato, A., and Inagaki, C. (1995). Atrial natriuretic peptide stimulates Cl- transport in retinal pigment epithelial cells. *Cur. Eye. Res.* **14,** 391–397.

Miller, S., and Steinberg, R. (1977). Active transport of ions across frog retinal pigment epithelium. *Exp. Eye Res.* **25,** 235–248.

Miller, S. S., and Edelman, J. L. (1990). Active ion transport pathways in the bovine retinal pigment epithelium. *J. Phy.* **424,** 283–300.

Mittag, T. W., Tormay, A., Ortega, M., and Severin, C. (1987). Atrial natriuretic peptide (ANP), guanylate cyclase, and intraocular pressure in the rabbit eye. *Cur. Eye. Res.* **6,** 1189–1196.

Mittag, T. W., Yoshimura, N., and Podos, S. M. (1987). Phorbol ester: Effect on intraocular pressure, adenylate cyclase, and protein kinase in the rabbit eye. *Inv. Oph. & Vis. Sci.* **28,** 2057–2066.

Mulders, S. M., Preston, G. M., Deen, P. M., Guggino, W. B., van Os, C. H., and Agre, P. (1995). Water channel properties of major intrinsic protein of lens. *J. Biol. Chem.* **270,** 9010–9016.

Nagelhus, E. A., Veruki, M. L., Torp, R., Haug, F. M., Laake, J. H., Nielsen, S., Agre, P., and Ottersen, O. P. (1998). Aquaporin-4 water channel protein in the rat retina and optic nerve: Polarized expression in Muller cells and fibrous astrocytes. *J. Neu.* **18,** 2506–2519.

Ni, Y., Wu, R., Xu, W., Maecke, H., Flammer, J., and Haefliger, I. O. (2006). Effect of cAMP on porcine ciliary transepithelial short-circuit current, sodium transport, and chloride transport. *Inv. Oph. & Vis. Sci.* **47,** 2065–2074.

Nielsen, S., Smith, B. L., Christensen, E. I., and Agre, P. (1993). Distribution of the aquaporin CHIP in secretory and resorptive epithelia and capillary endothelia. *Proc. Nat. Acad. Sci. USA* **90,** 7275–9.

Oen, H., Cheng, P., Turner, H. C., Alvarez, L. J., and Candia, O. A. (2006). Identification and localization of aquaporin 5 in the mammalian conjunctival epithelium. *Exp. Eye. Res.* **83,** 995–998.

Patil, R. V., Han, Z., and Wax, M. B. (1997). Regulation of water channel activity of aquaporin 1 by arginine vasopressin and atrial natriuretic peptide. *Bio. & Bio. Res. Com.* **238,** 392–396.

Patil, R. V., Han, Z., Yiming, M., Yang, J., Iserovich, P., Wax, M. B., and Fischbarg, J. (2001). Fluid transport by human nonpigmented ciliary epithelial layers in culture: a homeostatic role for aquaporin-1. *Am. J. Phy. - Cell Phy.* **281,** C1139–C1145.

Patil, R. V., Saito, I., Yang, X., and Wax, M. B. (1997). Expression of aquaporins in the rat ocular tissue. *Exp. Eye. Res.* **64,** 203–209.

Raina, S., Preston, G. M., Guggino, W. B., and Agre, P. (1995). Molecular cloning and characterization of an aquaporin cDNA from salivary, lacrimal, and respiratory tissues. *J. Biol. Chem.* **270,** 1908–1912.

Ren, G., Reddy, V. S., Cheng, A., Melnyk, P., and Mitra, A. K. (2001). Visualization of a water-selective pore by electron crystallography in vitreous ice. *Proc. Nat. Acad. Sci. USA* **98,** 1398–1403.

Ruiz-Ederra, J., and Verkman, A. S. (2006). Accelerated cataract formation and reduced lens epithelial water permeability in aquaporin-1-deficient mice. *Inv. Oph. & Vis. Sci.* **47,** 3960–3967.

Saparov, S. M., Kozono, D., Rothe, U., Agre, P., and Pohl, P. (2001). Water and ion permeation of aquaporin-1 in planar lipid bilayers. Major differences in structural determinants and stoichiometry. *J. Biol. Chem.* **276,** 31515–31520.

Shahidullah, M., and Delamere, N. A. (2006). NO donors inhibit Na,K-ATPase activity by a protein kinase G-dependent mechanism in the nonpigmented ciliary epithelium of the porcine eye. *Bri. J. Phar.* **148,** 871–880.

Shahidullah, M., Yap, M., and To, C. H. (2005). Cyclic GMP, sodium nitroprusside and sodium azide reduce aqueous humour formation in the isolated arterially perfused pig eye. *Bri. J. Phar.* **145,** 84–92.

Shiels, A., and Bassnett, S. (1996). Mutations in the founder of the MIP gene family underlie cataract development in the mouse. *Nat. Gene.* **12,** 212–215.

Stamer, W. D., Snyder, R. W., Smith, B. L., Agre, P., and Regan, J. W. (1994). Localization of aquaporin CHIP in the human eye: Implications in the pathogenesis of glaucoma and other disorders of ocular fluid balance. *Inv. Oph. Vis. Sci.* **35,** 3867–3872.

Stamer, W. D., Seftor, R. E. B., Snyder, R. W., and Regan, J. W. (1995). Cultured human trabecular meshwork cells express aquaporin-1 water channels. *Curr. Eye. Res.* **14,** 1095–1100.

Stamer, W., Peppel, K., O'Donnell, M., Roberts, B., Wu, F., and Epstein, D. (2001). Expression of Aquaporin-1 in human trabecular meshwork cells: Role in volume regulation. *Inv. Oph. Vis. Sci.* **42,** 1803–1811.

Stamer, W., Bok, D., Hu, J., Jaffe, G., and McKay, B. (2003). Aquaporin-1 channels in human retinal pigment epithelium: Role in transepithelial water movement. *Inv. Oph. Vis. Sci.* **44,** 2803–2808.

Sui, H., Han, B. G., Lee, J. K., Walian, P., and Jap, B. K. (2001). Structural basis of water-specific transport through the AQP1 water channel. *Nature* **414,** 872–878.

Thiagarajah, J. R., and Verkman, A. S. (2002). Aquaporin deletion in mice reduces corneal water permeability and delays restoration of transparency after swelling. *J. Biol. Chem.* **277,** 19139–19144.

Tietz, P., Jefferson, J., Pagano, R., Larusso, N. F. (2005) Membrane microdomains in hepatocytes: Potential target areas for proteins involved in canalicular bile secretion. *J. Lip. Res.* **46,** 1426–1432.

Toris, C.B., Yablonski, M.E., Wang, Y.L., Camras, C.B. (1999) Aqueous humor dynamics in the aging human eye. *Am. J. Oph.* **127,** 407–412.

Townsend, D.J., Brubaker, R.F. (1980) Immediate effect of epinephrine on aqueous formation in the normal human eye as measured by fluorophotometry. *Inv. Oph. & Vis. Sci.* **19,** 256–266.

Varadaraj, K., Kumari, S., Shiels, A., and Mathias, R. T. (2005). Regulation of aquaporin water permeability in the lens. *Inv. Oph. & Vis. Sci.* **46,** 1393–1402.

Vessey, J. P., Shi, C., Jollimore, C. A., Stevens, K. T., Coca-Prados, M., Barnes, S., and Kelly, M. E. (2004). Hyposmotic activation of ICl,swell in rabbit nonpigmented ciliary epithelial cells involves increased ClC-3 trafficking to the plasma membrane. *Biochem. & Cell Biol.* **82,** 708–718.

Wu, J., Marmorstein, A. D., and Peachey, N. S. (2006). Functional abnormalities in the retinal pigment epithelium of CFTR mutant mice. *Exp. Eye. Res.* **83,** 424–428.

Yool, A., Stamer, W., and Regan, J. (1996). Forskolin stimulation of water and cation permeability in Aquaporin 1 water channels. *Science* **273,** 1216–1218.

Yool, A., and Stamer, W. (2004). Novel roles for aquaporins as gated ion channels. Elsevier, Amsterdam.

Yu, J., Yool, A. J., Schulten, K., and Tajkhorshid, E. (2006). Mechanism of gating and ion conductivity of a possible tetrameric pore in aquaporin-1. *Structure* **14,** 1411–1423.

Yu, X. S., Yin, X., Lafer, E. M., and Jiang, J. X. (2005). Developmental regulation of the direct interaction between the intracellular loop of connexin 45.6 and the C terminus of major intrinsic protein (aquaporin-0). *J. Biol. Chem.* **280,** 22081–22090.

Zampighi, G. A., Eskandari, S., Hall, J. E., Zampighi, L., and Kreman, M. (2002). Microdomains of AQP0 in lens equatorial fibers. *Exp. Eye. Res.* **75,** 505–519.

Zampighi, G. A., Hall, J. E., Ehring, G. R., and Simon, S. A. (1989). The Structural Organization and Protein Composition of Lens Fiber Junctions. *The J. Cell Biol.* **108,** 2255–2275.

Zhang, D., Vetrivel, L., and Verkman, A. S. (2002). Aquaporin deletion in mice reduces intraocular pressure and aqueous fluid production. *J. Gen. Phy.* **119,** 561–569.

CHAPTER 3

The Role of Gap Junction Channels in the Ciliary Body Secretory Epithelium

Richard T. Mathias, Thomas W. White, and Peter R. Brink
Department of Physiology and Biophysics, State University of New York, Stony Brook, New York 11794

I. OVERVIEW

The secretory epithelium of the ciliary body is responsible for generating the aqueous humor (AH). The epithelium comprises two cell layers: the pigmented epithelium (PE) and non pigmented epithelium (NPE), whose

Current Topics in Membranes, Volume 62 1063-5823/08 $35.00

DOI: 10.1016/S1063-5823(08)00403-1

apical surfaces appose one another. At the apical–apical interface, the two layers are in communication through gap junctions formed from connexins Cx40 and Cx43. Cells of the PE are not in direct gap junctional communication with each other; however, the cells of the NPE are coupled by gap junctions made from Cx26 and Cx32. We first review the properties of the ciliary body, then those of gap junction channels in general, with special emphasis on channels made from the connexins present in the ciliary epithelium. An important question arises concerning the gap junction channels at the apical–apical interface: Do these channels conduct the water that is ultimately secreted? Model calculations are presented that suggest the channels could do so, but only if they provide a rather high degree of ion coupling. Finally, experimental animal models, which might help test this hypothesis, are reviewed.

II. INTRODUCTION

A. The Anatomy of the Ciliary Body

The ciliary body is an annular structure that attaches to the lens via the suspensory ligament and is connected to the sclera at the base of the cornea. The major tissue type within the ciliary body is smooth muscle, which has a role in the accommodation process. On the surface of the ciliary body, facing the posterior chamber, a number of fingerlike projections are found. These make up the secretory epithelium of the ciliary body. Each process has an underlying, capillary rich, connective tissue that is covered by a stratified cuboidal epithelium. The stratified epithelium represents the site of the blood–AH barrier. The epithelium consists of two cell layers: a pigmented layer that faces the interstitial and vascular space, and a nonpigmented layer that faces the posterior chamber. The apical surfaces of the pigmented cells are in intimate contact with the apical surfaces of the nonpigmented cells.

B. Ciliary Body Epithelial Function: Production of Aqueous Humor

The ciliary body secretory epithelium generates a slightly hyperosmotic to isosmotic fluid of ~300 mOsm (Hayward *et al.*, 1976; Gaasterland *et al.*, 1979) and is able to generate fluid flow rates from plasma to aqueous of 15–30 μl/hour in animal models (Candia *et al.*, 2005). Detailed experimental analysis and modeling of the stratified epithelium suggest the net flux of chloride and sodium ions as the motive force in generating the AH. The transport of fluid from the ciliary body into the posterior chamber is balanced by the drainage of fluid by Schlemm's canal. The autonomic nervous system is the major modulator of fluid transport and drainage (Uusitalo, 1972).

Various ion channels and transporters (Do and Civan, 2006) participate in the extraction of ions from the vascular space and subsequent transport to the AH. Water is thought to follow as a consequence of osmosis and is facilitated by the presence of aquaporins (Patil *et al.*, 1997). Gap junction channels are unique within the ensemble of channels and transporters (Coffey *et al.*, 2002). They are unique because they neither extract nor secrete from either epithelial surface. Rather they are presumed to allow the stratified epithelium to act as a monolayer, facilitating the movement of solutes from the pigmented to the nonpigmented layer. The blood–AH barrier is associated with the presence of tight junctions on the lateral surfaces of NPE cells. Given this anatomical barrier, the transepithelial movement of ions and other solutes is thought to be via the gap junction channels connecting the two cell layers. Pharmacological inhibition of gap junction channel activity reduces the net flux of ions across the epithelium (Wolosin *et al.*, 1997; Do and Civan, 2004), supporting the role of gap junctions in transepithelial solute flux.

These data implicate gap junctions as a necessary component of the ciliary body; gap junctions create a functional syncytium connecting the PE and NPE. The end result of such a linkage is a double-layered epithelia functioning similar to a simple monolayer with regard to ions and small molecules. To gain a better understanding of how gap junction channels within the ciliary body epithelium might participate in the production of AH, we will review first the subunit proteins forming the gap junction channels, second the distribution of different channel types, and third the functional properties of those gap junction channels both in general and in relation to the ciliary body. We shall seek to answer the question: are the known properties of ciliary body gap junctions of importance for the normal functioning of the epithelium?

C. Gap Junction Channels Formed by Connexins

Gap junction channels in vertebrates are formed from subunit proteins called connexins. A gap junction channel is composed of two hemichannels, each of which is formed from six connexins. When two cells are in close apposition, it is possible for a hemichannel from each cell to link together via the extracellular loops of the component connexins to form a cell-to-cell gap junction channel. This channel represents a unique intercellular pathway because it is the only form of intercellular communication that excludes the extracellular space. Gap junction channels tend to aggregate and form plaques containing tens to thousands of channels (Goodenough, 1975). The reasons for plaque formation are not completely understood, but have been attributed to lipid membrane domains including lipid rafts (Locke *et al.*, 2005).

There are over 20 identified connexins within the human genome and all are able to form gap junction channels between cells. Most cell types coexpress several connexins. For example, in the heart almost every cell type expresses at least two of the following three connexins, Cx40, Cx43, and Cx45 (Van Veen *et al.*, 2001), while liver hepatocytes and the lacrimal gland coexpress Cx26 and Cx32 (Kojima *et al.*, 1996; Walcott *et al.*, 2002; Ott *et al.*, 2006). The possibility therefore exists of channels being formed that contain more than one connexin type. If a hemichannel contains two connexin types and forms a gap junction channel with an adjacent cell also containing hemichannels made from two connexin types, then the channel is called a heteromeric. If all the subunit connexins from both cells are only of one type of connexin, then the channel is referred to as homotypic. If two adjacent cells are each making a single type of hemichannel, but by different connexins, then cell-to-cell gap junction channels are referred to as heterotypic. A number of connexins have been shown to form heteromeric and heterotypic gap junction channels (Brink *et al.*, 1997; Valiunas *et al.*, 2000, 2001; Cottrell *et al.*, 2002), but not all connexins are able to mix (White and Bruzzone, 1996; Gemel *et al.*, 2004).

D. Connexins in the Ciliary Body Epithelium

The ciliary body is no exception to the general rule that tissues express more than one connexin. In fact, four connexins are expressed in the nonpigmented cell layer whereas two are expressed in the pigmented layer (Coffey *et al.*, 2002). The PE and NPE both express Cx40 and Cx43, which colocalize on their apposing apical surfaces (Coffey *et al.*, 2002). This colocalization implies that these connexins could be forming mixed channels, but an explicit demonstration of heteromeric or heterotypic channels between these two cell types has not been shown. A number of studies using Cx40 and Cx43 have shown that these two connexins have the ability to form heteromeric and heterotypic channels (He *et al.*, 1999; Valiunas *et al.*, 2001, 2002), but other studies suggest that they prefer to not to do so (Bruzzone *et al.*, 1993; Rackauskas *et al.*, 2007). In addition, the cells of the NPE couple to each other with gap junction channels composed of either Cx26 or Cx31 (Fig. 1).

Cx26 forms heterotypic and heteromeric channels with Cx31 *in vitro* (Abrams *et al.*, 2006), but since Cx31 and Cx26 do not appear to colocalize in plaques in the nonpigmented cell layer (Coffey *et al.*, 2002), heterotypic or heteromeric channels from Cx31 and Cx26 seems unlikely.

Figure 1 depicts a schematic of the double-layered epithelium of the ciliary body showing the location of the various connexins based on immunostaining (Coffey *et al.*, 2002). Figure 2 illustrates sections from the ciliary body epithelium of the mouse, illustrating the pigmented layer (left-hand side) and immunostained for Cx43 (right-hand side). The majority of the Cx43 staining lies at the apical surfaces between the two epithelial layers.

III. GENERAL PROPERTIES OF CONNEXINS INCLUDING THOSE COMPOSING THE CILIARY BODY EPITHELIUM GAP JUNCTIONS

A. Voltage Dependence and Open Probability

All homotypic, heterotypic, and heteromeric gap junction channels studied display voltage dependence. For homotypic channels, the voltage dependence is characterized as a symmetric and time-dependent decline in conductance, where increasing voltage step amplitude induces larger and more rapid declines in conductance, but the polarity of the step is irrelevant (Wang *et al.*, 1992). For heterotypic channels, the voltage dependence is often asymmetric:

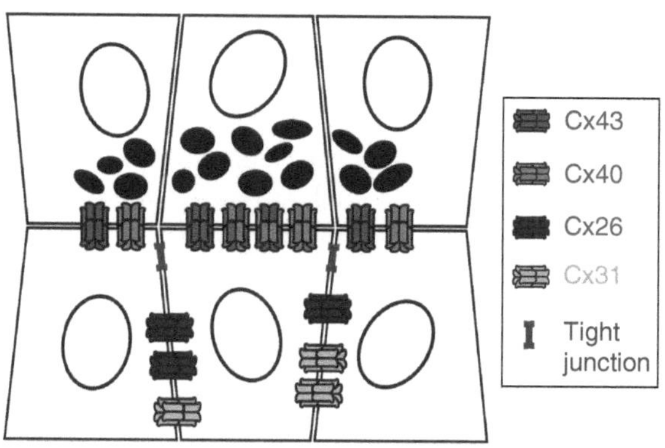

FIGURE 1 A schematic diagram showing connexin localization in the ciliary body. PE and NPE both express Cx40 and Cx43, which are concentrated in gap junction channels that mix within plaques. NPE also expresses Cx26 and Cx31, which provide homotypic gap junction channels that do not mix within the plaques. (See Color Insert.)

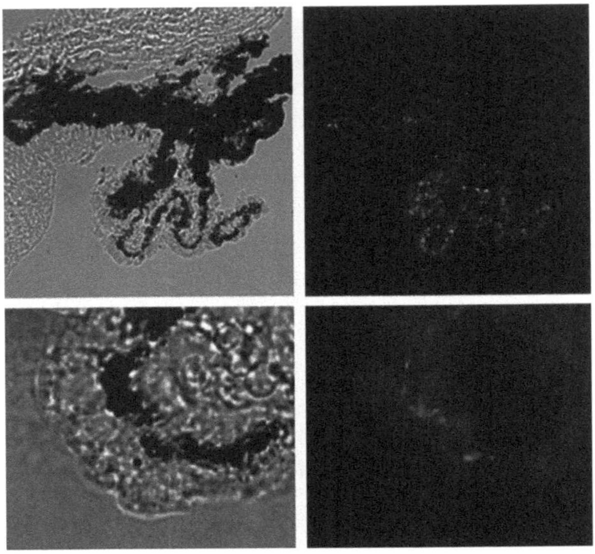

FIGURE 2 Connexin43 is expressed in ciliary epithelial cells. Phase contrast images (left) and immunolocalization of Cx43 (right) in frozen sections of the mouse ciliary body. At low power (upper) Cx43 is highly expressed in the ciliary body. At high power (lower), Cx43 staining is concentrated at the interfaces between pigmented and nonpigmented cells. Nuclei are counterstained with Dapi in lower right image.

the voltage step induced reduction in conductance is polarity dependent (White and Bruzzone, 1996; Brink *et al.*, 1997; Valiunas *et al.*, 2000). Heteromeric channels display symmetric and asymmetric behaviors (Brink *et al.*, 1997; He *et al.*, 1999; Valiunas *et al.*, 2001).

A number of studies have determined the open probability of specific homotypic gap junction channels using a variety of approaches. When the transjunctional voltage is near zero, the normal physiological state, the open probability is between 0.5 and 0.9. Standing transjunctional potentials result in reduced open probabilities (Brink *et al.*, 1996; Christ and Brink, 1999; Chen-Izu *et al.*, 2001; Ramanan *et al.*, 2005). The mean open and closed times for gap junction channels range in the tens to hundreds of milliseconds (Brink *et al.*, 1996) which is—ten to hundred times greater than that for specific cation channels such as sodium channels (Nav), potassium channels (Kv), or calcium channels (Cav). Determination of the open probability of heterotypic and heteromeric channels has not been as rigorously assessed as homotypic channels, but multichannel recordings are qualitatively similar to those of homotypic channels, suggesting analogous mean open and closed times.

B. Single Channel Conductance and Permeability/Selectivity

Single channel conductance for homotypic gap junction channels is connexin specific and ranges over an order of magnitude from ~10 pS for Cx36 (Srinivas *et al.*, 1999) to ~350 pS for Cx37 (Veenstra *et al.*, 1994), yet for all channels studied the sequence for monovalent cation selectivity is essentially the same. The generalized sequence is Cs~K>Na>TEA and roughly follows the mobility sequence for these species (Beblo and Veenstra, 1997; Wang and Veenstra, 1997; Brink *et al.*, 2000). In general, monovalent anions are less permeate than cations of similar mobility, but they too follow a sequence roughly equivalent to their own mobilities. For both monovalent cations and anions, gap junction channels appear to be poorly selective or nonselective. Of particular interest is the K^+ to Cl^- ratio, which was measured by Wang and Veenstra (1997) to be 0.13 for Cx43; a similar value was calculated for Cx40 by Beblo and Veenstra (1997).

The four connexins, Cx43, Cx40, Cx31, and Cx26, that are found in the ciliary body epithelium, all form homotypic channels *in vitro* and their single channel conductances are shown in Table I.

Combining the single channel conductance with the high open probability allows one to estimate the number of ions traversing a single channel from one cell to another. The estimated flux of a monovalent ion such as K^+ for a 10 mV steady-state voltage is between 10^6 and 10^7 ions/s per channel (Valiunas *et al.*, 2002), given single channel conductances like those shown in Table I. How does this affect ion concentration within a coupled cell? If a cell pair is coupled by a single gap junction channel and has a +10 mV transjunctional voltage applied, for every second the single channel is open and delivering 10^7 K^+ ions, the concentration would be elevated by ~1 μM, assuming a cell volume of 1 pl. This robust ability to move monovalent ions suggests gap junction channels are probably not rate limiting in the transepithelial movement of solutes destined for secretion by the ciliary body epithelium.

TABLE I
Single Channel Conductances

Connexin type	Homotypic unitary conductance (pS)
Cx40	140 (Valiunas *et al.*, 2002)
Cx43	90 (Valiunas *et al.*, 2002)
Cx26	130 (Kojima *et al.*, 1999)
Cx31	85 (Abrams *et al.*, 2006)

Exogenous probes such as Lucifer Yellow, with minor diameters of 1.0 nm, are able to permeate the homotypic channels listed in Table 1. In the case of Cx43 and Cx40, Lucifer Yellow permeability has been quantified relative to K^+ permeability (Valiunas *et al.*, 2002); the ratio of Lucifer Yellow to K^+ permeability is 1:400 for Cx40 and 1:40 for Cx43. The heteromeric and heterotypic forms have ratios between the two homotypic forms (Valiunas *et al.*, 2002). In addition to passing current and allowing the passage of exogenous probes, gap junction channels, including Cx43, Cx40, and Cx26, also display selective permeability to a variety of larger solutes including endogenous molecules such as IP_3, cAMP, and small polypeptides (Tsien and Weingart, 1976; Niessen *et al.*, 2000; Goldberg *et al.*, 2004; Neijssen *et al.*, 2005; Ayad *et al.*, 2006). Cx31 has also been shown to allow the passage of exogenous probes, but no data exist with regard to endogenous solutes (Abrams *et al.*, 2006). The expectation is that Cx31 will also be selectively permeable to endogenous solutes, but this will require experimental validation.

In addition to the conductivity/permeability properties of gap junction channels, their distribution within a tissue is another factor that can influence function. For example, within the ventricular myocardium, gap junction channels composed of Cx43 have their highest density at the intercalated discs. This is the most effective way to minimize longitudinal resistance within an array of cells and allow rapid action potential propagation. Gap junction channel localization in the ciliary body epithelium is another example where distribution appears to be as important to function as the properties of the gap junction channels themselves. Gap junction channels composed of Cx43 and Cx40 are principally distributed along the two apical surfaces that appose each other (Fig. 1). The properties of these channels are apparently of sufficient conductivity and present in adequate number to allow the two layers to function like a monolayer. But this distribution on its own would produce multiple two cell syncytia, with each pigmented cell coupled to one nonpigmented cell. To ensure more complete functional uniformity of either sheet of cells, lateral communication between cells, in at least one of the two layers, is also necessary. This is apparently achieved by the laterally situated gap junction plaques containing Cx26 and Cx31 in the nonpigmented layer. Thus, the properties and distribution of the gap junction channels in the ciliary body seem sufficient to allow the stratified epithelium to act as a unified monolayer. The aforementioned effects of gap junction channel blockers on ciliary body ion flux are also consistent with this model (Wolosin *et al.*, 1997; Do and Civan, 2004).

The properties of connexin-specific gap junctions and their distribution within the ciliary body epithelium are consistent with experimental evidence that disruption of gap junction channel-mediated coupling affects function in secretory epithelia. Up to this point however, we have dealt with the properties of solute permeation through gap junction channels. What about fluid secretion by this double-layered epithelium? Does the fluid move through the gap junction channels connecting the nonpigmented and pigmented layers? There are no data to address this question, so the next section presents model calculations on a secretory epithelium made from two cell layers coupled by gap junction channels.

IV. MODELING OF FLUID TRANSPORT BY THE CILIARY EPITHELIUM

The most widely accepted model is that secretion of the AH occurs through the active transport of salt causing fluid to follow by osmosis. Although the transporters responsible for salt secretion have mostly been characterized (reviewed in Civan and Macknight, 2004), the details of fluid secretion are not well understood. Here, using model predictions, we will describe some of the properties required for efficient fluid transport through gap junctions.

A. Derivation of Parameters

Mathias and Wang (2005) used several models of local osmosis and fluid transport across a simple one-layered epithelium, starting with the simplest three-compartment model (Curran, 1960; Curran and McIntosh, 1962), then the standing gradient model (Diamond and Bossert, 1967), and finally modeling fluid reabsorption by the proximal tubule of the kidney. The question of interest was how an epithelium could generate near isotonic fluid transport. Isotonic transport is the theoretical maximum rate at which fluid can be moved through osmosis. It occurs when the osmolarity of the fluid being transported is the same as that of the surrounding solutions; hence, all standing osmotic gradients go to zero. That is, if u (cm/s) is the rate of fluid flow, c_o (mol/cm^3) is the surrounding osmolarity, and j (mol/cm^2 s) is the rate of salt transport, isotonic transport occurs when $j/u = c_o$. Since this is the theoretical maximum water flow, it follows that $u < j/c_o$. This inequality can be used to bound the concentration change needed to generate an attainable water flow (Fig. 3).

Define the membrane osmotic permeability as RTL_m [(cm/s)/(mol/cm^3)], where L_m [(cm/s)/mm Hg] is the hydraulic permeability and $RT = 20$ mm Hg/mM, then $u = RTL_m\Delta c$. If L_m increases, so will u; however, it will not

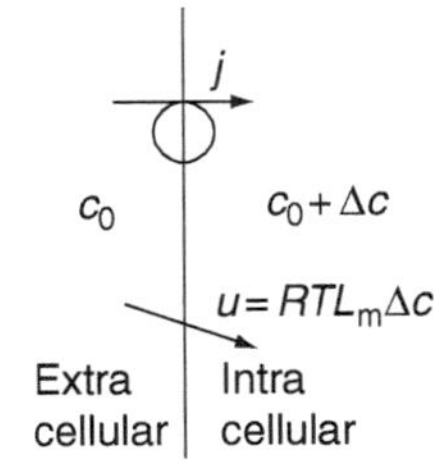

FIGURE 3 Water flow (u cm/s) and solute flux (j mol/cm^2 s) across a membrane. The solute flux j is due to active or secondary active transport processes and is independent of the small osmotic difference Δc (mol/cm^3) that is generated by j. However, the small osmotic pressure Δc creates the water flow u through passive osmosis.

increase as much as L_m because the increased water flow will carry away some of the solute gradient. Indeed, as $L_m \to \infty$, $\Delta c \to 0$, and u will achieve its isotonic limit. Since isotonic transport is a theoretical maximum, we know that $u = RTL_m \Delta c < j/c_o$. Dividing both sides by c_o, then rearranging the inequality leads to the condition:

$$\frac{\Delta c}{c_o} \leq \frac{j/c_o}{RTL_m c_o} = \epsilon \tag{1}$$

We have thus defined a parameter, ϵ, which is essentially the ratio of membrane salt permeability to water permeability, and as long as ϵ is small, osmotic gradients will be small, and transport has the possibility of approaching its isotonic limit.

Mathias and Wang (2005) used a perturbation approach [first used by Segel (1970)] to obtain approximate series solutions in powers of the small parameter ϵ. They concluded the complex "standing gradient models," as initiated by the analysis of Diamond and Bossert (1967), were flawed because they described the wrong experiment. The actual experiments were to collect the fluid transported by an epithelium and measure its osmolarity. These measurements were within experimental error of isotonic, hence the models fixed the osmolarity of the transported solution at exactly c_o. Mathias and Wang (2005) modeled the situation where the fluid is collected without imposing any conditions, and found the osmolarity will naturally be within O(ϵ) of isotonic [i.e., O(ϵ^n) means terms multiplied by ϵ^n or higher powers of ϵ]. For a typical cell, $\epsilon \leq 10^{-3}$–10^{-5}, hence the transported solution would indeed be within experimental error of isotonic. If one does the modeling without imposing the condition that the solutions on both side of the epithelium have exactly the osmolarity c_o, then diffusion gradients in the lateral spaces disappear, at least to within O(ϵ^2), and even the "standing gradient" model breaks down to the simple three-compartment model of Curran and McIntosh (1962). Hence in what follows, we will use an extension of Curran and McIntosh (1962) model, one in which there are four compartments (see Fig. 4): (i) the stroma, containing normal extracellular solution; (ii) the PE; (iii) the NPE; and (iv) the posterior chamber containing the AH.

Membranes of the NPE and PE actively transport salt, primarily NaCl, to generate the AH. These transport processes use the energy from hydrolysis of ATP, or the energy in the electrochemical gradients created by hydrolysis of ATP, to generate a transmembrane salt flux j. Hence j is not affected by the tiny osmotic gradients that generate fluid movement, so in models that focus on fluid transport, j can be considered an independent, fixed parameter that is established by the cell through expression of transporters, and adjustments in

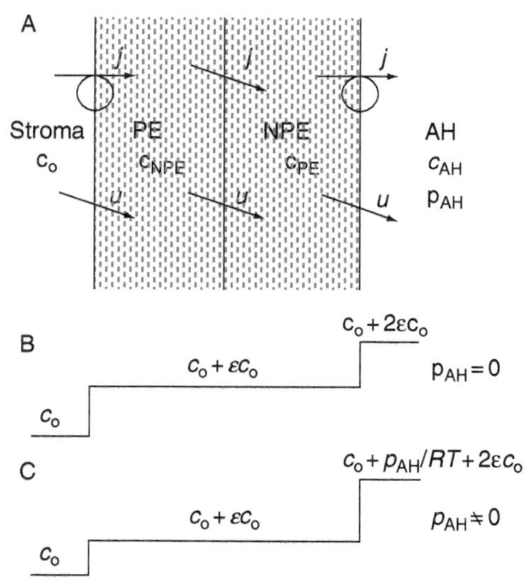

FIGURE 4 Transport by the ciliary epithelium when gap junctions are not a significant barrier to salt or water fluxes. (A) A four-compartment model of the ciliary epithelium illustrating the transport parameters. (B) Changes in osmolarity across a fluid transporting ciliary epithelium, assuming the gap junctions connecting the PE with the NPE provide no resistance to water or salt fluxes. In this panel, the assumption is that the intraocular hydrostatic pressure in the AH is zero. (C) Changes in osmolarity across a fluid transporting ciliary epithelium, assuming the gap junctions connecting the PE with the NPE provide no resistance to water or salt fluxes. In this panel, the assumption is that the intraocular hydrostatic pressure in the AH is not zero, but is about 10 mm Hg. Thus a transmembrane hydrostatic pressure opposes fluid transport across the basolateral membranes of the NPE cells.

the relative intracellular ion concentrations and voltages. Moreover, the salt flux is through membrane transport proteins (the Na/K ATPase, secondary active transport proteins, and membrane channel proteins) that are independent of and in parallel with the path for fluid movement (the aquaporins and lipid bilayer).

Figure 4B and C illustrate the predictions of the perturbation analysis when the gap junctions are sufficiently permeable to salt and water that they have no effect on transport. Figure 4B illustrates the concentrations in the absence of hydrostatic pressure in the AH, assuming the water permeability of the NPE layer is the same as that of the PE layer. Fluid is pulled from the stroma into the PE cells because transport of NaCl creates the small transmembrane osmotic gradient ϵc_o. Fluid is pulled from the NPE cells into the AH because transport of NaCl has again generated the small transmembrane osmotic gradient ϵc_o. Since the fluid transported from the stroma has to cross two membranes to reach the AH, the osmolarity of the AH is predicted to be $c_o(1+2\epsilon)$. Fluid movement is thus entirely generated through membrane transport of salt creating small transmembrane osmotic gradients and hydrostatic pressure is not a necessary component. Of course, there has to be small pressure gradients within the cells to drive the flow of fluid, but these are predicted to be very small and completely negligible in comparison to ϵc_o.

A significant hydrostatic pressure in the AH is necessary, however, to drive the exit of fluid through Schlemm's canal. Figure 4C illustrates the predicted concentrations when there is a significant pressure in the AH. Fluid transported from the NPE to the AH is now moving against the pressure p_{AH} (mm Hg). However, this does not significantly reduce the rate of fluid transport. As analyzed by Mathias (1985), it increases the osmolarity of the AH such that there are two components to the transmembrane osmotic gradient: one is given by ϵc_o and this drives the fluid movement; the other is given by p_{AH}/RT (mM) and this balances the effect of hydrostatic pressure, leaving water transport dependent on membrane salt transport. RT is 20 mm Hg/mM, so a typical intraocular pressure (IOP) of 10 mm Hg causes the AH to be about 0.5 mM hypertonic, hence the effect is small.

Fluid and salt fluxes through gap junctions differ fundamentally from those through the plasma membrane. Gap junctions are entirely passive devices, which do not generate electrochemical gradients, and the path for fluid flow is the same as that for salt flux, namely, through the cell-to-cell channels of the gap junction. With these simple ideas in mind, we can write down some fundamental relationships that need to hold if fluid moves through gap junctions to generate secretion of a nearly isotonic AH, which is of course the observation (Gaasterland *et al.*, 1979).

Assume the osmolarity of the extracellular solution in the stroma is fixed at c_o, whereas based on the analysis in Mathias and Wang (2005), the osmolarities of the NPE, PE, and AH will be slightly hypertonic. Thus:

$$\begin{array}{rr} \text{Stroma}: & c_o \\ \text{PE}: & c_{PE} = c_o + \Delta c_{PE} \\ \text{NPE}: & c_{NPE} = c_o + \Delta c_{NPE} \\ \text{AH}: & c_{AH} = c_o + \Delta c_{AH} \end{array} \quad (2)$$

Within the channels of the gap junctions, the osmolarity is less than that of cytoplasm because the impermeant anions of the cytoplasm are too large to enter. Since the hydro-osmotic pressure must be a continuous function of position across the epithelium, a negative hydrostatic pressure will exist within the channels. However, this pressure is not related to water flow, hence it is constant across the junction.

As illustrated in Fig. 5, we assume each gap junction channel has a length δ (cm) and a radius a (cm). There will be hydrostatic pressure (p_j mm Hg) and osmotic (c_j mol/cm^3) gradients that are associated with water flow through the junctional channels. These are related to water flow by

$$u = -RTL_j\delta\left(\frac{1}{RT}\frac{dp_j}{dx} - \frac{dc_j}{dx}\right) \quad (3)$$

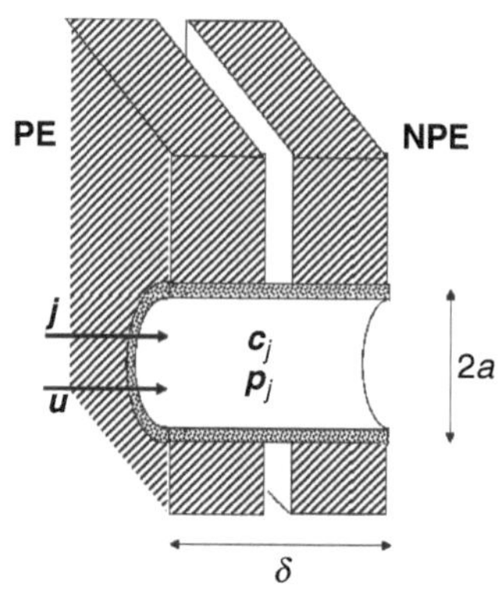

FIGURE 5 A cross-sectional view of a typical gap junction channel connecting the PE and NPE cells. A single channel would carry a small fraction of the total solute flux, j, and water flux, u, so the arrows are simply to indicate that both fluxes follow the same path. Each channel is assumed to be a right circular cylinder with radius a (cm) and length δ (cm). Within the channel, the hydrostatic pressure is indicated by p_j (mm Hg), and the concentration of solute by c_j (mol/cm^3).

The analysis by Mathias and Wang (2005) did not include hydrostatic pressure, because within a cell pressure gradients are predicted to be extremely small. The same cannot be assumed for a gap junction. The osmotic permeability of a pore, based on laminar flow, goes as the fourth power of the radius [see Eq. (14)], so a large number of small diameter pores in parallel will have much lower water permeability than one large pore of the same cross-sectional area. A significant pressure gradient is likely to exist across the junction.

Each parameter in Eq. (3) needs to be normalized so that its value is close to unity. The fluid flow is normalized to its isotonic limit, $U = u/(j/c_o)$, and we will seek to identify conditions that will ensure $U \approx 1$; position is normalized to junction width, $X = x/\delta$; osmolarity is normalized to that of the stroma, $C_j = c_j/c_o$; and pressure is normalized to $P_j = p_j/(RTc_o)$. In terms of normalized parameters, fluid flow is described by

$$\epsilon \Lambda_j U = -\frac{dP_j}{dX} + \frac{dC_j}{dX} \tag{4}$$

Thus, for $U \approx 1$ whereas the hydro-osmotic gradients are proportional to ϵ:

$$\Lambda_j = \frac{RTL_m}{RTL_j} \approx 1 \tag{5}$$

This is a rather reasonable condition that the junctional osmotic permeability must be of the same order of magnitude as the membrane osmotic permeability, then the hydrostatic and osmotic gradients will be small, leading to near isotonic transport.

Within the channels of the gap junction, salt will be carried by a combination of diffusion and convection.

$$j = -D_j \frac{dc_j}{dx} + uc_j \tag{6}$$

where D_j (cm^2/s) is the effective diffusion coefficient of the gap junction. Again, if the parameters in Eq. (6) are normalized, we can determine another constraint on parameter values. The normalized salt flux is $J = 1$, yielding

$$\frac{dC_j}{dX} = \epsilon k_j (UC_j - 1) \tag{7}$$

The parameter k_j is given by

$$k_j = \frac{RTL_m c_o}{D_j/\delta} \approx 1 \tag{8}$$

The condition $k_j \approx 1$ implies ϵk_j is small and hence concentration gradients will be small, which is necessary for near isotonic transport through the gap junction. In this situation, the flux of solute is carried mostly by convection.

B. Evaluation of Parameters

$\epsilon = (j/c_o)/(RTL_m c_o)$: The osmotic permeability of a membrane is typically around 0.3 (cm/s)/(mol/cm^3); however, for an epithelium, there are lateral membranes and infoldings, so that the permeability relative to the area of epithelial surface is about 30-fold greater (Whittembury and Reuss, 1982), hence we estimate:

$$RTL_m = 10(\text{cm/s})/(\text{mol/cm}^3) \tag{9}$$

For typical mammalian cells

$$c_o = 300 \times 10^{-6}\text{mol/cm}^3 \tag{10}$$

Based on several studies (reviewed in Do and Civan, 2004), the net secretion of Cl^- occurs at a rate of about 5×10^{-10} mol/s cm^2 of epithelial surface. Assuming Na^+ is secreted at the same rate gives:

$$j = 10^{-9}\text{mol/s cm}^2 \tag{11}$$

Inserting these numbers into the definition of ϵ yields:

$$\epsilon = 10^{-3} \tag{12}$$

$k_j = (RTL_m c_o)/(D_j/\delta)$: Based on the parameter values above, $RTL_m c_o = 3\times10^{-3}$ cm/s. The parameter D_j/δ (cm/s) is the gap junctional salt permeability, which can be estimated from the single channel permeability times the number of channels per area of epithelial surface, N_{GJ} (channels/cm^2 of epithelial surface). Assume each channel is a cylinder of length $\delta = 14\times10^{-7}$ cm and radius $a = 0.8\times10^{-7}$ cm, and that salt within each channel has a typical diffusion coefficient of $D = 10^{-5}$ cm^2/s. The predicted single channel permeability is given by $\pi a^2 D/\delta = 1.4 \times 10^{-13}$ cm^3/s. For $k_j = 1$, $N_{GJ}\, \pi a^2 D/\delta = 3 \times 10^{-3}$, yielding:

$$N_{GJ} = 2 \times 10^{10} \text{channels/cm}^2 \text{of epithelial surface} \tag{13}$$

This is a nominal value and implies that a range exists that would satisfy the requirement that ϵk_j be small so that diffusion gradients are small [see Eq. (7)]. There is no upper bound to N_{GJ} since the more channels, the smaller k_j; however, there is a lower bound of around 2×10^9, which would make $\epsilon k_j = 0.01$. This is still a rather small number, but if concentrations deviate from isotonic by $\epsilon k_j c_o$, this would be 3 mM, which is probably detectable.

$\Lambda_j = (RTL_m)/(RTL_j)$: For Λ_j to be near unity, we require $RTL_j \approx RTL_m = 10$ (cm/s)/(mol/cm^3). There are no data on the water permeability of a gap junction channel. However, if we assume laminar flow in a tube, the theoretical value of the osmotic permeability for a single gap junction channel can be calculated from standard physics:

$$\frac{RT\pi a^4}{8\eta\delta} = 3.3 \times 10^{-11} \frac{\text{cm}^3/\text{s}}{\text{mol/cm}^3} \tag{14}$$

The viscosity of water is $\eta = 7\times10^{-6}$ mm Hg s. For basis of comparison, the single channel water permeability of AQP1 has been estimated to be 2.2×10^{-13} (cm^3/s)/(mol/cm^3) (Chandry *et al.*, 1997). Thus gap junction channels are far better transporters of water than aquaporins, however gap junction channels allow ions to pass as well, whereas aquaporins do not. The independence of the water and ion pathways across the plasma membrane is essential in order for the flow of water to be controlled by salt transport.

The overall junctional water permeability is

$$RTL_j = N_{GJ} \frac{RT\pi a^2}{8\eta\delta} \frac{\text{cm/s}}{\text{mol/cm}^3} \tag{15}$$

Hence for $\Lambda_j = 1$:

$$N_{GJ} = 3 \times 10^{11} \text{channels/cm}^2 \text{of epithelial surface} \tag{16}$$

This is a stronger constraint than that imposed by $k_j = 1$, in that more channels are required to satisfy the need for high water permeability ($\Lambda_j = 1$). We therefore focus on the value of $\epsilon\Lambda_j$. Because ϵ is very small, it is not necessary for Λ_j to equal unity for $\epsilon\Lambda_j$ to be small enough to have a negligible effect. A limit on the value of Λ_j is estimated in the next section.

C. *Predictions of the Model*

It is not possible to model the gap junctions in isolation, since the concentrations and pressures in the PE and NPE cells, which set the boundary conditions for junctional fluxes, depend on the interactions of membrane and junctional transport. The complete model is a perturbation expansion in the small parameter ϵ, similar to the expansion used in Mathias and Wang (2005) for the three-compartment model shown in Fig. 2A of that paper, except that the analysis here has four compartments, and it includes hydrostatic pressure and fluxes of each individual ion. We do not present the complete model because it is beyond the scope of this review. The complete model is not yet published, but the analysis we present leads to some rather simple conclusions.

The predictions of a perturbation expansion in the small parameter ϵ (similar to that presented in Mathias and Wang, 2005) are shown in Fig. 6. To within $O(\epsilon^2)$, there is no concentration gradient across the junctions, whereas there is a hydrostatic pressure difference given by

$$p_j = -\epsilon\Lambda_j RTc_o \tag{17}$$

The concentration and pressure profiles are shown in Fig. 6B and C, respectively. The implication of these results is that the flux j is carried through the gap junctions by convection. This is not what our initial intuition would have predicted, but when one knows the result, intuition through hindsight works better. For example, it is intuitively obvious that for water transport to approach isotonic, there should be negligible concentration gradients within the cell, and the intracellular flux j will be carried predominantly by convection. At the entrance to the gap junction channels, the concentrations of Na^+ and Cl^- will be the same as in the PE; the water flow through the channels will be equal to u, as it is in the PE; thus convection of Na^+ and Cl^- will be the same as in the cell. Hence, convection will carry a solute flux j through the junction. In this situation, a transjunctional diffusion gradient does not develop, and in the absence of any transjunctional osmotic gradient, a hydrostatic gradient develops to drive the fluid flux.

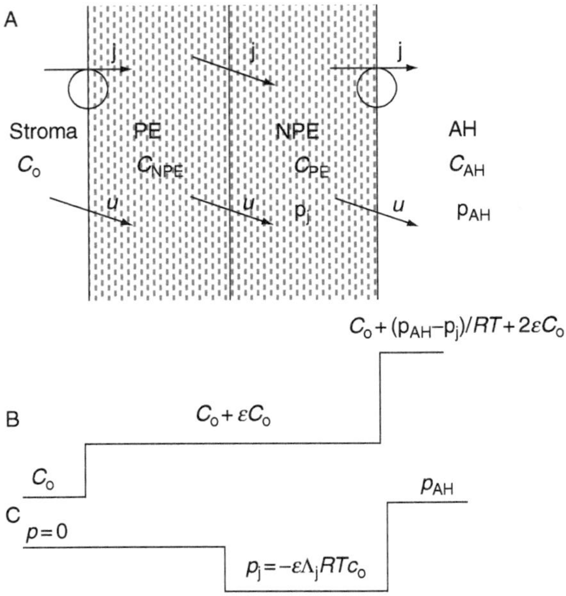

FIGURE 6 The effect of PE–NPE gap junctions on fluid transport by the ciliary epithelium. (A) A four-compartment schematic of the ciliary epithelium with definitions of transport parameters. (B) A profile of the changes in osmolarity across the transporting epithelium. (C) A profile of the changes in hydrostatic pressure across the transporting epithelium.

As a consequence of the transjunctional pressure drop, the transmembrane pressure difference between the NPE and AH is increased (see Fig. 6C). As discussed earlier for Fig. 4, the presence of a transmembrane pressure does not reduce fluid flow very much, but it makes the AH more hypertonic; namely,

$$c_{AH} = c_o + (p_{AH} - p_j)/RT + \epsilon c_o \tag{18}$$

The osmolarity of the AH is known to be close to that of normal extracellular solution (Gaasterland *et al.*, 1979). Assume that normal AH is no more than 3 mM hypertonic, or about 1%. The contributions to hypertonicity come from: $\epsilon c_o = 0.3$ mM, $p_{AH}/RT = 0.5$ mM, leaving the contribution of $-p_j/RT \leq 2.2$ mM. Given $\epsilon RTc_o = 6$ mm Hg [see Eq. (17)], our estimated limit implies $\Lambda_j \leq 7.3$ and

$$N_{GJ} \geq 4 \times 10^{10} \tag{19}$$

This constraint satisfies both Eqs. (5) and (8), so we next turn to what it implies for ciliary gap junctions relative to the known properties of other gap junctions.

D. Conductance and Structural Properties of Gap Junctions

For comparison with electrical data, the single gap junction channel conductance predicted by this model (Fig. 5) is 91 pS, which is a typical value. Based on the constraint in Eq. (18), for the number of channels, the overall junctional conductance would be at least 3.6 S/cm^2 of epithelial surface.

For comparison with other data, the area of the NPE–PE interface per area of epithelium needs to be estimated. The NPE and PE interface is certainly not a flat sheet, but we do not know the degree to which membrane undulations increase the area. In many fluid transporting epithelia, the area of apical membrane is actually about the same as the total area of basolateral membrane. But this is due to the presence of microvilli on the apical surface, and these are not apparent in the ciliary epithelium. Figure 7 illustrates the ciliary epithelium as made from simple cubic cells. The basolateral membrane area per area of epithelium is increased at least fivefold due to the presence of four lateral membranes for every basal membrane. In typical fluid transporting epithelia, the area is actually on the order of 30-fold greater than the apparent area of epithelium (reviewed in Whittembury and Reuss, 1982), which implies a 6-fold increase due to membrane undulations. We will therefore assume the surface area of the apical–apical interface between PE and NPE cells is at least sixfold greater than the apparent surface area of epithelium.

With this assumption, the PE to NPE junctional conductance is 0.6 S/cm^2 of cell-to-cell contact. Heart cells are coupled by about 0.3 S/cm^2 of cell-to-cell contact (Cohen *et al.*, 1982), whereas lens fiber cells are coupled by 1–10 S/cm^2 of cell-to-cell contact (reviewed in Mathias *et al.*, 1997). The ciliary epithelium is reported to have a relatively high density of gap junctions (reviewed in Do and Civan, 2004), so a value of 0.6 S/cm^2 may be reasonable. This implies that each NPE to PE cell pair is coupled with a conductance of about 0.6 μS or 1.7 MΩ (assuming the area of contact is about 100 μm^2). The heart and the lens are the only tissues in which coupling conductance has been measured in near *in vivo* conditions, and their values compare reasonably with that of 0.6 S/cm^2 for the ciliary epithelium.

Again, assuming the apical area is about sixfold greater than the apparent surface of ciliary epithelium, the number of channels per area of cell-to-cell contact will be about 0.7×1010 channels/cm^2, or 70 channels/μm^2. Within

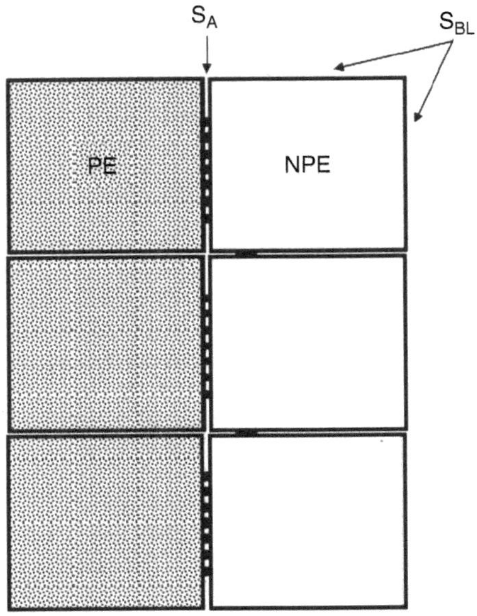

FIGURE 7 An idealized schematic of the ciliary epithelium. Each cell is assumed to be a cube, thus the area of basolateral membrane (S_{BL}) is at least fivefold greater than the apparent surface area of ciliary epithelium. However, as described in the text, in other fluid transporting epithelia, the area of basolateral membrane is at least 30-fold greater than the apparent surface of epithelium, owing to undulations in the membranes. Thus, we assume that the undulations increase the area sixfold, hence the area of apical membrane (S_A) is about sixfold greater than the apparent area of ciliary epithelial surface. The junctions between NPE cells are tight junctions, which include gap junctions. PE cells lack tight junctions. The NPE–PE connections represent gap junctions.

plaques, gap junction channels have a spacing of 8.5–9.5 nm when crystallized into closely packed hexagonal arrays. For the purpose of this calculation, assume a spacing of 10 nm in normal (uncrystallized) conditions. This implies there are about 10,000 channels/μm^2 of plaque. Our previous calculations have concerned the number of open channels. In a typical gap junction plaque, only about 10% of the channels are functional. If this is also true for the ciliary junctions, the area of plaque per area of apical surface would be about 7% when the conductance is 0.6 S/cm^2. Again, this is a very reasonable number. In the equatorial fiber cells of the lens, the junctional plaques occupy about 50% of the membrane, but this is the highest density of junctions reported, and the measured conductance of up to 10 S/cm^2 is significantly larger than that needed by the ciliary epithelium.

E. Summary

Membranes generate and regulate water transport by the regulated transport of salt in connection with expression of a relatively high water permeability, which ensures fluid follows salt transport almost isotonically. Conversely, neither water nor salt transport through gap junctions is regulated, as gap junctions are passive devices that must conduct whatever flux is delivered by the membranes. In order to conduct these fluxes, transjunctional gradients will be generated. A surprising conclusion of the analysis presented here is that transjunctional osmotic gradients do not develop, rather a transjunctional hydrostatic pressure develops to drive fluid transport, which convects the salt across the junction. Furthermore, the prediction is that this pressure is balanced by a small increase in the osmolarity of the AH. There are currently available Cx40 knockout mice, which are therefore predicted to generate a measurably hypertonic AH, so the model can be experimentally tested.

There are currently no experimental data on the water permeability of gap junction channels formed from any of the connexin isoforms. At the PE–NPE interface, the ciliary epithelium expresses Cx40 and Cx43. These may be particularly good water channels and our estimate of water permeability based on laminar flow in a pipe could be much too low. Conversely, these connexins could form channels that have little or no water permeability. If so, the modeling presented here would be inappropriate and water would have to follow another path.

One alternative possibility is that the PE cells are present only to increase salt flux, whereas they have a low membrane water permeability. If so, the water flow path would be into NPE cells through apical membranes and then into the AH through basolateral membranes. As shown in Fig. 6A, the PE cells have tight junctions isolating apical and basolateral membranes, so this seems *a priori* feasible, but would require data on membrane water permeabilities and new models of fluid transport by this epithelium. Swelling assays of isolated PE and NPE cells could provide the data on membrane water permeabilities and thus test this model.

Another possibility is that the water path is outward through the apical membranes of the PE cells then inward across the apical membranes of the NPE cells. This model, however, would require an isolated extracellular space at the NPE–PE interface, since the osmolarity between the cells would have to be $c_o(1+2\epsilon)$ to draw the water out of the NPE cells, and the osmolarity of the PE cells would be $c_o(1+3\epsilon)$ to draw the water into them. The problem is that there are no tight junctions between PE cells (see Fig. 6A), so there is no known structure to create this isolated space. One way to test this model is to see if lanthanum can penetrate into this space, as it does in the extracellular spaces between other gap junctions.

Despite the possibility of other mechanisms, the analysis presented here suggests that gap junctions can indeed transport significant water flow and thus are the prime candidate for the path leading to AH formation.

V. ANIMAL MODELS SUPPORT A ROLE FOR GAP JUNCTIONS IN FLUID TRANSPORT BY OCULAR EPITHELIA

Having reviewed the functional properties of ciliary body gap junctions and then hypothesizing how those properties could participate in secretion of the AH, we will now evaluate evidence from genetically engineered mice to see if the available experimental data are consistent with our model.

A recent study by Calera *et al.*, (2006) examined mice in which the Cx43 protein was eliminated by conditional gene knockout in the ciliary body. Immunohistochemical staining showed that Cx43 was eliminated from the PE but not the NPE, resulting in a ciliary epithelium that displayed areas of separation between the pigmented and nonpigmented layers. By 2 weeks of age, knockout mice had smaller eyes that were flaccid when dissected. The authors speculated that these flaccid eyes were due to a reduction in IOP, which in turn was due to a reduction in ion and water flow across the ciliary epithelium caused by the loss of Cx43 in the PE. Although IOP was not measured in this study, the authors provided some support for this hypothesis by showing that back-diffusion of plasma proteins into the AH was occurring in the conditional knockouts, leading to protein precipitates in the anterior chamber. By 5 weeks of age, the mice had a dramatically reduced vitreal space and a variety of other ocular defects. Due to the choice of Cre-expressing mice (nestin-Cre) used to generate these animals, Cx43 was removed from a number of ocular cell types in addition to the PE, complicating interpretation of the described phenotype. While further study will be required, it is clear that loss of Cx43 in the PE was correlated with a loss of morphologically recognizable gap junctions from the NPE/PE when examined by electron microscopy. The observed elevation in plasma protein in the AH and pathohistological changes consistent with loss in IOP are consistent with the model for fluid secretion described above.

A second example of an epithelium where gap junction channels and their distribution have been shown to affect the secretion of fluid is provided by the lacrimal gland, where a monolayer of acinar cells generates the fluid of tears. These cells express Cx32 and Cx26. In Cx32, knockout animals tear production was significantly reduced in female, but not male mice (Walcott *et al.*, 2002). The distribution of Cx26 in the Cx32 knockout mice was also determined and in the males extensive Cx26 containing gap junction plaques were found but in the females Cx26 staining was absent from the plasma membrane (Walcott *et al.*, 2002). Thus lateral coordination of lacrimal epithelial function was lost in the females and this disrupted fluid secretion. Gap junctions can therefore affect fluid transport either directly, by actually being the path of fluid flow as in the model in this review, or indirectly through signaling between cells.

The lens of the eye generates an internal circulation of ions and fluid (reviewed in Mathias *et al.*, 2007). This circulation enters the lens through the extracellular spaces between cells, moves into fiber cells, and then circulates to the equatorial surface cells by flowing through fiber cell gap junctions made from Cx46 and Cx50. Knockout of Cx46 (Gong *et al.*, 1997) caused loss of coupling between the central mature fiber cells (Gong *et al.*, 1998). This disrupted the egress pathway for the lens' circulation and caused calcium to accumulate in the central fiber cells (Gao *et al.*, 2004). Disruption of the path for ion flow would be sufficient to disrupt the circulation of fluid, so these data do not directly demonstrate a role of gap junctions in conducting fluid flow in the lens, but they are consistent with such a role.

There are additional connexin knockout/knockin mice that would be potentially useful for evaluating the role of gap junctions in the ciliary body. As noted above, Cx40 knockout animals might be expected to have diminished production of a hypertonic AH, which could be experimentally tested as these mice are viable (Simon *et al.*, 1998). Perhaps more intriguing would be knockin mice where the Cx43 gene has been replaced with either Cx32, Cx40, or Cx26 (Plum *et al.*, 2000; Winterhager *et al.*, 2007). These mice would explore the role of connexin specificity in AH production and allow comparison of the roles of relatively nonspecific ionic coupling (provided by all connexins) with the selective permeability for larger solutes such as cyclic nucleotides (which differ dramatically from one connexin to another), and perhaps water permeability, which potentially could vary between connexins. Many of the ideas expressed in this chapter could be easily tested with existing animal models.

References

Abrams, C. K., Freidin, M. M., Verselis, V. K., Bargiello, T. A., Kelsell, D. P., Richard, G., Bennett, M. V., and Bukauskas, F. F. (2006). Properties of human connexin 31, which is implicated in hereditary dermatolgical disease and deafness. *Proc. Natl. Acad. Sci. USA* **103,** 5213–5218.

Ayad, W. A., Locke, D., Koreen, I. V., and Harris, A. L. (2006). Heteromeric, but not homomeric, connexin channels are selectively permeable to inositol phosphates. *J. Biol. Chem.* **281,** 16727–16739.

Beblo, D. A., and Veenstra, R. D. (1997). Momovalent cation permeation through the connexin40 gap junction channel. Cs, Rb, K, Na, Li, TEA, TMA, TBA, and effects of anions Br, C1, F, acetate, aspartate, glutamate, and NO3. *J. Gen. Physiol.* **109**(4), 509–522.

Brink, P. R., Ramaman, S. V., and Christ, G. J. (1996). Human connexin43 gap junction channel gating: Evidence for mode shifts and/or heterogeneity. *Am. J. Physiol.* **271,** C321–C331.

Brink, P. R., Cronin, K., Banach, K., Peterson, E., Westphale, E., Seul, K. H., Ramanan, S. V., and Beyer, E. C. (1997). Evidence of rheteromeric gap junction channels formed from rat connexin43 and human Connexin37. *Am. J. Physiol.* **273,** C1386–C1396.

Brink, P. R., Ricotta, J., and Christ, G. J. (2000). Biophysical characteristics of gap junctions in vascular wall cells: Implication for vascular biology and disease. *Braz. J. Med. Biol. Res.* **33,** 415–422.

Bruzzone, R., Haefliger, J.-A., Gimlich, R. L., and Paul, D. L. (1993). Connexin40, a component of gap junctions in vascular endothelium, is restricted in its ability to interact with other connexins. *Mol. Biol. Cell* **4,** 7–20.

Calera, M. R., Topley, H. L., Liao, Y., Buling, B. R., Paul, D. L., and Goodenough, D. A. (2006). Connexin43 is required for production of the aqueous humor in the murine eye. *J. Cell Sci.* **119,** 4510–4519.

Candia, O. A., To, C., Gerometta, R. M., and Zamudio, A. C. (2005). Spontaneous fluid transport across isolated rabbit and bovine ciliary body preparations. *Invest. Ophthalmol. Vis. Sci.* **46,** 939–947.

Chandry, G., Zampighi, G. A., Kreman, M., and Hall, J. E. (1997). Comparison of the water transporting properties of MIP and AQP1. *J. Membr. Biol.* **159**(1), 29–39.

Chen-Izu, Y., Moreno, A. P., and Spangler, S. R. A. (2001). Opposing gates model for voltage gating of gap junction channels. *Am. J. Physiol.* **281,** C1604–C1613.

Christ, G. J., and Brink, P. R. (1999). Analysis of the presence and physiological relevance of subconducting states of Connexin43-derived gap junction channels in cultured human corporal vascular smooth muscle cells. *Circ. Res.* **84**(7), 797–803.

Civan, M. M., and Macknight, A. D. (2004). The ins and outs of aqueous humour secretion. *Exp. Eye Res.* **78**(3), 625–631.

Coffey, K. L., Krushinsky, A., Green, C. R., and Donaldson, P. (2002). Molecular profiling and cellular localization of connexin isoforms in the rat ciliary epithelium. *Exp. Eye Res.* **75,** 9–21.

Cohen, I. S., Falk, R. T., and Kline, R. P. (1982). Voltage-clamp studies on the canine Purkinje strand. *Proc. R. Soc. Lond. B Biol. Sci.* **217**(1207), 215–236.

Cottrell, G. T., Wu, Y., and Burt, J. M. (2002). Cx40 and Cx43 expression ratio Influences heteromeric/heterotypic gap junction Cchannel properties. *Am. J. Physiol. Cell Physiol.* **282** (6), C1469–C1482.

Curran, P. F. (1960). Na, Cl, and water transport by rat ileum in vitro. *J. Gen. Physiol.* **43,** 1137–1148.

Curran, P. F., and McIntosh, J. R. (1962). A model system for biological water transport. *Nature* **193,** 347–348.

Diamond, J. M., and Bossert, W. H. (1967). Standing-gradient osmotic flow. A mechanism for coupling of water and solute transport in epithelia. *J. Gen. Physiol.* **50**(8), 2061–2083.

Do, C. W., and Civan, M. M. (2004). Basis of chloride transport in ciliary epithelium. *J. Membr. Biol.* **200,** 1–13.

Do, C. W., and Civan, M. M. (2006). Swelling-activated chloride channels in aqueous humour formation, on the one side and the other. *Acta. Physiol. (Oxf)* **187**(1–2), 345–352.

Gaasterland, D. E., Pederson, J. E., MacLellan, H. M., and Reddy, V. N. (1979). Resus monker aqueous humor composition and a primate ocular perfusate. *Invest. Ophthalmol. Vis. Sci.* **18** (11), 1139–1150.

Gao, J., Sun, X., Martinez-Wittinghan, F. J., Gong, X., White, T. W., and Mathias, R. T. (2004). Connections between connexins, calcium, and cataracts in the lens. *J. Gen. Physiol.* **124**(4), 289–300.

Gemel, J., Valiunas, V., Brink, P. R., and Beyer, E. C. (2004). Connexin43 and connexin26 form gap junctions, but not heteromeric channels in co-expressing cells. *J. Cell Sci.* **117,** 24690–2480.

Goldberg, G., Valiunas, V., and Brink, P. R. (2004). Selectivity permeability of gap junction channels. *Biochem. Biophys. Acta* **662,** 96–101.

Gong, X., Li, E., Klier, G., Huang, Q., Wu, Y., Lei, H., Kumar, N. M., Horwitz, J., and Gilula, N. B. (1997). Disruption of alpha 3 connexin gene leads to proteolysis and cataractogenesis in mice. *Cell* **91**(6), 833–843.

Gong, X., Baldo, G. J., Kumar, N. M., Gilula, N. B., and Mathias, R. T. (1998). Gap junctional coupling in lenses lacking alpha3 connexin. *Proc. Natl. Acad. Sci. USA* **95**(26), 15303–15308.

Goodenough, D. (1975). The structure and permeability of isolated hepatocyte gap junctions. "Symposia on Quantitative Biology,"Vol. XL, pp. 37–44. Cold Spring Harbor.

Hayward, J. N., Pavasuthipiaisit, K., Perez-Lopez, F. R., and Sofreoview, M. V. (1976). Radioimmunoassay of arginine vasopressin in Rhesus monkey plasma. *Endocrinology* **98,** 975–981.

He, D. S., Jiang, J. X., Taffet, S. M., and Burt, J. M. (1999). Formation of heteromeric gap junction channels by connexins 40 and 43 in vascular smooth muscle cells. *Proc. Natl. Acad. Sci.* **96**(11), 6495–6500.

Kojima, T., Yamamoto, M., Tobioka, H., Mizuguchi, T., Mitaka, T., and Mochizuki, Y. (1996). Charges in cellular distribution of connexins 32 and 26 during formation of gap junctions in primary cultures of rat hepatocytes. *Exp. Cell Res.* **223,** 314–326.

Kojima, T., Srinivas, M., Fort, A., Hopperstand, M., Urban, M., Hertzberg, E. L., Mochizuki, Y., and Spray, D. C. (1999). TPA induced expression and function of human connexin 26 by post-translational mechanisms in stably transfected neuroblastoma cells. *Cell Struct. Funct.* **24,** 435–441.

Locke, D., Liu, J., and Harris, A. L. (2005). Lipid rafts prepared by different methods contain different connexin channels, but gap junctions are not lipid rafts. *Biochemistry* **44**(39), 13027–13042.

Mathias, R. T. (1985). Epithelial water transport in a balanced gradient system. *Biophys. J.* **47,** 823–835.

Mathias, R. T., and Wang, H. (2005). Local osmosis and isotonic transport. *J. Mem. Biol.* **208** (1), 39–53.

Mathias, R. T., Rae, J. L., and Baldo, G. J. (1997). Physiological properties of the normal lens. *Physiol. Rev.* **77**(1), 21–50.

Mathias, R. T., Kistler, J., and Donaldson, P. (2007). The lens circulation. *J. Mem. Biol.* **216,** 1–16.

Neijssen, J., Herberts, C., Drijfhout, J. W., Reits, E., Janssen, L., and Neefjes, J. (2005). Cross-presentation by intercellular peptide transfer through gap junctions. *Nature* **434,** 84–88.

Niessen, H., Harz, H., Bedner, P., Kramer, K., and Willecke, K. (2000). Selective permeability of different connexin channels to the second messenger IP3. *J. Cell Sci.* **113,** 1365–1372.

Ott, T., Jokwitz, M., Lenhard, D., Romualdi, A., Dombrowski, F., Ittrich, C., Schwarz, M., and Willecke, K. (2006). Ablation of gap junctional communication in hepatocytes of transgenic mice does not lead to disrupted cellular homeostasis or increased spontaneous tumourigenesis. *Eur. J. Cell Biol.* **85,** 717–728.

Patil, R. V., Han, Z., and Wax, M. B. (1997). Regulation of water channel activity of aquaporin 1 by arginine vasopressin and atrial natriuretic peptide. *Biochem. Biophys. Res. Commun.* **238,** 392–396.

Plum, A., Hallas, G., Magin, T., Dombrowski, F., Hagendorff, A., Schumacher, B., Wolpert, C., Kim, W. H., Lamers, W. H., Evert, M., Meda, P., Traub, O., *et al.* (2000). Unique and shared functions of different connexins in mice. *Curr. Biol.* **10,** 1083–1091.

Rackauskas, M., Kreuzberg, M. M., Pranevicius, M., Willecke, K., Verselis, V. K., and Bukauskas, F. F. (2007). Gating properties of heterotypic gap junction channels formed of connexins 40, 43, and 45. *Biophys. J.* **92,** 1952–1965.

Ramanan, S. V., Valiunas, V., and Brink, P. R. (2005). Non-stationary fluctuation analysis of macroscopic gap junction channel records. *J. Memb. Biol.* **205,** 81–88.

Segel, L. A. (1970). Standing-gradient flows driven by active solute transport. *J. Theor. Biol.* **29,** 233–250.

Simon, A. M., Goodenough, D. A., and Paul, D. L. (1998). Mice lacking connexin40 have cardiac conduction abnormalities characteristic of atrioventricular block and bundle branch block. *Curr. Biol.* **8,** 295–298.

Srinivas, M., Rozental, R., Kojima, T., Dermietzel, R., Mehler, M., Condorelli, D. F., Kessler, J. A., and Spray, D. C. (1999). Functional properties of channels formed by the neuronal gap junction protein connexin36. *J. Neurosci.* **19,** 9845–9855.

Tsien, R. W., and Weingart, R. (1976). Inotropic effect of cyclic AMP in calf ventricular muscle studied by a cut end method. *J. Physiol.* **260,** 117–142.

Uusitalo, R. (1972). Effect of sympathetic and parasympathetic stimulation on the secretion and outflow of aqueous humor in the rabbit eye. *Acta. Physiol. Scand.* **86,** 315–326.

Valiunas, V. R., Weingart, R., and Brink, P. R. (2000). Formation of heterotypic gap junction channels by connexins Cx40 and Cx43. *Circ. Res.* **86,** e42–e49.

Valiunas, V., Gemel, J., Brink, P. R., and Beyer, E. C. (2001). Gap junction channels formed by co-expressed Cx40 and Cx43. *Am. J. Physiol.* **281,** H1675–H1688.

Valiunas, V., Beyer, E. C., and Brink, P. R. (2002). Gap junction channels show a quantitative difference in selectivity. *Circ. Res.* **91,** 104–111.

Van Veen, A. A., van Rijen, H. V., and Opthof, T. (2001). Cardiac gap junction channels: Modulation of expression and channel properties. *Cardiovasc. Res.* **51,** 217–229.

Veenstra, R. D., Wang, H. Z., Beyer, E. C., Ramanan, S. V., and Brink, P. R. (1994). Connexin37 forms high conductance gap junction channels with subconductance state activity and selective dye and ionic permeabilities, *Biophy. J.* **66,** 1915–1928.

Walcott, B., Moore, L. C., Birzgalis, A., Claros, N., Valiunas, V., Ott, T., Willecke, C., and Brink, P. R. (2002). The role of gap junctions in fluid secretion of lacrimal glands. *Am. J. Physiol.* **282,** C501–C507.

Wang, H. Z., and Veenstra, R. D. (1997). Monovalent ion selectivity sequences of the rat connexin43 gap junction channel. *J. Gen. Physiol.* **109**(4), 491–507.

Wang, H. Z., Li, K., Lemanski, L. F., and Veenstra, R. D. (1992). Gating of mammalian cardiac gap junction channels by transjunctional voltage. *Biophys. J.* **63,** 139–151.

White, T. W., and Bruzzone, R. (1996). Multiple connexin proteins in single intercellular channels: Connexin compatibility and functional consequences. *J. Bioenerg. Biomembr.* **28** (4), 339–350.

Whittembury, G., and Reuss, L. (1982). Chapter 13: Mechanisms of coupling of solute and solvent transport in epithelia. *In* "The Kidney: Physiology. 2nd Addition" (D. W. Seldin and G. Giebisch, eds.). Raven Press, New York.

Winterhager, E., Pielensticker, N., Freyer, J., Ghanem, A., Schrickel, J. W., Kim, J. S., Behr, R., Grummer, K., Maass, K., Urschel, S., Lewalter, T., Tiemann, K., *et al.* (2007). Replacement of connexin43 by connexin26 in transgenic mice leads to dysfunctional reproductive organs and slowed ventricular conduction in the heart. *BMC Dev. Biol.* **7,** 26.

Wolosin, J. M., Candia, O. A., Peterson-Yantorno, K., Civan, M. M., and Shi, X.-P. (1997). Effect of heptanol on the short circuit currents of cornea and ciliary body demonstrates rate limiting role of heterocellular gap junctions in active ciliary body transport. *Exp. Eye Res.* **64,** 945–952.

CHAPTER 4

Regional Dependence of Inflow: Lessons from Electron Probe X-ray Microanalysis

Anthony D. C. Macknight* and Mortimer M. Civan†

*Department of Physiology, University of Otago Medical School, Dunedin, New Zealand
†Departments of Physiology and Medicine, University of Pennsylvania, Philadelphia, Pennsylvania 19104-6085

Current Topics in Membranes, Volume 62 1063-5823/08 $35.00

DOI: 10.1016/S1063-5823(08)00404-3

I. OVERVIEW

The ciliary epithelium secretes the aqueous humor into the posterior chamber of the eye. Despite the physiologic and pharmacologic importance of this epithelium, its complex structure and heterogeneous cell composition have impeded progress in studying the integrated functioning of the intact tissue. Electron probe X-ray microanalysis has provided an unusual opportunity to study the different cell populations in different regions of the intact ciliary epithelium. This chapter summarizes the advances in our understanding made possible by exploiting this technique.

II. INTRODUCTION

The mammalian ciliary epithelium is often studied as though it were a homogeneous preparation. However, both anatomical and histological studies show significant differences between different regions. The minor, flat *pars plana* is posterior to two well-defined anatomic regions, the posterior and anterior portions of the ciliary epithelium. The posterior region, the posterior *pars plicata*, displays long ciliary processes reaching regularly down to the iris. In contrast, in the rabbit anterior region, comprising the iridial portion of the primary ciliary processes (Weingeist, 1970), the folds are more tortuous. From histological studies, it is known that proteins and biologically active peptides are expressed nonuniformly in different regions of the ciliary epithelium (Flügel and Lütjen-Drecoll, 1988; Flügel *et al.*, 1989, 1993; Eichhorn *et al.*, 1990; Ghosh *et al.*, 1990, 1991; Eichhorn and Lütjen-Drecoll, 1993; Dunn *et al.*, 2001). However, a number of investigators have observed regional differences in the expression of Na^+, K^+-activated ATPase (Ghosh *et al.*, 1990) and additional proteins and biologically active peptides. For example, nonpigmented epithelial (NPE) cells of young calves display higher expression of $\alpha_1/\alpha_2/\alpha_3/\beta_1/\beta_2$ isoforms of Na^+,K^+-activated ATPase anteriorly than posteriorly, but pigmented epithelial (PE) cells expressed a constant relative concentration of α_1/β_1 throughout the epithelium (Coca-Prados and Sánchez-Torres, 1998). Another example is the localization of the Na^+-K^+-$2Cl^-$ cotransporter largely at the basolateral edge of the PE cell layer in the anterior region of young calves (Dunn *et al.*, 2001).

Until recently, however, it has not been possible to assess the functional significance of these variations in ciliary epithelial organization. Much has been learnt about overall aqueous humor secretion with isolated ciliary bodies from a variety of mammalian species studied in Ussing chambers under open- and short-circuited conditions. However, electrical measurements do not detect electroneutral solute movements and, given the leaky nature of the tissues, it is not always easy to quantify cellular solute movements from isotope studies. Other techniques that have been employed include microelectrodes, patch clamping, and fluorescence microscopy with isolated cells or groups of cells. None of these, however, enables one to determine the cellular ion composition of the transporting cells in an intact ciliary body. Without this, it is not possible to identify the ions and describe the major pathways involved in aqueous humor production.

We have applied the technique of electron-probe X-ray microanalysis (EPMA) to study this problem. This approach allows measurement of the elemental composition of individual epithelial cells within the intact tissue and has provided new insights into the transport properties of the ciliary epithelium.

III. REVIEW OF ELECTRON-PROBE X-RAY MICROANALYSIS

A. Theory

EPMA permits localization and quantification of intracellular elements within visualized cells. Using an electron microscope, a specific area within a cell is targeted with an electron beam. Incident electrons of sufficient energy can knock electrons out of the inner shell of an atom within the irradiated area. When an electron from an outer orbit relaxes into the vacated orbit, a quantum of X-ray energy is released that is characteristic of the atom bombarded. Measurement of the number of quanta at each characteristic energy permits quantification of the identified intracellular elements.

B. Technique

The application of this technique to cells requires the tissue to be mounted in an electron microscope with the elements remaining in the positions that they occupied in the living state. It has been shown that the cellular and extracellular locations of diffusible ions can be preserved by a combination of very rapid freezing of tissues to liquid N_2 temperatures followed by thin sectioning of the frozen tissues in a cryoultramicrotome at temperatures below –80 °C. The sections 0.2–0.4 μm in thickness must then be freeze-dried at these temperatures and at subatmospheric pressures (typically around 10^{-4} Pa or 7.5×10^{-7} Torr) to remove the water while preserving cell elemental contents.

The frozen-dried sections can then be visualized in a scanning electron microscope with a transmitted electron detector. X-rays are collected, usually with an energy-dispersive X-ray detector. The intracellular data are obtained by scanning the electron beam over a rectangular area within each selected cell. It is important to avoid the cell boundaries as, during the freeze-drying, extracellular elements tend to collect at the cell surface so that a cellular signal acquired near the plasma membrane often has artificially high Na^+ and Cl^- contributions. For this reason, we normally choose to sample an area within the nucleus. Direct measurements of *Chironomus* salivary gland cells have demonstrated that the intracellular activities of K^+ and Cl^- are the same in the nucleoplasm and cytosol (Palmer and Civan, 1975, 1977). In practice, the dimensions of the irradiated areas vary from $\sim$0.9 $\times$ 1.2 μm to $\sim$2.4 $\times$ 3.0 μm, depending on the size of the nucleus analyzed.

Elemental peaks can be quantified by filtered least-square fitting to a library of monoelemental peaks (Bowler *et al.*, 1991). The library spectra for Na, Mg, Si, P, S, Cl, K, and Ca are derived from microcrystals sprayed onto a Formvar film. In addition to the quantal element-specific X-rays, irradiating sections with an electron beam produces nonquantal continuous or white radiation, reflecting electron deceleration by coulombic interaction with atomic nuclei. The white counts (w), an index of tissue mass (Civan, 1983), are summed over the range 4.6–6.0 keV, and corrected for the nontissue contributions arising from the specimen holder and grid.

C. Application to the Ciliary Epithelium

In our studies, we use tissues from adult Dutch black-belted rabbits of either sex. The iris-ciliary body (ICB) is excised from each enucleated eye, cut into quarters, and each quarter bonded with cyanoacrylate to a Mylar support frame on its stromal border.

Quadrants of each ICB are mounted between the two halves of incubation chambers, so oriented as to occlude the common aperture. Each half-chamber is filled with 1.5 ml, and fresh solution constantly infused at 0.5 ml/min. A gas-lift in each half-chamber aerates and gently stirs the solution. Drugs, when used, are normally added to both sides of the tissue to maximize the effects.

Tissues are incubated for 1–2 hours at room temperature (18–22 °C), initially under control conditions. Pairs of quadrants (one from each eye) are then incubated for at least 30 min under either control or experimental conditions. After incubation, the tissues are blotted, and then plunged into liquid propane at –180 °C to freeze the preparation rapidly and preclude redistribution of ions and water. Blocks are fractured from the frozen tissue under a dissecting microscope (×7). Care is paid to the origin and orientation of the block so that, after transfer of a block to the cryoultramicrotome and subsequent trimming, we can identify and accurately select the region from which the sections were cut. Sections, 0.4- to 0.6-μm thick, are usually cut first from the pupillary side and perpendicular to the major plane of the ICB. The orientation of the block is then reversed and perpendicular sections taken from the side of the *pars plicata*. The sections are then freeze-dried and transferred for analysis to a scanning electron microscope equipped with an energy-dispersive X-ray spectrometer. Typical energy spectra are shown (Fig. 1). We analyze two well-defined regions, the posterior and anterior portions of the ciliary epithelium.

The Na, K, and Cl signals are normalized to the P signal obtained in the same scanned area of each cell, yielding molar ratios of these elements (McLaughlin *et al.*, 1998). This normalization corrects for variations in section thickness both within and between sections. The mean P content of the tissue is ~500 mmol/kg dry weight (Bowler *et al.*, 1996). P is chosen for normalization because of the constancy of its intracellular signal, almost entirely reflecting the covalently linked fraction in epithelial cells (e.g., Civan *et al.*, 1983). Normalization to P has been validated by the observed close linear relationship linking the two largely intracellular elements K and P (Fig. 3; Bowler *et al.*, 1996). NPE cells have the same P content anteriorly and posteriorly, as is true for the PE cells (Table I; McLaughlin *et al.*, 2004). However, the dry weight content of P is ~12% lower in NPE than in PE cells in both regions, so that the ion contents normalized to P are usually comparably higher in NPE than PE cells.

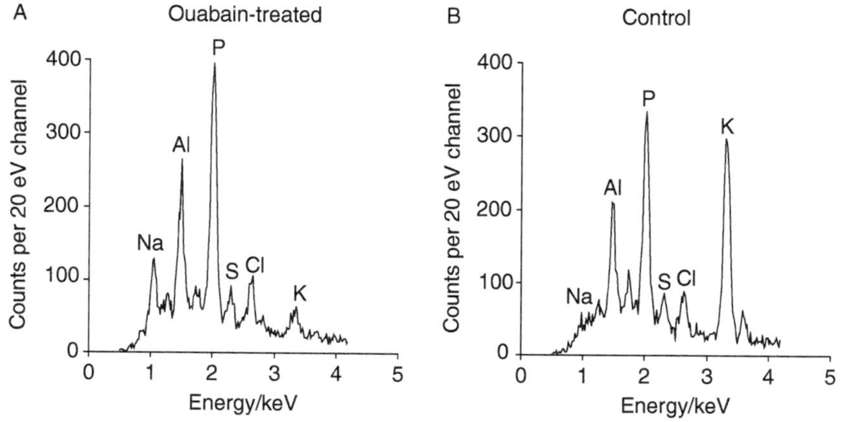

FIGURE 1 Energy spectra. Spectra were recorded from tissue incubated with 100 μM ouabain (A) and incubated in control Ringer's solution (B). The aluminum signal (Al) arises from the tissue holder and is used to align the spectra.

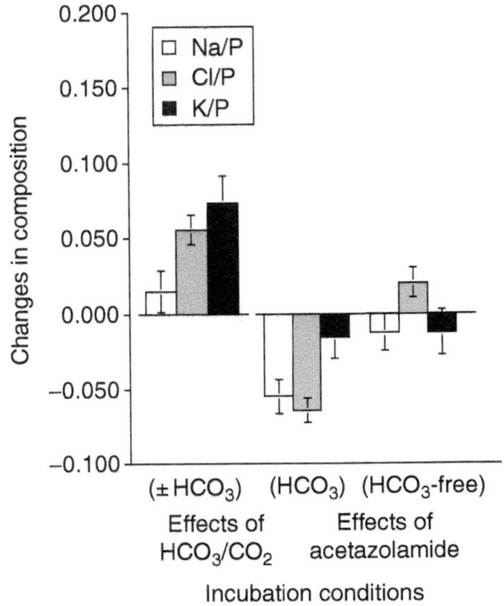

FIGURE 2 Dependence of intracellular ion content of ciliary epithelial cells on presence of CO_2/HCO_3^-. Analyses based on results from McLaughlin *et al.* (1998, 2001b) in form of bar graphs presenting means±1 SE. NPE and PE cells were affected to similar extents in these experiments, so that their values were combined.

The major strength of EPMA is the ability to obtain information about the elemental contents within individual cells under a wide variety of experimental conditions and manipulations. This has allowed us to identify relationships between the PE and NPE cells, to detect appreciable differences in behavior between the anterior and posterior regions, to extend our understanding of the role of CO_2/HCO_3^- in influencing cell electrolyte composition, and to reach a deeper understanding of the important ion transport pathways in these epithelial cells. In addition, it has been possible to obtain new information about how some of the drugs used in the treatment of glaucoma work at the cellular level.

The technique is not, of course, without its limitations. First, cell water cannot be measured directly. Therefore, it is the elemental contents and not concentrations that are measured and, in the absence of direct measurements of water content, the normalized ion contents cannot provide a direct estimate of intracellular ion concentrations. Nevertheless, we can use changes in the sum of the normalized contents of Na^+ and K^+[(Na+K)/P] as a useful indicator of changes in intracellular water content (Abraham

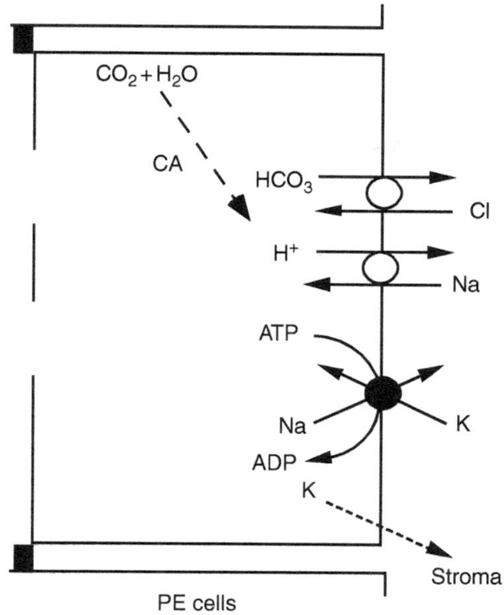

FIGURE 3 Stimulation of KCl uptake by CO_2/HCO_3^-. HCO_3^- and H^+ are formed from water and CO_2 through a reaction catalyzed by the enzyme carbonic anhydrase (CA). Both Cl^-/HCO_3^- and Na^+/H^+ exchangers operate on the basolateral membrane of the PE cells and remove HCO_3^- and H^+ from the cells, which gain Na^+ and Cl^- in exchange. This Na^+ is then extruded from the NPE and PE cells by the Na^+,K^+-activated ATPase, both at the stromal surface (illustrated here) and at the aqueous surface. The net effect is a gain of Cl^- with K^+ by the cells.

et al., 1985). Likewise, we can use the normalized anion gap, defined as (Na+K−Cl)/P, as an approximate index of changes in intracellular HCO_3^- content (McLaughlin *et al.*, 2001b), although other unmeasured anions can, of course, also contribute to this parameter.

IV. TOTAL INFLOW

A. Feasibility of EPMA

It was first necessary to establish that the technique of EPMA could be applied to the ciliary epithelium. We initially demonstrated that loss of K^+ and gain of Na^+ after tissue exposure to ouabain was readily detected and then went on to study cell elemental composition under a variety of situations.

B. Role of Gap Junctions Between PE and NPE Cells

We found that, under control conditions, there were no measurable elemental gradients across the gap junctions between paired PE and NPE cells for Na^+, K^+, or Cl^-, when allowance is made for the differences in P content between the two cell types. This indicates that, under normal conditions, the gap junctions between paired PE and NPE cells do not represent a rate-limiting step in transepithelial ion movements. When we used heptanol to block the gap junctions, at least partially, the elemental contents of the two cell types did not change appreciably unless transport inhibitors, such as ouabain, were used (McLaughlin *et al.*, 2004). This finding is consistent with our understanding that, when the cells' multiple transporters are all available, cells adjust the rates of movement of ions to maintain their steady-state compositions.

C. Cellular Chloride

Chloride plays a key role in transepithelial secretion and absorption of salt and water. For example, loss-of-function mutation of the *Clcnk* gene for the ClC-Kb Cl^- channel produces the type III form of Bartter's syndrome, with renal salt wasting (Hebert, 2003). However, an integrated model of the many mechanisms underlying net ciliary epithelial Cl^- secretion and its regulation has not been fully developed.

The electron microprobe was first utilized to study the known stimulatory effects of CO_2/HCO_3^- on the rabbit ciliary epithelium. Exposure to CO_2/HCO_3^- increases the magnitude of the intracellular ciliary epithelial potential (Carré *et al.*, 1992), restores the normally negative transepithelial potential (aqueous with respect to stroma) (Kishida *et al.*, 1981; Krupin *et al.*, 1984), and stimulates transepithelial secretion of fluid (Candia *et al.*, 2005). We demonstrated that the presence of CO_2/HCO_3^- in the bath resulted in increased cell Cl^- content, together with an increase in cell K^+ content, which was prevented by the carbonic anhydrase inhibitor acetazolamide (Fig. 2). The increase in Cl^- content likely reflects exchange for intracellular HCO_3^- through a Cl^-/HCO_3^- antiport. The increase in K^+ content can be understood as arising from two sequential steps (Fig. 3), an initial exchange of extracellular Na^+ for intracellular proton through a Na^+/H^+ antiport, and subsequent exchange of intracellular Na^+ for extracellular K^+ through Na^+, K^+-activated ATPase.

Another widely distributed Cl^- transport pathway, the Na^+-K^+-$2Cl^-$ cotransporter was also studied. In secretory epithelia, it is this transporter that is usually responsible for accumulation of Cl^- to levels greater than would be seen for a passive distribution of the ion. If this cotransporter were supporting uptake of Cl^- by the cells under our experimental conditions, blocking the transporter with bumetanide, a 5-sulfamoylbenzoic acid loop diuretic, would be expected to reduce cell Cl^- and K^+ contents (Haas and McManus, 1983; Fig. 4). In fact, as shown in Fig. 5, cells actually gained significant amounts of

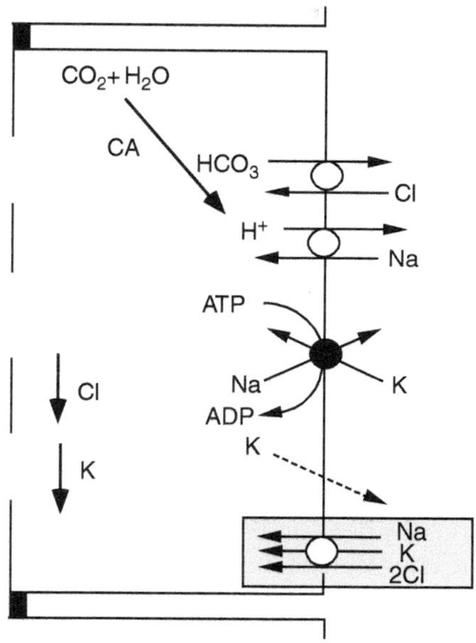

FIGURE 4 The predicted response of cell Cl^- and K^+ contents to the cotransporter inhibitor bumetanide under conditions of baseline uptake of Na^+, K^+, and Cl^- through the cotransporter.

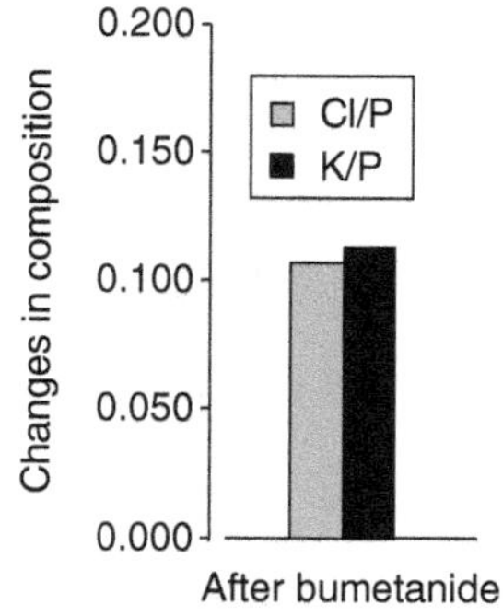

FIGURE 5 Measured changes in cell Cl^- and K^+ contents in the presence of the Na^+-K^+-$2Cl^-$ cotransporter inhibitor bumetanide (10^{-4} mol/liter).

Cl^- with K^+. The most direct interpretation is that, rather than net Cl^- uptake through this pathway, the PE cells in the anterior epithelium may be actually losing Cl^- through it to the surrounding extracellular fluid.

The net direction of solute transport by this electrically neutral transporter is determined by the combined chemical potential driving force. It is instructive to calculate the driving force using reasonable assumptions for the cell ion concentrations, based on the measured cell composition (Fig. 6). These calculations suggest that cell Cl^- might be accumulated in the cells not only beyond its electrochemical equilibrium concentration, but also to a higher concentration than the Na^+-K^+-$2Cl^-$ cotransporter can generate. This accumulation would be a consequence of the combined contributions of the Cl^-/HCO_3^- and Na^+/H^+ exchangers. In contrast, in a HCO_3^--free solution, where the contributions of the Cl^-/HCO_3^- and Na^+/H^+ exchangers will be minimal, bumetanide actually results in a fall in cell Cl^- content and, in the presence of acetazolamide and bumetanide, the cells lose rather than gain Cl^- (McLaughlin *et al.*, 1998). Under these conditions, therefore, the Na^+-K^+-$2Cl^-$ cotransporter would surely contribute to net cell Cl^- accumulation. The view that net movement of solute through the symport can be bidirectional is consistent with flux and volumetric measurements of rabbit ciliary epithelial preparations. Under the experimental conditions of Crook *et al.* (2000), stromal bumetanide inhibited net Cl^- transport across the rabbit ciliary epithelium. Furthermore, Edelman *et al.* (1994) observed that reducing external ionic concentrations to reverse the thermodynamic force driving uptake through the symport led to bumetanide-inhibitable reduction in cell volume.

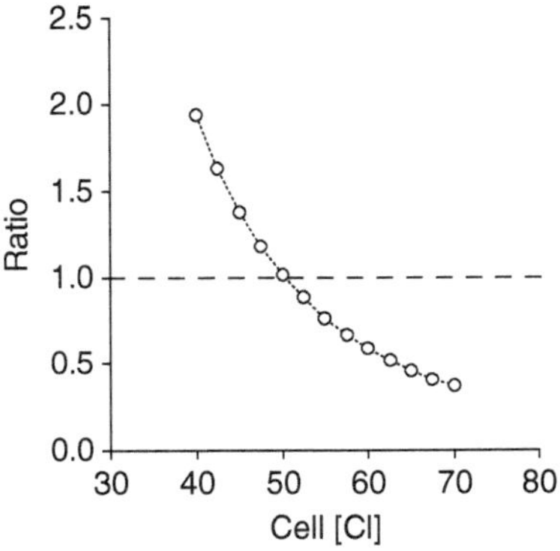

FIGURE 6 The calculated net thermodynamic driving forces for ion movements on the Na^+-K^+-$2Cl^-$ cotransporter. Ratios above 1 indicate that the driving force would be for net Cl^- entry to the cells, ratios below 1 indicate that the driving force favors net Cl^- loss from the cells on the cotransporter.

D. Role of the Na^+, K^+-Activated ATPase in Aqueous Humor Production in Rabbit Ciliary Epithelium

Na^+,K^+-activated ATPase is found on the basolateral membranes of both the PE and NPE cells. We have studied the effects of inhibiting the pump with the cardiac gycoside ouabain, both in the absence and presence of the gap junction uncoupler heptanol. We have found that ouabain causes significant loss of K^+ and uptake of Na^+ by the cells (Fig. 7). With ouabain only on the

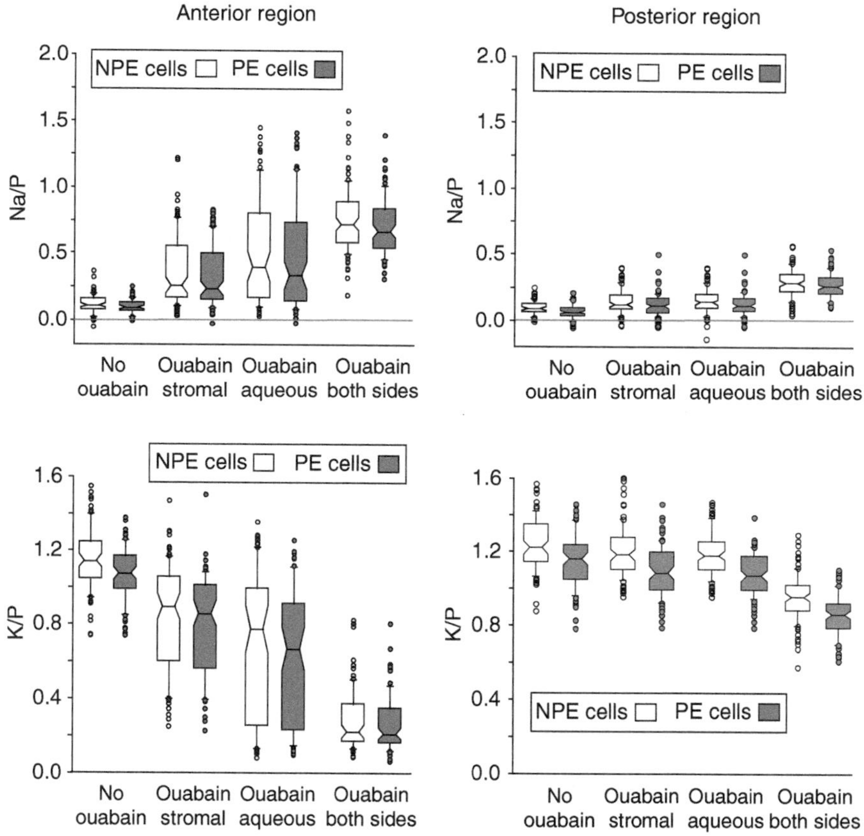

FIGURE 7 The effects of 0.1 mM ouabain on the ion contents of cells incubated in CO_2/HCO_3^- Ringer's solution (McLaughlin *et al.*, 2004). In these box plots, the medians are indicated by the central horizontal lines, the notch indicates the 95% confidence intervals, the lower and upper lines include all data between the 25th and 75th percentiles, and the "whiskers" display the data range between the 10th and 90th percentiles. Changes in Na^+ and K^+ are far greater in the anterior than posterior region. Since the rate-limiting step in cation changes after ouabain is the membrane Na^+ permeability, the anterior cells must have a higher Na^+ permeability than the posterior cells. Used with the permission of the American Physiological Society.

stromal or aqueous side of the tissue, its effects are less than when it is present simultaneously on both sides, indicating that indeed the Na^+,K^+-activated ATPase is active on the basolateral membranes of both the PE and NPE cells. As illustrated by Fig. 8, ouabain-induced changes in composition are shared by contiguous NPE and PE cells. The loss of K^+ and gain of Na^+ are strongly linked within a given NPE or PE cell (Fig. 9). Furthermore, the compositions of contiguous NPE and PE cells are also consistently and strongly linked to one another even when neighboring NPE–PE cell couplets are affected to very different extents by ouabain (Fig. 10). These findings indicate that, although under physiological conditions cells are linked through gap junctions, different pairs of cells differ in the activities of their membrane pathways.

E. Relationship of the EMPA Findings to the Consensus Model for Aqueous Humor Secretion

The results from the EMPA studies are consistent with the model that has emerged from a variety of studies of ciliary epithelial function. At the basolateral membrane of the PE cells, there is uptake of Na^+ and Cl^- through electroneutral Cl^-/HCO_3^- and Na^+/H^+ exchangers. Carbonic anhydrase both catalyzes the production of HCO_3^- (Meldrun and Roughton, 1933) and stimulates Cl^-/HCO_3^- (Sterling *et al.*, 2001) and Na^+/H^+

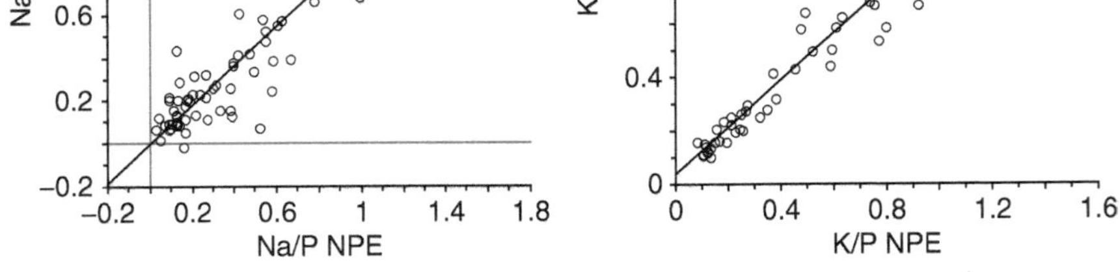

FIGURE 8 Relationship between elemental compositions of contiguous paired NPE and PE cells in the anterior epithelium after aqueous ouabain (McLaughlin *et al.*, 2004). The correlation coefficient (*r*) is 0.95 for both least-squares fits, indicating a strong correlation between ionic changes in adjoining the NPE and PE cells. Used with the permission of the American Physiological Society.

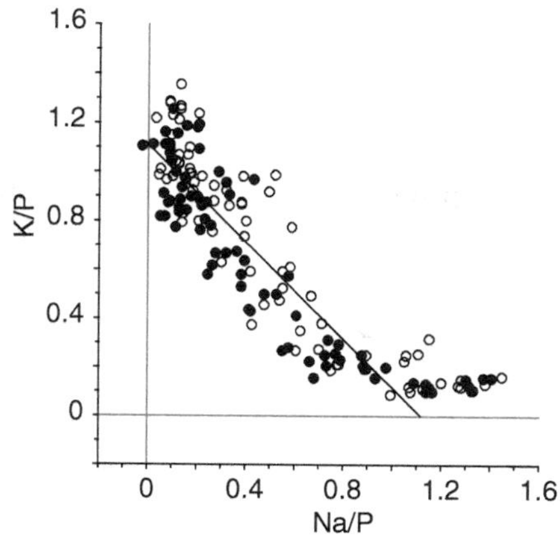

FIGURE 9 Relationship between Na/P and K/P in paired anterior NPE (open circles) and PE cells (filled circles) after aqueous ouabain (McLaughlin *et al.*, 2004). At the highest Na^+ and lowest K^+ contents, cell Cl^- content was beginning to increase showing that cells have begun to swell. The fall in K^+ content is tightly linked with the increase in Na^+ content within the same cell. Used with the permission of the American Physiological Society.

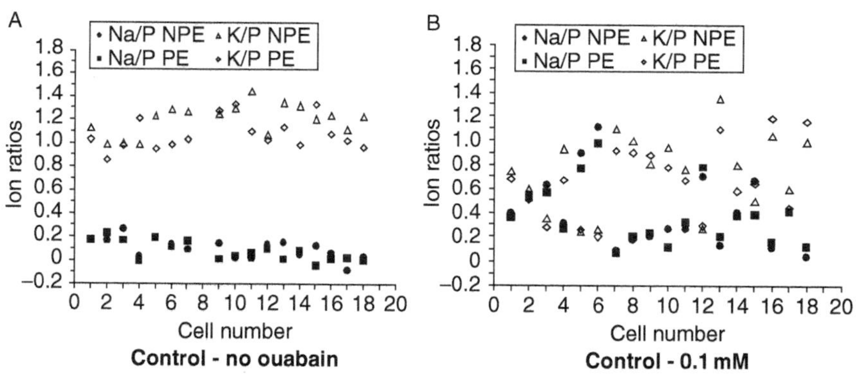

FIGURE 10 Na^+ and K^+ contents in neighboring anterior NPE cells within a single section (McLaughlin *et al.*, 2004). (A) Under control conditions, the elemental contents are similar. (B) In contrast, ouabain alters the composition dramatically of neighboring pairs of PE-NPE cells. Some cell pairs are hardly affected, whereas others show marked changes in composition. Used with the permission of the American Physiological Society.

exchangers (Li *et al.*, 2002). The Na^+ and Cl^- diffuse through the gap junctions between the contiguous PE and NPE cells and are then released to the aqueous through Na^+,K^+-activated ATPase and Cl^- channels, respectively.

V. TOPOGRAPHY OF INFLOW

So far, we have discussed the ciliary epithelium as though it were a homogeneous tissue. However, as emphasized in Section II, both anatomical and histological studies show significant differences between different regions. In our initial EPMA studies, we analyzed sections obtained from the iridial portion of the primary ciliary processes. On occasion, we sectioned more posteriorly and found that such areas appeared to have very different elemental cell responses to drugs. For this reason, we began a systematic study of the composition of the cells from different regions. Initially, we identified three regions to analyze (McLaughlin *et al.*, 2001b), which we called anterior, middle, and posterior (Fig. 11). However, it became apparent that cells in the middle region tended to be similar to those in the anterior region, while still displaying properties intermediate between the two well-defined anterior and posterior regions. In later studies (McLaughlin *et al.*, 2004, 2007), we focused on the elemental compositions of the two very well-defined structural areas, the anterior iridial part of the primary ciliary processes and the posterior *pars plicata*.

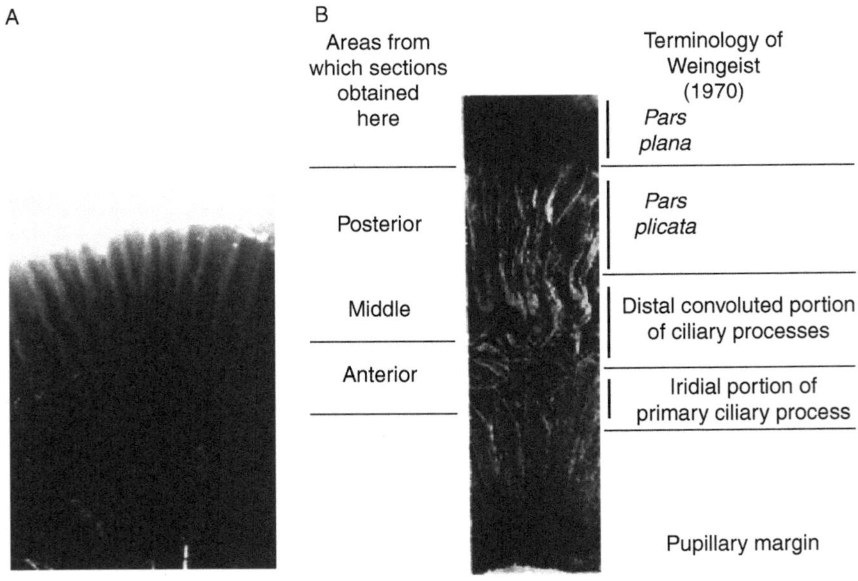

FIGURE 11 Rabbit iris-ciliary body (ICB) photographs showing the gross morphology of the structure. (A) A portion of the isolated ICB. (B) A photograph (from Weingeist, 1970) of the whole rabbit ICB showing the relationship between the areas analyzed and the terminology of Weingeist. Modified from Fig. 2 of McLaughlin *et al.* (2001b).

When compared to the posterior region, the more anterior region shows greater changes in Na^+, K^+, and Cl^- in response to inclusion of CO_2/HCO_3^- in the bath (Fig. 12), inhibition of the Na^+-K^+-$2Cl^-$ cotransporter with bumetanide (Fig. 13), and inhibition of the Na^+,K^+-activated ATPase by ouabain (Fig. 7).

These regional variations might indicate that the anterior region is the more active in secreting the aqueous humor. However, when we used heptanol to block the gap junctions between contiguous PE and NPE cells, we found that the largest ion changes after ouabain were in the anterior NPE cells. After heptanol, the anterior PE cells show relatively little Na^+ uptake and K^+ loss after ouabain in either the stromal or aqueous solutions. Indeed, their composition is very similar to that of the posterior cells incubated under these conditions (Fig. 14). This finding indicates that the anterior NPE cells must gain Na^+ from the aqueous humor. Recent studies with acylguanidines [amiloride, benzamil, and dimethylamiloride (DMA)]

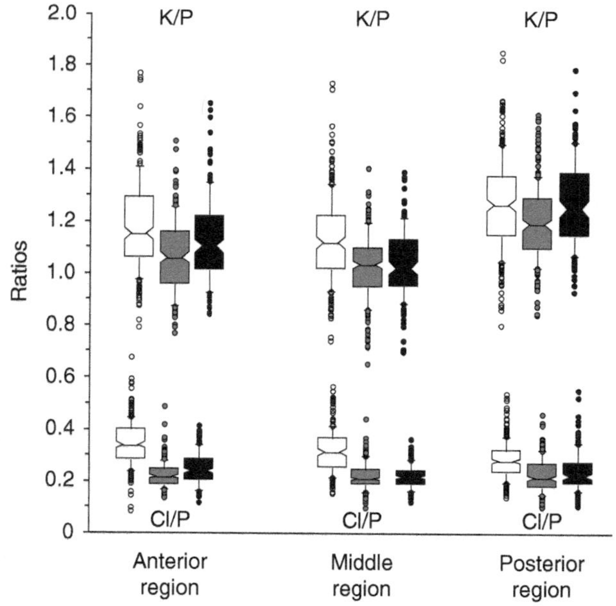

FIGURE 12 The effects of incubation in CO_2/HCO_3^- Ringer's (open symbols), HCO_3^--free solution (gray symbols), or CO_2/HCO_3^- Ringer's+acetazolamide (0.5 mmol/liter) (black symbols) on cell ion contents. Note that the effects on cell ions of the CO_2/HCO_3^- Ringer's are greater in the anterior and middle regions than in the posterior region of the epithelium and were largely prevented by the carbonic anhydrase inhibitor, acetazolamide (0.5 mmol/liter).

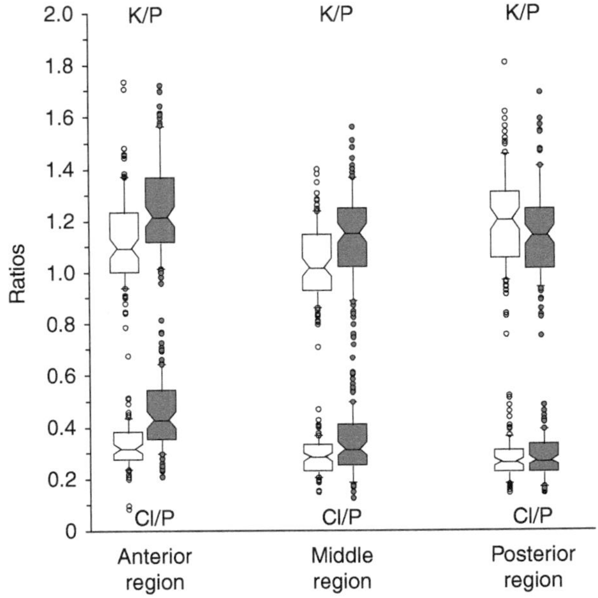

FIGURE 13 The effects of incubation in CO_2/HCO_3^- Ringer's on cell ion contents in the presence (gray symbols) or absence (open symbols) of bumetanide (0.5 mmol/liter). Note that in the anterior and middle regions, cells gained Cl^- and K^+ after bumetanide, whereas in the posterior region there was loss of K^+ and little change in Cl^-.

indicate that this gain in Na^+ may be mediated both by Na^+ channels and Na^+/H^+ exchangers (McLaughlin *et al.*, 2007). Thus, the anterior NPE cells appear to have a greater capacity than the posterior NPE cells to reabsorb ions from the aqueous humor.

In order to explore the possibility that different regions also display differing capacities to secrete aqueous humor, we have made use of the recent observation that hypotonic swelling of the entire bovine ciliary epithelium stimulates secretion of Cl^- (Do *et al.*, 2006). The time course of the swelling-activated Cl^- secretion follows that of the regulatory volume decrease (RVD) displayed by NPE cells freshly isolated from the same preparation (Do *et al.*, 2006). The swelling-activated Cl^- channels of NPE cells are functionally similar to those activated by agonists of NPE-cell A_3 adenosine receptors, suggesting that a single population of Cl^- channels may provide the final conduits for much of the Cl^- secretion by ciliary epithelium (Carré *et al.*, 2000).

We incubated tissues in a solution in which the NaCl concentration was reduced by 50%. This gave a solution whose total osmolarity was 60% of normal. As shown in Fig. 15, this leads to rapid losses of Cl^- and K^+ from

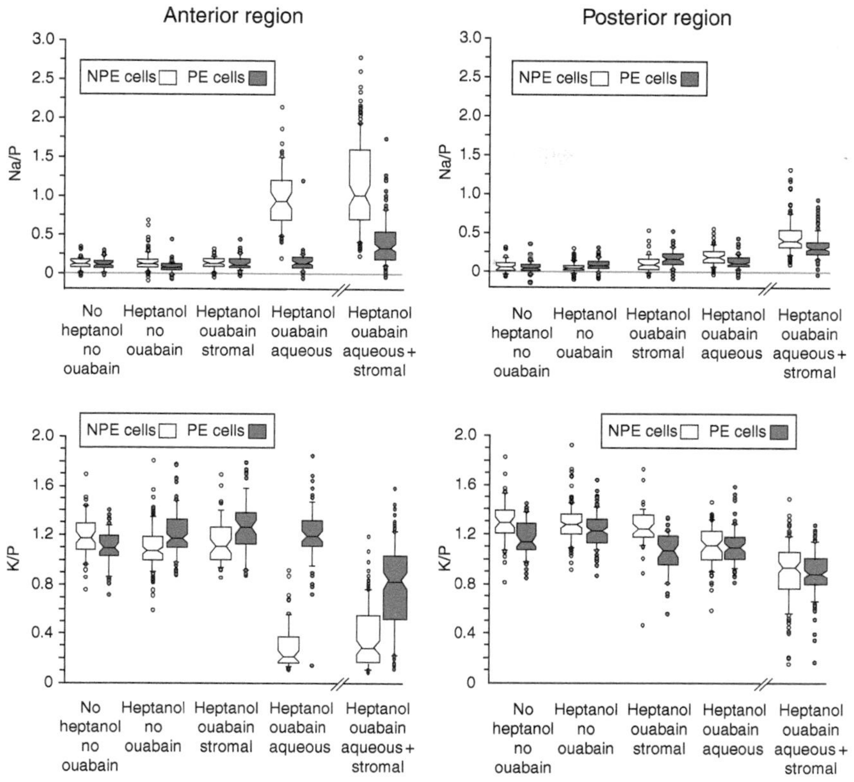

FIGURE 14 Effects of ouabain (0.1 mmol/liter) on epithelial cells in the presence of heptanol (3 mmol/liter) (McLaughlin *et al.*, 2004). Note the much greater effects of ouabain on the anterior NPE cells. Used with permission of the American Physiological Society.

the posterior cells, with little change in Na^+ content. The effects on the anterior cells are very much smaller. In the posterior epithelial cells, the loss of K^+ greatly exceeds the loss of Cl^- (Figs. 15 and 16). Cell electroneutrality requires that these cells must either have lost significant quantities of HCO_3^- ions with K^+ or gained appreciable H^+ ions that would be buffered on cellular macromolecules, thus decreasing cellular electronegativity. Although our results cannot distinguish between these two possibilities, it is known that Cl^- channels are also permeable to the HCO_3^- ion (Tabcharani *et al.*, 1989; Nicholl *et al.*, 2002). In addition, a Cl^-/HCO_3^- exchanger could recycle Cl^- across the membrane, so that the net effect was loss of HCO_3^- and Cl^-, rather than loss only of Cl^-.

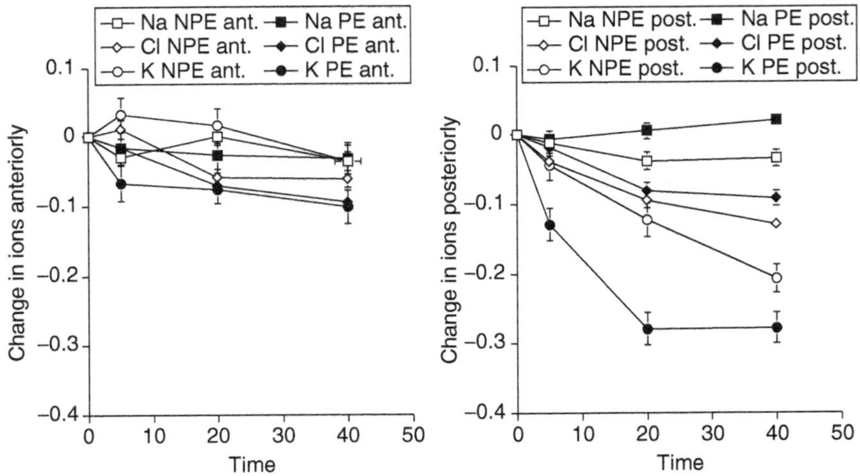

FIGURE 15 Effects of incubation in hypo-osmotic solution. The differences are shown ±1 SE for anterior (ant.) and posterior (post.) regions of rabbit ciliary epithelium.

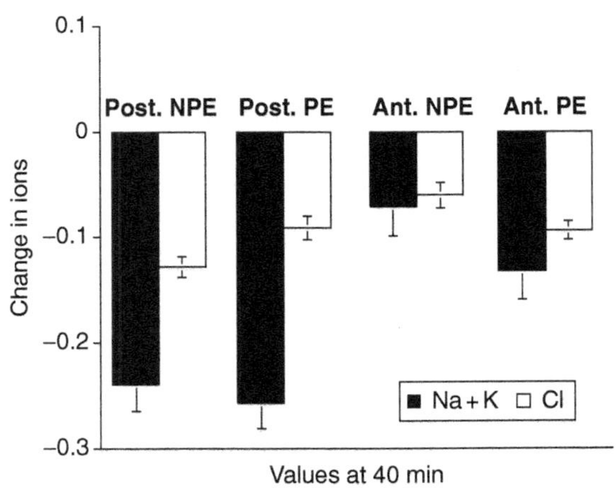

FIGURE 16 The relationship between losses of measured cations (Na+K) and losses of measured anion, Cl. Differences are shown ±1 SE for anterior (ant.) and posterior (post.) regions of rabbit ciliary epithelium.

VI. A NEW MODEL FOR AQUEOUS HUMOR PRODUCTION

The results obtained with ouabain and hypotonic challenge suggest that the posterior epithelium may be the primary region for continued net secretion of aqueous humor (Fig. 17A). The recent data indicate that the anterior

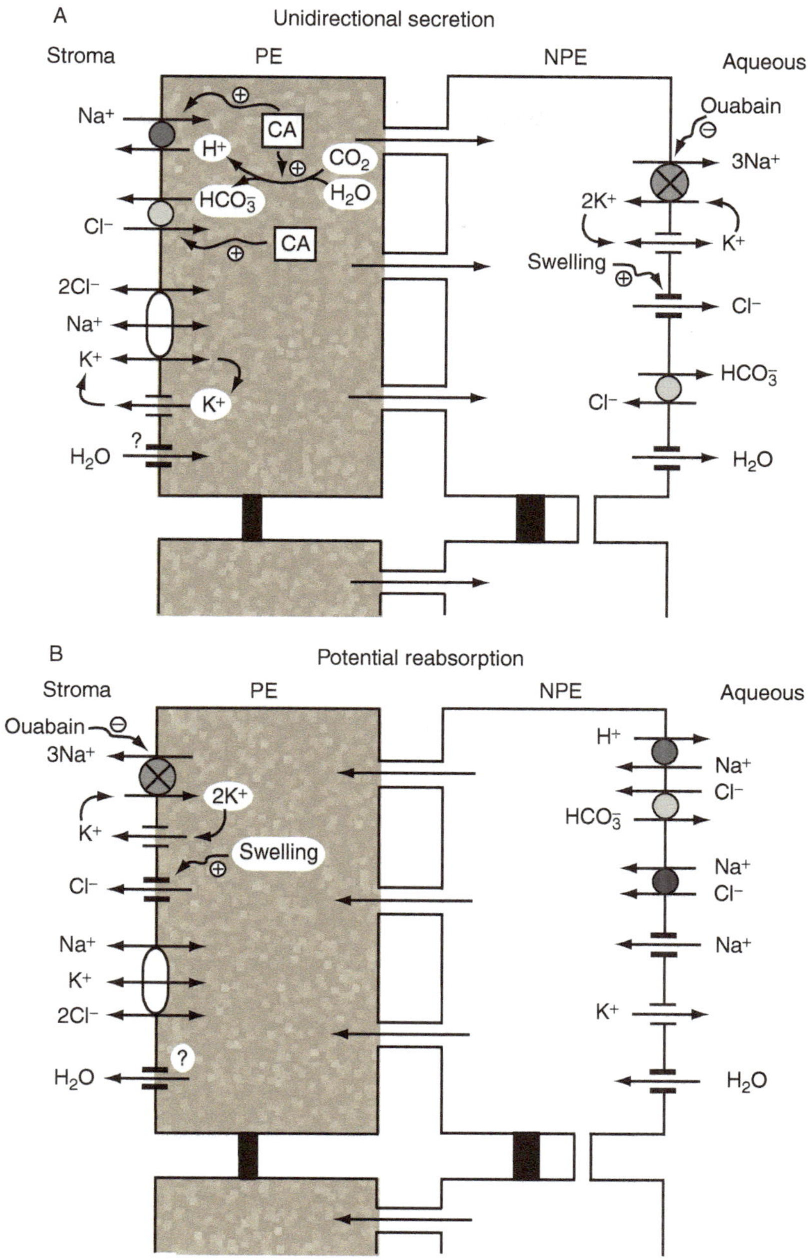

FIGURE 17 Pathways for unidirectional secretion (A) and possible reabsorption (B) across the ciliary epithelium (McLaughlin *et al.*, 2007). Used with permission of the American Physiological Society.

region can be a site for substantial recycling at the aqueous surface (Fig. 17B). High rates of NaCl reabsorption from the aqueous humor with secondary uptake of water would lead to swelling of both the NPE and, through the gap junctions, the PE cells. Swelling (Zhang and Jacob, 1997) and/or cyclic adenosine monophosphate (cAMP) activates PE-cell maxi-Cl^- channels, particularly at high intracellular Cl^- concentration (Do *et al.*, 2004), releasing Cl^- into the stroma. Na^+ is extruded by PE-cell Na^+,K^+-activated ATPase (Krupin *et al.*, 1984). Although the Na^+-K^+-$2Cl^-$ cotransporter commonly mediates cellular uptake of solute, reversal of the net thermodynamic driving force can lead to solute release that is bumetanide sensitive (Dong and Delamere, 1994; Edelman *et al.*, 1994). Thus, the solute recycling at the aqueous surface could lead to net transepithelial reabsorption back into the stroma of the anterior region. In contrast, the posterior ciliary epithelium could be the site for continuous, albeit regulated secretion.

This proposed functional topography is consonant with the finding that the cell Na^+/K^+ ratio in the anterior ciliary region (0.10) is double that in the posterior (0.05) region in control cells. This would be expected if Na^+ recycling from the aqueous humor were higher in the anterior epithelium even under baseline conditions, and not solely after exposure to ouabain.

Data from cells incubated in the absence of any drugs provides semiquantitative support to this model in documenting that the driving force favoring solute uptake through the PE-cell Na^+-K^+-$2Cl^-$ cotransporter is significantly different in the anterior and posterior regions. The Na^+-K^+-$2Cl^-$ cotransporter has been localized largely to the PE cells at the stromal surface (Dunn *et al.*, 2001), and at least under certain conditions, contributes significantly to transepithelial Cl^- secretion (Crook *et al.*, 2000; Do and To, 2000) and aqueous humor formation (Shahidullah *et al.*, 2003). For the Na^+-K^+-$2Cl^-$ cotransporter to contribute to net secretion, the net thermodynamic force must favor delivery of solute from stroma into PE cell. That condition is met when:

$$[Na^+]_o[K^+]_o[Cl^-]_o^2 > [Na^+]_c[K^+]_c[Cl^-]_c^2 \quad (1)$$

where the subscripts "o" and "c" refer to the extracellular and cellular phases. Without knowledge of the absolute water contents of the PE cells in the anterior and posterior regions, we cannot calculate the intracellular concentrations of Eq. (1). However, using [(Na+K)/P] as an index of water content, the baseline PE-cell water contents are the same in the two regions ($\Delta = 0.014 \pm 0.015$). In this case, the baseline intracellular concentrations are directly proportional to the elemental contents. We calculate that the right-hand product of the inequality [Eq. (1)] above is 3.3×10^{-3} posteriorly and approximately threefold higher (9.1×10^{-3}) anteriorly. This supports the idea that the net thermodynamic driving force favors cotransport entry of Na^+, K^+, and Cl^- into the PE cells of the posterior epithelium, while favoring their release from the anterior PE cells. We do not know whether the posterior and anterior products for the intracellular concentrations bracket that for the extracellular concentrations [Eq. (2)], as would be expected from the model:

$$\{[Na^+]_c[K^+]_c[Cl^-]_c^2\}_{anterior} \geq \{[Na^+]_o[K^+]_o[Cl^-]_o^2\} > \{[Na^+]_c[K^+]_c[Cl^-]_c^2\}_{posterior} \quad (2)$$

However, bumetanide inhibition of the Na^+-K^+-$2Cl^-$ cotransporter results in an increase in Cl/P only anteriorly, consistent with the notion that the thermodynamic force on the Na^+-K^+-$2Cl^-$ favors release of solute in that region.

In summary, the baseline elemental contents and their response to hypotonic challenge suggest that the posterior epithelium is the major site of secretion. The effects of ouabain and acylguanidines point to fine-tuning of aqueous humor formation by the anterior epithelium through substantial solute recycling at the aqueous surface and net transepithelial reabsorption. This organization is analogous to the common functional integration of glandular secretion by acinar cells and subsequent processing of the secretion by ductal cells (e.g., Luo *et al.*, 2001). Selective stimulation of reabsorption by the ductal analogue (the anterior epithelium) might provide a novel approach for reducing the rate of net aqueous humor formation, and thereby, intraocular pressure (IOP).

VII. EFFECT OF TIMOLOL ON INFLOW

Our main thrusts in applying the technique of EPMA of ciliary epithelium have been to (i) monitor the chemical driving forces across plasma and gap-junctional membranes, (ii) examine different ciliary epithelial regions to assess functional topography, and (iii) examine how some of the drugs commonly used in the treatment of glaucoma may regulate inflow.

We began by examining the effects of timolol on epithelial cell composition, focusing at the time on the anterior region. Timolol both acts as a β-adrenergic receptor antagonist and lowers IOP, but whether these are causal or parallel actions has been unclear (Yorio, 1985). Timolol (10 μM) produced similar K^+ and Cl^- losses from rabbit ciliary epithelia in CO_2/HCO_3^- solution, but had no effect in CO_2/HCO_3^--free solution, or in CO_2/HCO_3^- solution containing the carbonic anhydrase inhibitor acetazolamide (McLaughlin *et al.*, 2001a). If timolol were to act solely by reducing intracellular cAMP, adding cyclic-AMP to the bath should reverse its effects. This was not the case. In further experiments, we found that inhibition of Na^+/H^+ exchange by DMA in CO_2/HCO_3^- solution reduced Cl and K contents comparably to timolol (Fig. 18).

The results documented a previously unrecognized cAMP-independent transport effect of timolol. One possibility is that inhibition of Cl^-/HCO_3^- exchange may mediate timolol's inhibition of aqueous humor formation. Alternatively, timolol's ocular hypotensive effect might be mediated by reducing cAMP, but that action is exerted at a compartmentalized membrane site, so that flooding the entire cell with cAMP triggers many additional unrelated and confounding effects. As noted earlier (Civan, 2008), another possibility is that timolol's actions on the ciliary epithelium may be mediated by antagonism of β-adrenergic receptors, but through the arachidonic acid signaling cascade by coupling to G_i proteins.

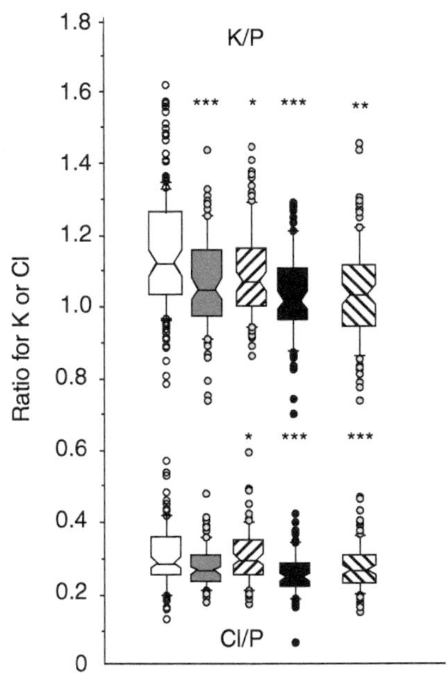

FIGURE 18 Effects of timolol (10 μM) and/or cAMP (1 mM) on ciliary epithelial Cl/P or K/P ratios in CO_2/HCO_3^- solution. The open symbols at the extreme left represent control conditions. Proceeding sequentially to the right, the next four columns present data obtained with timolol, cAMP, cAMP and timolol, and DMA, respectively. Stars indicate significant differences from the control data (*$P<0.05$, **$P<0.01$, ***$P<0.001$).

VIII. FUTURE DIRECTIONS

Many questions remain to be answered. In particular, it is essential that we understand whether the co- and countertransporters and channels that we have identified as playing major roles in the formation of the aqueous humor in rabbits are also those involved in other species. Equally, it is essential that we determine whether the new model of functional topography of rabbit ciliary epithelium that we propose here, applies to inflow of other species.

References

Abraham, E. H., Breslow, J. L., Epstein, J., Chang-Sing, P., and Lechene, C. (1985). Preparation of individual human diploid fibroblasts and study of ion transport. *Am. J. Physiol.* **248** (1 Pt. 1), C154–C164.

Bowler, J. M., Purves, R. D., and Macknight, A. D. (1991). Effects of potassium-free media and ouabain on epithelial cell composition in toad urinary bladder studied with X-ray microanalysis. *J. Membr. Biol.* **123**(2), 115–132.

Bowler, J. M., Peart, D., Purves, R. D., Carré, D. A., Macknight, A. D., and Civan, M. M. (1996). Electron probe X-ray microanalysis of rabbit ciliary epithelium. *Exp. Eye Res.* **62**(2), 131–139.

Candia, O. A., To, C. H., Gerometta, R. M., and Zamudio, A. C. (2005). Spontaneous fluid transport across isolated rabbit and bovine ciliary body preparations. *Invest. Ophthalmol. Vis. Sci.* **46**(3), 939–947.

Carré, D. A., Tang, C. S., Krupin, T., and Civan, M. M. (1992). Effect of bicarbonate on intracellular potential of rabbit ciliary epithelium. *Curr. Eye Res.* **11**(7), 609–624.

Carré, D. A., Mitchell, C. H., Peterson-Yantorno, K., Coca-Prados, M., and Civan, M. M. (2000). Similarity of A(3)-adenosine and swelling-activated Cl(-) channels in nonpigmented ciliary epithelial cells. *Am. J. Physiol. Cell Physiol.* **279**(2), C440–C451.

Civan, M. M. (1983). "Epithelial Ions & Transport: Application of Biophysical Techniques." Wiley-Interscience, John Wiley & Sons, New York.

Civan, M. M. (2008). Formation of the aqueous humor: Transport components and their integration. *In* "The Eye's Aqueous Humor" (M. M. Civan, ed.), 2nd edn. Elsevier, San Diego.

Civan, M. M., Peterson-Yantorno, K., DiBona, D. R., Wilson, D. F., and Erecinska, M. (1983). Bioenergetics of Na^+transport across frog skin: Chemical and electrical measurements. *Am. J. Physiol.* **245**(6), F691–F700.

Coca-Prados, M., and Sánchez-Torres, J. (1998). Molecular approaches to the study of the Na^+, K^+-ATPase and chloride channels in the ocular ciliary epithelium. *In* "The Eye's Aqueous Humor: From Secretion to Glaucoma" (M. M. Civan, ed.), Vol. 45, pp. 25–53. Academic Press, San Diego.

Crook, R. B., Takahashi, K., Mead, A., Dunn, J. J., and Sears, M. L. (2000). The role of NaKCl cotransport in blood-to-aqueous chloride fluxes across rabbit ciliary epithelium. *Invest. Ophthalmol. Vis. Sci.* **41**(9), 2574–2583.

Do, C. W., and To, C. H. (2000). Chloride secretion by bovine ciliary epithelium: A model of aqueous humor formation. *Invest. Ophthalmol. Vis. Sci.* **41**(7), 1853–1860.

Do, C. W., Peterson-Yantorno, K., Mitchell, C. H., and Civan, M. M. (2004). cAMP-activated maxi-Cl^- channels in native bovine pigmented ciliary epithelial cells. *Am. J. Physiol. Cell Physiol.* **287**(4), C1003–C1011.

Do, C. W., Peterson-Yantorno, K., and Civan, M. M. (2006). Swelling-activated Cl^- channels support Cl^- secretion by bovine ciliary epithelium. *Invest. Ophthalmol. Vis. Sci.* **47,** 2576–2582.

Dong, J., and Delamere, N. A. (1994). Protein kinase C inhibits Na(+)-K(+)-2Cl- cotransporter activity in cultured rabbit nonpigmented ciliary epithelium. *Am. J. Physiol.* **267**(6 Pt. 1), C1553–C1560.

Dunn, J. J., Lytle, C., and Crook, R. B. (2001). Immunolocalization of the Na-K-Cl cotransporter in bovine ciliary epithelium. *Invest. Ophthalmol. Vis. Sci.* **42**(2), 343–353.

Edelman, J. L., Sachs, G., and Adorante, J. S. (1994). Ion transport asymmetry and functional coupling in bovine pigmented and nonpigmented ciliary epithelial cells. *Am. J. Physiol.* **266** (5 Pt. 1), C1210–C1221.

Eichhorn, M., and Lütjen-Drecoll, E. (1993). Distribution of endothelin-like immunoreactivity in the human ciliary epithelium. *Curr. Eye Res.* **12**(8), 753–757.

Eichhorn, M., Flügel, C., and Lütjen-Drecoll, E. (1990). Regional differences in the ciliary body of cattle. An electron microscopy and histochemical study. *Fortschr. Ophthalmol.* **87**(3), 241–246.

Flügel, C., and Lütjen-Drecoll, E. (1988). Presence and distribution of Na^+/K^+-ATPase in the ciliary epithelium of the rabbit. *Histochemistry* **88**(3–6), 613–621.

Flügel, C., Lütjen-Drecoll, E., Zadunaisky, J. A., and Wiederholt, M. (1989). Regional differences in the morphology and enzyme distribution of the spiny dogfish (Squalus acanthias) ciliary epithelium. *Exp. Eye Res.* **49**(6), 1097–1114.

Flügel, C., Liebe, S., Voorter, C., Bloemendal, H., and Lütjen-Drecoll, E. (1993). Distribution of alpha B-crystallin in the anterior segment of primate and bovine eyes. *Curr. Eye Res.* **12**(10), 871–876.

Ghosh, S., Freitag, A. C., Martin-Vasallo, P., and Coca-Prados, M. (1990). Cellular distribution and differential gene expression of the three alpha subunit isoforms of the Na,K-ATPase in the ocular ciliary epithelium. *J. Biol. Chem.* **265**(5), 2935–2940.

Ghosh, S., Hernando, N., Martin-Alonso, J. M., Martin-Vasallo, P., and Coca-Prados, M. (1991). Expression of multiple Na^+,$K(^+)$-ATPase genes reveals a gradient of isoforms along the nonpigmented ciliary epithelium: Functional implications in aqueous humor secretion. *J. Cell. Physiol.* **149**(2), 184–194.

Haas, M., and McManus, T. J. (1983). Bumetanide inhibits (Na+K+2Cl) co-transport at a chloride site. *Am. J. Physiol.* **245**(3), C235–C240.

Hebert, S. C. (2003). Bartter syndrome. *Curr. Opin. Nephrol. Hypertens* **12**(5), 527–532.

Kishida, K., Sasabe, T., Manabe, R., and Otori, T. (1981). Electric characteristics of the isolated rabbit ciliary body. *Jpn. J. Ophthalmol.* **25,** 407–416.

Krupin, T., Reinach, P. S., Candia, O. A., and Podos, S. M. (1984). Transepithelial electrical measurements on the isolated rabbit iris-ciliary body. *Exp. Eye Res.* **38**(2), 115–123.

Li, X., Alvarez, B., Casey, J. R., Reithmeier, R. A., and Fliegel, L. (2002). Carbonic anhydrase II binds to and enhances activity of the Na+/H+ exchanger. *J. Biol. Chem.* **277**(39), 36085–36091.

Luo, X., Choi, J. Y., Ko, S. B., Pushkin, A., Kurtz, I., Ahn, W., Lee, M. G., and Muallem, S. (2001). HCO_3^- salvage mechanisms in the submandibular gland acinar and duct cells. *J. Biol. Chem.* **276**(13), 9808–9816.

McLaughlin, C. W., Peart, D., Purves, R. D., Carré, D. A., Macknight, A. D., and Civan, M. M. (1998). Effects of HCO_3^- on cell composition of rabbit ciliary epithelium: A new model for aqueous humor secretion. *Invest. Ophthalmol. Vis. Sci.* **39**(9), 1631–1641.

McLaughlin, C. W., Peart, D., Purves, R. D., Carré, D. A., Peterson-Yantorno, K., Mitchell, C. H., Macknight, A. D., and Civan, M. M. (2001a). Timolol may inhibit aqueous humor secretion by cAMP-independent action on ciliary epithelial cells. *Am. J. Physiol. Cell Physiol.* **281**(3), C865–C875.

McLaughlin, C. W., Zellhuber-McMillan, S., Peart, D., Purves, R. D., Macknight, A. D., and Civan, M. M. (2001b). Regional differences in ciliary epithelial cell transport properties. *J. Membr. Biol.* **182**(3), 213–222.

McLaughlin, C. W., Zellhuber-McMillan, S., Macknight, A. D., and Civan, M. M. (2004). Electron microprobe analysis of ouabain-exposed ciliary epithelium: PE-NPE cell couplets form the functional units. *Am. J. Physiol. Cell Physiol.* **286,** C1376–C1389.

McLaughlin, C. W., Zellhuber-McMillan, S., Macknight, A. D., and Civan, M. M. (2007). Electron microprobe analysis of rabbit ciliary epithelium indicates enhanced secretion posteriorly and enhanced absorption anteriorly. *Am. J. Physiol. Cell Physiol.* **293,** C1455–C1466.

Meldrun, N. U., and Roughton, R. J. W. (1933). Carbonic anhydrase. Its preparation and properties. *J. Physiol.* **80,** 113–142.

Nicholl, A. J., Killey, J., Leonard, M. N., and Garner, C. (2002). The role of bicarbonate in regulatory volume decrease (RVD) in the epithelial-derived human breast cancer cell line ZR-75-1. *Pflügers Arch.* **443**(5–6), 875–881.

Palmer, L. G., and Civan, M. M. (1975). Intracellular distribution of free potassium in Chironomus salivary glands. *Science* **188**(4195), 1321–1322.

Palmer, L. G., and Civan, M. M. (1977). Distribution of Na^+, K^+ and Cl^- between nucleus and cytoplasm in Chironomus salivary gland cells. *J. Membr. Biol.* **33**(1–2), 41–61.

Shahidullah, M., Wilson, W. S., Yap, M., and To, C. H. (2003). Effects of ion transport and channel-blocking drugs on aqueous humor formation in isolated bovine eye. *Invest. Ophthalmol. Vis. Sci.* **44**(3), 1185–1191.

Sterling, D., Reithmeier, R. A., and Casey, J. R. (2001). A transport metabolon. Functional interaction of carbonic anhydrase II and chloride/bicarbonate exchangers. *J. Biol. Chem.* **276**(51), 47886–47894.

Tabcharani, J. A., Jensen, T. J., Riordan, J. R., and Hanrahan, J. W. (1989). Bicarbonate permeability of the outwardly rectifying anion channel. *J. Membr. Biol.* **112**(2), 109–122.

Weingeist, T. A. (1970). The structure of the developing and adult ciliary complex of the rabbit eye: A gross, light, and electron microscopic study. *Doc. Ophthalmol.* **28**(2), 205–375.

Yorio, T. (1985). Cellular mechanisms in the actions of antiglaucoma drugs. *J. Ocul. Pharmacol.* **1**(4), 397–422.

Zhang, J. J., and Jacob, T. J. (1997). Three different Cl^- channels in the bovine ciliary epithelium activated by hypotonic stress. *J. Physiol.* **499**(Pt. 2), 379–389.

CHAPTER 5

Functional Modulators Linking Inflow with Outflow of Aqueous Humor

Miguel Coca-Prados and Sikha Ghosh

Department of Ophthalmology and Visual Science, Yale University School of Medicine, New Haven, Connecticut 06510

I. OVERVIEW

The gene expression program of the human ocular ciliary epithelium (CE) indicates that this tissue is the source of multiple neuroendocrine factors found in the aqueous humor (AqH). In particular, the biosynthesis

Current Topics in Membranes, Volume 62

1063-5823/08 $35.00
DOI: 10.1016/S1063-5823(08)00405-5

of enzymes involved in the processing and maturation of neuropeptides, vasoactive peptides, prostaglandins, steroid hormones, and retinoids suggests that they are likely functional components of multiple local interactive and metabolic endocrine loops. The diversity of biological activities assigned to regulatory peptides identified in the AqH predicts that they are engaged in neuroendocrine circuitries, including inflow and outflow of AqH, immune homeostasis, and circadian rhythms. Emerging evidence demonstrates that the trabecular meshwork (TM) cells, the main cell type of the conventional AqH outflow pathway, express receptors for many of the endocrine factors secreted by the CE. Therefore, the CE should be considered a multifunctional and interactive tissue with the avascular tissues of the anterior segment such as the cornea, the lens, and the TM. In this chapter, we review the molecular and physiological basis of a potential local neuroendocrine circuitry mechanism linking inflow with outflow of AqH.

II. INTRODUCTION

Intraocular pressure (IOP) reflects a balance between the rate of AqH inflow by the CE and the rate of its drainage through the outflow pathways. The CE is a bilayer of polarized, secretory, neuroepithelial cells, [pigmented (PE) and nonpigmented (NPE)], forming with the underlying ciliary muscle (CM) and stroma, as a multicellular unit. The ciliary body (CB) exhibits multiple and important functions in the physiology of the eye (Coca-Prados and Escribano, 2007; Yorio *et al.*, 2007). The CE is responsible for the transport and secretion of AqH, a fluid that nourishes the avascular tissues of the anterior segment of the eye, such as the cornea, the lens, and the TM. The CM, with its contraction–relaxation properties, is known to independently influence accommodation of the lens and outflow of AqH (Kaufman, 1984; Wiederholt *et al.*, 2000).

The AqH upon secretion moves from the posterior chamber to the anterior chamber, through the pupil, leaving the eye through multiple outflow pathways, including the TM, Schlemm's canal, uveoscleral and episcleral veins, by mechanisms not yet fully understood. The study of AqH dynamics, linking inflow with outflow, has clinical relevance to the treatment of glaucoma. As a condition characterized by the progressive, neurodegeneration of the optic nerve, and loss of the visual field due to death of retinal ganglion cells, glaucoma can often lead to blindness. An abnormal elevation in IOP is the best known risk factor in the development of glaucoma, and reduction of AqH secretion is the most effective approach in lowering IOP, and stopping the progression of the disease.

The mechanisms underlying the transport of fluid across the CE have been extensively studied and a working cellular model has emerged in recent years (Civan, 1998). Key transport proteins with important roles in AqH secretion by the CE, and in the maintenance of IOP in conjunction with the TM are: the chloride channels; the paired Na^+/H^+ exchanger (NHE) and Cl^-/HCO_3^- exchanger (AE); the bumetanide-sensitive Na^+-K^+-$2Cl^-$ cotransporter; the Na^+/HCO_3^- cotransporter; and multiple α and β subunit Na^+,K^+-ATPase isoforms. Additional transporters identified in the CE include: (1) the SVCT2, involved in the active transport of ascorbate (Tsukaguchi *et al.*, 1999); and (2) several GLUT isoforms, involved in glucose transport (Escribano and Coca-Prados, 2002; Chan *et al.*, 2007); and members of the OATP family, which mediate the uptake of a wide range of steroid hormone precursors, drugs, and toxins (Schuster *et al.*, 1997; Ito *et al.*, 2003; Hagenbuch and Meier, 2004; Gao *et al.*, 2005). In spite of their possible relevance in AqH secretion and cell signaling, little is known as to how these transporters are regulated.

Over the years, work on AqH dynamics has examined inflow and outflow as two independent physiological processes. However, evidence supports that inflow and outflow are likely linked, since genetic changes affecting AqH secretion can alter IOP (Civan and Macknight, 2004; Libby *et al.*, 2005). Mutations of key transport components involved in AqH secretion, including the Na^+/HCO_3^- cotransporter of the CE, can cause hypertension and glaucoma (Igarashi *et al.*, 1999, 2001). Mice, either lacking adenosine receptors A3 or carrying mutations in the water channel *Aqp1* and *Aqp4*, present in the CE, have reduced AqH inflow and reduced IOP (Zhang *et al.*, 2002; Avila *et al.*, 2002a).

In this chapter, we introduce the "inflow–outflow link hypothesis," which proposes that inflow and outflow of AqH are linked by endocrine factors. These factors synthesized and secreted by the neuroendocrine CE function as modulators. They represent a major amplification system, establishing communication via the AqH with cell receptors in the inflow and outflow pathways, and by classical endocrine mechanisms (autocrine and endocrine/paracrine) regulate AqH secretion and IOP.

III. SOURCES OF NEUROPEPTIDES AND PEPTIDE HORMONES IN THE AqH

It is known that the CB and the TM are innervated by postganglionic peptidergic nerves from the autonomic peripheral and sensory nervous systems. The terminals of these peptidergic fibers lie near the blood vessels in the stroma of ciliary processes, in the CM, and in the proximities of the TM.

The neuropeptides released at the terminals are synthesized by their respective ganglia [i.e., ciliary ganglion (CG), pterygopalatine ganglion (PPG), superior cervical ganglion (SCG), and trigeminal ganglion (TG)], located at a considerable distance from their target tissues in the eye (Troger *et al.*, 2007). Upon synthesis, neuropeptides are delivered through their axons to their terminal ends, and then released upon neural stimulus. Under certain experimental conditions, such as induced ocular inflammation or the electrical stimulation of the cervical sympathetic nerves, neuropeptides are released at high levels and get accumulated in the AqH (Unger, 1990; Gallar and Liu, 1993). Peptides, of neural origin, are found in AqH, and are believed to influence AqH dynamics and IOP (Sears, 1984; Mittag *et al.*, 1987).

Interestingly, a group of peptides distinct from those of autonomic and sensory origin have been detected in the CE at the transcriptional level suggesting that they are locally synthesized. These peptides have also been detected by radioimmunoassay (RIA), at nanomolar range, in the culture medium of NPE cells grown in serum-free medium conditions. The peptides identified are: (1) secretogranin II (SgII) (Ortego *et al.*, 1996a); (2) neurotensin (NT) (Ortego and Coca-Prados, 1999); (3) galanin (Ortego and Coca-Prados, 1998); and (4) somatostatin (SST) (Ghosh *et al.*, 2006). Since these peptides have also been found to be expressed by distinct cells in the mammalian retina (Linden *et al.*, 2005; Troger *et al.*, 2007), it suggests that the bilayered CE may potentially harbor functionally distinct cells (NPE and PE cells), with the capacity to synthesize and release different peptides. This interpretation is consistent with the finding that along the NPE cell layer of the CE, NPE cells express distinct patterns of gene and protein expression of α ($\alpha1$, $\alpha2$, $\alpha3$) and β ($\beta1$, $\beta2$, $\beta3$) subunit isoforms of the $Na^{+}K^{+}$-ATPase (Ghosh *et al.*, 1990, 1991; Wetzel and Sweadner, 2001). This diversity in expression of $Na^{+}K^{+}$-ATPase α and β isoforms, although not totally understood, may have important physiological implications regarding AqH secretion by the CE. For example, the expression of the $\alpha2\beta2$ and $\alpha2\beta3$ isoforms in distinct NPE cells could lead to the secretion of Na^{+} and fluid, whereas transport by the expression of $\alpha1\beta1$ isoforms in PE cells could lead to reabsorption of Na^{+} and fluid (Wetzel and Sweadner, 2001). Thus, distinct combinations of α and β subunit isoforms in NPE cells may underscore different cellular functions along the distinct anatomical regions of the CE (i.e., *pars plicata, pars plana,* and *ora serrata*).

The secretory CE also contributes to the protein composition of the AqH by its expression, among others, of (1) the plasma protein $\alpha2$-macroglobulin (Escribano *et al.*, 1995), transferrin (Bertazolli-Filho *et al.*, 2003), transthyretin (Kawaji *et al.*, 2005), and ceruloplasmin (Bertazolli-Filho *et al.*, 2006); (2) the proteases cathepsins D and O (Ortego *et al.*, 1997); (3) the atrial and brain natriuretic peptides (NPs) (Ortego and Coca-Prados, 1999); (4) angiotensin

(Savaskan *et al.*, 2004); (5) the anti-angiogenic factors, pigment epithelium-derived factor (PEDF) (Ortego *et al.*, 1996b), and chondromodulin-I (Funaki *et al.*, 2001); and (6) the growth factors transforming growth factor (TGFβ2) and epidermal growth factor (EGF) (Escribano *et al.*, 1994; unpublished results). This extraordinary capacity of the CE to synthesize so many secretory factors can only be explained within the context of a complex multicellular, multifunctional, and interactive tissue.

Cells releasing peptides in other tissues, including in various regions of the brain, occur in single or in small groups of cells. However, in the gastrointestinal tract where more than 18 different types of endocrine cells have been identified, they are widely distributed (Solcia *et al.*, 1987). These neuroendocrine cells receive chemical signals and then transduce them into hormonal signals for target cells.

One possible interpretation of the capacity of the CE to express and secrete "neural-like" peptides may rely on its common embryological origin with the multiple cell layers of the retina and the retinal pigment epithelium (RPE). The PE cell layer of the CE is continuous with the RPE, and the NPE is related to the multiple neural cells of the retina (Beebe, 1986). This relationship of the NPE cell layer with the neural and sensory cell layers of the retina is manifested by the expression of a large pool of retina-specific genes and retinal transcription factors along the CE. For example, genes restricted to rod-phototransduction continue to be expressed in the adult mammalian eye in the NPE (Bertazolli-Filho *et al.*, 2001; Ghosh *et al.*, 2004). In contrast, genes encoding proteins and enzymes that are components of the visual cycle, also known as the retinoid cycle, are expressed in the PE cell layer of the CE (Salvador-Silva *et al.*, 2005). In the RPE, the components of the visual cycle provide the necessary enzymatic machinery for regeneration of visual pigment. This permits the photoreceptor cells in the retina to initiate the phototransduction signaling cascade upon absorption of light by rhodopsin. The functional significance of the expression of genes restricted to photoreception, or the retinoid cycle in the physiology and metabolism of the CE, is presently unknown, although a number of potential roles have been suggested including a novel nonvisual phototransduction mechanism, the regulation of gene expression, and circadian tasks, including AqH secretion and IOP (Escribano and Coca-Prados, 2002; Salvador-Silva *et al.*, 2005).

Thus, the expression and secretion of peptides and transmitters by the CE indicates that there is an alternative source of peptide in the AqH that is distinct from the autonomic or sensory nerves. Since many of the factors secreted by the CE fulfill functions in fluid homeostasis and cell-to-cell communication, it suggests an endocrine microenvironment by which the CE signals to the avascular tissues in the anterior segment of the eye. Because of the hypotensive and/or hypertensive effects of some of the peptides synthesized and released by the CE (i.e., NPs, endothelin), when tested either in the eye or in the cardiovascular system, it also suggests that these same factors may participate in the mechanism of regulation of AqH secretion and IOP.

The expression of neuropeptides by the human CE and cultured NPE cells is supported by the identification of neuropeptide-processing enzymes. These are involved in the endoproteolytic cleavage and maturation of larger propeptides into biologically active peptides. For example, prohormone convertases (PCs), which are usually restricted to neuroendocrine cells, have been found distinctly in NPE and PE cells (see following section). Table I shows the expression of neuroendocrine markers identified in the human CE.

IV. NEUROENDOCRINE CHARACTERISTICS OF THE BILAYERED CE

In the course of an extensive analysis of a large number of subtracted cDNAs isolated from a human CB library, our laboratory identified one clone (CBS-294) encoding the neuropeptide-processing enzyme carboxypeptidase E (CPE), [Escribano *et al.*, 1995; National Eye Institute Bank

TABLE I

Expression of Neuroendocrine Markers and Hormones in the Human CE

Neuropeptide-processing enzymes
Prohormone convertases: PC1, PC2, PACE4, PC5, Furin
Carboxypeptidase E (CPE)
Peptidylglycine α-amidating monooxygenase (PAM)
Neuropeptides and peptide hormones
7B2
Secretogranin II (SgII)
Neurotensin (NT)
Galanin (Gal)
Somatostatin (SST)
Natriuretic peptides (ANP, BNP)
NPY
Angiotensin II
Endothelin
Substance-P

(NEIBank) http://neibank.nei.nih.gov]. This enzyme is involved in the processing and maturation of neuropeptide precursors (pro-neuropeptides) into biologically active peptides, and is selectively found in neuroendocrine cells (Fricker *et al.*, 1986; Fricker, 1988). This finding led us to further explore whether additional neuropeptide-processing enzymes, and specific neuroendocrine neuropeptide precursors, were expressed in the CE. Typically, the processing and maturation of pro-neuropeptides into biologically active peptides involves the coordinated participation of a set of proteolytic enzymes such as the prohormone convertases PC1 and PC2 which introduce cleavages on the C-terminal side of dibasic amino acids (Lys-Arg). This cleavage generates a C-terminal dibasic extension that is removed by CPE before the mature peptides can be secreted extracellularly. Often, if the peptide after CPE intervention ends in glycine (Gly), another enzyme, the peptidyl-glycine-α-amidating monoxigenase (PAM), removes the Gly residue generating an amidated carboxyl group. Approximately 50% of all the biologically active neuropeptides known so far are amidated including calcitonin-gene-related peptide (CGRP), neuropeptide Y (NPY), and galanin, all found in the AqH. In addition to CPE, the human CE expresses the following neuropeptide-processing enzymes: furin, PC1, PC2, 7B2 (a cofactor of PC2), and PAM. PC1 and PC2 are found exclusively in neural and endocrine cells including the gut and the brain where they are involved in the intracellular processing of neuropeptides including pro-neurotensin (pro-NT) and pro-somatostatin (pro-SST). PC1 and PC2 have been localized distinctly along the PE and the NPE cell layers, respectively, suggesting that both cells layers might be capable of posttranslational-processing pro-neuropeptides in a cell-specific fashion (Ghosh *et al.*, 2006).

Because of the cell-to-cell communication between PE and NPE cells through gap junctions, intermediate forms of immature neuropeptides (lower than 5000 in molecular mass) likely move from one cell to an adjacent cell where they undergo further processing. Gap junctions represent a major conduit in the transfer of ions and water from the PE cells to NPE in AqH secretion, and in the transmission of cell signaling along the entire CE. Some neuropeptides in the AqH are detected in relatively high levels, including SgII and its processed form SN (Stemberger *et al.*, 2004; Troger *et al.*, 2005). Transcripts encoding SgII have been identified in NPE cells, and SgII or a SgII-like propeptide have been immunolocalized along the NPE cell layer (Ortego *et al.*, 1996a).

Proteolytic processing of neuropeptide precursors into biologically active peptides is a major mechanism of regulation. The endoproteolytic cleavage of pro-NT, a 169-amino acid polypeptide, gives rise to several peptides of different sizes, including the tridecapeptide NT and the hexapeptide neuromedin N (NN), respectively. The biologically active peptides are then stored in secretory vesicles before they are secreted by either the constitutive secretory pathway (CSP) or the regulated secretory pathway (RSP). Earlier studies have documented the identification and separation of NT and NN from tissue extracts prepared from the iris-ciliary complex (Elbadri *et al.*, 1991; Hayes *et al.*, 1992). Interestingly, in these tissues the level of NT was higher than NN, which contrasted the retina where the level of NN is higher than NT (Hayes *et al.*, 1992). This distinct, tissue-specific, posttranslational processing suggests that the processed NT and NN forms may display different functions within the eye. NT is released in the gastrointestinal track by specialized neuroendocrine cells (N cells) and enteric neurons, and its action is mediated by multiple NT receptors (NTRs). In the gut, NT is a pro-inflammatory peptide, and it is associated with fatty acid translocation, gut motility, and stimulation of growth of normal gut mucosa. However, in the cardiovascular system, it has been shown to exhibit both hypotensive and hypertensive effects (Gully *et al.*, 1996). Under serum-free conditions, cultured ciliary-derived cells secrete and accumulate NT, or an NT-like immunoreactive material, in the culture medium (Ortego and Coca-Prados, 1997). The function of NT in the CE or in the AqH has not been studied. NTRs are found in the CE and in the TM, suggesting putative autocrine, and paracrine functions of NT in these tissues (Ortego and Coca-Prados, 1997). In the human colon, NTRs have been found to mediate chloride secretion in response to NT, and this effect is blocked by antagonists of adenosine receptors (Riegler *et al.*, 2000).

Another neuroendocrine peptide identified in the CE is SST. This peptide is widely distributed in the central nervous system (CNS), gastrointestinal tract, and endocrine cells. In the rat gut, SST accounts for approximately 65% of the total SST-like immunoreactivity, whereas 25% in the brain, 5% in the pancreas, and 5% in the remaining organs (Patel and Reichlin, 1978). SST regulates many cellular functions, including the secretion of hormones, neuronal excitability, and vascular smooth muscle contractility. In the gut, SST inhibits the release of every hormone that has been tested. In addition, SST is known for its inhibitory effect on the secretion of growth hormone and thyroid-stimulating hormone. In the mammalian retina, SST-like immunoreactivity has been detected sparsely (Johnson *et al.*, 2000). It has been suggested to serve as a trophic factor in the development of the retina, and possibly as an important regulator of synaptic communication in the retina (Cristiani *et al.*, 2002; Thermos, 2003). SST-like immunoreactivity has also been detected along the axons innervating the iris sphincter cells and CM cells (Firth *et al.*, 2002).

SST derives from a 92-amino acid polypeptide precursor, known as pro-SST, which undergoes processing at both the N- and C-terminal sites of specific pair basic amino acids into multiple shorter forms, including the biologically active peptides SST-28 and SST-14 (Patel, 1999). The endoprotease furin and the proprotein convertases PC1 and PC2 cleaved pro-SST into SST-28 and SST-14 at specific residues with different efficiency (Galanopoulou *et al.*, 1993, 1995; Brakch *et al.*, 1995). Since both PC1 and PC2 are expressed in the CE (Ortego and Coca-Prados, 1997, 1999; Ortego *et al.*, 2002; Ghosh *et al.*, 2006), and they are distinctly localized along the PE cells and NPE cells, it suggests that pro-SST is likely to be processed differently within the CE. The inhibitory activity exerted by SST on endocrine and exocrine secretions is mediated through at least five distinct SSTRs (1–5), of which SSTR1 and SSTR2 are expressed in the CE (Klisovic *et al.*, 2001). These receptors are also expressed in the iris and retina, and they are coupled to G_i-proteins, and, upon activation, can negatively couple to the adenylyl cyclase-cAMP pathway to inhibit stimulated but not basal cAMP production, and result in inhibition of ion exchangers (Patel, 1999). SST elicits an attenuation of the NHE activity in the intact CE by multiple intracellular signaling pathways, including the PI3 K/Akt pathway and phosphorylation of the endothelial nitric oxide synthase (eNOS) at a specific residue within the calmodulin regulatory site of the enzyme (Ghosh *et al.*, 2006). SST has been detected in the AqH, and it has been shown also to exhibit immunosuppressive properties that may contribute to ocular immune privilege (Taylor and Yee, 2003).

Galanin, a 30-amino acid neuropeptide originally isolated from porcine small intestine (Tatemoto *et al.*, 1983), is widely distributed throughout the mammalian neural and endocrine systems (Bedecs *et al.*, 1995). It has also been found (mRNA and peptide) in excised ciliary processes, and in a human cell line established from the NPE (Ortego and Coca-Prados, 1998). Interestingly, galanin receptor type 1 is also expressed in human NPE cells, predicting an autocrine mechanism of action in these cells (Ortego and Coca-Prados, 1998). These studies also indicate that pro-galanin in the AqH is locally synthesized, processed, and secreted by the CE. Therefore, galanin in the CE exhibits a distinct origin in contrast to galanin derived from the sensory nerve terminals in the uvea (Firth *et al.*, 2002). Galanin exerts multiple biological functions in the gastrointestinal tract, including the inhibition of gastric acid secretion and the release of numerous pancreatic peptides such as insulin, glucagons, and SST. In the hypothalamus, it is involved in osmotic regulation, as a survival- and growth-promoting factor for different types of neurons *in vitro* and *in vivo*, and in neurogenesis (Lang *et al.*, 2007). In the rabbit eye, galanin induces mydriasis by attenuating cholinergic neurotransmitter release (Yamaji *et al.*, 2003). However, when injected in the monkey eye, galanin has little effect on either outflow of AqH or IOP (Almegård and Anderson, 1990). Galanin, or a galanin-like product, has been detected in the AqH of human eye donors and in the culture medium of human NPE cells. Its secretion by cultured NPE is regulated by the action of catecholamines (Ortego and Coca-Prados, 1998).

Finally, in support of the neuroendocrine phenotype of the ocular CE is also its ability to metabolize steroid hormones and cortisone. The enzymes involved in this metabolism have been identified as members of the 17β- and 11β-hydroxysteroid dehydrogenase families. These enzymes are rate limiting in the synthesis of sex steroid hormones and cortisol (Coca-Prados *et al.*, 2003; Rauz *et al.*, 2003).

A. NPs: Ocular Modulators of Inflow and Outflow of AqH

Little is known on the nature of the endogenous effectors that regulate AqH and IOP. In clinical studies, NPs have been shown to display an ocular hypotensive effect (Wolfensberger *et al.*, 1994; Fernández-Durango *et al.*, 1999; Goldmann and Waubke, 1999). It has been suggested that NPs mediate this effect by either reducing the rate of AqH secretion by the CE and/or by increasing outflow through the TM (Pang *et al.*, 1994; Chang *et al.*, 1996). The natriuretic system (NS) is expressed in the ciliary processes. The NS consists of three NPs and three natriuretic peptide receptors (NPRs). The interaction of NPs with their cognate receptors plays important physiological roles in hypertension and cardiovascular pathophysiological disorders. NPs are composed of three peptide hormones: atrial, brain, and C-type NPs (ANP, BNP, and CNP); exhibit multiple functions in the cardiovascular system including lowering blood pressure; and contribute to the maintenance of fluid volume homeostasis. Although originally described in the heart, the NS is also expressed in many extracardiac tissues, including the kidney, adrenal glands, brain, and the eye. NPs are involved in cardiac endocrine function by establishing cross-talk communication between endocrine and contractility function of the heart. Within the ciliary processes, ANP and BNP are colocalized in the NPE cell layer, whereas CNP is distinctively distributed in the vascular endothelium (Fidzinski *et al.*, 2004).

The three NPs, ANP, BNP, and CNP, exhibit the ocular hypotensive effect by lowering IOP. The NPs action is mediated by three different NP-specific cell surface receptors: NPR-A, NPR-B, and NPR-C, of which the A- and B-types are associated with intrinsic guanylyl cyclase (GC) activities producing the second messenger cGMP. In contrast, the NPR-C does not have GC activity, functioning as a clearance receptor regulating the level of NPs in AqH. The three NPs have been detected in the AqH by RIA at levels higher than in plasma with the exception of ANP that is found in lower levels. Among the NPs, BNP is the most abundant in the AqH of human and rabbit eyes, followed by CNP and ANP in very low levels (Salzmann *et al.*, 1998; Fernández-Durango *et al.*, 1999; Ortego and Coca-Prados, 1999; DeBold and Bruneau, 2000; Potter *et al.*, 2004). ANP receptors are downregulated in rabbits with experimental glaucoma, and immunoreactive ANP levels in AqH are increased as the IOP is increased (Fernández-Durango *et al.*, 1990).

Recently, the potential role of NPs as paracrine modulators of AqH outflow has been suggested (Potter *et al.*, 2004). Bremazocine, a relatively selective agonist of κ opioid receptors, lowers IOP by increasing the level of CNP in AqH, and increasing outflow. This is consistent with the presence of B-type as the predominant NP receptor in the TM (Chang *et al.*, 1996; Fernández-Durango *et al.*, 1999) and the potent hypotensive effect exerted by CNP in the eye (Takashima *et al.*, 1998).

Cardiac ANP lowers blood pressure and stimulates diuresis and natriuresis by mechanisms involving vasodilatation, inhibition of renal Na^+ reabsorption, and inhibition of the sympathetic and renin-angiotensin-aldosterone system (Kuhn, 2003). In the ventricular myocardium, ANP and BNP production and secretion is regulated by complex interactions with the neurohormonal and immune systems, including endothelin-1, angiotensin-II, glucocorticoids, sex steroid hormones, thyroid hormones, growth factors, and cytokines (especially TNF-α, interlukin-1, and interlukin-6). Whether endocrine communications exist between NPs, and the aforementioned neurohormonal or immune systems, has yet to be explored in the ciliary processes.

B. *Inhibition of the NHE by NPs and Their Possible Role on IOP*

NPs influence ion transport activities of the NHE and $Na^+/K^+/Cl^-$ cotransporter in the kidney and the heart. These transport systems are believed to contribute to AqH secretion by the CE. The NHE is considered a key player in Na^+ absorption in the CE, the first step in AqH secretion (Civan, 1998), and it is distributed along the basal plasma membrane of PE and NPE cells (Fidzinski *et al.*, 2004). NHE is considered a molecular sensor of intracellular pH (pH_i) and cell volume regulation (Fig. 1). It serves as the principal alkalinizing mechanism in many cell types to protect them from the harmful effects of excessive acidification from metabolic acid generation or from H^+ accumulation. The NHE functions in parallel with bicarbonate transport systems, including the Cl^--HCO_3^- and the Na^+-HCO_3^- cotransporters, to maintain a cytoplasmic acid-base balance. The concerted action of these transporters results in cellular uptake of NaCl and water, and to cell swelling. Activation of the NHE by osmotic shrinkage has been described in many cells and tissues. It has been suggested that under hypotonic conditions, NHE activity could be inhibited whereas under hypertonic conditions, NHE can be fully activated (Rotin and Grinstein, 1989; Lang *et al.*, 1998).

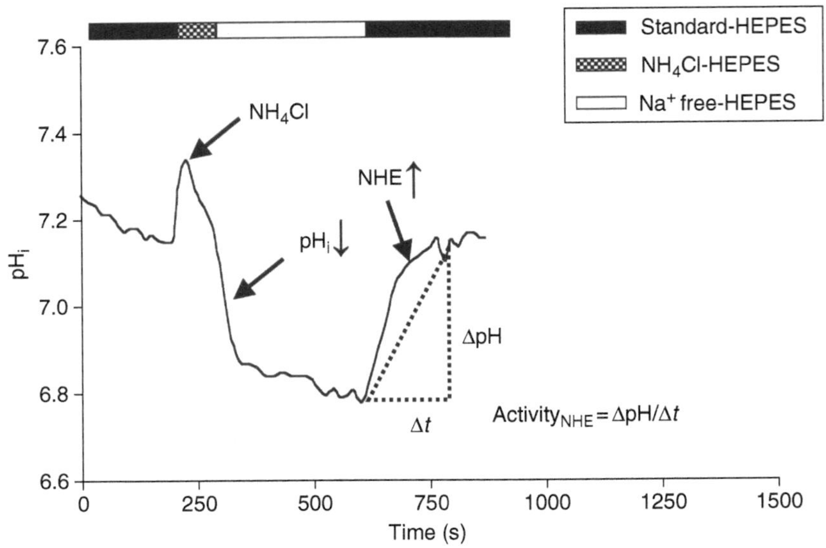

FIGURE 1 Typical intracellular pH (pH_i) recording of an NPE cell in an excised bovine ciliary process. NPE cells in the intact ciliary epithelium (CE) were loaded with protons by adding NH_4Cl as described (Fidzinski *et al.*, 2004). Following the removal of NH_4^+ and in the absence of Na^+ the cell acidified (pH_i ↓). Addition of Na^+ resulted in the cell realkalinization, a function of the Na^+/H^+ exchanger (NHE) activity, to a level observed before the NH_4^+ pulse. The rate of realkalinization (NHE ↑) was determined as $\Delta pH/\Delta t$. Horizontal bars indicate length of exposure in seconds (s) to buffer solutions as indicated.

Nine different NHE isoforms have been cloned so far, of which NHE1 is ubiquitous and the best studied. The NHE isoforms differ on the basis of their differences in amiloride sensitivity, regulation by several kinases, and their distinct distribution in polarized epithelial cells (apical or basal plasma membrane). In the CE, so far, the NHE1 is the only isoform identified, and has been immunolocalized along the basal plasma membrane of PE and NPE cells (Fidzinski *et al.*, 2004). In contrast, NHE2 and NHE3, in the intestinal and kidney epithelial cells, are localized along the apical plasma membrane, and they are involved in Na^+ absorption (Tse *et al.*, 1992, 1993). Because of the distinctive function of NHE1 as a molecular sensor of pH_i, cell volume, and transcellular transport of Na^+across epithelia exhibits a potential role in the regulation of AqH secretion. The NHEs are plasma membrane transport proteins that mediate the electroneutral exchange of extracellular Na^+ with intracellular H^+, with a stoichiometry of 1:1 (Aronson, 1985).

The NHE1 is an integral membrane protein composed of 12 membrane-spanning domains at the N-terminus and an intracellular hydrophilic C-terminus containing numerous canonical sites for phosphorylation and regulatory binding sites. NHE1 activity is inhibited by the diuretic amiloride and its analogues including benzoylguanidinium-based derivatives (e.g., HOE642, HOE694). NHE serves as the principal alkalinizing mechanism in many cell types to guard against the damaging effects of excess acidification from metabolic acid generation or from H^+ accumulation by different pathways. NHE maintains the cytoplasmic acid-base balance by working in parallel with bicarbonate-transporting systems (i.e., Na^+-HCO_3 cotransporters, Na^+-dependent HCO/Cl exchangers, and Cl^-/HCO_3 exchangers). NHE also provides a major conduit for Na^+ influx, coupled to Cl^-, and H_2O uptake which is required to restore cell volume to steady-state levels following cell shrinkage induced by acute elevations in external osmolality. A large number of factors modulate NHE activity by phosphorylation of a number of serine/threonine (Ser/Thr) kinases on residues localized on the distal region of the cytosolic C-terminus (Baumgartner *et al.*, 2004). A number of regulatory proteins are known to bind the cytosolic domain of the NHE, and either inhibit (i.e., tescalcin) or stimulate (i.e, calmodulin and carbonic anhydrase II) its activity.

In the first step in AqH secretion, in the uptake by the PE cells of NaCl, NHE1 plays a predominant role together with the AE. It is predicted that blocking one or the other antiport should reduce inflow, and thereby lower IOP. The NHE1 activation leads to a cellular influx of Na^+ ions, coupled to Cl^- and H_2O, and extrusion of H^+ ions. The osmostic influx of water will consequently lead to cell swelling. This makes the NHE a suitable candidate as a key regulator of IOP because it demonstrates powerful mechanisms for increasing cell volume. This interpretation is supported by experimental evidence that NPs lower IOP in experimental animals, as well as function as inhibitors of NHE (Avila *et al.*, 2002a,b). Although the current understanding of ionic transport mechanisms of AqH secretion also includes the bumetanide-sensitive Na^+-K^+-$2Cl^-$ cotransporter, and the Na^+,K^+-ATPase, the regulation of the NHE1 exchanger appears to be pivotal in the mechanism of action of the endogenous peptides. The NHE2 isoform is also capable of regulating pH_i, and cellular volume in a manner similar to NHE1. However, the activation of this isoform differs in its apparent sensitivity to pH_i, and differs in its sensitivity (NHE1>NHE2) to inhibition by amiloride, and by benzoylguanidium compounds (i.e., HOE694). NHE2 is highly sensitive to inhibition by extracellular protons. It is strongly upregulated by the increase of extracellular pH.

NPs have been shown to either activate NHE or to inhibit it. In the CE, NPs effects have been examined by recording the pH_i responses of the NPE cells in excised ciliary processes. NPs exhibit a profound inhibition of NHE activity in the CE, which may explain their ocular hypotensive effect by a reduction in AqH secretion. Using the pH_i sensitive dye 2′7′-bis

(2-carboxyethyl)-5(6)-carboxyfluorescein, it has been determined that the rate of realkalinization (a function of NHE activity) of NPE cells, following their acidification with NH_4Cl, was attenuated by the three NPs in a pharmacological order of potency, indicating that CNP was the most potent, followed by ANP and BNP (Fidzinski *et al.*, 2004). This pharmacological profile is identical to their ability to activate GC in cultured NPE cells, suggesting that NPR-B receptors mediate these effects *in vivo* and *in vitro* (Ortego and Coca-Prados, 1999). This interpretation is also consistent with the ability of lysophosphatidic acid (LPA), an NPR-B receptor blocker, to reverse the CNP effect on the NHE (Abbey and Potter, 2003) (Fig. 2).

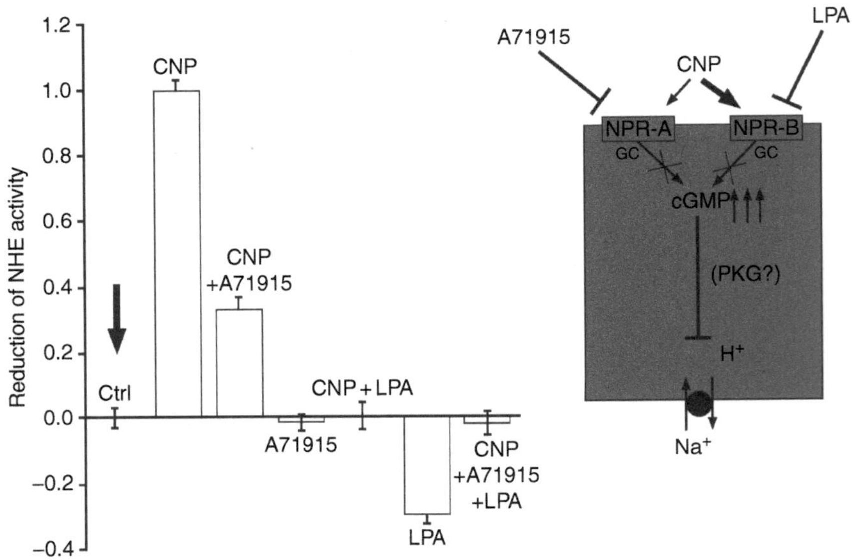

FIGURE 2 Mechanism of action of CNP in the ciliary epithelium (CE). CNP is the most potent natriuretic peptide to block NHE activity (given the value of 1) in the intact CE. When A71915 (1 μM), an NPR-A blocker, was added in the presence of CNP, it reversed over 60% the inhibitory effect elicited by CNP (100 nM) on NHE activity. However, lysophosphatidic acid (LPA, 10 μM), an NPR-B blocker, completely reversed the effect of CNP. The partial reversal by A71915 was complete when added together with LPA (CNP + A71915 + LPA) supporting the view that NPR-B is the main natriuretic receptor in the CE. LPA but not A71915 when added alone exhibited a stimulatory effect on NHE activity (Fidzinski *et al.*, 2004). The scheme on the right summarizes the cell mechanism by which CNP attenuates the Na^+/H^+ exchanger (NHE) activity in the CE. CNP acting preferentially on NPR-B receptors in the NPE cells (and in a lesser level on NPR-A) results in an increased intracellular cGMP production. Whether a cGMP-dependent protein kinase (PKG) is involved downstream to attenuate NHE activity is not known. Blockers of NPR-A (A71915) or NPR-B (LPA) receptors prevent with different efficacy the attenuation of NHE activity.

Concerted NHE inhibition in the CE by NPs predicts a reduction of AqH secretion and IOP. This interpretation is also consistent with the ocular hypotensive effect of NHE inhibitors on anaesthetized mice (Avila *et al.*, 2002b). Since NHE is ubiquitous and NPR-B receptors are also expressed in the TM, NPs can modulate NHE activity in these cells in a similar fashion as in the CE and influence inflow and outflow of AqH and IOP.

V. NEUROENDOCRINE PHENOTYPE OF THE TM

Studies on the gene program in the human TM have revealed that this tissue expresses genes that encode molecular markers restricted to neuroendocrine cells (González *et al.*, 2000a; Borrás, 2003). Among these markers, figured: SgII and the neuroendocrine processing enzyme neuron-specific enolase. SgII is a prototype of secretory peptide present in dense-core vesicles of many neuroendocrine and nervous tissues (Fischer-Colbrie *et al.*, 1995). Earlier studies have shown the localization of a discrete cluster of cells that lie circumferentially in the TM of the rhesus monkey, and rat eyes that are labeled with the neuron-specific enolase, a marker restricted to neurons and neuroendocrine tissues (Stone *et al.*, 1984; Nucci *et al.*, 1992). The potential capability of the human TM to exhibit neuroendocrine characteristics is also supported by the expression of neuropeptide receptors including SST receptor type 2 (SSTR2), NTRs types 1 and 3, the neuropeptide-processing enzymes (PC1), the neuroendocrine protein (7B2), which confer neuroendocrine specificity, and the neuroendocrine peptide SgII (Fig. 3 and Table II).

The neuroendocrine features of the TM, as those proposed for the CE, are best explained within an endocrine microenvironment in the anterior segment. The neuroendocrine factors synthesized and secreted by these cells may target primarily their cognate receptors on their own peptide-producing cells by autocrine mechanisms. However, the possibility that neuropeptides released by the CE, and carried out by the AqH, which may target receptors on cells of the conventional outflow pathway by endocrine/paracrine mechanisms, is predictable. The activation of the neuropeptide receptors found in TM cells elicits cellular responses including contraction/relaxation, a property that is characteristic of the TM cells (see review by Wiederholt *et al.*, 2000). A characteristic of many neuroendocrine cells is the presence of secretory granules in their cytoplasm. The formation of secretory granules is a critical step in the packaging of peptides, and their peptide-processing enzymes before their release by exocytosis following the RSP. Further studies are required to characterize the putative neuroendocrine phenotype of TM cells and to determine whether these cells exhibit characteristic organelles to support storage of neuropeptide and peptide hormones.

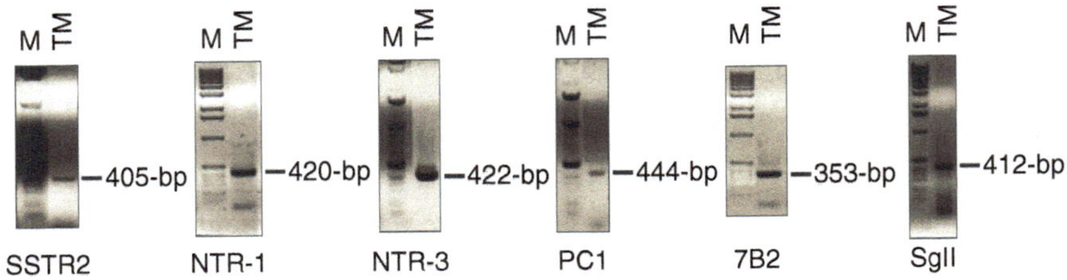

FIGURE 3 Expression of neuroendocrine markers and peptide receptors in TM cells. Complementary DNA synthesized *in vitro* from polyA$^+$ mRNA from cultured TM cells was used as template to amplify by RT-PCR DNA fragments for somatostatin receptor type 2 (SSTR2), neurotensin receptors type 1 (NTR-1) and type 3 (NTR-3), the prohormone convertase 1 (PC1), the neuroendocrine cofactor 7B2, and the neuroendocrine peptide secretogranin II (SgII). Oligonucleotide primers shown in Table II were annealed to cDNA template from TM cells and the predicted DNA products amplified by PCR. The DNA fragments amplified were gel purified, sequenced, and their nucleotide verified to share 100% homology with the respective cDNA sequences published in the GenBank database.

VI. REGULATION OF NEUROENDOCRINE SIGNALS: THE POTENTIAL ROLE OF NEUTRAL ENDOPEPTIDASE 24.11 (NEPRELYSIN)

In many neuroendocrine systems there is a feedback mechanism to terminate a specific endocrine signal. In the context of inflow-outflow of AqH, we have suggested a neuroendocrine communication between CE and TM. In the AqH, neuropeptides and hormones could survive degradation by proteases if bound to carrier proteins that protect them from inactivation. A number of peptide carriers are present in the AqH that could fulfill this function including α2-macroglobulin and albumin. These proteins are known to render protection to cytokines and peptides from degradation.

One effective mechanism to regulate communication between cells and tissues is by regulating the level of extracellular peptide. This could be accomplished in part by enzymatic clearance. In the case of NPs, which are potential endocrine signals between inflow and outflow, they can be internalized by the NPR-C receptors, expressed in TM cells (Chang *et al.*, 1996), or they could be inactivated by the membrane bound neutral endopeptidase 24.11 (NEP, EP24.11) also known as neprilysin. This enzyme, a metallopeptidase, is abundant in the brain (Akiyama *et al.*, 2001) and in the tissues associated to inflow and outflow of AqH. It is known by its specificity in degrading extracytoplasmic peptides including NPs, NT, SST, NPY, and the cleavage of the amyloid β-protein (Aβ), all of which are detected in the AqH. It has been observed that SST regulates neuronal neprilysin activity by up regulating its gene expression, resulting in an increased degradation of the specific substrates including Aβ (Saito *et al.*, 2005). Interestingly, by

TABLE II
Oligonucleotide Primers of Neuroendocrine Markers Tested by RT-PCR in TM Cells

Gene	Forward/Reverse	Product length (bp)	Annealing temp (°C)	GenBank Acc. No.
Somatostatin receptor type 2 (SSTR2)	5′-TCGGCCAAGTG GAGGAGAC-3′ 5′-GATGGCCACTGA GACCGAAGAGAC-3′	405	57.9	L06613
Neurotensin receptor type 1 (NRT-1)	5′-CATGTCCC-GAAGCCGCACCAA GAA-3′ 5′-TAGGGCAGCCAG CAGACCACAAAG-3′	420	61	NM_002531
Neurotensin receptor type 3 (NTR-3)	5′-CACAGGCGGGGA GACGGACTTTAC-3′ 5′-CAATGGCCACGAT GATGC-3′	422	56.9	AF175279
Prohormone convertase1 (PC1)	5′-TGGCTGAAAGA GAACGGGAT ACAT-3′ 5′-ATTGCTTTGGCGGT GAGTTTTTAC-3′	444	57.8	M90753
7B2 (Chaper-on protein of PC2)	5′-CACCAGGCCAT GAATCTTG-3′ 5′-CTCCGCTTGCG TCTCTGTCCTC-3′	353	55.8	M23654
Secretogranin II (SgII)	5′-CTACCAGACGGGCT CAGTGTTG-3′ 5′-GGGCCAGCTTGT CAGTCTCCT-3′	412	56	M25756

upregulating neprilysin activity the proteolytic degradation of SST also occurs, suggesting a negative feedback mechanism of regulation of endocrine peptides (Iwata *et al.*, 2004; Fig. 4).

Selective neprilysin inhibitors prevent the degradation of NPs, both *in vitro* and *in vivo*, resulting in the increased biological activity of NPs. Neprilysin is also involved in the enzymatic conversion of big endothelin (ET-1) to its active form, the vasoconstrictor peptide ET-1. Thus, the balance of effects of neprilysin inhibition will depend on whether the predominant endocrine signals in the AqH degraded by neprilysin are vasodilator (i.e., NPs) or vasoconstrictors (i.e., angiotensin II, ET-1). Another important observation is that SST has been found to modulate the proteolytic activity of neprilysin

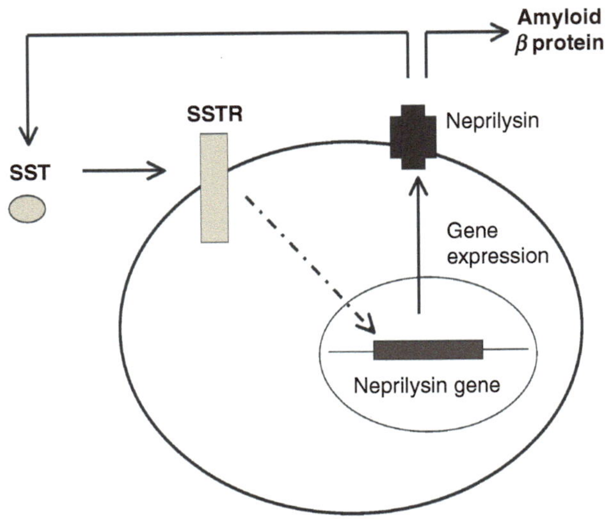

FIGURE 4 Neprilysin, a potential regulator of endocrine signals in the aqueous humor (AqH). Neprilysin (NEP, EP24.11), a metallopeptidase highly abundant in cells of the inflow and outflow pathways of AqH is highly specific in the cleavage and degradation of extracytoplasmic peptides, including amyloid β protein [Aβ, natriuretic peptides, neurotensin, and somatostatin (SST)] secreted by the ciliary epithelium (CE). SST regulates neprilysin activity by upregulating its gene expression, resulting in an increased degradation of Aβ and of other specific peptides including SST, suggesting a negative feedback mechanism of regulation of endocrine peptides.

by making the enzyme more effective in degrading its substrates, as suggested in the brain (Barnes *et al.*, 1995; Iwata *et al.*, 2004). This indicates a feedback endocrine signaling between peptide level and endocrine signaling in the target cell. Thus, the expression on the one hand of SST, neprilysin, and NPs in the CE, and on the other hand of SST receptors, and neprilysin in the TM, suggests cell-to-cell communication between inflow and outflow by endocrine signals. The role of neprilysin in the regulation of endocrine peptide and hormone levels, and its influence on the outflow of AqH could be highly significant.

VII. NEUROENDOCRINE SIGNALING IN THE CE AND TM

The L-arginine-nitric oxide (NO) pathway has been shown to participate in the regulation of AqH outflow (Wiederholt *et al.*, 1994). NO donors display hypotensive effects by lowering IOP (Kotikoski *et al.*, 2002). NO is synthesized by NO-synthases (endothelial eNOS, inducible iNOS, and neuronal nNOS) from L-arginine (L-Arg). NO formation activates the soluble guanylate cyclase (sGC) enzyme, which results in increased levels of cyclic guanosine 3′,5′-monophosphate (cGMP). Earlier studies have shown that tissues involved in inflow and outflow of AqH are enriched with sites of NO synthesis (Nathanson and McKee, 1995; Geyer *et al.*, 1997).

There is general agreement that increased cGMP production is associated with a reduction in IOP by NO donors, by decreasing AqH secretion or by increasing the outflow facility. A number of ligands synthesized and released by the CE act on their autoreceptors probably in the same peptide-producing cells, and in cells of the outflow system including the TM cells via the NO/cGMP signaling pathway. Endothelin (ET-1), a potent vasoactive peptide, is one of these ligands released by the human NPE cells, and acting on ETA and ETB receptors, expressed in the tissues involved in inflow and outflow AqH regulation (Osborne *et al.*, 1993; Tao *et al.*, 1998; Fernández-Durango *et al.*, 2003). ET-1 induces iNOS expression in cultured human NPE cells (Prassana *et al.*, 2000) and increases perfusion pressure in human eyes, which results in an upregulation of iNOS gene expression, and suggests an increase in outflow facility (Schneemann *et al.*, 2003).

Glutamate has also been shown to modulate the production of NO (Kosenko *et al.*, 2003), and preliminary studies indicate that glutamate receptors are expressed in the CE and TM cells. So far, there is no work addressing the intracellular signaling of glutamate on the intrinsic control of eNOS to modulate the production of NOS in CE or TM. Physiologically, the TM cells are exposed to the hemodynamic forces of AqH, metabolic and oxidative stress, and to endocrine peptides and growth factors, released by the CE. Shear stress, certain peptides, and steroids mediate their action via the PI3 K/Akt signaling pathway leading to eNOS activation via phosphorylation of specific Ser and Thr amino acid residues in the calmodulin (CaM) and reductase domains of the enzyme. Recent studies have investigated the time course and rate of protein phosphorylation of eNOS in the CE, and provided evidence of unique sites of phosphorylation on eNOS Ser^{617} but not on Ser^{1179} by SST (Ghosh *et al.*, 2006). These studies raise the possibility that mediators released by the CE might influence eNOS activity in cells of inflow and outflow pathways.

VIII. PUTATIVE GLUTAMATERGIC SYSTEM IN THE INFLOW-OUTFLOW AXIS: GLUTAMATE AS A FUNCTIONAL ENDOCRINE/PARACRINE SIGNAL BETWEEN CE AND TM CELLS

L-Glutamate has many important physiological functions including its role as neurotransmitter in the retina and in the CNS. In the retina, glutamate is the most prevalent neurotransmitter for excitatory synaptic transmission.

It serves additional functions as a metabolic substrate for other neurotransmitters [i.e., γ-amino butyric acid (GABA)], and an amino acid for general cellular metabolism. On the other hand, glutamate transporters maintain low extracellular glutamate and influence the kinetics of glutamate receptor activation.

Recent studies have revealed that several of the genes associated with the function of L-glutamate as a neurotransmitter in the CNS are also expressed in a number of peripheral tissues, including bone, pancreas, gastrointestinal tract, testes, and RPE, where it may play a role as an autocrine or paracrine signal molecule (Miyamoto and Del Monte, 1994; Skerry and Genever, 2001; Hayashi *et al.*, 2003; McGahan *et al.*, 2005). Bone cells express functional glutamate receptors and glutamate transporters which are electrophysiologically and pharmacologically similar to those in the CNS (Kalariti and Koutsilieris, 2004). Although the significance of glutamate signaling in these non-neuronal tissues is not fully understood, the regulation of glutamate transporters in response to mechanical pressure has implicated glutamate signaling in the transmission of mechanical stimuli in osteocytes, and in the response of these cells to their mechanical environment. In the stomach, glutamate has been shown to modulate histamine-induced acid secretion, and contractility in different parts of the intestine (Shannon and Sawyer, 1989; Tsai *et al.*, 1999). In pancreatic α and β cells, glutamate has been suggested to modulate the secretion of insulin and glucagon.

Preliminary studies indicate that glutamate receptors, glutamate transporters, and enzymes (glutaminase and glutamine synthase) of the glutamate cycle are also expressed in the human CE and TM cells (unpublished results). The characterization of a putative glutamatergic system within tissues involved in inflow-outflow of AqH will be of unique relevance since extracellular functions of glutamate as an excitatory neurotransmitter in cell-to-cell communication and as an intracellular signal is well documented.

Studies by 3′-*r*apid *a*mplification of *c*DNA *e*nds (RACE) of a human CB cDNA library have revealed the identification of an alternative splicing variant of the metabotropic glutamate receptor mGluR1 (Fig. 5). The wild mGluR1 receptor is abundant in the retina, it is a GTP-binding protein that can couple positively to phospholipase C, usually leading to mobilization of Ca^{2+}. However, the human variant identified in the CB, and in cultured human NPE cells exhibited 100% similarity with the human splicing form mGluR1b (Lin *et al.*, 1997). This splice variant contains an 85-bp insertion at the intracellular C-terminus, between nucleotides 2660 and 2661, of the human wild-type mGluR1 form. The insertion is in frame, but it introduces a premature stop codon (TGA), within this extra 85 bp, that results in a replacement of the last 313 C-terminal residues in the mGluR1 wild form by only 20 residues (Fig. 5). The mGluR1b is one of four other alternatively

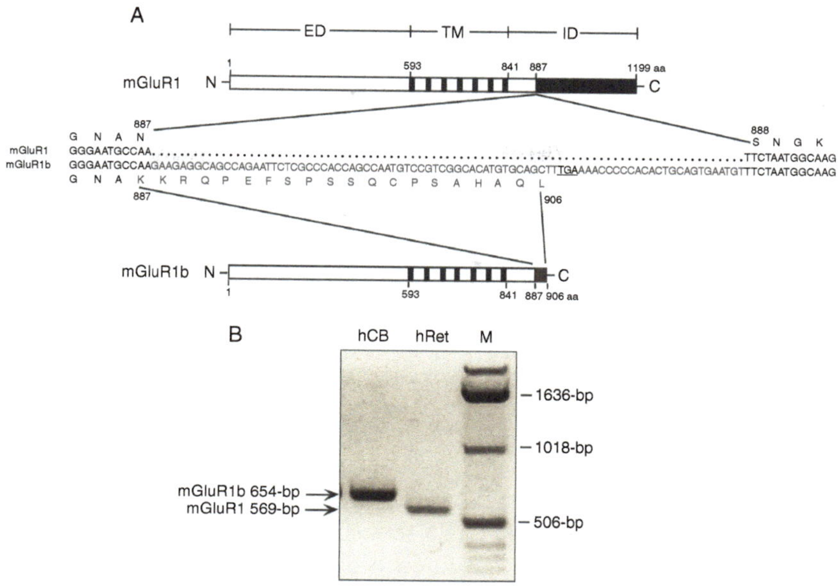

FIGURE 5 (A) Expression of the metabotropic glutamate receptor splice variant 1 b (mGluR1b) in the human ciliary body. Schematic representation of the coding region of mGluR1 [1199 amino acids (aa)] and of the splicing variant mGluR1b (906 aa). mGluR1b conserves both the extracellular domain (ED), and the transmembrane (TM) domains of mGluR1. However, at the C-terminus within the intracellular domain (ID), mGluR1b contains an 85 nucleotide insertion (indicated below in red). This insertion is in frame, but it introduces a premature STOP codon (TGA) within this extra 85-bp that results in a replacement of the 313 C-terminal amino acids (887–1199) by only 20 amino acids (887–906). (B) PCR amplification of the 3′-end region comprising the 85-bp insert present in mGluR1b from a human ciliary body (hCB) cDNA library (Escribano *et al.*, 1995). Primers encompassing the 85-bp insertion were annealed to DNA from an hCB library or to cDNA synthesized *in vitro* from polyA$^+$ RNA from a human retina (hRet). At the optimal annealing temperature of 59 °C a DNA product of 654-bp was amplified in the hCB library and a DNA product of 569-bp in the hRet. Gel extraction, purification, and DNA sequencing of both DNA size products indicated that they exhibited 100% homology with the splice variant mGluR1b (654-bp) and with the wild form mGluR1 (569-bp). Oligonucleotide primers used forward: 5′-CAACGTGCCCGCCAACTTCAAC-3′; reverse: 5′-GCTGGGCATCCTCCTCCTCCTCTA-3′ from mGluR1 cDNA (GenBank accession number NM_000838). (See Color Insert.)

spliced variants of mGluR1 described thus far (Pin *et al.*, 1992; Tanabe *et al.*, 1992). The functional significance of the splicing variant mGluR1b in the CE awaits further study. However, several differences have been described, including slower and more prolonged Ca^{2+} responses, distinct intracellular targeting, and regulation by different cellular kinases (Chan *et al.*, 2001; Mundell *et al.*, 2004).

A. *Expression in the Human CB of Glutamate Transporters of the Excitatory Amino Acid Transporters Family*

In the retina L-glutamate is released by photoreceptors, bipolar cells, and retinal ganglion cells (Lam, 1997; Thoreson and Witkovsky, 1999). Whether L-glutamate in the AqH derives from the retina is presently unknown. However, based on preliminary studies, it is not premature to speculate whether a putative glutamatergic system is expressed in the inflow–outflow pathways of AqH. Glutamate transporters, members of the excitatory amino acid transporters (EAAT) family, mediate the uptake of glutamate and exhibit currents that depend on extracellular Na^+. Studies suggest that this complex process also involves the cotransport of protons (H^+) and gates the passive flux of Cl^- ions (Arriza *et al.*, 1997).

Screening of a human CB cDNA library, by PCR, with specific sets of oligonucleotide primers spanning coding regions of genes encoding distinct members of the EAAT family, has revealed the surprising amplification of EAAT1 and EAAT5 sequences. The expression of EAAT5 in the human CB was further examined by Northern blot, and the mRNA (3.1-kb) size of EAAT5 in this tissue was identical to the one described in the human retina (Arriza *et al.*, 1997; Ghosh and Coca-Prados, unpublished results). In the rat, rabbit, and monkey retinas, EAAT5 is restricted to rod photoreceptor cells (Pow and Barnett, 2000), in cone and rod photoreceptor terminals, and in axon terminals of rod bipolar cells in the mouse retina (Wersinger *et al.*, 2006). The large Cl^- conductance of EAAT5 suggests that it functions more like a ligand-gated Cl^- channel than a glutamate transporter. Cl^- is mainly translocated in the presence of glutamate or related substrates, and Cl^- movement is not thermodynamically coupled to substrate transport.

In certain forms of glaucoma, the expression of the glutamate transporter, EAAT1 (GLAST) has been shown to be expressed at reduced levels (Naskar *et al.*, 2000) and suggested to be associated with retinal ganglion cell death (Vorwerk *et al.*, 2000). Whether the glutamate receptors and glutamate transporters expressed in the CB are functional or play a role in disease, awaits further studies at the cellular, molecular, and electrophysiological levels. Finally, two enzymes known to be associated with the glutamate cycle were also detected in the human CB and TM cells. Of these enzymes, glutamine synthase, involved in the conversion of glutamate into glutamine, was described to be present in the NPE of the CE, and suggested to be involved in the synthesis of multipolysaccharides (Riepe and Norenberg, 1978). The second glutamate enzyme in the CB is glutaminase that converts glutamine into glutamate.

IX. IMPLICATIONS OF A NEUROENDOCRINE SIGNALING IN THE ANTERIOR SEGMENT OF THE EYE

A. *Potential Neuroendocrine Entrainment of Circadian Rhythms: AqH Secretion and IOP*

There is abundant evidence indicating that in the mammalian eye, AqH flow (Ericson, 1958; Reiss *et al.*, 1984; Topper and Brubaker, 1985; Smith and Gregory, 1989; Maus *et al.*, 1994), and IOP (Liu, 2001) exhibit well-defined diurnal rhythms. The rate of secretion of AqH in humans falls 50–60% during sleep (Brubaker, 1998), and IOP is higher during the night than during the day (Liu, 2001), suggesting distinct circadian signals modulating inflow and outflow during the day and during the night. Clarifying the mechanisms and regulation of circadian AqH secretion can lead to novel strategies for lowering IOP, the only known intervention in slowing the onset and progression of blindness in glaucoma. There is supporting evidence that factors found in AqH follow a circadian rhythm (Nii *et al.*, 2001; Liu, 2000, 2002). Although these factors remain to be determined, other factors known to be present in the AqH and in the CE are presumably circadian (Fig. 6). Among these factors are melatonin, and the melatonin rhythm-generating enzymes, arylalkylamine *N*-acetylytransferase (AA-NAT), and hydroxyindole-*O*-methyltransferase (HIOMT) (Martin *et al.*, 1992). Melatonin has been demonstrated to be circadian in the tissues where it is synthesized including the pineal gland and retina. A number of additional endocrine factors in the AqH are predictably circadian including SST and cortisol, both synthesized and secreted by the CE (Rauz *et al.*, 2003; Ghosh *et al.*, 2006). SST, for example, has been shown to be circadian in the cerebrospinal fluid and chemically similar to the AqH (Berelowitz *et al.*, 1981). In contrast, cortisol is the main circulating human glucocorticoid in plasma, exhibiting a diurnal rhythm, and a potent natural immunosuppressant activity. It has been suggested that cortisol may impose diurnal variation on immune responsiveness, similar to melatonin.

The circadian rhythmicity of AqH flow and IOP has been suggested to be synchronized by the master pacemaker (clock), the suprachiasmatic nucleus (SCN). The central clock in the SCN receives photic information from photoreceptor cells in the retinal inner layer through the retinohypothalamic tract (RHT) in the form of neural signals (the so-called input signals). The input signals (i.e., glutamate, nitric oxide) turn on the circadian clock genes and reset the clock (Ding *et al.*, 1997). The clock genes turn on the expression of other genes, the so-called clock-controlled genes, encoding output signals. Among the circadian output signals figure: endocrine/paracrine factors,

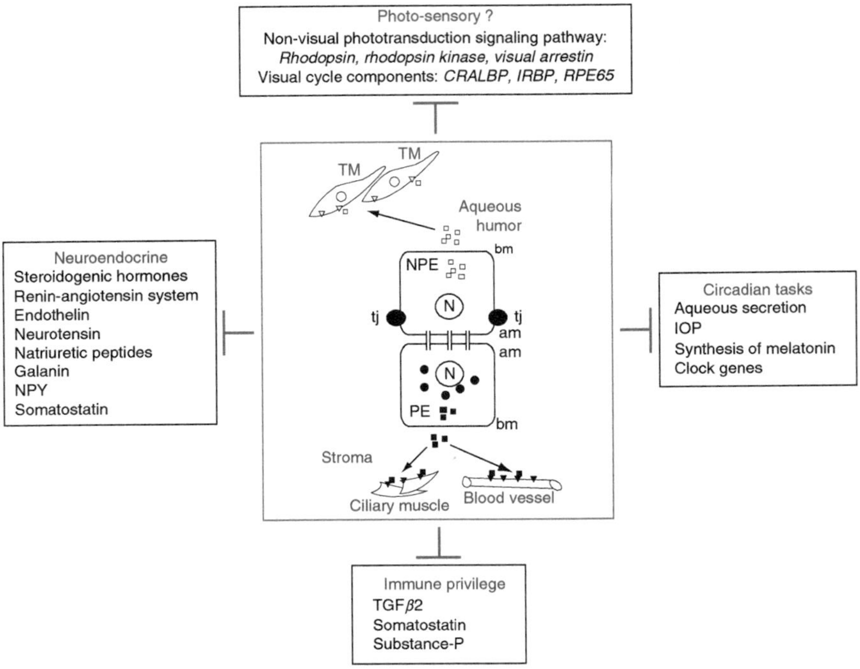

FIGURE 6 Potential endocrine systems linking inflow and outflow of aqueous humor (AqH). The multiple neuropeptides and hormones released by the ciliary epithelium, CE (PE and NPE cells) either into the AqH or toward the stroma may serve as endocrine messengers to establish cell-to-cell communication with tissues within the anterior segment of the eye. Of interest is the putative interaction between the CE and the trabecular meshwork (TM) cells in the outflow, the blood vessels in the stroma, and the ciliary muscle. Several steroid-metabolizing enzymes, including members of the 17β- and 11β-hydroxysteroid dehydrogenases, regulatory peptides, and hormone peptides expressed in the CE, confer steroidogenic and neuroendocrine functions to the CE. Some of the peptides released by the CE including TGFβ2, somatostatin, and substance-P are known to exhibit immunomodulatory properties and therefore they may contribute to maintain the immune privilege status of the anterior segment of the eye. On the other hand, the multiple components of rod-phototransduction (i.e., rhodopsin, rhodopsin kinase, and visual arrestin) and of the visual cycle (i.e., CRALBP, IRBP, and RPE65) found in the adult mammalian CE may underlie a novel local nonvisual photosensory system to regulate circadian tasks including the secretion of AqH and intraocular pressure (IOP). It is predicted that neuropeptides and hormones released by the CE in the AqH might be under the control of an internal circadian clock. The expression of melatonin-synthesizing enzymes arylalkylamine *N*-acetyltransferase (AA-NAT) and hydroxyindole-*O*-methyltransferase (HIOMT), melatonin receptors including Mel 1A and 1B, and the detection of melatonin in the CE strongly supports that circadian tasks such as AqH secretion and IOP may be regulated by endocrine factors. N = nucleus; tj = tight junctions; NPE = nonpigmented ciliary epithelial cell; PE = pigmented ciliary epithelial cell; TM = trabecular meshwork cells; am = apical membrane; bm = basal membrane; Δ, $\blacktriangle$ = receptors; $\square$, $\blacksquare$ = peptides. Filled circles in the cytoplasm of PE cell represent melanin granules.

norepinephrine (NE), cAMP, arginine-vasopressin, adenosine, and dopamine. These humoral and neural signals then synchronize circadian rhythm activities in peripheral tissues expressing circadian oscillators. In the adrenal gland, numerous genes, and entire pathways associated to catecholamine and steroid metabolism are transcriptionally regulated by an endogenous circadian clock in this tissue (Oster *et al.*, 2006). In recent years, the existence of circadian clocks in peripheral tissues (Zylka *et al.*, 1998; Balsalobre, 2002), and in dissociated cells (Balsalobre *et al.*, 1998), has been demonstrated. These peripheral clocks are self-sustained and cell-autonomous (Balsalobre *et al.*, 1998; Yamazaki *et al.*, 2000; Balsalobre, 2002). Most importantly, the circadian clock in SCN and in peripheral tissues share the same molecular components. However, there is a significant difference between oscillations in peripheral tissues, and in cells *in vitro* which damped after time, and those recorded from SCN neurons which maintain circadian rhythmicity for a longer time.

It has been suggested that the human CB contains a peripheral clock involved in circadian rhythm of AqH secretion. This hypothesis is supported, in part, by the expression in the CB of the clock genes: Period (*Per*), Clock (*Clock*), Cryptochrome (*Cry*), *BMAL 1*, and Timeless (*Tim*). In peripheral tissues, clock genes oscillate as in the SCN (Ripperger and Schibler, 2001); form feedback loops, resulting in oscillations of expression levels of clock, and clock-controlled genes with 24-hour cycles (King and Takahashi, 2000). Disruption of clock genes (i.e., *Cry1*) affects the diurnal rhythm of IOP (Maeda *et al.*, 2006). Cultured human ciliary epithelial cells (NPE) entrained to a 24-hour light-dark (LD) cycle (i.e., 12-hour L and 12-hour D) resets the circadian rhythm of the clock gene *Per1* by the action of the input signal NE, and exhibit circadian expression of cAMP, SST mRNA, and AA-NAT (unpublished results).

We also have hypothesized that the neuroendocrine functions of the CB are most likely linked to an endogenous peripheral circadian oscillator. The secretion of hormones and neuropeptides by the CE in the AqH are output signal candidates for clock-controlled genes, as shown in many peripheral clocks (Hastings *et al.*, 2003; Oster *et al.*, 2006). These output signals can potentially regulate AqH flow and IOP. Although the SCN may be important in the synchronization of circadian physiology in peripheral tissues, it has been shown that the input of the SCN or of photoreceptor (rods and cones) cells of the retina are not essential for maintaining the circadian rhythm in peripheral clocks. For example, the circadian AA-NAT expression and activity in the retina occur in the absence of SCN (Tosini and Menaker, 1998; Iuvone *et al.*, 2005). These studies suggest a high degree of autonomy in peripheral clocks which run with some degree of independence from the SCN.

The local influence of peripheral clocks on circadian physiology has two clear advantages. First, the central clock would need to maintain and/or reset the peripheral clocks through a limited number of ligands. And second, the local control of circadian rhythmicity would allow it to adapt rapidly. It has been suggested that key circadian proteins determine circadian periodicity in gene expression and physiology for many organs (Reppert and Weaver, 2002), and up to 10% of genes in different tissues are directly or indirectly regulated by the circadian clock system (Panda *et al.*, 2002; Storch *et al.*, 2002). Since different tissues have characteristic sets of genes with clock-regulated timing and amplitude of expression, it is likely that the circadian clock in the CB may be involved in the control of aqueous inflow and outflow with distinct signals.

The CB may be capable to autosustain circadian activity due to the expression of many components of phototransduction and of the retinoid cycle (Bertazolli-Filho *et al.*, 2001; Ghosh *et al.*, 2004; Salvador-Silva *et al.*, 2005). Melatonin is viewed as a hormonal message (output signal) for the duration of the dark phase in circadian rhythm. Melatonin receptors are also functionally expressed in the CE (Osborne and Chidlow, 1994), and under 24-hour LD cycle, the AA-NAT activity in CB explants is higher in the dark than in the light. Recent studies indicate that melatonin receptors (MT3) activation expressed in the CB could modulate IOP reduction (Serle *et al.*, 2004). Table III summarizes several biological functions assigned to the human CE.

B. Neuroendocrine-Immune Circuitry

Neuropeptides in the AqH may exhibit biological activities associated with a putative neuroendocrine-immune circuit in the anterior chamber (Fig. 6). It has been well recognized that the anterior segment of the eye is an immune-privileged site, and that antigens injected into the anterior chamber elicit deviant systemic immune responses that are devoid of immunogenic inflammation. This distinctive response known as anterior chamber-associated immune deviation (ACAID) arises in part by the soluble immunosuppressive and anti-inflammatory factors released by the surrounding tissues into the AqH. The AqH is able to suppress interferon-γ (INF-γ) production by effector T cells, and neuropeptides present in the AqH including α-MSH, CGRP, VIP, and SST, which are capable of suppressing pathogen-induced inflammation in the anterior chamber of the eye. SST, for example, contributes to the immunosuppressive properties of the AqH by promoting the production of the potent immunosuppressive cytokine α-MSH, and by inducing the activation of regulatory T cells (Taylor and Yee, 2003).

TABLE III

Biological Functions Assigned to the Human Ciliary Epithelium

- Synthesis and secretion of plasma proteins, proteases, anti-proteases, anti-angiogenic, and growth factors.
- Expression of neuropeptides, transmitters, peptide hormones, and cognate receptors.
- Expression of immunosuppressive factors TGFβ2, somatostatin, substance-P
- Steroidogenic: Steroid-converting enzymes (17β-HSD2, 4, 5 and 7; 11βHSD1)
- Prostaglandins: Cyclooxygenase-2 (COX-2); prostaglandin D2 synthase
- Expression of components of rod-phototransduction including rhodopsin, rhodopsin kinase, arrestin
- Expression of components of the visual cycle including CRALBP, IRBP, CRBP, 11-*cis*-RDH, LRAT
- Expression of glutamate metabolizing enzymes, glutamate receptors, and glutamate transporters
- Circadian clock: Expression of melatonin rhythm-generating enzymes AA-NAT, HIOMT, melatonin receptors, and clock genes
- Expression of glaucoma genes including *MYOC, CYP1B1, OPTN*

The presence of pro-inflammatory (i.e., substance-P and CGRP) and anti-inflammatory (SST) neuropeptides in AqH indicates that there is a regulated balance between these factors to influence the ocular immune-privileged microenviroment of the anterior chamber. TGFβ2, a cytokine that promotes immune deviation and immunosuppressive activities in the anterior chamber has been detected in high levels in the AqH in approximately 50% of glaucoma patients with primary open-angle glaucoma (POAG) (Picht *et al.*, 2001).

Neuropeptides with immunosuppressive and anti-inflammatory properties in the AqH may derive either from the autonomic and sensory systems and/or from the ocular tissues in the anterior segment. The pigmented cells of the iris, the bilayered CE, and possibly the TM cells are capable of synthesizing and releasing pro-inflammatory and immunosuppressive factors. As indicated earlier, SST expression has been shown in the iris and the ciliary processes, and SST receptors have been detected in the NPE cells of the CE (Ghosh *et al.*, 2006) and in TM (Fig. 3). The human TM expresses VIP and SgII (González *et al.*, 2000b). Immunoregulatory actions of VIP are mediated by VIP receptors that are expressed on NPE cells. In mice, VIP induces the production of interleukins and displays anti-inflammatory actions. High level of TGFβ2 mRNA is expressed in human ciliary processes and in cultured NPE cells (Escribano *et al.*, 1994), and in the TM as well (unpublished results), suggesting that these tissues have the potential ability to produce immunosuppressive factors into the AqH. However, recent studies have documented that T cells exposed to cultured PE cells of the iris-CB were induced to secrete large amounts of active and latent TGF-β2, a response that is mimicked by exposing T cells to AqH (Knisley *et al.*, 1991). These observations indicate that TGF-β released into the AqH could be from multiple sources, and confirm that cells from the CE exhibit immunomodulatory functions of their own, since they have the ability to synthesize, and release cytokine factors independent of the involvement of T cells. The finding that the CE and TM cells express cognate receptors for many of the neuropeptides and cytokine released in the AqH is not coincidental, and it suggests that these factors may act on autocrine and paracrine based mechanisms to promote immune privilege in the anterior chamber of the eye.

Finally, neuropeptides with immunosuppressive properties in AqH are potentially under a 24-hour LD circadian cycle. There is evidence that ambient light is an important factor in inducing the ocular immune privilege in the anterior chamber (Ferguson *et al.*, 1988) and that green light (500–510 nm) within the visual spectrum is the relevant photic energy which influences intraocular immune reactions (Ferguson *et al.*, 1992). Injection of antigen into the anterior chamber of mice, raised in complete darkness from birth, failed to induce ACAID whereas mice raised in the same manner and exposed to ambient light for 48 hours restored the capacity of the eyes to support ACAID. Interestingly, severance of the optic nerve had no effect on the ocular immune privilege of the anterior chamber, suggesting that the presence of photoreceptors in the anterior segment of the eye may be involved in the modulation ACAID. It has been suggested that circadian factors including melatonin, expressed and released by the CE in the AqH, might contribute to the light-dependent modulation of ACAID. Thus, neuropeptides released by the CE could exhibit multiple endocrine functions, including communication with cells of the AqH outflow pathways and the immune system, to maintain the immune privilege condition of the anterior segment of the eye (Fig. 6). Finally, the release of peptides in AqH could follow a 24-hour LD circadian rhythm cycle as described for AqH secretion and IOP.

X. SUMMARY

This chapter introduces the "inflow-outflow link hypothesis." It proposes that inflow and outflow of AqH are linked by neuroendocrine factors released by the CE. The neuropeptides released by the CE and carried out by the AqH may represent a major amplification system which establishes communication with cognate receptors expressed in cells of the outflow pathways by classical endocrine mechanisms. The unique neural origin of many of the neuropeptides and neurotransmitters, synthesized and secreted by the NPE cells, may reflect its common embryological origin with the retina, and its neural-like function. The diversity of functions assigned to neuropeptides in the AqH is best explained by their interaction in biological systems of endocrine tissues (CB and TM). Neuropeptides with immunosuppressive properties (SST, substance-P, TGFβ2) could be establishing communication with the immune system and maintaining the immune privilege status of the anterior segment of the eye. The role of angiotensin, endothelin, and NPs in the cardiovascular system could explain their involvement in fluid homeostasis and regulation of IOP in AqH. The synthesis of melatonin and the expression of clock genes could also explain the presence of a peripheral circadian clock in the CB, synchronizing with the SCN the circadian rhythm control of IOP in AqH.

Acknowledgments

This work was supported by National Eye Institute NIH grants EY04873, EY00785, Research to Prevent Blindness, and Alcon Laboratories. The Connecticut Lions Foundation.

References

Abbey, S. E., and Potter, L. R. (2003). Lysophosphatidic acid inhibits C-type natriuretic peptide activation of guanylyl cyclase-B. *Endocrinology.* **144,** 240–246.

Akiyama, H., Kondo, H., Ikeda, K., Kato, M., and McGeer, P. L. (2001). Immunohistochemical localization of neprilysin in the human cerebral cortex: Inverse association with vulnerability to amyloid beta-protein (Abeta) deposition. *Brain Res.* **902,** 277–281.

Almegård, B., and Anderson, S. E. (1990). Outflow facility in the monkey eye: Effects of calcitonin gene-related peptide, cholecystokinin, galanin, substance P and capsaicin. *Exp. Eye Res.* **51,** 685–689.

Aronson, P. S. (1985). Kinetic properties of the plasma membrane Na+/H+ exchanger. *Annu. Rev. Physiol.* **47,** 545–560.

Arriza, J. L., Eliasof, S., Kavanaugh, M. P., and Amara, S. G. (1997). Excitatory amino acid transporter 5, a retinal glutamate transporter coupled to a chloride conductance. *Proc. Natl. Acad. Sci. USA* **94,** 4155–4160.

Avila, M. Y., Stone, R. A., and Civan, M. M. (2002a). Knockout of A3 adenosine receptors reduces mouse intraocular pressure. *Invest. Ophthalmol. Vis. Sci.* **43,** 3021–3026.

Avila, M. Y., Seidler, R. W., Stone, R. A., and Civan, M. M. (2002b). Inhibitors of NHE-1 Na^+/H^+ exchange reduce mouse intraocular pressure. *Invest. Ophthalmol. Vis. Sci.* **43,** 1897–1902.

Balsalobre, A. (2002). Clock genes in mammalian peripheral tissues. *Cell Tissue Res.* **309,** 193–199.

Balsalobre, A., Damiola, F., and Schibler, U. (1998). A serum shock induces circadian gene expression in mammalian tissue culture cells. *Cell* **93,** 929–937.

Barnes, K., Doherty, S., and Turner, A. J. (1995). Endopeptidase-24.11 is the integral membrane peptidase initiating degradation of somatostatin in the hippocampus. *J. Neurochem.* **64,** 1826–1832.

Baumgartner, M., Patel, H., and Barber, D. L. (2004). Na+/H+ exchanger NHE1 as plasma membrane scaffold in the assembly of signaling complexes. *Am. J. Cell Physiol.* **28,** 844–850.

Bedecs, K., Langel, U., and Bartfai, T. (1995). Metabolism of galanin and galanin (1–16) in isolated cerebrospinal fluid and spinal cord membranes from rat. *Neuropeptides* **29,** 137–143.

Beebe, D. C. (1986). Development of the ciliary body: A brief review. *Trans. Ophthalmol. Soc UK* **105,** 123–130 (Review).

Berelowitz, M., Perlow, M. J., Hoffman, H. J., and Frohman, L. A. (1981). The diurnal variation of immunoreactive thyrotropin-releasing hormone and somatostatin in the cerebrospinal fluid of the rhesus monkey. *Endocrinology* **109,** 2102–2109.

Bertazolli-Filho, R., Ghosh, S., Huang, W., Wollmann, G., and Coca-Prados, M. (2001). Molecular evidence that human ocular ciliary epithelium expresses components involved in phototransduction. *Biochem. Biophys. Res. Commun.* **284,** 317–325.

Bertazolli-Filho, R., Laicine, E. M., and Haddad, A. (2003). Synthesis and secretion of transferrin by isolated ciliary epithelium of rabbit. *Biochem. Biophys. Res. Commun.* **305,** 820–825.

Bertazolli-Filho, R., Laicine, E. M., Haddad, A., and Rodrigues, M. L. (2006). Molecular and biochemical analysis of ceruloplasmin expression in rabbit and rat ciliary body. *Curr. Eye Res.* **31,** 155–161.

Borrás, T. (2003). Gene expression in the trabecular meshwork and the influence of intraocular pressure. *Prog. Retin. Eye Res.* **22,** 435–463 (Review).

Brakch, N., Galanopoulou, A. S., Patel, Y. C., Boileau, G., and Seidah, N. G. (1995). Comparative proteolytic processing of rat prosomatostatin by the convertases PC1, PC2, furin, PACE4 and PC5 in constitutive and regulated secretory pathways. *FEBS Lett.* **362,** 143–146.

Brubaker, R. F. (1998). Clinical measurement of aqueous humor dynamics: Implications for addressing glaucoma. *In* "Eye's Aqueous Humor: From Secretion to Glaucoma" (M. M. Civan, ed.), Vol. 45, pp. 234–284. Academic Press, San Diego.

Chan, W. Y., Soloview, M. M., Ciruela, F., and McIlḥinnery, R. A. (2001). Molecular determinants of metabotropic glutamate receptor 1B trafficking. *Mol. Cell. Neurosci.* **17,** 577–588.

Chan, C. Y., Guggenheim, J. A., and To, C. H. (2007). Is active glucose transport present in the bovine ciliary body epithelium? *Am. J. Physiol. Cell Physiol.* **292,** C1087–C1093.

Chang, A. T., Polansky, J. R., and Crook, R. B. (1996). Natriuretic peptide receptors on human trabecular meshwork cells. *Curr. Eye Res.* **15,** 137–143.

Civan, M. M. (1998). Transport components of net secretion of the aqueous humor and their integrated regulation. *In* "Current Topics in Membranes *The Eye's Aqueous Humor. From secretion to glaucoma*" (A. Kleinzeller, D. M. Fambrough, and D. J. Benos, eds.), Vol. 45, pp. 1–24. Academic Press, New York.

Civan, M. M., and Macknight, A. D. (2004). The ins and outs of aqueous humor secretion. *Exp. Eye Res.* **78,** 625–631.

Coca-Prados, M., Ghosh, S., Wang, Y., Escribano, J., Herrala, A., and Vihko, P. (2003). Sex steroid hormone metabolism takes place in human ocular cells. *J. Steroid Biochem. Mol. Biol.* **86,** 207–216.

Coca-Prados, M., and Escribano, J. (2007). New perspectives in aqueous humor secretion and in glaucoma: The ciliary body as a multifunctional endocrine gland. *Prog. Retin. Eye Res.* **26,** 239–262 (Review).

Cristiani, R., Petrucci, C., Dal Monte, M., and Bagnoli, P. (2002). Somatostatin (SRIF) and SRIF receptors in themouse retina. *Brain Res.* **936,** 1–14.

DeBold, A. J., and Bruneau, B. G. (2000). Natriuretic peptides. *In* "Handbook of Physiology, Section 7: The Endocrine System" (J. C. S. Fray and H. M. Goodman, eds.), Vol III, pp. 377–409. Oxford University Press, New York.

Ding, J. M., Faiman, L. E., Hurst, W. J., Kuriashkina, L. R., and Gillete, M. U. (1997). Resetting the biological clock: Mediation of nocturnal CREB phosphorylation via light, glutamate and nitric oxide. *J. Neurosci.* **17,** 667–675.

Elbadri, A. A., Shaw, C., Johnston, C. F., Archer, D. B., and Buchanan, K. D. (1991). The distribution of neuropeptides in the ocular tissues of several mammals: A comparative study. *Comp. Biochem. Physiol. C.* **100,** 625–627.

Ericson, L. A. (1958). Twenty-four hourly variations on the aqueous humor flow: Examination with perilimbal suction cup. *Acta Ophthalmol. (Copenhagen)* **50**(Suppl.), 1.

Escribano, J., and Coca-Prados, M. (2002). Bioinformatics and reanalysis of subtracted expressed sequence tags from the human ciliary body: Identification of novel biological functions. *Mol. Vis* **8,** 315–332. http://www.molvis.org/molvis/v8/a39/

Escribano, J., Hernando, N., Ghosh, S., Crabb, J., and Coca-Prados, M. (1994). cDNA from human ocular ciliary epithelium homologous to βig-h3 is preferentially expressed as an extracellular protein in the corneal epithelium. *J. Cell. Physiol.* **160,** 511–521.

Escribano, J., Ortego, J., and Coca-Prados, M. (1995). Isolation and characterization of cell-specific cDNA clones from a subtractive library of the ocular ciliary body of a single normal human donor: Transcription and synthesis of plasma proteins. *J. Biochem. (Tokyo).* **118,** 921–931.

Fernández-Durango, R., Triviño, A., Ramirez, J. M., García de la Coba, M., Ramirez, A., Salazar, A., Fernandez-Cruz, A., and Gutkowska, J. (1990). Immunoreactive atrial natriuretic factor in aqueous humor: Its concentration is increased with high intraocular pressure in rabbit eyes. *Vis. Res.* **30,** 1305–1310.

Fernández-Durango, R., Moya, F. J., Ripodas, A., de Juan, J. A., Fernández-Cruz, A., and Bernal, R. (1999). Type B and type C natriuretic peptide receptors modulate intraocular pressure in the rabbit eye. *Eur. J. Pharmacol.* **364,** 107–113.

Fernández-Durango, R., Rollin, R., Mediero, A., Roldan-Pallares, M., Garcia Feijo, J., Garcia Sanchez, A., Fernandez-Cruz, A., and Ripodas, A. (2003). Localization of endothelin-1 mRNA expression and immunoreactivity in the anterior segment of human eye: Expression of ETA and ETB receptors. *Mol. Vis.* **9,** 103–109.

Ferguson, T. A., Hayashi, J. D., and Kaplan, H. J. (1988). Regulation of the systemic immune response by visible light and the eye. *FASEB J.* **2,** 3017–3021.

Ferguson, T. A., Mahendra, S. L., Hooper, P., and Kaplan, H. J. (1992). The wavelength of light governing intraocular immune reactions. *Invest. Ophthalmol. Vis. Sci.* **33,** 1788–1795.

Fidzinski, P., Salvador-Silva, M., Choritz, L., Geibel, J., and Coca-Prados, M. (2004). Inhibition of the NHE-1 Na^+/H^+-exchanger by natriuretic peptides in the ocular nonpigmented ciliary epithelium. *Am. J. Physiol. Cell Physiol.* **287,** C655–C663.

Firth, S. I., Kaufman, P. L., De Jean, B. J., Byers, J. M., and Marshak, D. W. (2002). Innervation of the uvea by galanin and somatostatin immunoreactive axons in macaques and baboons. *Exp. Eye Res.* **75,** 49–60.

Fischer-Colbrie, R., Laslop, A., and Kirchmair, R. (1995). Secretogranin II: Molecular properties, regulation of biosynthesis and processing to the neuropeptide secretoneurin. *Prog. Neurobiol.* **46,** 49–70.

Fricker, L. D. (1988). Activation and membrane binding of carboxypeptidase E. *J. Cell. Biochem.* **38,** 279–289.

Fricker, L. D., Evans, C. J., Esch, F. S., and Herbert, E. (1986). Cloning and sequence analysis of cDNA for bovine carboxypeptidase E. *Nature* **323,** 461–464.

Funaki, H., Sawaguchi, S., Yaoeda, K., Koyama, Y., Yaoita, E., Funaki, S., Shirakashi, M., Oshima, C., Shukunami, C., Hiraki, Y., Abe, H., and Yamamoto, T. (2001). Expression and localization of angiogenic inhibitory factor, chondromodulin-I, in adult rat eye. *Invest. Ophthalmol. Vis. Sci.* **42,** 1193–1200.

Galanopoulou, A. S., Kent, G., Rabbani, S. N., Seidah, N. G., and Patel, Y. C. (1993). Heterologous processing of prosomatostatin in constitutive and regulated secretory pathways. Putative role of the endoproteases furin, PC1, and PC2. *J. Biol. Chem* **268,** 6041–6049.

Galanopoulou, A. S., Seidah, N. G., and Patel, Y. C. (1995). Direct role of furin in mammalian prosomatostatin processing. *Biochem. J.* **309,** 33–40.

Gallar, J., and Liu, J. H. K. (1993). Stimulation of the cervical sympathetic nerves increases intraocular pressure. *Invest. Ophthalmol. Vis. Sci.* **34,** 596–605.

Gao, B., Huber, R. D., Wenzel, A., Vavricka, S. R., Ismair, M. G., Reme, C., and Meier, P. J. (2005). Localization of organic anion transporting polypeptides in the rat and human ciliary body epithelium. *Exp. Eye Res.* **80,** 61–72.

Geyer, O., Podos, S. M., and Mittag, T. (1997). Nitric oxide synthase activity in tissues of the bovine eye. *Graefes Arch. Clin. Exp. Ophthalmol.* **235,** 786–793.

Ghosh, S., Choritz, L., Geibel, J., and Coca-Prados, M. (2006). Somatostatin modulates PI3 K-Akt, eNOS and NHE activity in the ciliary epithelium. *Mol. Cell. Endocrinol.* **253,** 63–75.

Ghosh, S., Freitag, A. C., Martin-Vasallo, P., and Coca-Prados, M. (1990). Cellular distribution and differential gene expression of the three alpha subunit isoforms of the Na^+,K^+-ATPase genes reveals a gradient of isoforms along the nonpigmented ciliary epithelium. *J. Biol. Chem.* **265,** 2935–2940.

Ghosh, S., Hernando, N., Martin-Alonso, J. M., Martin-Vasallo, P., and Coca-Prados, M. (1991). Expression of multiple Na^+, K(+)-ATPase genes reveals a gradient of isoforms along the nonpigmented ciliary epithelium: Functional implications in aqueous humor secretion. *J. Cell. Physiol.* **149,** 184–194.

Ghosh, S., Salvador-Silva, M., and Coca-Prados, M. (2004). The bovine iris-ciliary epithelium expresses components of rod phototransduction. *Neurosci. Lett.* **370,** 7–12.

Goldmann, D. B., and Waubke, N. (1999). A pilot study on the effect of atrial natriuretic peptide on intraocular pressure in the human. *Fortschr. Ophthalmol.* **86,** 494–496.

González, P., Epstein, D. L., and Borrás, T. (2000a). Characterization of gene expression in human trabecular meshwork using single-pass sequencing of 1060 clones. *Invest. Ophthalmol. Vis. Sci.* **41,** 3678–3693.

González, P., Epstein, D. L., and Borrás, T. (2000b). Genes upregulated in the human trabecular meshwork in response to elevated intraocular pressure. *Invest. Ophthalmol. Vis. Sci.* **41,** 352–361.

Gully, D., Lespy, L., Canton, M., Rostene, W., Kitabgi, P., Le Fur, G., and Maffrand, J. P. (1996). Effect of the neurotensin receptor antagonist SR48692 on rat blood pressure modulation by neurotensin. *Life Sci.* **58,** 665–674.

Hayashi, M., Morimoto, R., Yamamoto, A., and Moriyama, Y. (2003). Expression and localization of vesicular glutamate transporters in pancreatic islets, upper gastrointestinal tract and testis. *J. Histochem. Cytochem.* **51,** 1375–1390.

Hagenbuch, B., and Meier, P. J. (2004). Organic anion transporting polypeptides of the OATP/SLC21 family: Phylogenetic classification as OATP/*SLCO* superfamily, new nomenclature and molecular/functional properties. *Pflugers Arch. Eur. J. Physiol.* **447,** 653–665.

Hayes, R. G., Shaw, C., Kitabgi, P., and Buchanan, K. D. (1992). Different relative abundance of neurotensin and neuromedin N in bovine ocular tissues. *Regul. Pept.* **39,** 147–155.

Hastings, M. H., Reddy, A. B., and Maywood, E. S. (2003). A clockwork web: Circadian timing in brain and periphery, in health and disease. *Nat. Rev. Neurosci.* **4,** 649–661.

Igarashi, T., Inatomi, J., Sekine, T., Cha, S. H., Kanai, Y., Kunimi, M., Tsukamoto, K., Satoh, M., Shimadzu, M., Tozawa, F., Mori, T., Shiobara, M., *et al.* (1999). Mutations in SLC4A4 cause permanent isolated proximal renal tubular acidosis with ocular abnormalities. *Nat. Genet.* **23,** 264–266.

Igarashi, T., Inatomi, J., Sekine, T., Seki, G., Shimadzu, M., Tozawa, F., Takeshima, Y., Takumi, T., Takahashi, T., Yoshikawa, N., Nakamura, H., and Endou, H. (2001). Novel nonsense mutation in the Na+/HCO3- cotransporter gene (SLC4A4) in a patient with permanent isolated proximal renal tubular acidosis and bilateral glaucoma. *J. Am. Soc. Nephrol.* **12,** 713–718.

Ito, A., Yamaguchi, K., Tomita, H., Suzuki, T., Onogawa, T., Sato, T., Mizutamari, H., Mikkaichi, T., Nishio, T., Suzuki, T., Unno, M., Sasano, H., *et al.* (2003). Distribution of rat organic transporting polypeptide-E (oatp-E) in the rat eye. *Invest. Ophthalmol. Vis. Sci.* **44,** 4877–4884.

Iuvone, P. M., Tosini, G., Pozdeyev, N., Haque, R., Klein, D. C., and Chaurasia, S. S. (2005). Circadian clocks, clock networks, arylalkylamine N-acetyltransferase, and melatonin in the retina. *Prog. Retin. Eye Res.* **24,** 433–456(Review).

Iwata, N., Mizukami, H., Shirotani, K., Takaki, Y., Muramatsu, S., Lu, B., Gerard, N. P., Gerard, K., Ozawa, K., and Saido, T. C. (2004). Presynaptic localization of neprilysin contributes to efficient clearance of amyloid-beta peptide in mouse brain. *J. Neurosci.* **24,** 991–998.

Johnson, J., Rickman, D. W., and Brecha, N. C. (2000). Somatostatin and somatostatin Subtype 2A expression in the mammalian retina. *Microsc. Res. Tech.* **50,** 103–111(Review).

Kalariti, N., and Koutsilieris, M. (2004). Glutamatergic system in bone physiology. *In Vivo* **18,** 621–628.

Kaufman, P. (1984). Aqueous humor outflow. *Curr. Top. Eye Res.* **4,** 97–138.

Kawaji, T., Ando, Y., Nakamura, M., Yamamoto, K., Ando, E., Takano, A., Inomata, Y., Hirata, A., and Tanihara, H. (2005). Transthyretin synthesis in rabbit ciliary pigment epithelium. *Exp. Eye Res.* **81,** 306–312.

King, D. P., and Takahashi, J. S. (2000). Molecular genetics of circadian rhythms in mammals. *Annu. Rev. Neurosci.* **23,** 713–742.

Klisovic, D. D., O'Dorisio, M. S., Katz, S. E., Sall, J. W., Balster, D., O'Dorisio, T. M., Craig, E., and Lubow, M. (2001). Somatostatin receptor gene expression in human ocular tissues: RT-PCR and immunohistochemical study. *Invest. Ophthalmol. Vis. Sci.* **42,** 2193–2201.

Knisley, T., Bleicher, P., Vibbard, C., and Granstein, R. (1991). Production of latent transforming growth factor-beta and other inhibitory factors by cultured murine iris and ciliary body cells. *Curr. Eye Res.* **10,** 761–772.

Kosenko, E., Llansola, M., Montoliu, C., Monfort, P., Rodrigo, R., Hernandez-Viadel, M., Erceg, A. M., Sanchez-Perez, A. M., and Felipo, V. (2003). Glutamine synthase activity and glutamine content in brain: Modulation by NMDA receptors and nitric oxide. *Neurochem. Int.* **43,** 493–499.

Kotikoski, H., Alajuuma, P., Moilanen, E., Salmenpera, P., Oksala, O., Laippala, P., and Vapaatalo, H. (2002). Comparison of nitric oxide donors in lowering intraocular pressure in rabbits: Role of cyclic GMP. *J. Ocul. Pharmacol. Ther.* **18,** 11–23.

Kuhn, M. (2003). Structure, regulation, and function of mammalian membrane guanylyl cyclase receptors, with a focus on guanylyl cyclase-A. *Circ. Res.* **93,** 700–709.

Lam, D. M. (1997). Neurotransmitters in the vertebrate retina. *Invest. Ophthalmol. Vis. Sci.* **38,** 553–556 (Review).

Lang, F., Busch, G. L., and Völkl, H. (1998). The diversity of volume regulatory mechanisms. *Cell. Physiol. Biochem.* **8,** 1–45.

Lang, R., Gundlach, A. L., and Kofler, B. (2007). The galanin peptide family: Receptor pharmacology, pleiotropic biological actions, and implications in health and disease. *Pharmacol. Ther.* **115,** 177–207.

Libby, R. T., Gould, D. B., Anderson, M. G., and John, S. W. M. (2005). Complex genetics of glaucoma susceptibility. *Annu. Rev. Genomics Hum. Genet.* **6,** 15–44.

Lin, F. F., Varney, M., Sacaan, A. I., Jachec, C., Daggett, L. P., Rao, S., Flor, P., Kuhn, R., Kerner, D., Standaert, D., Young, A. B., and Velicelebi, G. (1997). Neuropharmacology Cloning and stable expression of the mGluR1b subtype of human metabotropic receptors. *Neuropharmacology* **36,** 917–931.

Linden, R., Martins, R. A., and Silveira, M. S. (2005). Control of programmed cell death by Neurotransmitters and neuropeptides in the developing mammalian retina. *Prog. Retin. Eye Res.* **24,** 457–491. (Review).

Liu, J. H. (2000). Circadian variations of prostaglandins in the rabbit aqueous humor. *J. Ocul. Pharmacol. Ther.* **16,** 49–54.

Liu, J. H. (2001). Diurnal measurement of intraocular pressure. *J. Glaucoma* **10,** S39–S41 (Review).

Liu, J. H. (2002). Circadian variations of transforming growth factor-beta2 and basic fibroblast growth factor in the rabbit aqueous humor. *Curr. Eye Res.* **24,** 75–80.

Maeda, A., Tsujiya, S., Higashide, T., Toida, K., Todod, T., Ueyama, T., Okamura, H., and Sugiyama, K. (2006). Circadian intraocular pressure rhythm is generated by clock genes. *Invest. Ophthalmol. Vis. Sci.* **47,** 4050–4052.

Martin, X. D., Malina, H. Z., Brennan, M. C., Hendrickson, P. H., and Lichter, P. R. (1992). The ciliary body—the third organ found to synthesize indoleamines in humans. *Eur. J. Ophthalmol.* **2,** 67–72.

Maus, T. L., Young, W. F., Jr., and Brubaker, R. F. (1994). Aqueous flow in humans after adrenalectomy. *Invest. Ophthalmol. Vis. Sci.* **35,** 3325–3331.

McGahan, M. C., Harned, J., Mukunnemkeril, M., Goralska, M., Fleisher, L., and Ferrell, J. B. (2005). Iron alters glutamate secretion by regulating cytosolic aconitase activity. *Am. J. Physiol. Cell Physiol.* **288,** 1117–1124.

Mittag, T. W., Tormay, A., and Podos, S. M. (1987). Vasoactive intestinal peptide and intraocular pressure: Adenylate cyclase activation and binding sites for vasoactive intestinal peptide in membranes of ocular ciliary processes. *J. Pharmacol. Exp. Ther.* **241,** 230–235.

Miyamoto, K., and Del Monte, M. A. (1994). A Na^+-dependent glutamate transporter in human retinal pigment epithelial cells. *Invest. Ophthalmol. Vis. Sci.* **35,** 3589–3598.

Mundell, S. J., Pula, G., More, J. C., Jane, D. E., Roberts, P. J., and Kelly, E. (2004). Activation of cyclic AMP-dependent protein kinase inhibits the desensitization and internalization of metabotropic glutamate receptors 1a and 1b. *Mol. Pharmacol.* **65,** 1507–1516.

Naskar, R., Vorwerk, C. K., and Dreyer, E. B. (2000). Concurrent downregulation of a glutamate transporter and receptor in glaucoma. *Invest. Ophthalmol. Vis. Sci.* **41,** 1940–1944.

Nathanson, J. A., and McKee, M. (1995). Identification of an extensive system of nitric oxide-producing cells in the ciliary muscle and outflow pathway of the human eye. *Invest. Ophthalmol. Vis. Sci.* **36,** 1765–1773.

Nii, H., Ikeda, H., Okada, K., Yoshitomi, T., and Gregory, D. S. (2001). Circadian change of adenylate cyclase activity in rabbit ciliary processes. *Curr. Eye Res.* **23,** 248–255.

Nucci, P., Tredici, G., Manitto, M. P., Pizzini, G., and Brancato, R. (1992). Neuron-specific enolase and embryology of the trabecular meshwork of the rat eye: An immunohistochemical study. *Int. J. Biol. Markers* **7,** 253–255.

Ortego, J., and Coca-Prados, M. (1997). Molecular characterization and differential gene induction of the neuroendocrine-specific genes neurotensin, neurotensin receptor, PC1, PC2, and 7B2 in the human ocular ciliary epithelium. *J. Neurochem.* **69,** 1829–1839.

Ortego, J., and Coca-Prados, M. (1998). Molecular identification and coexpression of galanin and GalR-1 galanin receptor in the human ocular ciliary epithelium: Differential modulation of their expression by the activation of alpha2- and beta2-adrenergic receptors in cultured ciliary epithelial cells. *J. Neurochem.* **71,** 2260–2270.

Ortego, J., and Coca-Prados, M. (1999). Functional expression of components of the natriuretic peptide system in human ocular nonpigmented ciliary epithelial cells. *Biochem. Biophys. Res. Commun.* **258,** 21–28.

Ortego, J., Escribano, J., Crabb, J., and Coca-Prados, M. (1996a). Identification of a neuropeptide and neuropeptide-processing enzymes in aqueous humor confers neuroendocrine features to the human ocular ciliary epithelium. *J. Neurochem.* **66,** 787–796.

Ortego, J., Escribano, J., Becerra, S. P., and Coca-Prados, M. (1996b). Gene expression of the neurotrophic pigment epithelium-derived factor in the human ciliary epithelium. Synthesis and secretion into the aqueous humor. *Invest. Ophthalmol. Vis. Sci.* **37,** 2759–2767.

Ortego, J., Escribano, J., and Coca-Prados, M. (1997). Gene expression of proteases and protease inhibitors in the human ciliary epithelium and ODM-2 cells. *Exp. Eye Res.* **65,** 289–299.

Ortego, J., Wollmann, G., and Coca-Prados, M. (2002). Differential regulation of gene expression of neurotensin and prohormone convertases PC1 and PC2 in the bovine ocular ciliary epithelium: Possible implications on neurotensin processing. *Neurosci. Lett.* **333,** 49–53.

Osborne, N. N., and Chidlow, G. (1994). The presence of functional melatonin receptors in the iris-ciliary processes of the rabbit eye. *Exp. Eye Res.* **59,** 3–9.

Osborne, N. N., Barnett, N. L., and Luttmann, W. (1993). Endothelin receptors in the cornea, iris, and ciliary processes. Evidence from binding, secondary messenger and PCR studies. *Exp. Eye Res.* **56,** 721–728.

Oster, H., Damerow, S., Hut, R. A., and Eichele, G. (2006). Transcriptional profiling in the adrenal gland reveals circadian regulation of hormone biosynthesis genes and nucleosome assembly genes. *J. Biol. Rhythms* **21,** 350–361.

Panda, S., Antoch, M. P., Miller, B. H., Su, A. I., Schook, A. B., Straume, M., Schultz, P. G., Kay, J. S., Takahashi, J. S., and Hogenesch, J. B. (2002). Coordinated transcription of key pathways in the mouse by the circadian clock. *Cell* **109,** 307–320.

Pang, I. H., Shade, D. L., Clark, A. F., Steely, H. T., and DeSantis, L. (1994). Preliminary characterization of a transformed cell strain derived from human trabecular meshwork. *Curr. Eye Res.* **13,** 51–63.

Patel, Y. C. (1999). Molecular pharmacology of somatostatin receptor subtypes. *J. Endocrinol. Invest.* **20,** 348–367 (Review).

Patel, Y. C., and Reichlin, S. (1978). Somatostatin in hypothalamus, extrahypothalamic brain and peripheral tissues of the rat. *Endocrinology* **102,** 523–530.

Picht, G., Welge-Luessen, U., Grehn, F., and Lutjen-Drecoll, E. (2001). Transforming growth factor beta 2 levels in the aqueous humor in different types of glaucoma and the relation to filtering bleb development. *Graefes Arch. Clin. Exp. Ophthalmol.* **239,** 199–207.

Pin, J. P., Waeber, C., Prezeau, L., Bockaert, J., and Heinemann, S. F. (1992). Alternative splicing generates metabotropic glutamate receptors inducing different patterns of calcium release in Xenopus oocytes. *Proc. Natl. Acad. Sci. USA* **89,** 10331–10335.

Potter, D. E., Russell, K. R., and Manhiani, M. (2004). Bremazocine increases C-type natriuretic peptide levels in aqueous humor and enhances outflow facility. *J. Pharmacol. Exp. Ther.* **309,** 548–553.

Pow, D. V., and Barnett, N. L. (2000). Developmental expression of excitatory amino acid transporter 5: A photoreceptor and bipolar cell glutamate transporter in rat retina. *Neurosci. Lett.* **280,** 21–24.

Prassana, G., Krishnamoorthy, R., Hulet, C., Zhang, H., Zhang, X., and Yorio, T. (2000). Endothelin-1 induces nitric oxide synthase-2 expression in human non-pigmented ciliary epithelial cells. *Exp. Eye Res.* **71,** 535–539.

Rauz, S., Cheung, C. M. G., Wood, P. J., Coca-Prados, M., Walker, E. A., Murray, P. I., and Stewart, P. M. (2003). Inhibition of 11β-hydroxysteroid dehydrogenase type 1 lowers intraocular pressure in patients with ocular hypertension. *Q. J. Med.* **96,** 481–490.

Reiss, G. R., Lee, D. A., Topper, J. E., and Brubaker, R. F. (1984). Aqueous humor flow during sleep. *Invest. Ophthalmol. Vis. Sci.* **25,** 776–778.

Reppert, S. M., and Weaver, D. R. (2002). Coordination of circadian timing in mammals. *Nature* **418,** 935–941.

Riegler, M., Castagliuolo, I., Wang, C., Wlk, M., Sogukoglu, T., Wenzl, E., Matthews, J. B., and Pothoulakis, C. (2000). Neurotensin stimulates Cl(−) secretion in human colonic mucosa *In vitro*: Role of adenosine. *Gastroenterology* **119,** 348–357.

Riepe, R. E., and Norenberg, M. D. (1978). Glutamine synthetase in the developing rat retina: An immunohistochemical study. *Exp. Eye Res.* **27,** 435–444.

Ripperger, J. A., and Schibler, U. (2001). Circadian regulation of gene expression in animals. *Curr. Opin. Cell Biol.* **13,** 357–362.

Rotin, D., and Grinstein, S. (1989). Impaired cell volume regulation in Na^{+}-H^{+} exchange-deficient mutants. *Am. J. Physiol.* **257,** C1158–C1165.

Saito, T., Iwata, N., Tsubuki, S., Takaki, Y., Takano, J., Huang, S. M., Suemoto, T., Higuchi, M., and Saido, T. C. (2005). Somatostatin regulates brain amyloid beta peptide Abeta42 through modulation of proteolytic degradation. *Nat. Med.* **11,** 434–439.

Salzmann, J., Flitcroft, D., Bunce, C., Gordon, D., Wormald, R., and Migdal, C. (1998). Brain natriuretic peptide: Identification of a second natriuretic peptide in human aqueous humor. *Br. J. Ophthalmol.* **82,** 830–834.

Salvador-Silva, M., Ghosh, S., Bertazolli-Filho, R., Boatright, J. H., Nickerson, J. M., Garwin, J. C., Saari, J. C., and Coca-Prados, M. (2005). Retinoid processing proteins in the ocular ciliary epithelium. *Mol. Vis.* **11,** 356–365. http://www.molvis.org/molvis/v11/a42/.

Savaskan, E., Loffler, K. U., Meier, F., Muller-Spahn, F., Flammer, J., and Meyer, P. (2004). Immunohistochemical localization of angiotensin-converting enzyme, angiotensin II and AT1 receptor in human ocular tissues. *Ophthalmic Res.* **36,** 312–320.

Schneemann, A., Leusink-Muis, A., van den Berg, T., Hoyng, P. F., and Kamphuis, W. (2003). Elevation of nitric oxide production in human trabecular meshwork by increased pressure. *Graefes Arch. Clin. Exp. Ophthalmol.* **241,** 321–326.

Schuster, V. L., Lu, R., and Coca-Prados, M. (1997). The prostaglandin transporter is widely expressed in ocular tissues. *Surv. Ophthalmol.* **41**(Suppl. 2), S41–S45.

Sears, M. L. (1984).*In* "Handbook of Experimental Pharmacology" (M. L. Sears, ed.), pp. 193–248. Springer, Berlin.

Serle, J. B., Wang, R. F., Peterson, W. M., Plourde, R., Yerxa, B. R., Serle, J. B., Wang, R. F., Peterson, R., Plourde, R., and Yerxa, B. R. (2004). Effect of 5-MCA-NAT, a putative melatonin MT3 receptor agonist, on intraocular pressure in glaucomatous monkey eyes. *J. Glaucoma* **13,** 385–388.

Shannon, H. E., and Sawyer, B. D. (1989). Glutamate receptors of the N-methyl-D-aspartate subtype in the myenteric plexus of the ginea pig ileum. *J. Pharmacol. Exp. Ther.* **251,** 518–523.

Skerry, T. M., and Genever, P. G. (2001). Glutamate signaling in non-neuronal tissues. *Trends Pharmacol. Sci.* **22,** 174–181.

Smith, S. D., and Gregory, D. S. (1989). A circadian rhythm of aqueous flow underlies the circadian rhythm of IOP in NZW rabbits. *Invest. Ophthalmol. Vis. Sci.* **30,** 775–778.

Solcia, E., Usellini, L., Buffa, R., Rindi, G., Villani, L., Zampatti, C., and Sillini, E. (1987). Endocrine cells producing regulatory peptides. *Experientia.* **43,** 839–850. (Review)

Stone, R. A., Kuwayama, Y., Laties, A. M., and Marangos, P. J. (1984). Neuron-specific enolase-containing cells in the rhesus monkey trabecular meshwork. *Invest. Ophthalmol. Vis. Sci.* **25,** 1331–1334.

Stemberger, K., Pallhuber, J., Doblinger, A., Troger, J., Kirchmair, R., Kralinger, M., Fischer-Colbrie, R., and Kieselbach, G. (2004). Secretoneurin in the human aqueous humor and the absence of an effect of frequently used eye drops on the levels. *Peptides* **25,** 2115–2118.

Storch, K. F., Lipan, O., Leykin, I., Viswanathan, N., Davis, F. C., Wong, W. H., and Weitz, C. J. (2002). Extensive and divergent circadian gene expression in liver and heart. *Nature* **417,** 78–83.

Takashima, Y., Taniguchi, T., Yoshida, M., Haque, M. S., Igaki, T., Itoh, H., Nakao, K., Honda, Y., and Yoshimura, N. (1998). Ocular hypotension induced by intravitreally injected C-type natriuretic peptide. *Exp. Eye Res.* **66,** 89–96.

Tanabe, Y., Masu, M., Ishii, T., Shigemoto, R., and Nakanishi, S. (1992). A family of metabotropic glutamate receptors. *Neuron* **8,** 169–179.

Taylor, A. W., and Yee, D. G. (2003). Somatostatin is an immunosuppressive factor in aqueous humor. *Invest. Ophthalmol. Vis. Sci.* **44,** 2644–2649.

Tao, W., Prasanna, G., Dimitrijevich, S., and Yorio, T. (1998). Endothelin receptor A is expressed amd mediates the [Ca^{2+}] immobilization of cells in human ciliary smooth muscle, ciliary nonpigmented epithelium, and trabecular meshwork. *Curr. Eye Res.* **17,** 31–38.

Tatemoto, K., Rökaeus, A., Jornvall, H., McDonald, T. J., and Mutt, V. (1983). Galanin- a novel biologically active peptide from porcine intestine. *FEBS Lett.* **164,** 124–128.

Thermos, K. (2003). Functional mapping of Somatostatin receptors in the retina. A review. *Vision Res.* **43,** 1805–1815.

Thoreson, W. B., and Witkovsky, P. (1999). Glutamate receptors and circuits in the vertebrate retina. *Prog. Retin. Eye Res.* **18,** 765–810.

Topper, J. E., and Brubaker, R. F. (1985). Effects of timolol, epinephrine, and acetazolamide on aqueous flow during sleep. *Invest. Ophthalmol. Vis. Sci.* **26,** 1315–1319.

Tosini, G., and Menaker, M. (1998). The clock in the mouse retina: Melatonin synthesis and photoreceptor degeneration. *Brain Res.* **789,** 221–228.

Troger, J., Doblinger, A., Leierer, J., Laslop, A., Schmid, E., Teuchner, B., Opatril, M., Philipp, L., Klimaschewski, L., Pfaller, K., Göttinger, W., and Fischer-Colbrie, R. (2005). Secretoneurin in the peripheral ocular innervation. *Invest. Ophthalmol. Vis. Sci.* **46,** 647–654.

Troger, J., Kieselbach, G., Teuchner, B., Kralinger, M., Nguyen, Q. A., Haas, G., Yayan, J., Gottinger, W., and Schmid, E. (2007). Peptidergic nerves in the eye, their source and potential pathophysiological relevance. *Brain Res. Rev.* **53,** 39–62(Review).

Tsai, L. H., Lee, Y. J., and Wu, J. (1999). Effect of excitatory amino acid neurotransmitters on acid secretion in the rat stomach. *J. Biomed. Sci.* **6,** 36–44.

Tse, C. M., Brant, S. R., Walker, M. S., Pouyssegur, J., and Donowitz, M. (1992). Cloning and sequencing of a rabbit cDNA encoding an intestinal and kidney-specific Na^+/H^+ exchanger isoform (NHE-3). *J. Biol. Chem.* **267,** 9340–9346.

Tse, C. M., Levine, S. A., Chris Yun, C. H., Montrose, M. H., Little, P. J., Pouyssegur, J., and Donowitz, M. (1993). Cloning and expression of a rabbit cDNA encoding a serum-activated ethylisopropylamiloride-resistant epithelial Na^+/H^+ exchanger isoform (NHE-2). *J. Biol. Chem.* **268,** 11917–11924.

Tsukaguchi, H., Tokui, T., Mackenzie, B., Berger, U. V., Chen, X. Z., Wang, Y., Brubaker, R. F., and Hediger, M. A. (1999). A family of mammalian Na+-dependent L-ascorbic acid transporters. *Nature* **399,** 70–75.

Unger, W. G. (1990). Review: Mediation of the ocular response to injury. *J. Ocul. Pharmacol.* **6,** 337–353.

Vorwerk, C. K., Naskar, R., Schuettauf, F., Quinto, K., Zurakowski, D., Gochenauer, G., Robinson, S. A., Mackler, S. A., and Dreyer, E. B. (2000). Depression of retinal glutamate transporter function leads to elevated intravitreal glutamate levels and ganglion cell death. *Invest. Ophthalmol. Vis. Sci.* **41,** 3615–3621.

Wersinger, E., Schwab, Y., Sahel, J. A., Rendon, A., Pow, D. V., Picaud, S., and Roux, M. J. (2006). The glutamate transporter EAAT5 works as a presynaptic receptor in mouse rod bipolar cells. *J. Physiol.* **577,** 221–234.

Wetzel, R. K., and Sweadner, K. J. (2001). Immunochemical localization of Na^{+}-K^{+}-ATPase isoforms in the rat and mouse ocular ciliary epithelium. *Invest. Ophthalmol. Vis. Sci.* **42,** 763-769.

Wiederholt, M., Sturm, A., and Lepple-Wienhues, A. (1994). Relaxation of trabecular meshwork and ciliary muscle by release of nitric oxide. *Invest. Ophthalmol. Vis. Sci.* **35,** 2515–2520.

Wiederholt, M., Thieme, H., and Stumpff, F. (2000). The regulation of trabecular meshwork and ciliary muscle contractility. *Prog. Retin. Eye Res.* **19,** 271–295(Review).

Wolfensberger, T. J., Singer, D. R., Freegard, T., Markandu, N. D., Buckley, M. G., and MacGregor, G. A. (1994). Evidence for a new role of natriuretic peptides: Control of intraocular pressure. *Br. J. Ophthalmol.* **78,** 446–448.

Yamaji, K., Yoshitomi, T., and Usui, S. (2003). Effect of somatostatin and galanin on isolated rabbit iris sphincter and dilator muscles. *Exp. Eye Res.* **77,** 609–614.

Yamazaki, S., Numano, R., Abe, M., Hida, A., Takahashi, R., Ueda, M., Block, G. D., Sakaki, M., Menaker, M., and Tei, H. (2000). Resetting central and peripheral circadian oscillators in transgenic rats. *Science* **288,** 682–685.

Yorio, T., Prasanna, G., and Coca-Prados, M. (2008). The Ciliary Body: A potential multifaceted functional neuroendocrine unit. *In* " Ocular Therapeutics" (T. Yorio, A. F. Clark, and M. B. Wax, eds.). Chapter 4, pp. 69–85. Academic Press, New York.

Zhang, D., Vetrivel, L., and Verkman, A. S. (2002). Aquaporin deletion in mice reduces intraocular pressure and aqueous fluid production. *J. Gen. Physiol.* **119,** 561–569.

Zylka, M. J., Shearman, L. P., Weaver, D. R., and Reppert, S. M. (1998). Three period homologs in mammals: Differential light responses in the suparachiasmatic circadian clock and oscillating transcripts outside of brain. *Neuron* **20,** 1103–1110.

CHAPTER 6

Aqueous Humor Outflow Resistance

Thomas F. Freddo* and Mark Johnson†
*School of Optometry, University of Waterloo, Ontario N2L 3G1, Canada
†Departments of Biomedical Engineering and Ophthalmology, Northwestern University, Illinois 60208

I. OVERVIEW

This chapter provides a summary of the work that has led to our current day understanding of conventional aqueous outflow. The anatomy and basic physiology of aqueous humor outflow through the trabecular meshwork, via Schlemm's canal, to the episcleral venous system is reviewed. Various postulates concerning the manner in which aqueous outflow resistance is generated are reviewed as well.

Current Topics in Membranes, Volume 62 1063-5823/08 $35.00

DOI: 10.1016/S1063-5823(08)00406-7

II. INTRODUCTION

It was already recognized in the seventeenth century by Rikchard Banister that glaucoma was associated with an elevated intraocular pressure (IOP) (Sorsby, 1963), although this was not widely accepted until the nineteenth century. Recognition that this elevated pressure was related to a blockage of the aqueous humor outflow pathway began with the studies of Leber, Schwalbe, Knies, and Smith (Schwalbe, 1870; Leber, 1873; Knies, 1875; Smith, 1888). The flow resistance of this pathway, in both normal and glaucomatous eyes, has been the focus of study for the past 130 years. Despite this effort medical treatments for glaucoma, which are directed at the principal outflow pathway, remain few. And none of these are among the current first line of therapy.

Aqueous humor is secreted by the ciliary body into the posterior chamber of the eye. Aqueous humor cannot traverse the intact iris and thus it passes through the pupil to reach the anterior chamber of the eye. Since the pupillary margin rests on the crystalline lens of the eye, the pupil acts as a one-way valve, preventing reflux of aqueous into the posterior chamber. Changes in the relationship between the pupil and the lens can alter aqueous flow. In some eyes, a mid-size pupil diameter can produce a relative blockage of aqueous movement into the anterior chamber. In the face of intraocular inflammation, the pupillary margin may become bound to the lens (aka posterior synechia). The result may be an elevation of IOP.

Aqueous humor that has reached the anterior chamber circulates in a convective current driven by the temperature difference between the warm iris and the cooler cornea. Aqueous humor rises in the back of the anterior chamber and falls along the inner surface of the cornea, while at the same time, flowing toward the "angle," where the iris and ciliary body insert into the sclera. At the angle, the bulk of this flow enters a pathway known as the conventional aqueous outflow pathway composed of the trabecular meshwork, the juxtacanalicular connective tissue (JCT), the endothelial lining of the inner wall of Schlemm's canal, Schlemm's canal itself, and the collecting channels that lead to the episcleral veins (Fig. 1).

Since this is a bulk flow, driven by a passive pressure gradient, it is clinically important to remember that anything leading to elevation of episcleral venous pressure will require IOP to rise to whatever level is required to surpass episcleral venous pressure in order for outflow to resume.

A small fraction of the aqueous humor flows out of a second pathway know as the "unconventional" pathway. The fluid traveling along this pathway also enters at the angle of the eye, but then travels posteriorly through the ciliary body and ciliary muscle, to the supraciliary and suprachoroidal spaces (Fig. 2; Bill, 1964a, b, 1965; Bill and Hellsing, 1965). The route by

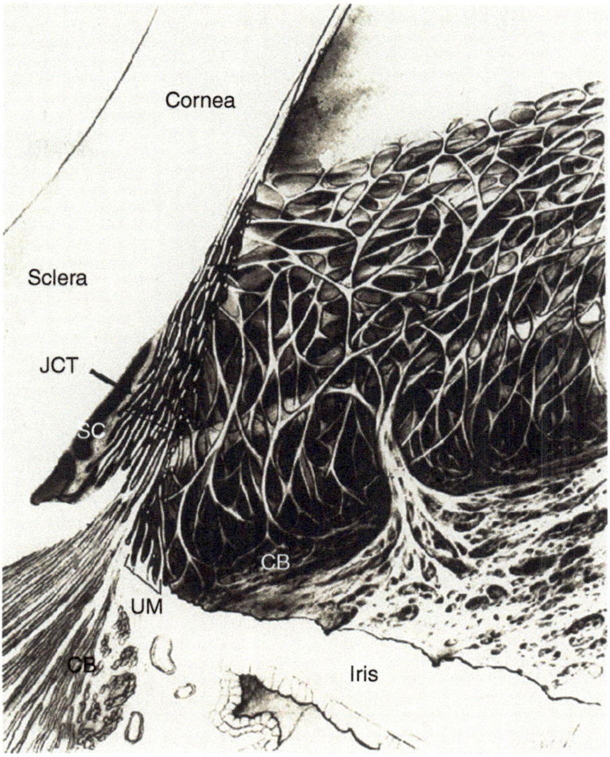

FIGURE 1 Sketch of the angle region of the human eye showing the relationship between the trabecular meshwork and surrounding structures. CB, ciliary body; UM, uveal meshwork; SC, Schlemm's canal; JCT, juxtacanalicular meshwork. [Adapted from Hogan, M., *et al.*, (1971). "Histology of the Human Eye" (W.B. Saunders), Figs. 4–16.]

which the fluid exits the eye from here is still debated. Some studies suggest that this fluid passes through the sclera into the episcleral tissue (Bill, 1966, 1975; Bill and Phillips, 1971), while other studies suggest that this fluid is absorbed osmotically into the choroidal vessels and vortex veins (Pederson *et al.*, 1977; Sherman *et al.*, 1978; Ethier *et al.*, 2004). This question is complicated by the difficulty of measuring flow through this pathway (Johnson and Erickson, 2000), and by the need to make this measurement *in vivo* (Wagner *et al.*, 2004). This pathway carries less than 10% of the total flow in the older adult human eye, but is important in the treatment of glaucoma as the mechanism of action of $PGF_{2\alpha}$ and commercially available prostanoids is on this pathway (Crawford and Kaufman, 1987; Gabelt and Kaufman, 1989).

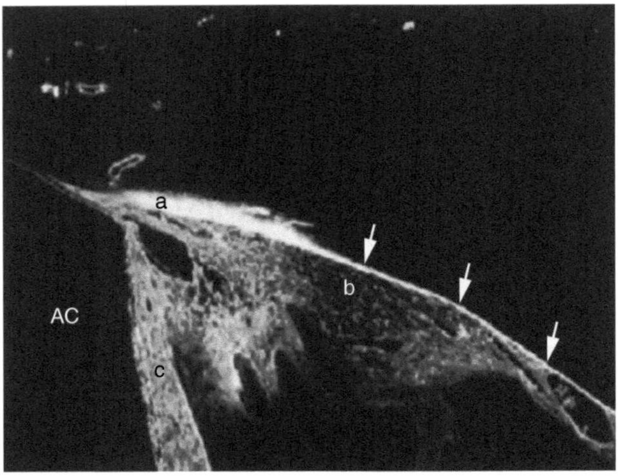

FIGURE 2 Sagittal section of a rabbit eye intracamerally injected with fluorescent tracer that has highlighted both the conventional outflow pathway (a) and the unconventional outflow pathway through the ciliary muscle (b) to the supraciliary space (arrows). AC, anterior chamber; c, iris. (Tripathi, 1977a; Fig. 2).

Our focus in this chapter is on the conventional outflow pathway. We begin with an examination of the aqueous humor itself. Then, we briefly review the flow resistance of those aspects of the outflow pathway that are thought to have minor roles in the generation of aqueous outflow resistance. Finally, we turn our focus to the region near the inner wall of Schlemm's canal where the bulk of outflow resistance is thought to be generated. However, we first begin by defining what we mean by flow resistance.

Fluid flow through a tissue, in the absence of active pumping, is driven by a gradient in hydrostatic and osmotic pressures. As there is no significant osmotic pressure difference between the fluid in the anterior chamber and the blood into which it empties (Bárány, 1963), it is simply the pressure difference ($\triangle P$, typically 5 mm Hg) between the IOP and the episcleral venous pressure that drives flow through the aqueous outflow network. The ratio of this pressure difference to the flow of aqueous humor passing through this system (Q, typically 2 μl/min) is the flow resistance:

$$R = \frac{\Delta P}{Q} \tag{1}$$

The inverse of the outflow resistance is known as the outflow facility.

III. THE AQUEOUS HUMOR

The question first arises as to what are the flow properties of the fluid passing through the aqueous outflow network. Aqueous humor is secreted by the ciliary epithelium from an ultrafiltrate of blood. It has a protein concentration roughly 0.5–1% of that in serum, depending on the specie (Gaasterland *et al.*, 1979; Dernouchamps, 1982; Pavao *et al.*, 1989). The most important hydrodynamic property of aqueous humor is its viscosity, and this has been measured to be essentially the same as that of saline (Beswick and McCulloch, 1956; Balazs *et al.*, 1959).

Johnson *et al.* (1986) reported that ultracentrifuged aqueous humor, when passed through microporous filters with pore sizes similar to the smallest openings found in the outflow pathway, can obstruct flow through these filters. Of interest, serum diluted to the same protein concentration as aqueous humor did not obstruct the filters. This obstruction with aqueous humor appeared to be due to a hydrophobic interaction between the filter surface and the proteins in the aqueous humor and could be relieved with a protease but not a GAGase (Johnson *et al.*, 1986; Ethier *et al.*, 1989; Pavao *et al.*, 1989). More recent work indicates that myocilin, a protein associated with juvenile and primary open-angle glaucoma (POAG) (Polansky *et al.*, 1997; Stone *et al.*, 1997; Fingert *et al.*, 2002), may be involved in this process of filter obstruction by aqueous humor (Russell *et al.*, 2001). It is not known whether this filter-blocking behavior of aqueous humor is of physiological significance.

IV. REGIONS OF LOW OUTFLOW RESISTANCE

Upon entering the conventional aqueous outflow pathway, the aqueous humor enters the trabecular meshwork, an avascular tissue composed of the uveal meshwork, the deeper corneoscleral meshwork, and the still deeper JCT (Fig. 1). The uveal meshwork consists of a set of beams organized into an irregular netlike structure (Fig. 3). This is a very open and porous network and negligible flow resistance is expected in the region, an observation confirmed experimentally by Grant (1963).

The corneoscleral meshwork extends from the uveal meshwork ~100 μm in the direction of flow toward Schlemm's canal. It consists of a number of interconnected sheets or trabeculae that extend from the peripheral cornea to the scleral spur. These sheets, like the cores of the uveal meshwork beams, have an avascular core of collagen and elastin covered with a basal lamina and finally a single layer of endothelial cells (Fig. 3; Tripathi, 1974). The number of trabecular endothelial cells decreases with age and is further decreased in glaucoma (Alvarado *et al.*, 1981).

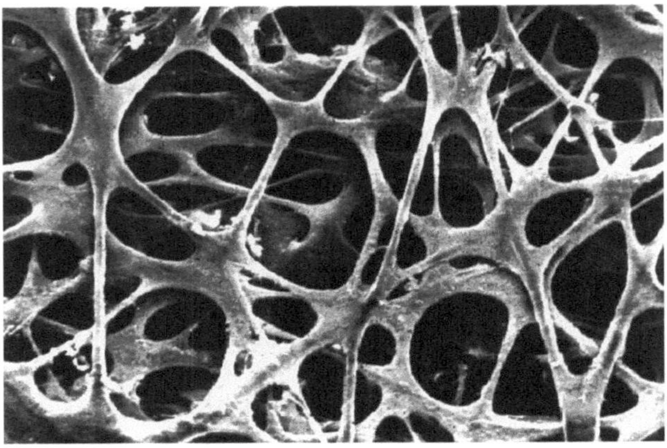

FIGURE 3 Scanning electron micrograph shows the uveal face of the trabecular meshwork showing branching and anastomosing trabecular beams covered in a thin layer of endothelial cells. Note how the open spaces get smaller in layers beneath the surface. [Freddo, T. F., *et al.* (1984)].

The trabecular meshwork exhibits openings that get progressively smaller as the deeper layers of the meshwork are reached (Fig. 3). Its design is like that of a filter. The trabecular endothelial cells on the surface of this filter are phagocytic and can thus ingest materials trapped by this filter (Rohen and van der Zypen, 1968). McEwen (1958) used Poiseuille's law to show that there is negligible flow resistance in the region.

The aqueous humor next passes through the JCT. The JCT and the endothelial lining of the inner wall of Schlemm's canal (and its basement membrane) are the regions where the bulk of outflow resistance is thought to be generated in the normal eye (Fig. 4). We defer our discussion of this region until the following section.

Upon passing through the inner wall of Schlemm's canal, the aqueous humor enters the canal itself. When cut in cross section, the canal has the appearance of a highly elongated ellipse [Freddo, 1993 (and revised 1999)], with its major axis having a diameter varying between 150 and 350 μm (Ten Hulzen and Johnson, 1996); while the minor axis (the distance between the inner and outer wall of the canal) can vary between roughly 1 and 30 μm, depending on the IOP (Fig. 5). As IOP increases, the canal progressively collapses (Johnstone and Grant, 1973; Johnson and Kamm, 1983). As it collapses, the outflow resistance it generates is increased (Moses, 1979; Van Buskirk, 1982). Preventing collapse of Schlemm's canal is likely the mechanism by which muscarinic agents (e.g., pilocarpine) act to decrease outflow resistance (Johnson and Erickson, 2000). These agents cause the longitudinal fibers of the ciliary muscle to contract, thus pulling on a system of elastic

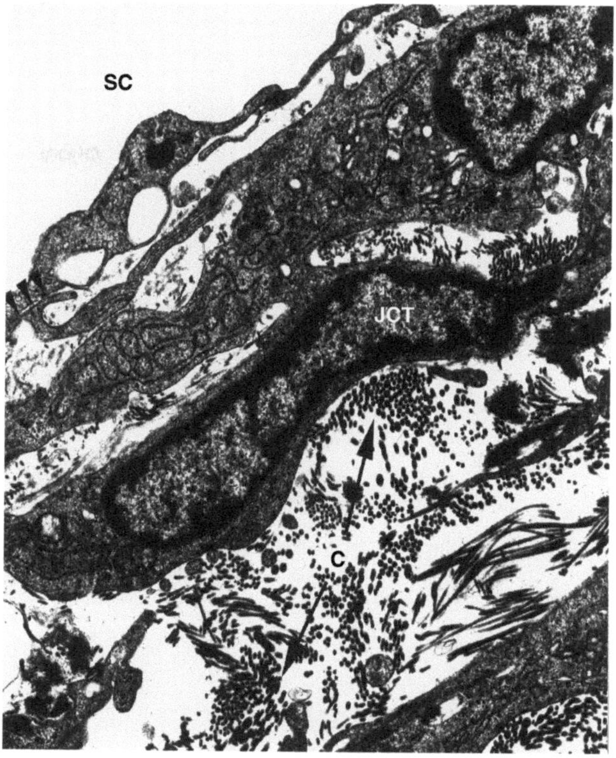

FIGURE 4 Transmission electron micrograph shows juxtacanalicular region (JCT) of the trabecular meshwork and both the endothelial lining and lumen of Schlemm's canal (SC). The JCT region contains an open extracellular matrix including collagen (C) and elastin. The fibroblast-like cells of the region extend filipodial connections to the endothelial cells lining Schlemm's canal (arrows) (Freddo, 1993).

fibers termed the cribriform plexus, which makes connections into the JCT region and the endothelial lining of inner wall of Schlemm's canal (Rohen, 1983; Gong *et al.*, 1989; Figs. 6 and 7).

A similar effect on outflow facility occurs experimentally when the lens is pushed posteriorly in the eye, thus pulling on the zonules that in turn pull the ciliary muscle in the posterior direction. This opens the canal in a fashion analogous to the action of muscarinic agents. Lens depression only decreases outflow resistance at elevated IOP when Schlemm's canal is collapsed (Fig. 8), confirming that the mechanism of action of muscarinic agents is largely one of preventing collapse of Schlemm's canal, despite earlier assumptions that the principal action of miotics was to pull the scleral spur posteriorly, thus "opening up" the trabecular meshwork (Van Buskirk, 1976; Rosenquist *et al.*, 1988).

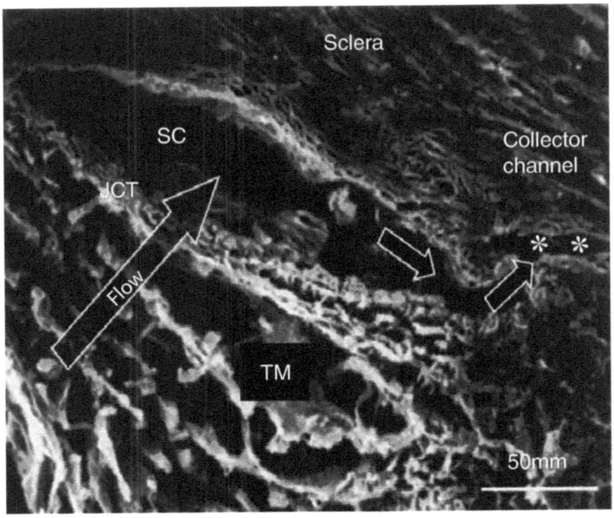

FIGURE 5 Scanning electron micrograph showing a sagittal section through the trabecular meshwork (TM), the juxtacanalicular region (JCT), Schlemm's canal (SC), and one of the external collector channels that leads from Schlemm's canal to the episcleral venous system (Freddo, 1993).

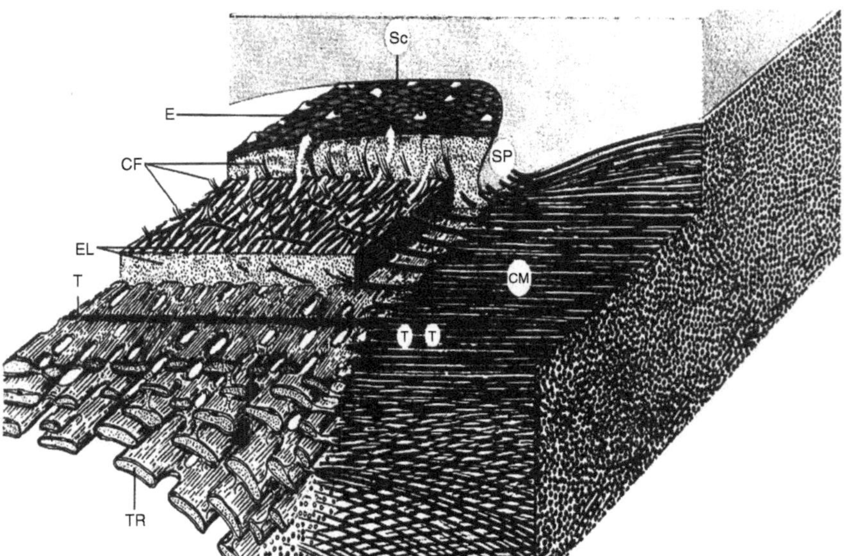

FIGURE 6 Tendons (T) extending from the longitudinal bundle of the ciliary muscle (CM) attach to the scleral spur (SP) but also extend elastic (EL) connecting fibrils (CF) to attach to the endothelial cells (E) lining Schlemm's canal (SC) (Rohen, 1983).

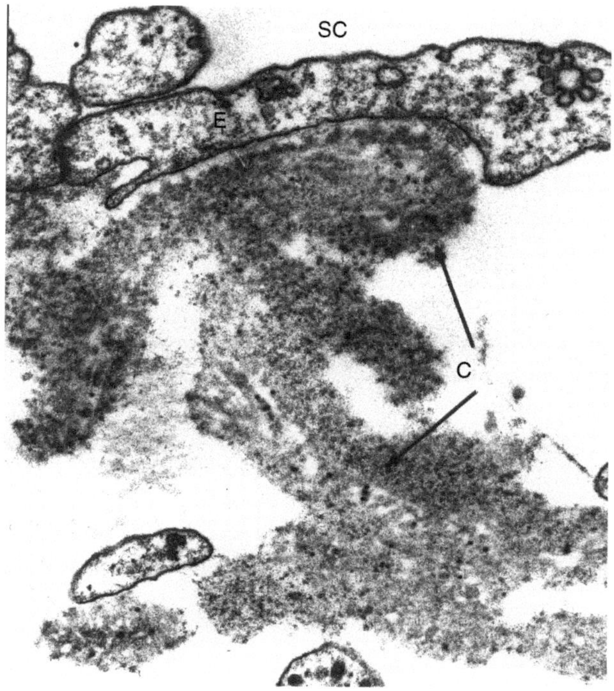

FIGURE 7 High magnification electron micrograph demonstrating detail of an elastic connecting fibril (C) of the cribriform plexus making connection with an endothelial cell (E) of the inner wall of Schlemm's canal (SC) (Gong *et al.*, 1989; Fig. 3).

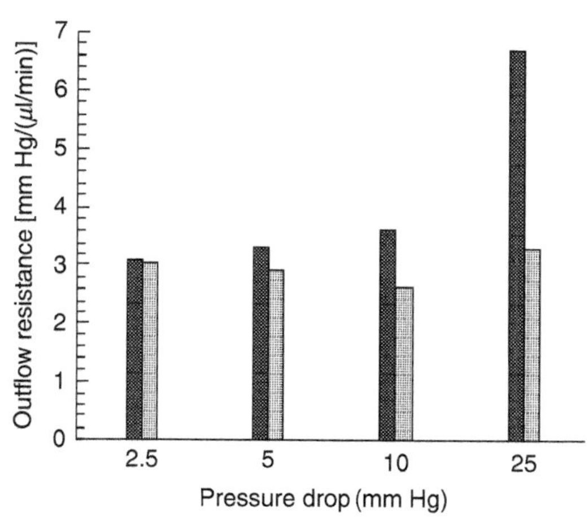

FIGURE 8 Effect of pressure drop (IOP, episcleral venous pressure) on outflow resistance in enucleated eyes without (▩) or with lens depression (▤); data from (Van Buskirk 1976).

The potential for collapse of Schlemm's canal as IOP increases has led some to speculate that this might be a cause of POAG (Nesterov, 1970). However, the outflow resistances of normal enucleated eyes, perfused at pressures that lead to extensive collapse of Schlemm's canal (Johnstone and Grant, 1973), were not as high as those of typical glaucomatous eyes that range from 10 to 100 mm Hg/(μl/min) in untreated glaucoma (Grant, 1951). Taken together, these findings suggest that while collapse of Schlemm's canal might make a glaucomatous condition worse, it is not likely to be a primary cause of the disease (Johnson and Kamm, 1983).

Leading from the outer wall of Schlemm's canal are approximately 30 collector channels spaced around the circumference of Schlemm's canal. These collector channels connect to the deep scleral plexus, then the intrascleral venous plexus, and finally, the episcleral veins where aqueous humor mixes with the venous blood. In some eyes, a smaller number of vessels lead directly from Schlemm's canal to the episcleral veins, bypassing the scleral plexi. These are termed aqueous veins and are identified clinically by the fact that aqueous humor and blood are seen to run within them in a laminar flow (Fig. 9). Both the collector channels and the channels to which they lead,

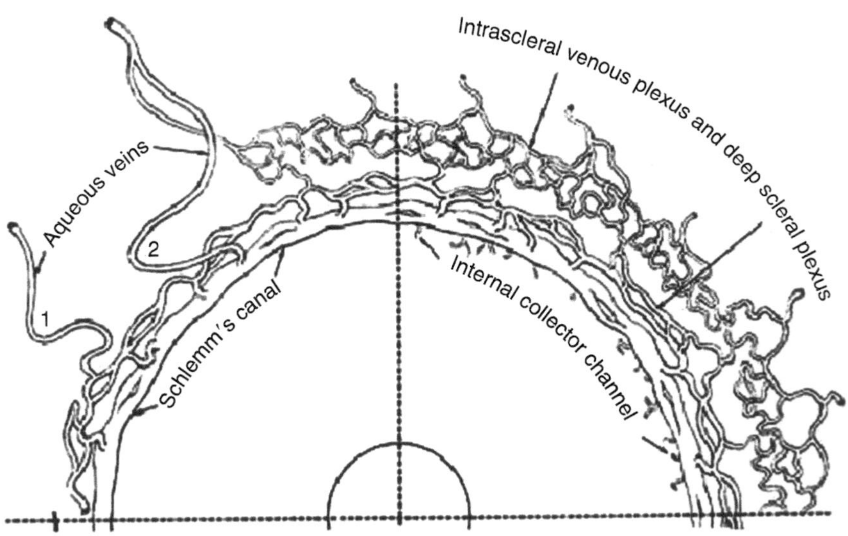

FIGURE 9 Diagram showing the aqueous outflow pathways from Schlemm's canal to the episcleral vessels. External collector channels emerging from the outer wall of Schlemm's canal (upper right) lead to deep and intrascleral plexuses and then to the episcleral vessels. Aqueous veins (upper left) arising from either external collector channels or the outer wall of Schlemm's canal, bypass this more convoluted pathway to the episcleral veins. [Modified from: Hogan, M., *et al.*, (1971). Figs. 4–19.]

enroute to the venous blood, have diameters that are tens of micrometers in size, and calculations indicate that these vessels should have negligible flow resistance (Dvorak-Theobald, 1934; Batmanov, 1968; Rohen and Rentsch, 1968; Rosenquist *et al.*, 1988).

The experimental evidence regarding this question is less clear. Mäepea and Bill (1992) using micropipettes measured the pressure in Schlemm's canal in living monkeys and received results in agreement with the theoretical calculations, namely, that the collector channels and vessels leading to the episcleral veins generated less than 10% of the total outflow resistance (Mäepea and Bill, 1989). However, a number of investigators have perfused enucleated primate and human eyes following a 360° trabeculotomy that should remove all flow resistance proximal to the collector channels and aqueous veins. These studies show that at high IOP, 25% of the flow resistance resides in collector channels and the vessels leading to the episcleral veins, while at low IOP (7 mm Hg in an enucleated eye), ~50% of the flow resistance resides in these vessels.

This discrepancy has not yet been resolved, perhaps partially because other findings indicate that the increased outflow resistance characteristic of POAG is not caused by an increased flow resistance of the collector channels or the vessels leading to the episcleral veins. Grant (1963) found that, in eight enucleated eyes from patients with POAG, a 360° trabeculotomy eliminated all of the elevated outflow resistance of these eyes, indicating that the enhanced flow resistance in POAG is proximal to the collector channels. This conclusion is further supported by the success of laser trabeculoplasty (LTP) in reducing the outflow resistance of such glaucomatous eyes (Wise and Witter, 1979). While it is not clear exactly how LTP works to lower IOP, the site of action appears to be in the trabecular meshwork rather than acting upon the collector channels or the vessels leading to the episcleral veins (Van Buskirk *et al.*, 1984; Bradley *et al.*, 2000; Johnson and Erickson, 2000).

The considerations above suggest that both in the normal eye and in the glaucomatous eye, the only tissues capable of generating a significant fraction of outflow resistance are those tissues in the immediate and vicinity of the JCT, the basement membrane of the inner wall endothelium of Schlemm's canal, and the endothelium itself. Experimental measurements attempting to localize the pressure drop in the outflow pathway lead to this same conclusion (at least in normal eyes) (Mäepea and Bill, 1992; Johnson, 2006).

V. REGIONS OF POTENTIAL SIGNIFICANT OUTFLOW RESISTANCE

Seidel (1921), examining the outflow pathway in 1921, stated that "the inner wall of Schlemm's canal stands in open communication with the anterior chamber, and that the aqueous humor directly washes around the inner wall endothelium of Schlemm's canal and is only separated from the lumen by a thin, outer membrane". It is either in or around this location that the bulk of outflow resistance likely resides in both the normal and the glaucomatous eye. We today tend to refer to this inner wall region as including the JCT, the basal lamina of the inner wall endothelium of Schlemm's canal, and the endothelium itself.

A. Extracellular Matrix and the JCT

The JCT, also called the endothelial meshwork or cribriform region, is the portion of the meshwork positioned between the beams of the corneoscleral meshwork and the basal lamina of the inner wall of Schlemm's canal. It varies in thickness between a few micrometers in some locations to perhaps 10 μm in others. It is not nearly as well ordered as is the corneoscleral meshwork. There is nothing resembling a beamlike structure. It is, instead, composed of a loose connective matrix that is very porous (30–40% open space) under typical flow conditions (Figs. 4, 10, and 11; Ten Hulzen and Johnson, 1996).

The cells in this region, whose type has not been definitively determined, are fibroblastic in appearance and lack a basal lamina (Gong *et al.*, 1996). They are connected to one another and also to the collagen and elastic fibrils in this tissue (Tervo *et al.*, 1995). These cells exhibit thin processes that make connections with the endothelium of the inner wall of Schlemm's canal (Fig. 4).

The extracellular matrix in the JCT region includes collagen types I, III, IV, V, and VI (but not type II) (Lütjen-Drecoll *et al.*, 1989; Marshall *et al.*, 1990, 1991) elastin, (Gong *et al.*, 1989); laminin (Marshall *et al.*, 1990); fibronectin (Gong *et al.*, 1996); and glycosaminoglycans (GAGs),

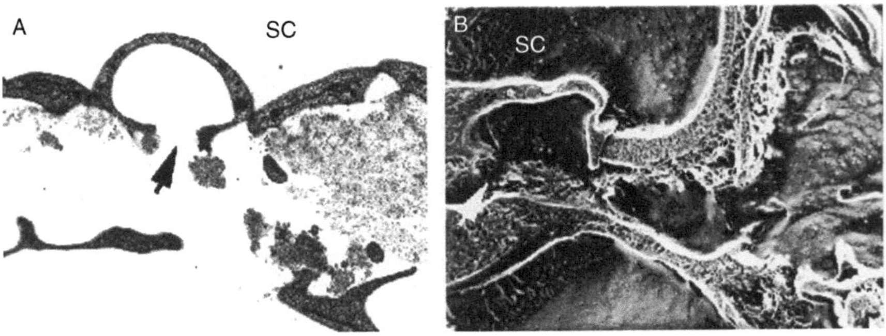

FIGURE 10 Micrographs of inner wall of Schlemm's canal (SC) and JCT from an 81-year-old eye. (A) Conventional TEM and (B) QFDE image preparation. Both show vacuoles with discontinuities in their basal lamina at the basal opening into the vacuole (arrows) (×11,100) (Gong *et al.* 2002).

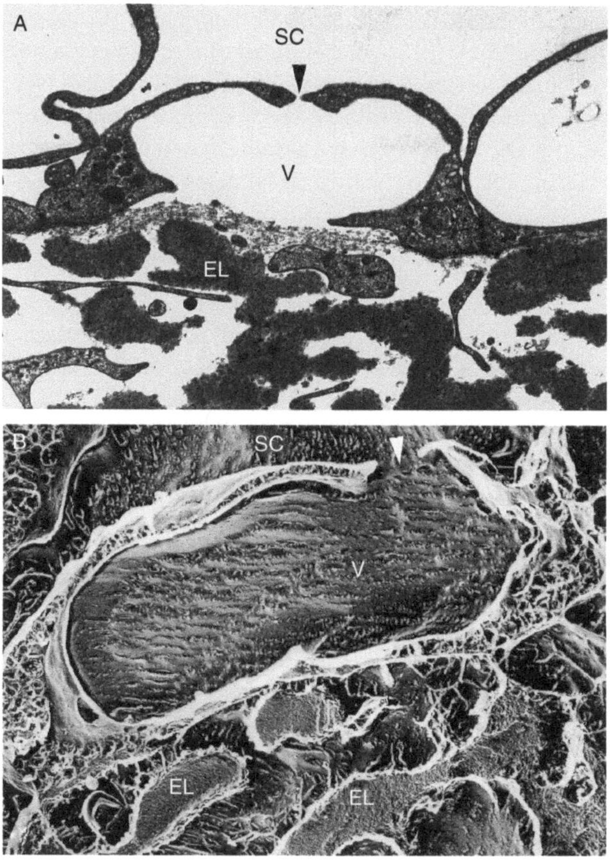

FIGURE 11 Transmission electron micrograph and quick-freeze, deep-etch micrograph of giant vacuoles exhibiting "pores" connecting the lumen of the giant vacuole (V) with the lumen of Schlemm's canal (SC). EL, elastic fibers in JCT region. (Courtesy of Haiyan Gong.).

particularly chondroitin sulfate, dermatan sulfate, and hyaluronic acid (Gong *et al.*, 1996). In glaucoma, there is a loss of hyaluronic acid from this region (Knepper *et al.*, 1996b). There is also an accumulation of a material called plaque (Lütjen-Drecoll *et al.*, 1981), although it does not appear to have any hydrodynamic consequences (Alvarado *et al.*, 1986; Murphy *et al.*, 1992).

The tortuous and relatively small flow pathways through the JCT were once attractive candidates for the generation of significant outflow resistance. Surprisingly, these expectations were not supported by hydrodynamic considerations (Kamm *et al.*, 1983; Seiler and Wollensak 1985; Ethier *et al.*, 1986). Fluid flow through complicated structures such as soils, polymer networks, and connective tissues are frequently modeled by using porous media theory. In such an approach, the specific hydraulic conductivity (K) of the medium is the property that characterizes its intrinsic capacity to carry flow. It is closely related to the flow resistance of that tissue:

$$R = \frac{\mu L}{KA} \tag{2}$$

where μ is the viscosity of the fluid passing through the tissue, L is the flow-wise length, and A is the cross-sectional area facing flow. For most connective tissue, K ranges in value from 10^{-13} to 10^{-15} cm^2 (Johnson, 2006), although the very loose vitreous humor has a specific hydraulic conductivity of 10^{-11} cm^2.

If we assume that the entire pressure drop in the aqueous outflow pathway (roughly 5 mm Hg) occurs across the JCT, then we can use Eqs. (1) and (2) to estimate the K of the JCT. As noted above, aqueous humor has a viscosity similar to that of saline (0.007 g/cm s) and flows through the outflow pathway at a rate of ~2 μl/min. The approximate cross-sectional area facing flow can be determined by multiplying the width of Schlemm's canal (150–350 μm) (Ten Hulzen and Johnson, 1996) by it length around the eye of ~3.6 cm.

The only parameter that is not well known is the length over which the pressure drop occurs. Mäepea and Bill (1992) used micropressure measurements in the outflow pathway to find that this pressure drop occurs within 14 μm of the inner wall of Schlemm's canal. We can then use Eqs. (1) and (2) to conclude that if all or most of the pressure drop in the outflow pathway occurs across the JCT, then the K of this tissue must be less than 9×10^{-13} cm^2 (Johnson and Erickson, 2000; Johnson, 2006).

K for a tissue can also be estimated from micrographs showing the ultrastructure of that tissue (Johnson, 2006). By morphometrically characterizing the open spaces in that tissue, K can be determined. Carmen-Kozeny theory relates the structure of a porous medium to K as:

$$K = \frac{\varepsilon D_h^2}{80} \tag{3}$$

where D_h is the hydraulic diameter characterizing the open spaces in the porous medium and ϵ is the porosity or fraction of open space in the medium. D_h can be found by determining both the porosity of a tissue and the surface area per unit volume of its open spaces (α) as $D_h = 4\varepsilon^3/\alpha^2$.

Several groups have used such an approach to estimate K of the JCT based on its morphological appearance as seen by conventional electron microscopy (EM) (Kamm *et al.*, 1983; Seiler and Wollensak, 1985; Ethier *et al.*, 1986; Murphy *et al.*, 1992; Ten Hulzen and Johnson, 1996). In these studies, D_h was typically 1–1.5 μm, and more importantly, K of the JCT was calculated to be $\sim 200–1000 \times 10^{-13}$ cm^2. This is at least 20 times greater than the maximum value of K of this tissue based on experimental measurements.

Based on these findings, Ethier *et al.* (1986) concluded that the JCT region, as visualized using conventional EM techniques, could not generate an appreciable fraction of aqueous outflow resistance. They further concluded that this region must either be filled with an extracellular matrix gel that is poorly visualized using conventional EM techniques, or that this region is not the primary site of outflow resistance.

To evaluate the first possibility, Gong *et al.* (2002) used the quick-freeze/deep-etch (QFDE) method to examine the apparent open spaces in the JCT region in detail. QFDE is a technique that allows exquisite preservation of tissue ultrastructure at nanometer length scales. Using this technique, a much more elaborate and extensive extracellular matrix was seen in the JCT than seen using conventional techniques; however, openings nearly a micrometer in size were still seen in this region casting doubts as to whether a significant fraction of outflow resistance could be generated by this tissue (Figs. 10 and 11).

An important caveat pointed out by Gong *et al.* (2002) was that it was not clear whether and to what extent GAGs would be well preserved using their methods, and this reservation leaves the question of generation of significant outflow resistance in the JCT region in doubt. There is conflicting evidence in the literature as to whether GAGs and other extracellular moieties contribute to aqueous humor outflow resistance.

B. Possible Role of Glycosaminoglycans

Proteoglycans consist of a protein core to which negatively charged GAGs side chains are attached. The resulting structure is space filling as a consequence of the highly charged GAGs, and this gives rise to the potential to generate significant flow resistance. This characteristic also leads to these structures being difficult to preserve during morphological examination since the counterions used as stains for conventional TEM and the salt that remains after the sublimation step in QFDE would each be expected to collapse the GAG structures.

Along with GAGs, other extracellular moieties such as small nonfibrillar collagens and fibronectin have also been shown in other connective tissues to be associated with the generation of flow resistance. However, unlike GAGs, these other extracellular macromolecules do not collapse when tissues are prepared for EM examination. Thus, as the appearance of the JCT as seen by both conventional EM and by QFDE preparation techniques appears not to be consistent with the generation of appreciable flow resistance, only GAGs are candidates as extracellular molecules in the JCT that might generate significant flow resistance.

Early studies by Bárány (1953, 1956) showed that testicular hyaluronidase dramatically decreased outflow resistance in enucleated bovine eyes. Pedler (1956) confirmed these findings. This was consistent with the role of GAGs in other tissues (Meyer, 1953; Comper and Laurent, 1978). Testicular hyaluronidase has been reported to increase outflow facility in a guinea pigs (Melton and DeVille, 1960), dogs (Van Buskirk and Brett, 1978), and rabbits (Knepper *et al.*, 1984). Knepper (1980) found that chondroitinase AC, chondroitinase ABC, and *Streptomyces* hyaluronidase increased outflow facility in the enucleated rabbit eye in a dose-dependent manner.

While the outflow pathways of nonprimates appear to be sensitive to these agents that degrade GAGs, the evidence is far less clear in primate and humans. Peterson and Jocson (1974) found a significant effect of testicular hyaluronidase on enucleated primates eye and Sawaguchi *et al.* (1992) reported that chondroitinase ABC decreased IOP in living cynomolgus monkeys as compared with control eyes receiving heat-inactivated enzymes. Hubbard *et al.* (1997) found no effect of chondroitinase ABC or *Streptomyces* hyaluronidase on IOP or outflow facility, either chronically or acutely, in living monkeys. Furthermore, studies on human eyes have shown no effects of GAGase on outflow resistance (Pedler, 1956; Grant, 1963).

Indeed, biochemical studies show a decrease in hyaluronan in the meshwork in glaucoma (Knepper *et al.*, 1996a), and additional studies also have shown a decrease in sulfated proteoglycans with age in normal human trabecular meshwork (Gong *et al.*, 1992). A decrease in these extracellular matrix components with age would be inconsistent with attribution of an increase in outflow resistance to excess accumulation of GAGs in an age-related disease such as glaucoma.

More recently, the role of the other extracellular matrix components in contributing to outflow resistance in human eyes has been supported by work of Acott's group showing that matrix metalloproteinases (MMPs) reversibly increase outflow facility in perfused human anterior segment organ culture (Bradley *et al.*, 1998). However, MMPs are relatively nonspecific in their action, and the locus of their activity was not determined in this study. It is possible that the MMPs were acting not on extracellular matrix in the JCT, but instead on the basement membrane of the cells of the inner wall endothelium of Schlemm's canal, the topic we address in the following section.

C. *Possible Role of the Basement Membrane of the Endothelial Lining of Schlemm's Canal*

The fundamental role of the basement membrane is as a structural support of the epithelial tissue it supports. Vascular endothelium requires a strong substrate to support it against the load of the vascular transmural pressure. In the case of the aqueous outflow pathway, attachments of the Schlemm's canal endothelial cells to this underlying substrate may assist the cell to "hold on" against the flow passing through this endothelium and entering Schlemm's canal.

The basement membrane can be a source of significant flow resistance in some tissues. Typically, the flow resistances of physiological membranes are described in term of their hydraulic conductivity (L_p) which is defined as the flow per unit area per pressure drop. This can be related to *K* as follows:

$$L_p \equiv \frac{Q/A}{\Delta P} = \frac{K}{\mu L} \tag{4}$$

Since the flow per unit area (Q/A) and the pressure drop of the outflow pathways are properties that are well known, it is straightforward to determine that L_p is between 4000×10^{-11} and 9000×10^{-11} cm^2 s/g (Johnson, 2006).

Table I shows measured value of L_p for several basement membranes. L_p for the aqueous outflow pathway is comparable to that of other basement membranes also involved in water transport, namely, Bruch's membrane through which the retinal pigment epithelium pumps fluid, and of course, the kidneys.

This supports the possibility that the basement membrane of the inner wall endothelium of Schlemm's canal might generate a significant flow resistance. Further supporting this possibility is the composition of basement membranes. The type IV collagen, heparan sulfate, fibronectin, and laminin that make up basement membrane would be expected to be degraded by the MMPs of the types shown to affect outflow resistance by Acott's group (Bradley *et al.*, 1998).

Morphological examination of the inner wall basement membrane using conventional methods of tissue preparation for EM do not preserve the basement membrane in sufficient detail to allow a morphometric analysis of the flow resistance such as has been done on the JCT. However, it has been reported (Gong *et al.*, 1996) that unlike vascular basement membranes, the basement membrane of the inner wall endothelium is discontinuous. Studies using QFDE appear to confirm this conclusion (Fig. 10; Gong *et al.*, 2002).

TABLE I

Hydraulic Conductivity of Basement Membranes.

Tissue	$L_p \times 10^{11}$ (cm^2 s/g)
Descement's membrane (Fatt, 1969)	15–37
Lens capsule (Fisher, 1982)	17–50
Bruch's membrane (eyes under 40 years old) (Starita *et al.*, 1996)	2000–12,500
Kidney tubule basement membrane (Bentzel and Reczek, 1978; Welling and Welling, 1978)	6300–13,700
Renal glomerulus basement membrane (Daniels *et al.*, 1992)	7600–25,000

It is hard to see how the basement membrane can be a significant source of flow resistance in the aqueous outflow pathway if it is a discontinuous layer, as the fluid would flow through the breaks rather than passing through the membrane itself. While the flow resistance of these breaks have not been explicitly calculated, these breaks are ubiquitous and typically are a fraction of a micrometer or larger in size. These breaks would not be expected to generate significant outflow resistance.

It is important to mention that Hann *et al.* (2001) found no difference in ultrastructural labeling for fibronectin, laminin, or type IV collagen comparing normal to glaucomatous eyes in the basal lamina of Schlemm's canal. This suggests that even though the basement membrane was found to be a significant source of outflow resistance in the normal eye, it is not likely to be responsible for the elevated flow resistance characteristic of the glaucomatous eye.

Changes have been found in glaucomatous eyes in the cells of their inner wall endothelium as compared to normal eyes. We now turn our examination to that tissue.

VI. ENDOTHELIAL LINING OF SCHLEMM'S CANAL

A. *How Does Aqueous Humor Cross the Continuous Endothelial Barrier Presented by the Endothelial Lining of Schlemm's Canal?*

The debate on how or even whether aqueous humor crosses the inner wall of Schlemm's canal was engaged early in the study of glaucoma. No less prominent luminaries than Schwalbe and Leber were diametrically opposed on this issue, the former contending that open communications across

Schlemm's canal must exist and the latter insisting that the inner wall of Schlemm's canal was a continuous membrane requiring either passive filtration or active transport (Leber, 1895). More than 100 years later the debate has been refined but certainly not resolved.

The inner wall of Schlemm's canal is composed of endothelial cells that appear to be vascular in their embryological origin (Krohn, 1999). These cells encounter a direction of flow that is inward, toward the lumen of the canal. As such, they are presumed to share more characteristics with postcapillary venules and lymphatics rather than with arterioles. A recent comprehensive comparison of known characteristics of blood capillaries, lymphatics, and the wall of Schlemm's canal concluded that the inner wall of Schlemm's canal is unique, sharing some but not all of the features of either vascular or lymphatic endothelia (Ramos *et al.*, 2007).

One of the clearly distinguishing features that makes the inner wall of Schlemm's canal unique is the way in which this endothelium responds to changes in pressure. When human eyes are placed into fixative fluid at zero (i.e., atmospheric) pressure, rather than being fixed under conditions of flow, the endothelium of Schlemm's canal is generally flat and featureless. When fixative is introduced under conditions of flow, however, remarkable endothelial blebs are found to bulge into the lumen of Schlemm's canal. These have been termed "giant vacuoles" (Figs. 10 and 11).

Giant vacuoles are discernable at the light microscopic level. Initially there was debate as to whether these structures were truly vacuoles, meaning transcytoplasmic channels that enveloped an aliquot of aqueous humor on the abluminal side of Schlemm's canal and conveyed it to the lumen (Tripathi, 1968, 1971, 1974). Electron microscopic studies using serial sections have since shown that the vacuoles all have an opening on their basal aspect and thus these structures appear to be outpouchings, bullous separations, or invaginations of the inner wall cells caused by the pressure drop across the inner wall, rather than intracellular structures (Inomata *et al.*, 1972; Johnson and Erickson, 2000).

Given the remarkable appearance of these structures, initial skepticism arose as to whether they were physiological or artifactual (Shabo *et al.*, 1973; Grierson and Johnson, 1981). In this regard, it is reassuring to appreciate that virtually identical "vacuoles" are found in the arachnoid villi, which reabsorb cerebrospinal fluid (Tripathi, 1977b).

A careful analysis of the baboon outflow pathway clearly distinguished giant vacuoles from postmortem changes in this tissue (Grierson and Johnson, 1981). Adding more credibility to the existence of giant vacuoles as a physiological phenomenon are findings that the number and size of giant vacuoles increases with increasing IOP. The number of vacuoles increases linearly with pressure while the increase in size with increasing pressure (and therefore the volume of fluid contained within them) is nonlinear (Grierson and Lee, 1977, 1978a).

Current evidence points to the process of giant vacuole formation as being entirely passive and not requiring either the expenditure of cellular energy resources or protein synthesis. These vacuoles are found in greater number near the ostia of collector channels (Parc *et al.*, 2000). As an indication of their longevity, within 3 min of discontinuing perfusion, Schlemm's canal returns to its featureless state, exhibiting no giant vacuoles (Brilakis and Johnson, 2001).

One line of investigation explored the possibility that a loss of passive deformability of the endothelium could create a resistive element in the eye with glaucoma, possibly making it more difficult to passively form giant vacuoles. It has been shown that inner wall cells from eyes with glaucoma exhibit fewer sialic acid moieties, inferring that this led to the membranes being less deformable, making the process of vacuole formation more difficult (Tripathi*et al.*, 1987).

B. How Does Aqueous Humor Cross the Inner Wall of Schlemm's canal?: Pores or Paracellular Flow or Both?

The fact that the number of giant vacuoles increases with increasing pressure suggests a relationship between their formation and outflow. But how does aqueous humor actually cross the endothelial barrier presented by Schlemm's canal? Years ago, it was assumed that the expanding giant vacuole compressed the cytoplasm of the distended endothelial cells, ultimately resulting in development of a pore through which its aliquot of aqueous humor was discharged into Schlemm's canal (Tripathi, 1977a). This implied that giant vacuoles form, burst, and collapse, a process referred to as its "life cycle." Although nonvacuolar openings were identified in the inner wall in these studies, they were found to be occupied by wandering cells (Tripathi, 1974), implying that any pores found in the inner wall should be associated with giant vacuoles as illustrated in Fig. 11. By scanning electron microscopy (SEM), however, numerous pores are found when viewing the luminal surface of the inner wall of Schlemm's canal (Fig. 12) and not all of these are associated with giant vacuoles. Some of these have ragged edges, leading investigators to suspect this population as preparation artifacts (Ethier and Chan, 2001). But pores with smooth edges found both within the walls of giant vacuoles and in areas of the inner wall show no evidence of the typical bulging associated with giant vacuoles. The apparent lack of a one-to-one correspondence between pores and vacuoles, as seen by SEM, merits further

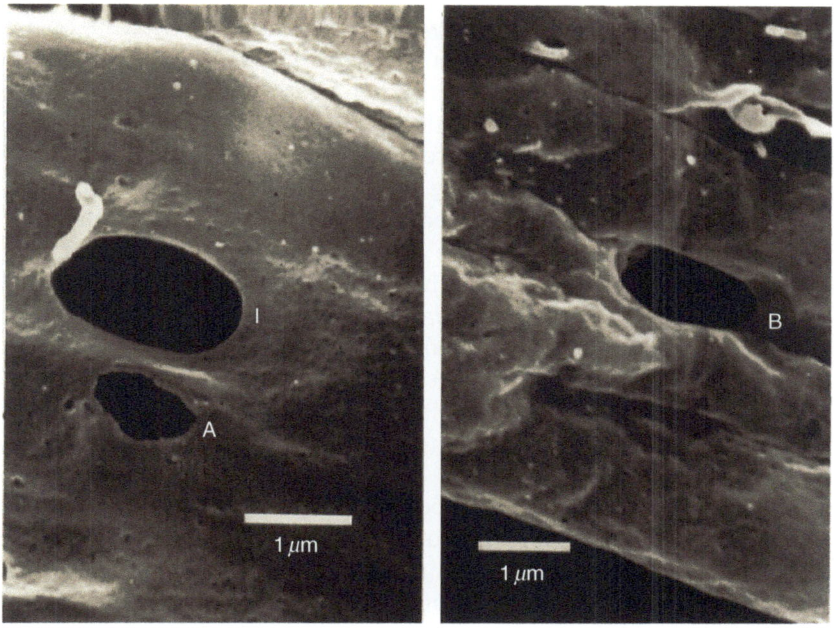

FIGURE 12 Scanning electron micrograph of pores in the inner wall endothelium. Left, an intracellular or I-pore (I) and an artifactual pore (A); and right, an intercellular or B-pore (B) (Ethier *et al.*, 1998).

study to better understand the role of the vacuole, if its purpose is not to give rise to a pore that releases the aliquot of aqueous contained within it. Equally important is the issue of whether pores associated with vacuoles and those not associated with vacuoles might form in different ways.

Allingham *et al.* (1992) reported that the density of pores in the inner wall endothelium was inversely correlated with outflow resistance and that fewer were found in eyes with POAG. This raised the question as to whether a reduced capacity to form pores might contribute to the added resistance in the outflow pathway of the glaucomatous eye. In that study, however, eyes were fixed at a constant pressure, resulting in much lower flow rates in glaucomatous eyes than in normal eyes.

In more recent studies (Sit *et al.*, 1997; Ethier *et al.*, 1998; Johnson *et al.*, 2002), in which fixation was completed under conditions of constant flow rather than constant pressure, pore density was not correlated with outflow facility but did increase with increasing volume of fixative passed through the system. Importantly, the fundamental notion that glaucomatous eyes exhibit fewer pores than normal eyes was confirmed.

Two subtypes of pores have more recently been described. Some of these occur at the border between adjacent endothelial cells and are termed "B" pores. Still others occur away from areas of cell borders. These are termed "I" (intracellular pores) (Ethier *et al.*, 1998). Studies attempting to better define these two types of pores have demonstrated that I-pores decrease with postmortem time suggesting that they are not artifactual (Johnson *et al.*, 2002). Both B-pore and I-pore density correlated with volume of fixative perfused, but only the I-pore result was statistically significant. Complicating this analysis was the fact that the density of B-pores correlated strongly ($p < 0.01$) with the number of I-pores, suggesting that either both were artifacts or neither were artifacts.

C. Paracellular Flow?

Whether either or both types of pores proves to be artifactual or real, other investigators favor paracellular flow, between inner wall endothelial cells, as the principal pathway for entry of aqueous humor into the lumen of Schlemm's canal. Epstein and Rohen (1991) perfused monkey eyes with cationized ferritin at normal and elevated IOPs. Remarkably, little tracer was found within or lining the giant vacuoles. Instead, the tracer was found to decorate the interendothelial clefts between inner wall cells and accumulate in paracellular channels that became more distended under conditions of elevated pressure. Such distentions would presumably have an impact on the permeability properties of intercellular junctions between adjacent endothelial cells. Subsequently, (Ethier *et al.* (2001) showed that cationized ferritin dramatically reduced outflow facility compared with its anionic counterpart, even with tenfold lower concentrations. Unlike anionic ferritin, cationized ferritin was shown to cluster and to distribute itself along the interendothelial clefts but especially around the openings of pores. These studies did not, however, reconcile the relative importance of the paracellular pathway versus pores.

In thin-sectioned material, several investigators reported various forms of intercellular junctions between cells forming the inner wall of Schlemm's canal (Vegge, 1967; Grierson *et al.*, 1978b). In one such study, horseradish peroxidase was perfused into the anterior chambers of normal human and monkey eyes. The junctions of the inner wall of Schlemm's canal blocked the passage of this material, suggesting the presence of tight junctions (MacRae and Sears, 1970).

Where tight junctions (zonulae occludentes) exist, they are invariably accompanied by zonular junctions of the adherens type. These latter junctions serve to provide adhesion, a prerequisite for junction formation and for repair following major disruptions of tight junction integrity. But the adherens junction itself does not represent the occluding element of the junctional complex. Screens of cDNAs from inner wall cells have been shown to exhibit PECAM-1 and VE-cadherin (Heimark *et al.*, 2002). Their presence suggests the presence of adherens junctions, and this expression is also consistent with Schlemm's canal being embryologically of vascular origin (Heimark *et al.*, 2002). The protein ZO-1 has also been documented in inner wall cells (Alvarado *et al.*, 2004), but this protein is also intracellular and does not directly influence flow in the paracellular space (McNeil *et al.*, 2006).

Definitive demonstration of tight junctions (zonulae occludentes) between the endothelial cells of the inner wall of Schlemm's canal came with publication of freeze-fracture replicas showing a simple zonulae occludentes represented by as many as four discontinuous tight junctional strands in monkey eyes, with corridors through the junctional matrix that were termed "slit-pores" (Raviola and Raviola, 1981). It is possible that these "slit-pores" could be the freeze-fracture correlate of the distentions in the paracellular channels observed by Epstein and Rohen (1991) in sectioned material. Using their freeze-fracture replicas, the Raviolas completed exhaustive morphological studies to calculate potential flow through the "slit-pores" they documented, concluding that paracellular flow would be negligible. Unfortunately, these studies were performed on eyes fixed by immersion and not under physiological conditions of flow, a point raised by Epstein and Rohen to suggest that the Raviola calculations could have underestimated the potential for flow.

When freeze-fracture studies were first completed on the inner wall of Schlemm's canal in human eyes, immersion fixed material was again used. In these studies, the interendothelial junctions of Schlemm's canal inner wall cells were even more robust than those seen in monkey eyes, but pathways across the junctional matrix, similar to those described by Raviola and Raviola (1981) as "slit-pores" were still found (Bhatt *et al.*, 1995).

Following on from these studies, the same group repeated these freeze-fracture studies but on human eyes, now fixed at increasing pressures (0, 15, 45 mm Hg). In these eyes, increasing pressure was found to result in simplification of tight junctional structure but not wholesale disruption or distortion of the junctional matrix (Fig. 13). In these same studies, in sectioned material, a reduction in overlap of the adjoined endothelial cells was observed (Ye *et al.*, 1997). Equally important, it was demonstrated that where an interendothelial cleft existed within the wall of a giant vacuole, a *focal* reduction in junctional complexity resulted (Fig. 14). This finding raised the prospect that if "B" pores are real, they might be the result of focal simplification of tight junctions from a normal strand number of 3 to 0. Importantly, small disruptions of tight junctions have been shown to repair very rapidly, and without

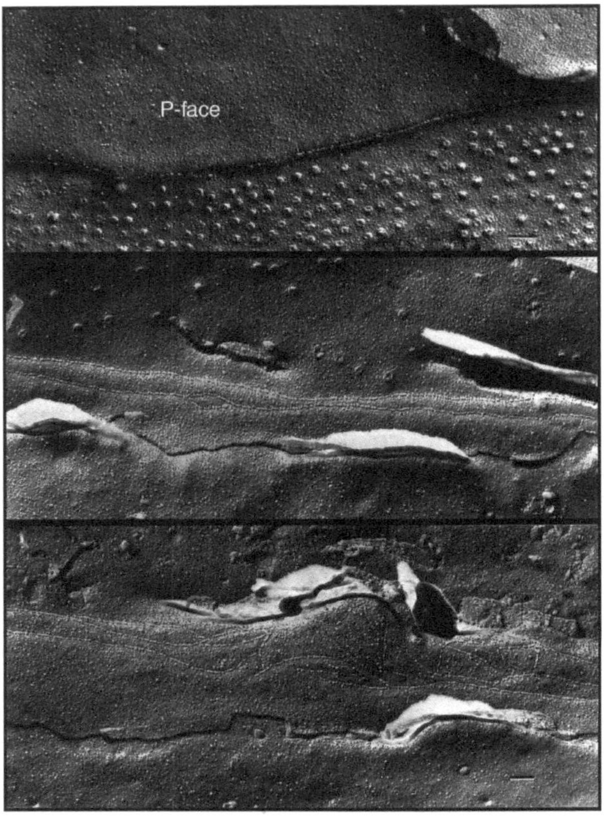

FIGURE 13 Freeze-fracture electron micrographs showing representative reduction in complexity of tight junctions between cells lining the inner wall of Schlemm's canal as pressure is increased from 0 mm Hg (bottom) to 15 mm Hg (middle) to 45 mm Hg (top). Magnification scale bar = 0.2 μm (Ye *et al.*, 1997, Fig. 2).

dependence upon either cellular energy or protein synthesis (Meyer *et al.*, 2001). As such, a focal reduction of tight junction strand number to zero would be expected to self-repair rapidly.

Junctional simplification with increasing pressure may also lie at the heart of the distentions in the paracellular pathway found using cationized ferritin, but Epstein and Rohen (1991) do not mention the reduction in length of the paracellular cleft found by Ye *et al.* (1997). We know from numerous studies that as IOP increases, facility of outflow decreases. Facility of outflow is increased, however, when separations are produced between cells lining the inner wall of Schlemm's canal. Such separations have been produced with an array of chemical agents, including ethacrynic acid (Epstein *et al.* 1987);

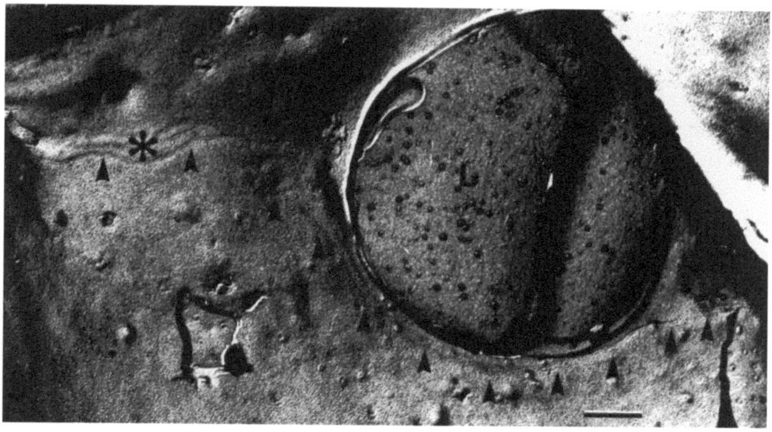

FIGURE 14 Freeze-fracture electron micrograph showing fracture plane through the lumen (L) of a giant vacuole. The resulting distention of the endothelial cells membranes diverted the course of the interendothelial tight junction joining two cells involved in formation of the vacuole (arrowheads). In the distended area, substantial simplification of the junctional structure is evident. Areas of the same junction on either side of the distended area continued to exhibit a more complex, multistranded junctional architecture (*). Magnification scale bar = 0.5 μm (Ye *et al.*, 1997, Fig. 4).

alpha-chymotrypsin (Hamanaka and Bill, 1988); and EDTA (Bill *et al.*, 1980; Hamanaka and Bill, 1987). Clearly in all such instances, the intercellular junctions of the inner wall have been disrupted, but in a more dramatic way than the focal changes discussed above in possible relation to border pores.

Further studies of these matters are clearly imperative if the roles of pores and paracellular pathway are to be understood and possibly unified into an encompassing model of aqueous flow across the inner wall of Schlemm's canal. The findings from studies of "B" pores and the studies of the paracellular pathway would suggest that an improved understanding of tight junction regulation in the inner wall endothelium should be a priority because these junctions could be the common element limiting flow through each of these. Clearly, if "I" pores are not artifactual, however, a broader concept will be required in order to unify these findings into a single physiological model.

Understanding the actual mechanisms underlying normal aqueous outflow, and the altered outflow in glaucoma, would lead to the development of medical therapies to treat glaucoma at the source of the problem rather than by reducing the formation of aqueous humor—the nutritive fluid upon which the cornea, lens, and trabecular meshwork depend for metabolic support.

VII. SUMMARY

Unraveling the mystery of most diseases often begins with a simple comparison at the light microscopic level between the affected tissue in its normal and diseased state. The differences that are found serve to guide investigators to the ultimate cause of the disease. At this point, the differences found between the trabecular meshworks of age-matched normals and glaucomatous human eyes are very few—regardless of the method of analysis. This provides ample room for further investigation but is simultaneously a great source of frustration for both the basic scientist and the clinician. Much as glaucoma is a disease of increased resistance, the disease process itself remains resistant to giving up its secrets. Outflow resistance is the inverse of facility of outflow. Although our search for the source of outflow resistance in both the normal and glaucomatous eye must continue, the fluid mechanics of the process can still be simplified to their clinical essence using the formula developed by Goldmann: $F = C_{tm} (P_i - P_e) + F_u$, where F equals aqueous flow in μl/min; P_i and P_e represent intraocular and episcleral venous pressure, respectively, in mm Hg; F_u represents the component of outflow traveling the unconventional route; and the value C_{tm} remains the measurable but elusive facility of outflow—the mirror of outflow resistance.

References

Allingham, R. R., de Kater, A. W., Ethier, C. R., Anderson, P. J., Hertzmark, E., and Epstein, D. L. (1992). The relationship between pore density and outflow facility in human eyes. *Invest. Ophthalmol. Vis. Sci.* **33**(5), 1661–1669.

Alvarado, J., Murphy, C., Polansky, J., and Juster, R. (1981). Age-related changes in trabecular meshwork cullularity. *Invest. Ophthalmol. Vis. Sci.* **21,** 714.

Alvarado, J. A., Yun, A. J., and Murphy, C. (1986). Juxtacanalicular tissue in primary open angle glaucoma and in nonglaucomatous normals. *Arch. Ophthalmol.* **104**(10), 1517–1528.

Alvarado, J. A., Betanzos, A., Franse-Carman, L., Chen, J., and Gonzalez-Mariscal, L. (2004). Endothelia of Schlemm's canal and trabecular meshwork: Distinct molecular, functional, and anatomic features. *Am. J. Physiol. Cell Physiol.* **286,** C621–C634.

Balazs, E., Laurent, T., Laurent, U. B., Deroche, M. H., and Bunney, D. M. (1959). Studies on the structure of the vitreous body. VIII. Comparative biochemistry. *Arch. Biochem. Biophys.* **81,** 464.

Bárány, E. (1963). A mathematical formulation of intraocular pressure as dependent on secretion, ultrafiltration, bulk outflow, and osmotic readsorption of fluid. *Invest. Ophthalmol.* **2**(6), 584–590.

Bárány, E. H. (1953). *In vitro* studies of the resistance to flow through the angle of the anterior chamber. *Acta Soc. Med. Ups.* **59,** 260–276.

Bárány, E. H. (1956). The action of different kinds of hyaluronidase on the resistance to flow through the anterior chamber. *Acta Ophthalmol.* **34,** 397–403.

Batmanov, I. (1968). The structure of the drainage system of the human eye. [Russian]. *Vestn. Oftalmol.* **81,** 27–31.

Bentzel, C., and Reczek, P. (1978). Permeability changes in *Necturus* proximal tubule during volume expansion. *Am. J. Physiol.* **234**(3), F225–F234.

Beswick, J., and McCulloch, C. (1956). Effect of hyaluronidase on the viscosity of aqueous humor. *Br. J. Ophthalmol.* **40,** 545.

Bhatt, K., Gong, H., and Freddo, T. F. (1995). Freeze-fracture studies of interendothelial junctions in the angle of the human eye. *Invest. Ophthalmol. Vis. Sci.* **36**(7), 1379–1389.

Bill, A. (1964a). The albumin exchange in the rabbit. *Acta Physiol. Scand.* **60,** 18.

Bill, A. (1964b). The drainage of albumin from the uvea. *Exp. Eye Res.* **3,** 179.

Bill, A. (1965). The aqueous humor drainage mechanism in the cynomolgus monkey (*Macaca irus*) with evidence for unconventional routes. *Invest. Ophthalmol.* **4,** 911.

Bill, A. (1966). Conventional and uveo-scleral drainage of aqueous humour in the cynomolgus monkey (*Macaca irus*) at normal and high intraocular pressures. *Exp. Eye Res.* **5,** 45.

Bill, A. (1975). Blood circulation and fluid dynamics in the eye. *Physiol. Rev.* **55,** 383–416.

Bill, A., and Hellsing, K. (1965). Production and drainage of aqueous humor in the cynomolgus monkey (*Macaca irus*). *Invest. Ophthalmol.* **4,** 920.

Bill, A., and Phillips, C. I. (1971). Uveoscleral drainage of aqueous humor in human eyes. *Exp. Eye Res.* **12,** 275–281.

Bill, A., Lutjen-Drecoll, E., and Svedbergh, B. (1980). Effects of intracameral Na2EDTA and EGTA on aqueous outflow routes in the monkey eye. *Invest. Ophthalmol. Vis. Sci.* **19,** 492–504.

Bradley, J. M. B., Vranka, J., Colvis, C. M., Conger, D. M., Alexander, J. P., Fisk, A. S., Samples, J. R., and Acott, T. S. (1998). Effect of matrix metalloproteinases activity on outflow in perfused human organ culture. *Invest. Ophthalmol. Vis. Sci.* **39,** 2649–2658.

Bradley, J. M., Anderssohn, A. M., Colvis, C. M., Parshley, D. E., Zhu, X. H., Ruddat, M. S., Samples, J. R., and Acott, T. S. (2000). Mediation of laser trabeculoplasty-induced matrix metalloproteinase expression by IL-1beta and TNFalpha. *Invest. Ophthalmol. Vis. Sci.* **41** (2), 422–430.

Brilakis, H. S., and Johnson, D. H. (2001). Giant vacuole survival time and implications for aqueous humor outflow. *J. Glaucoma* **10,** 277–283.

Comper, W. D., and Laurent, T. C. (1978). Physiological function of connective tissue polysaccharides. *Physiol. Rev.* **58,** 255–315.

Crawford, K., and Kaufman, P. L. (1987). Pilocarpine antagonizes prostaglandin F2 alpha-induced ocular hypotension in monkeys. Evidence for enhancement of Uveoscleral outflow by prostaglandin F2 alpha. *Arch. Ophthamol.* **105**(8), 1112–1116.

Daniels, B. S., Hauser, E. B., Deen, W. M., and Hostetter, T. H. (1992). Glomerular basement membrane: *In vitro* studies of water and protein permeability. *Am. J. Physiol.* **262,** F919–F926.

Dernouchamps, J. P. (1982). The proteins of the aqueous humor. *Doc. Ophthalmol.* **53,** 193.

Dvorak-Theobald, G. (1934). Schlemm's canal: Its anastomoses and anatomic relations. *Trans. Am. Ophthalmol. Soc.* **32,** 574–585.

Epstein, D. L., and Rohen, J. W. (1991). Morphology of the trabecular meshwork and inner-wall endothelium after cationized ferritin perfusion in the monkey eye. *Invest. Ophthalmol. Vis. Sci.* **32,** 160–171.

Epstein, D. L., Freddo, T. F., Bassett-Chu, S., Chung, M., and Karageuzian, L. (1987). Influence of ethacrynic acid on outflow facility in the monkey and calf eye. *Invest. Ophthalmol. Vis. Sci.* **28,** 2067–2075.

Ethier, C. R., and Chan, D. W. (2001). Cationic ferritin changes outflow facility in human eyes whereas anionic ferritin does not. *Invest. Ophthalmol. Vis. Sci.* **42,** 1795–1802.

Ethier, C. R., Kamm, R. D., Palaszewski, B. A., Johnson, M. C., and Richardson, T. M. (1986). Calculations of flow resistance in the juxtacanalicular meshwork. *Invest. Ophthalmol. Vis. Sci.* **27**(12), 1741–1750.

Ethier, C. R., Kamm, R. D., Johnson, M., Pavao, A. F., and Anderson, P. J. (1989). Further studies on the flow of aqueous humor through microporous filters. *Invest. Ophthalmol. Vis. Sci.* **30**(4), 739–746.

Ethier, C. R., Coloma, F. M., Sit, A. J., and Johnson, M. (1998). Two pore types in the inner wall endothelium of Schlemm's canal. *Invest. Ophthalmol. Vis. Sci.* **39,** 2041–2048.

Ethier, C., Johnson, M., and Ruberti, J. (2004). Ocular biomechanics and biotransport. *Annu. Rev. Biomed. Eng.* **6,** 249–273.

Fatt, I. (1969). Permeability of Descemet's membrane to water. *Exp. Eye Res.* **8,** 34–354.

Fingert, J. H., Stone, E. M., Sheffield, V. C., and Alward, W. L. M. (2002). Myocilin glaucoma. *Surv. Ophthalmol.* **47**(6), 547–561.

Fisher, R. F. (1982). The water permeability of basement membrane under increasing pressure: Evidence for a new theory of permeability. *Proc. R. Soc. Lond.* **B216,** 475–496.

Freddo, T. F. (1993 and revised 1999). Anatomy and physiology related to aqueous humor production and outflow. *In* "Primary Care of the Glaucomas" (M. Fingeret and T. Lewis, eds.). Chapter 3. Appleton and Lange, Stamford, CT.

Freddo, T., Patterson, M., Scott, D., and Epstein, D. (1984). Influence of mercurial sulfhydryl agents on aqueous outflow pathways in enucleated eyes. *Invest. Ophthalmol. Vis. Sci.* **25,** 278–285.

Gaasterland, D. E., Pederson, J. E., McLellan, H. M., and Reddy, V. N. (1979). Rhesus monkey aqueous humor composition and a primate ocular perfusate. *Invest. Ophthalmol. Vis. Sci.* **18,** 1139.

Gabelt, B., and Kaufman, P. (1989). Prostaglandin F increases uveoscleral outflow inthe cynomolgus monkey. *Exp. Eye Res.* **49,** 389–402.

Gong, H. Y., Trinkaus-Randall, V., and Freddo, T. F. (1989). Ultrastructural immunocytochemical localization of elastin in normal human trabecular meshwork. *Curr. Eye Res.* **8**(10), 1071–1082.

Gong, H., Freddo, T. F., and Johnson, M. (1992). Age-related changes of sulfated proteoglycans in the normal human trabecular meshwork. *Exp. Eye Res.* **55**(5), 691–709.

Gong, H., Tripathi, R. C., and Tripathi, B. J. (1996). Morphology of the aqueous outflow pathway. *Microsc. Res. Tech.* **33,** 336–367.

Gong, H., Ruberti, J., Overby, D., Johnson, M., and Freddo, T. F. (2002). A new view of the human trabecular meshwork using quick-freeze, deep-etch electron microscopy. *Exp. Eye Res.* **75,** 347–358.

Grant, W. M. (1951). Clinical measurements of aqueous outflow. *Arch. Ophthalmol.* **46,** 113–131.

Grant, W. M. (1963). Experimental aqueous perfusion in enucleated human eyes. *Arch. Ophthalmol.* **69,** 783–801.

Grierson, I., and Johnson, N. F. (1981). The post-mortem vacuoles of Schlemm's canal. *Graefes Arch. Clin. Exp. Ophthalmol.* **215,** 249–264.

Grierson, I., and Lee, W. R. (1977). Light microscopic quantitation of the endothelial vacuoles in Schlemm's canal. *Am. J. Ophthalmol.* **84,** 234–246.

Grierson, I., and Lee, W. R. (1978a). Pressure effects on flow channels in the lining endothelium of Schlemm's canal. *Acta Ophthamol.* **56,** 935–952.

Grierson, I., Lee, W. R., Abraham, S., and Howes, R. C. (1978b). Associations between the cells of the walls of Schlemm's canal. *Albrecht Von Graefes Arch. Klin. Exp. Ophthalmol.* **208,** 33–47.

Hamanaka, T., and Bill, A. (1987). Morphological and functional effects of Na2EDTA on the outflow routes for aqueous humor in monkeys. *Exp. Eye Res.* **44,** 171–190.

Hamanaka, T., and Bill, A. (1988). Effects of alpha-chymotrypsin on the outflow routes for aqueous humor. *Exp. Eye Res.* **46,** 323–341.

Hann, C. R., Springett, M. J., Wang, X., and Johnson, D. H. (2001). Ultrastructural localization of collagen IV, fibronectin, and laminin in the trabecular meshwork of normal and glaucomatous eyes. *Ophthalmic Res.* **33**(6), 314–324.

Heimark, R. L., Kaochar, S., and Stamer, W. D. (2002). Human Schlemm's canal cells express the endothelial adherens proteins, VE-cadherin and PECAM-1. *Curr. Eye Res.* **25**(5), 299–308.

Hogan, M. J., Alvarado, J. A., and Weddell, J. E. (1971). *Histology of the Human Eye.* Saunders, Philadelphia.

Hubbard, W. C., Johnson, M., Gong, H., Gabelt, B. T., Peterson, J. A., Sawhney, R., Freddo, T., and Kaufman, P. L. (1997). Intraocular pressure and outflow facility are unchanged following acute and chronic intracameral chondroitinase ABC and hyaluronidase in monkeys. *Exp. Eye Res.* **65,** 177–190.

Inomata, H., Bill, A., and Smelser, G. (1972). Aqueous humor pathways through the trabecular meshwork and into Schlemm's canal in cynomologus monkey (Macca Irus). *Am. J. Ophthalmol.* **73,** 760–789.

Johnson, M. (2006). What controls aqueous humour outflow resistance? *Exp. Eye Res.* **82**(4), 545–557.

Johnson, M., and Erickson, K. (2000). Mechanisms and routes of aqueous humor drainage. *In* "Principles and Practice of Ophthalmology" (D. M. Albert and F. A. Jakobiec, eds.), Vol. 4, pp. 2577–2595. WB Saunders Co., Philadelphia, Glaucoma, Chapter 193B.

Johnson, M., and Kamm, R. D. (1983). The role of Schlemm's canal in aqueous outflow from the human eye. *Invest. Ophthalmol. Vis. Sci.* **24,** 320–325.

Johnson, M., Ethier, C. R., Kamm, R. D., Grant, W. M., Epstein, D. L., and Gaasterland, D. (1986). The flow of aqueous humor through micro-porous filters. *Invest. Ophthalmol. Vis. Sci.* **27,** 92–97.

Johnson, M., Chen, D., Read, A. T., Christensen, C., Sit, A., and Ethier, C. R. (2002). Glaucomatous eyes have a reduced pore density in the inner wall endothelium of Schlemm's canal. *Invest. Ophthalmol. Vis. Sci.* **43,** 2950–2955.

Johnstone, M. A., and Grant, W. M. (1973). Pressure dependent changes in the structures of the aqueous outflow system of human and monkey eyes. *Am. J. Ophthalmol.* **75,** 365.

Kamm, R. D., Palaszewski, B. A., *et al.* (1983). Calculation of flow resistance in the juxtacanalicular meshwork. ARVO Abstracts. *Invest. Ophthalmol. Vis. Sci.* **24,** 135.

Knepper, P. A. (1980). *Glycosaminoglycans and aqueous outflow resistance.* Ph.D. Thesis. Northwestern University, Ann Arbor, MI.

Knepper, P. A., Farbman, A. I., and Telser, A. G. (1984). Exogenous hyaluronidases and degradation of hyaluronic acid in the rabbit eye. *Invest. Ophthalmol. Vis. Sci.* **25,** 286–293.

Knepper, P. A., Goossens, W., Hvizd, M., and Palmberg, P. F. (1996a). Glycosaminoglycans of the human trabecular meshwork in primary open-angle glaucoma. *Invest. Ophthalmol. Vis. Sci.* **37,** 1360–1367.

Knepper, P. A., Goossens, W., and Palmberg, P. F. (1996b). Glycosaminoglycan stratification of the juxtacanalicular tissue in normal and primary open-angle glaucoma. *Invest. Ophthalmol. Vis. Sci.* **37,** 2414–2425.

Knies, M. (1875). Die resorption von blut in der vorderen augenkammer. *Archiv für Pathologische Anatomie und Physiologie klinische Medicin.* **62,** 537–553.

Krohn, J. (1999). Expression of Factor VIII-related antigen in human aqueous drainage channels. *Acta Ophthalmol. Scand.* **77,** 9–12.

Leber, T. (1873). Studien über den Flüssigkeitswechsel im Auge. *Albrecht Von. Graefes Arch. Ophthalmol.* **19,** 87–106.

Leber, T. (1895). Der circulus venosus Schlemii steht nicht in offener verbindung mit der vorderen Augenkammer. *Albrecht Von. Graefes Arch. Ophthalmol.* **41,** 235–280.

Lütjen-Drecoll, E., Futa, R., and Rohen, J. W. (1981). Ultrahistochemical studies on tangential sections of trabecular meshwork in normal and glaucomatous eyes. *Invest. Ophthalmol. Vis. Sci.* **21,** 563–573.

Lütjen-Drecoll, E., Rittig, M., Rittig, M., Rauterberg, J., Jander, R., and Mollenhauer, J. (1989). Immunomicroscopical study of type VI collagen in the trabecular meshwork of normal and glautomatous eyes. *Exp. Eye Res.* **48,** 139–147.

MacRae, D., and Sears, M. L. (1970). Peroxidase passage through the outflow channels of human and rhesus eyes. *Exp. Eye Res.* **10,** 15–18.

Mäepea, O., and Bill, A. (1989). The pressures in the episcleral veins, Schlemm's canal and the trabecular meshwork in monkeys: Effects of changes in intraocular pressure. *Exp. Eye Res.* **49,** 645–663.

Mäepea, O., and Bill, A. (1992). Pressures in the juxtacanalicular tissue and Schlemm's canal in monkeys. *Exp. Eye Res.* **54,** 879–883.

Marshall, G. E., Konstas, A. G., and Lee, W. R. (1990). Immunogold localization of type IV collagen and laminin in the aging human outflow system. *Exp. Eye Res.* **51,** 691–699.

Marshall, G. E., Konstas, A. G., and Lee, W. R. (1991). Immunogold ultrastructural localization of collagen in the aged human outflow system. *Ophthalmology* **98,** 692–700.

McEwen, W. K. (1958). Application of Poiseuille's law to aqueous outflow. *Arch. Ophthamol.* **60,** 290.

McNeil, E., Capaldo, C. T., and Macara, I. G. (2006). Zonula occludens-1 Function in the assembly of tight Junctions in Madin-Darby Canine Kidney epithelial cells. *Mol. Biol. Cell* **17,** 1922–1932.

Melton, C. E., and DeVille, W. B. (1960). Perfusion studies on eyes of four species. *Am. J. Ophthalmol.* **50,** 302–308.

Meyer, K. (1953). The biological significance of hyaluronic acid and hyaluronidase. *Physiol. Rev.* **27,** 335–359.

Meyer, T. N., Schwesinger, C., Ye, J., Denker, B. M., and Nigam, S. K. (2001). Reassembly of the tight junction after oxidative stress depends on tyrosine kinase activity. *J. Biol. Chem.* **276,** 22048–22055.

Moses, R. A. (1979). Circumferential flow in Schlemm's canal. *Am. J. Ophthalmol.* **88,** 585–591.

Murphy, C. G., Johnson, M., and Alvarado, J. A. (1992). Juxtacanalicular tissue in pigmentary and primary open angle glaucoma. The hydrodynamic role of pigment and other constituents. *Arch. Ophthamol.* **110,** 1779–1785.

Nesterov, A. P. (1970). Role of blockade of Schlemm's canal in pathogenesis of primary open angle glaucoma. *Am. J. Ophthalmol.* **70,** 691–696.

Parc, C. E., Johnson, D. H., and Brilakis, H. S. (2000). Gian vacuoles are found preferentially near collector channels. *Invest. Ophthalmolmol. Vis. Sci.* **41,** 2894–2890.

Pavao, A. F., Lee, D. A., Ethier, C. R., Johnson, M. C., Anderson, P. J., and Epstein, D. L. (1989). Two-dimensional gel electrophoresis of calf aqueous humor, serum, and filter-bound proteins. *Invest. Ophthalmol. Vis. Sci.* **30,** 731–738.

Pederson, J. E., Gaasterland, D. E., and MacLellan, H. M. (1977). Uveoscleral aqueous outflow in the rhesus monkey: Importance of uveal reabsorption. *Invest. Ophthalmol. Vis. Sci.* **16,** 1008.

Pedler, C. (1956). The relationship of hyaluronidase to aqueous outflow resistance. *Trans. Ophthalmol. Soc. U. K.* **76,** 51–63.

Peterson, W. S., and Jocson, V. L. (1974). Hyaluronidase effects on aqueous outflow resistance. *Am. J. Ophthalmol.* **77,** 573–577.

Polansky, J. R., Fauss, D. J., Chen, P., Chen, H., Lütjen-Drecoll, E., Johnson, D., Kurtz, R. M., Ma, E., Bloom, E., and Nguyen, T. D. (1997). Cellular pharmacology and molecular biology of the trabecular meshwork inducible glucocorticoid response gene product. *Ophthalmologica* **211,** 126–139.

Ramos, R. F., Hoying, J. B., Witte, M. H., and Stamer, W. D. (2007). Schlemm's canal endothelia, lymphatic or blood vasculature? *J. Glaucoma* **16,** 391–405.

Raviola, G., and Raviola, E. (1981). Paracellular route of aqueous outflow in the trabecular meshwork and canal of Schlemm. *Invest. Ophthalmol. Vis. Sci.* **21,** 52–72.
Rohen, J. W. (1983). Why is intraocular pressure elevated in chronic simple glaucoma? Anatomical considerations. *Ophthalmology* **90,** 758–765.
Rohen, J. W., and Rentsch, F. J. (1968). Morphology of Schlemm's canal and related vessels in the human eye. [German]. *Albrecht Von Graefes Arch. Klin. Exp. Ophthalmol.* **176,** 309–329.
Rohen, J. W., and van der Zypen, E. (1968). The phagocytic activity of the trabecular meshwork endothelium. *Graefes Arch. Clin. Exp. Ophthalmol.* **175,** 143.
Rosenquist, R. C., Jr., Melamed, S., and Epstein, D. L. (1988). Anterior and posterior axial lens displacement and human aqueous outflow facility. *Invest. Ophthalmol. Vis. Sci.* **29,** 1159–1164.
Russell, P., Tamm, E. R., Grehn, F. J., Picht, G., and Johnson, M. (2001). The presence and properties of myocilin in the aqueous humor. *Invest. Ophthalmol. Vis. Sci.* **42,** 983–986.
Sawaguchi, S., Yue, B. Y., Yeh, P., and Tso, M. O. (1992). Effects of intracameral injection of chondroitinase ABC *in vivo*. *Arch. Ophthamol.* **110**(1), 110–117.
Schwalbe, G. (1870). Untersuchen über die Lymphbahnen des Auges und ihre Begrenzungen. *Arch. mikrosk. Anat.* **6,** 261–362.
Seidel, E. (1921). Weitere experimentelle Untersuchungen über die Quelle und den Verlauf der intraokularen Saftströmung. IX Mitteilung. Über den Abfluss des Kammerwassers aus der vorderen AugenKammer. *Graefe Arch. Clini. Exp. Ophthalmol.* **104,** 357–402.
Seiler, T., and Wollensak, J. (1985). The resistance of the trabecular meshwork to aqueous humor outflow. *Graefe Arch. Clin. Exp. Ophthalmol.* **223,** 88–91.
Shabo, A. L., Reese, T. S., and Gaasterland, D. (1973). Postmortem formation of giant vacuoles in Schlemm's canal of the monkey. *Am. J. Ophthalmol.* **76,** 896–905.
Sherman, S. H., Green, K., and Laties, A. M. (1978). The fate of anterior chamber fluorescein in the monkey eye. I. The anterior chamber outflow pathways. *Exp. Eye Res.* **27,** 159.
Sit, A. J., Coloma, F. M., Ethier, C. R., and Johnson, M. (1997). Factors affecting the pores of the inner wall endothelium of Schlemm's canal. *Invest. Ophthalmol. Vis. Sci.* **38,** 1517–1525.
Smith, P. (1888). On the escape of fluid from the aqueous and vitreous chambers under different pressure. *Ophthalmol. Rev.* **7,** 193–208.
Sorsby, A. (1963). "Modern Ophthalmology." Butterworth, Washington, DC.
Starita, C., Hussain, A., Pagliarini, S., and Marshall, L. (1996). Hydrodynamics of ageing Bruch's membrane: Implications for macular disease. *Exp. Eye Res.* **62,** 565–572.
Stone, E. M., Fingert, J. H., Alward, W. L., Nguyen, T. D., Polansky, J. R., Sunden, S. L., Nishimura, A. F., Clark, A. F., Nystuen, A., Nichols, B. E., Mackey, D. A., Ritch, R., *et al.* (1997). Identification of a gene that causes primary open angle glaucoma. *Science* **275,** 668–670.
Ten Hulzen, R. D., and Johnson, D. H. (1996). Effect of fixation pressure on juxtacanalicular tissue and Schlemm's canal. *Invest. Ophthalmol. Vis. Sci.* **37,** 114–124.
Tervo, K., Päällysaho, T., Virtanen, I., and Tervo, T. (1995). Integrins in human anterior chamber angle. *Graefe Arch. Clin. Exp. Ophthalmol.* **233,** 291–292.
Tripathi, R. C. (1968). Ultrastructure of Schlemm's canal in relation to aqueous outflow. *Exp. Eye Res.* **7,** 335–341.
Tripathi, R. C. (1971). Mechanism of the aqueous outflow across the trabecular wall of Schlemm's canal. *Exp. Eye Res.* **11,** 116–121.
Tripathi, R. C. (1974). Ch 3: Comparative physiology and anatomy of the aqueous outflow pathway. *In* "The Eye: Comparative Physiology" (H. Davson and L. T. Graham, eds.), Vol. 5, pp. 163–356. Academic Press, London, England; New York, NY.
Tripathi, R. C. (1977a). Uveoscleral drainage of aqueous humour. *Exp. Eye Res.* **25**(suppl.), 305–308.

Tripathi, R. C. (1977b). The functional morphology of the outflow systems of ocular and cerbrospinal fluids. *Exp. Eye Res.* **25**(suppl.), 65–116.

Tripathi, R. C., Tripathi, B. J., and Spaeth, G. L. (1987). Localization of sialic acid moieties in the endothelial lining of Schlemm's canal in normal and glaucomatous eyes. *Exp. Eye Res.* **44,** 293–306.

Van Buskirk, E. M. (1976). Changes in the facility of aqueous outflow induced by lens depression and intraocular pressure in excised human eyes. *Am. J. Ophthalmol.* **82,** 736–740.

Van Buskirk, E. M. (1982). Anatomic correlates of changing aqueous outflow facility in excised human eyes. *Invest. Ophthalmol. Vis. Sci.* **22,** 625–632.

Van Buskirk, M. S., and Brett, J. (1978). The canine eye: *In vitro* dissolution of the barriers to aqueous outflow. *Invest. Ophthalmol. Vis. Sci.* **17,** 258–263.

Van Buskirk, E. M., Pond, V., Resenquist, R. C., and Acott, T. S. (1984). Argon laser trabeculoplasty. Studies of mechanism of action. *Ophthalmology* **91,** 1005–1010.

Vegge, T. (1967). The fine structure of the trabeculum cribriforme and the inner wall of Schlemm's canal in the normal human eye. *Z. Zellforsch. Mikrosk. Anat.* **77,** 267–281.

Wagner, J. A., Edwards, A., and Schuman, J. S. (2004). Characterization of uveoscleral outflow in enucleated porcine eyes perfused under constant pressure. *Invest. Ophthalmol. Vis. Sci.* **45,** 3203–3206.

Welling, L., and Welling, D. (1978). Physical properties of isolated perfused basement membranes from rabbit loop of Henle. *Am. J. Physiol.* **234,** F54–F58.

Wise, J. B., and Witter, S. L. (1979). Argon laser therapy for open-engle glaucoma. A pilot study. *Arch. Ophthalmol.* **97,** 319–322.

Ye, W., Gong, H., Sit, A., Johnson, M., and Freddo, T. F. (1997). Interendothelial junctions in normal human Schlemm's canal respond to changes in pressure. *Invest. Ophthalmol. Vis. Sci.* **38,** 2460–2468.

CHAPTER 7

Aqueous Humor Dynamics I

Measurement Methods and Animal Studies

Carol B. Toris
Department of Ophthalmology, University of Nebraska Medical Center, Omaha, Nebraska, 68198-5840

I. OVERVIEW

Intraocular pressure (IOP) is maintained by the dynamics of ocular aqueous humor that involves its secretion, circulation throughout the anterior chamber, and drainage into the iridocorneal angle. The measurable components of

Current Topics in Membranes, Volume 62

1063-5823/08 $35.00
DOI: 10.1016/S1063-5823(08)00407-9

aqueous humor dynamics include aqueous flow, outflow facility, uveoscleral outflow, and episcleral venous pressure. Multiple methods are available to assess these components. Interpretation of data collected by these methods requires an understanding of the inherent assumptions and limitations of each method applicable to the species of animal under investigation. Despite the inevitable problems associated with each method, invaluable information has been collected regarding normal circadian rhythms and interspecies differences in aqueous humor dynamics. Additionally, studies of animal models of spontaneous and induced glaucoma have enhanced our understanding of human glaucoma and facilitated the design of improved pharmacological treatments and surgical procedures. This chapter describes the various methods to assess aqueous humor dynamics and summarizes findings from the animal species that have contributed the most to our understand of aqueous humor dynamics.

II. COMPONENTS OF AQUEOUS HUMOR DYNAMICS AND MEASUREMENT TECHNIQUES

The major components of aqueous humor dynamics include the production rate of aqueous humor, resistance to outflow through the trabecular meshwork, rate of outflow from the anterior chamber angle other than through the trabecular meshwork (most commonly termed uveoscleral outflow, although other descriptive names exist), and the pressure in the aqueous humor collection vessels (episcleral venous pressure).

The original Goldmann equation described the relationship between the IOP and the components of aqueous humor dynamics as they were understood half a century ago:

$$\mathrm{IOP} = \frac{F_{\mathrm{p}}}{C_{\mathrm{trab}}} + P_{\mathrm{v}} \tag{1}$$

where IOP is intraocular pressure, F_{p} is the rate of aqueous humor production and drainage, C_{trab} is the facility of outflow through the trabecular meshwork, and P_{v} is the pressure in the vessels that drain the aqueous humor from the trabecular meshwork (episcleral veins). In the early studies describing the relationship between IOP and components of aqueous humor dynamics, it was believed that the aqueous humor secreted by the ciliary processes entered the anterior chamber and drained exclusively through the trabecular meshwork. Tracer originally injected into the anterior chamber that was occasionally observed in the uvea was thought to be inconsequential. Since the original equation was written, the importance of uveoscleral outflow has become more apparent and the Goldmann equation was changed accordingly. In a steady state, aqueous humor production (F_p) is the sum of aqueous humor drainage through the two outflow pathways: trabecular (F_{trab}) and uveoscleral (F_u).

$$F_p = F_u + F_{trab} \quad (2)$$

F_{trab} is pressure dependent, that is, as IOP increases, fluid flow through the trabecular meshwork increases. F_{trab} can be rewritten as the product of the facility of trabecular outflow (C_{trab}) and the pressure difference across the trabecular meshwork (IOP$-P_v$). Hence, trabecular outflow is described as given below:

$$F_{trab} = C_{trab}(\mathrm{IOP} - P_v) \quad (3)$$

Substitution of Eq. (3) into Eq. (2) yields:

$$F_p = F_u + C_{trab}(\mathrm{IOP} - P_v) \quad (4)$$

Assessment of aqueous humor dynamics requires the evaluation of each of these variables. Studies of aqueous humor dynamics in mice, rats, rabbits, cats, dogs, and monkeys are summarized in Tables I through VI. Clinical studies of aqueous humor dynamics are described in Chapter 8.

A. *Intraocular Pressure*

The health of the eye and normal visual function require a firm control of the IOP. When the IOP deviates from a narrow healthy range for an extended period of time, irreversible damage may occur to the optic nerve and retina, resulting in permanent vision loss. Transient changes in the IOP occur by seemingly benign daily activities including blinking, wearing a tight necktie, stooping, a valsalva maneuver, or a bowel movement. Eyelid pressure, tension on extraocular muscles, arterial and venous blood pressure, and gravity are additional factors that affect the IOP. Intraocular pressure under steady state conditions occurs at a time when aqueous humor inflow and outflow are equal and the pressure in the eye is relatively stable. Normal values of steady-state IOP in research animals are surprisingly similar among species, 15–17 mmHg in mice and rats (Tables I and II), 10–25 mmHg in rabbits (Table III), 14–25 mmHg in cats (Table IV), 9–26 mmHg in dogs (Table V), and 14–23 mmHg in monkeys (Table VI). Acute measurements of the IOP are made indirectly by tonometry and directly by manometry. Continuous IOP measurements over a period of weeks to months are made by telemetry.

TABLE I

Aqueous Humor Dynamics in Mice

Reference	Animal	Anesthesia	IOP (mmHg)	F_a (μl/min)	C (μl/min/mmHg)	P_{ev} (mmHg)	F_u(μl/min)
Fan *et al.*, 2007	CD1	Ketamine-xylazine		0.09 ± 0.01[a]			
		Avertin[d]		0.20 ± 0.03[a]			
Crowston *et al.*, 2004	Swiss white	Ketamine-xylazine	15.7 ± 1.0	0.14 ± 0.04[b]	0.0053 ± 0.0014		0.115
Aihara *et al.*, 2003a	Swiss white	Ketamine-xylazine	15.7 ± 2.0	0.18 ± 0.05[b]	0.0051 ± 0.0006	9.5 ± 1.2	0.148
Aihara *et al.*, 2003b	Swiss white	Ketamine-xylazine	16.5 ± 0.6			9.6 ± 1.3	
Zhang *et al.*, 2002	CD1	Ketamine	16.0 ± 0.4	0.06 ± 0.03[c]	0.006 ± 0.001		

Reported are mean values from untreated or vehicle-treated control groups ± standard errors or standard deviations as provided in the original papers.

C, outflow facility measured by two-level constant pressure perfusion method; F_a, aqueous flow measured by [a]fluorophotometry, [b]aqueous humor sampling method, [c]confocal microscopy; F_u, uveoscleral outflow calculated from the modified Goldmann equation; IOP, intraocular pressure measured by manometry; P_{ev}, episcleral venous pressure measured by the intracameral microneedle method, [d]2,2,2 tribromoethanol.

TABLE II

Aqueous Humor Dynamics in Rats

References	Animal	Anesthesia	IOP (mmHg)	F_a (μl/min)	C (μl/min/mmHg)
Nguyen *et al.*, 2007	Sprague-Dawley	Ketamine-xylazine	15.5 ± 0.29[a]		0.034 ± 0.001
Kee, Hong and Choi, 1997	Sprague-Dawley	Sodium pentobarbital			0.051 ± 0.004
Kee and Seo, 1997	Sprague-Dawley	Sodium pentobarbital			0.040 ± 0.004
Mermoud *et al.*, 1996	Lewis rats	Ketamine-xylazine	17.2 ± 1.8[b]	0.35 ± 0.11	0.044 ± 0.015

Reported are mean values from untreated or vehicle-treated control groups ± standard errors or standard deviations as provided in the original papers.

C, outflow facility measured by the two-level constant pressure perfusion method; F_a, aqueous flow measured by the aqueous humor sampling method; IOP, intraocular pressure measured by [a]Tono-Pen, [b]manometry.

TABLE III
Recent Studies of Aqueous Humor Dynamics in Rabbits

References	Animal	Treatments	IOP (mmHg)	F_a (μl/min)	C (μl/min/mmHg)	F_u (μl/min)	P_{ev} (mmHg)
Ocular normotensive rabbits							
Kiel and Reitsamer, 2007	New Zealand white	Pentobarbital	18.9 ± 1.3	3.3 ± 0.2[d]			
Oka *et al.*, 2006	Japanese albino	Vehicle	17–18	2.8 ± 0.4[d]	0.15 ± 0.02[b]	0.53 ± 0.05[f]	
Toris, Zhan and McLaughlin, 2003	Dutch belted	Vehicle	23.7 ± 2.2	2.7 ± 0.6[d]			
Kotikoski, Vapaatalo and Oksala, 2003	New Zealand white	Ketamine/xylazine indomethacin	10–12		0.45–0.75[b]		
Wang *et al.*, 2003	Rabbit	Saline				0.18 ± 0.05[f]	
Zhan *et al.*, 2002	New Zealand white	None	20.1 ± 1.5	2.8 ± 0.2[d]		0.32 ± 0.03[f]	
		None	21.7 ± 0.9	2.5 ± 0.2[d]	0.24 ± 0.03[a]		
		None	17.2 ± 0.9	1.7 ± 0.2[d]	0.21 ± 0.01[a]	0.37 ± 0.06[f]	
Reitsamer and Kiel, 2002	New Zealand white	Pentobarbital	15.8 ± 0.8				9.6 ± 0.9
			16.4 ± 1.1				
			17.2 ± 1.2				
Puras *et al.*, 2002a	New Zealand white	Vehicle	20.7 ± 0.7	2.9 ± 0.4[e]	0.38 ± 0.05[a]		

(*Continued*)

TABLE III (*Continued*)

References	Animal	Treatments	IOP (mmHg)	F_a (μl/min)	C (μl/min/mmHg)	F_u (μl/min)	P_{ev} (mmHg)
Chu and Potter, 2002	New Zealand white	None	21.0 ± 1.5	2.3 ± 0.3[d]			
Ogidigben and Potter, 2001	New Zealand white	None	25.5	2.3 ± 0.6[d]			
Chidlow *et al.*, 2001	New Zealand white	None	17.2 ± 0.3		0.15 ± 0.01[b]		
Inoue *et al.*, 2001	New Zealand white	Saline			0.29 ± 0.08[b]	1.84 ± 0.07[f]	
Kiel *et al.*, 2001	New Zealand white	Pentobarbital	15.8 ± 1.2	2.8 ± 0.2[d]			
			15.7 ± 1.1	3.1 ± 0.4[d]			
Sugiyama *et al.*, 2001	Japanese white	None	14–17		0.21 ± 0.03[b]	0.42 ± 0.04[f]	
Honjo *et al.*, 2001	Japanese white	None			0.12 ± 0.01[b]	0.47 ± 0.03[f]	
Crosson, 2001	New Zealand white	Vehicle		2.0 ± 0.3[d]	0.22 ± 0.02[c]		
Russell, Wang and Potter, 2000	New Zealand white	None	25–26	2.4–3.3[d]			
Muta *et al.*, 2000	Albino	None	26.0 ± 1.3	2.7 ± 0.3[d]	0.22 ± 0.06[c]	0.04 ± 0.42[g]	
Melena *et al.*, 1999	New Zealand white	None	20.1 ± 0.8	3.1 ± 0.2[e]	0.24 ± 0.01[a]		
Crosson and Petrovich, 1999	New Zealand white	Vehicle			0.23 ± 0.01[b]		

Artru and Momota, 1999	New Zealand white	Sevoflurane	13 ± 3	4.0 ± 1.0[e]	0.24 ± 0.13[b]	0.5 ± 0.1[g]
Kanno *et al.*, 1998	New Zealand white	Vehicle		2.9 ± 0.1[d]	0.19 ± 0.01[c]	0.16 ± 0.01[f]
Zhan *et al.*, 1998	New Zealand white	Vehicle	19.4 ± 1.3	2.6 ± 0.1[d]	0.25 ± 0.03[b]	0.31 ± 0.09[f]
Ocular hypertensive rabbits (α-chymotrypsin model)						
Puras *et al.*, 2002b	New Zealand white	α-chymotrypsin	35.0 ± 0.9	4.9 ± 1.0[e]	0.25 ± 0.04[a]	
Melena *et al.*, 1999	New Zealand white	α-chymotrypsin	39.3 ± 2.5	5.4 ± 0.6[e]	0.21 ± 0.02[a]	

Values are means ± standard error or standard deviation as provided in the original paper. The selected articles are those published within the past decade that reported one or mores value of aqueous humor dynamics in untreated or vehicle-treated ocular normotensive or hypertensive rabbits.

C, outflow facility measured by [a]tonography, [b]two-level constant pressure perfusion method, [c]flow to blood method; F_a, aqueous flow measured by [d]fluorophotometry, [e]calculation; F_u, uveoscleral outflow measured by [f]intracameral tracer method, [g]mathematical calculation; IOP, intraocular pressure measured by pneumatonometry.

TABLE IV

Aqueous Humor Dynamics in Cats

References	Anesthesia	IOP (mmHg)	F_a (μl/min)	C (μl/min/mmHg)	F_u (μl/min)	P_{ev} (mmHg)
Hayashi, Yablonski and Bito, 1987	Ketamine-xylazine ($n = 15$)	17 ± 0.8	4.4 ± 0.5[d]	1.37 ± 0.34[a]		12.0 ± 0.8
		16 ± 2.3	3.5 ± 0.4[d]	1.14 ± 0.11[a]		13.6 ± 0.8
Goh, Oshima and Araie, 1994	Ketamine ($n = 10$)	24.5 ± 1.1	12.1 ± 2.9[d]	0.48 ± 0.12[a]	0.88 ± 0.07[f]	
		23.0 ± 1.3	12.4 ± 1.1[d]		1.13 ± 0.25[f]	
Toris *et al.*, 1995	Ketamine ($n = 12$)	14.0 ± 3.5	6.0 ± 1.4[d]	0.48 ± 0.36[b]	2.1 ± 2.2[g]	
					1.5 ± 0.6[f]	
Wang *et al.*, 1999	Ketamine ($n = 10$–12)	22.2 ± 1.2	6.0 ± 0.3[d]	0.26 ± 0.04[b]	2.98 ± 0.58[g]	
				0.27 ± 0.02[a]	1.42 ± 0.48[f]	
Higginbotham *et al.*, 1988	Ketamine-acepromazine ($n = 4$)	22.8 ± 11.4	6.6 ± 3.3[d]			
		23.0 ± 11.5	6.4 ± 3.2[d]			
Rosenberg *et al.*, 1996	Ketamine-xylazine	24.6 ± 3.1	4.0 ± 1.8[e]	0.25[a]		8
Oksala and Stjernschantz, 1988	Ketamine-xylazine	16.7 ± 1.6		0.74 ± 0.22[b]		
Colasanti, 1990	Sodium pentobarbital	21.9 ± 1.0	10.6 ± 2.4[c]	0.98 ± 0.06[b]		
		24.2 ± 0.6	12.5 ± 2.4[c]	0.87 ± 0.10[b]		
			8.25 ± 0.6[c]	0.70 ± 0.14[b]		

Values are means ± standard error or standard deviation (when available) as provided in original publications. The selected articles are those published since 1980 that reported one or more values of aqueous humor dynamics in healthy untreated or vehicle-treated cats.

C, outflow facility measured by [a]tonography, [b]two-level constant pressure perfusion method; F_a, aqueous flow measured by [c]an aqueous humor sampling method, [d]fluorophotometry, [e]mathematical calculation; F_u, uveoscleral outflow measured by [f]intracameral tracer method, [g]mathematical calculation; IOP, intraocular pressure measured by pneumatonometry; P_{ev}, episcleral venous pressure measured by venomanometry.

TABLE V

Aqueous Humor Dynamics in Dogs

References	Animal	Anesthesia	IOP (mmHg)	F_a (μl/min)	C (μl/min/mmHg	P_{ev} (mmHg)
Ocular normotensive dogs						
Toris *et al.*, 2006a	Normal beagles (4)	Butorphenol	12.8 ± 2.1	6.8 ± 2.4		
Skorobohach, Ward and Hendrix, 2003	Normal beagles (5)	Tiletamine-zolazepam, butorphanol	14–17 diurnal range	5.1 ± 2.0		
Ward *et al.*, 2001	Normal beagles (15)	Tiletamine-zolazepam, butorphanol		5.2 ± 1.9		
Cawrse, Ward and Hendrix, 2001	Normal beagles (15)	Tiletamine-zolazepam, butorphanol	16–18 diurnal range	5.9		
Kurata *et al.*, 1998	Normal beagles (6)	Not reported	17.7–19.0		0.23–0.28[a]	
Artru, 1995	Mongrel dogs (10–12)	Desflurane	11.3 ± 3.8	1.6 ± 0.6	0.097 ± 0.062[a]	
		Halothane	9.4 ± 2.8	1.5 ± 0.5	0.091 ± 0.054[a]	
Gelatt *et al.*, 1982	Normal beagles (6)	Acepromazine-ketamine	23.4 ± 1.5			11.7 ± 0.3
		Thiamylal sodium	19.9 ± 0.6			11.6 ± 0.4
		Halothane				
		Ketamine-xylazine	26.0 ± 2.1			11.4 ± 0.7
Peiffer *et al.*, 1980	Normal beagles (18)	Sodium pentobarbital			0.22 ± 0.03[a]	
Gelatt *et al.*, 1977	Normal beagles (37)	Acepromazine-ketamine			0.24 ± 0.07[b]	

(*Continued*)

TABLE V (*Continued*)

References	Animal	Anesthesia	IOP (mmHg)	F_a (μl/min)	C (μl/min/mmHg	P_{ev} (mmHg)
Peiffer *et al.*, 1976	Normal beagles (36)	Sodium pentobarbital			0.21 ± 0.14^b	
		Ketamine-acetyl-promazine				
Glaucomatous dogs						
Gelatt *et al.*, 1982	Glaucomatous beagles (12)	Acepromazine-ketamine	34.2 ± 3.7			10.6 ± 0.3
		Thiamylal sodium	27.4 ± 0.4			12.1 ± 0.4
		Halothane				
		Ketamine-xylazine	37.8 ± 4.2			12.5 ± 0.5
Peiffer *et al.*, 1980	Glaucomatous beagles (17)	Sodium pentobarbital			0.09 ± 0.01^a	
Gelatt *et al.*, 1977	Glaucomatous beagles (35)	Acepromazine-ketamine			0.09 ± 0.04^b	
Peiffer *et al.*, 1976	Glaucomatous beagles (35)	Sodium pentobarbital			0.15 ± 0.09^b	
		Ketamine-acetyl-promazine				

Values are a range or means±standard error or standard deviation as provided in the original papers. The articles are those reporting one or more values of aqueous humor dynamics in glaucomatous or healthy untreated or vehicle-treated dogs.

C, outflow facility measured by [a]two-level constant pressure perfusion method, [b]tonography; F_a, aqueous flow measured by fluorophotometry; IOP, intraocular pressure measured by pneumatonometry; P_{ev}, episcleral venous pressure measured by force-displacement method.

TABLE VI

Recent Studies of Aqueous Humor Dynamics in Monkeys

References	Species	IOP mmHg	F_a (μl/min)	C (μl/min/ mmHg)	F_u (μl/min)
Ocular normotensive monkeys					
Nilsson *et al.*, 2006	Cynomolgus	18 ± 1	1.9 ± 0.3	0.69 ± 0.13[a]	0.53 ± 0.18[e]
Okka, Tian and Kaufman, 2004	Cynomolgus	19.3 ± 0.8		0.51 ± 0.08[a]	
		18.8 ± 0.7			
Peterson *et al.*, 2000b	Cynomolgus	16–17	1.8 ± 0.2		
			2.0 ± 0.2		
Peterson *et al.*, 2000a	Cynomolgus			Multiple studies with results between 0.27 ± 0.04[a] and 0.59 ± 0.12[a]	
Toris *et al.*, 2006b	Cynomolgus	22.5 ± 0.7	1.5 ± 0.1	0.11 ± 0.02[b]	0.47 ± 0.17[f]
Gabelt *et al.*, 2005	Cynomolgus	14–16	2.0 ± 0.3	0.90 ± 0.26[a]	0.48 ± 0.13[e]
				0.40 ± 0.06[c]	
Toris *et al.*, 2005	Cynomolgus	23.4 ± 5.3	1.7 ± 0.3	0.15 ± 0.07[b]	0.35 ± 0.72[f]
Tian and Kaufman, 2005	Cynomolgus and Rhesus			0.36 ± 0.04[a]	
Tian *et al.*, 2004	Cynomolgus	12–13		0.50 ± 0.07[a]	
Takagi *et al.*, 2004	Cynomolgus	21–23	1.5 ± 0.1	0.45 ± 0.08[a]	1.01 ± 0.22[f]
Gabelt *et al.*, 2004	Cynomolgus	15–18	1.4 ± 0.3	0.26 ± 0.02[a]	0.32 ± 0.12[f]
				0.27 ± 0.04[a]	
Chien *et al.*, 2003	Cynomolgus	15–18	1.8 ± 0.2	0.55 ± 0.07[d]	

(*Continued*)

TABLE VI (*Continued*)

References	Species	IOP mmHg	F_a (μl/min)	C (μl/min/ mmHg)	F_u (μl/min)
Gabelt *et al.*, 2001	Cynomolgus	14.9 ± 1.0	Multiple studies with results between 1.4 ± 0.18 and 2.1 ± 0.1	0.35 ± 0.05[a]	
		16.0 ± 1.7		0.50 ± 0.09[a]	
Toris *et al.*, 2003	Cynomolgus	22.3 ± 3.6	1.7 ± 0.3	0.15 ± 0.09[b]	0.35 ± 0.92[f]
Gabelt *et al.*, 2003	Rhesus	17.3 ± 0.8	1.7 ± 0.1		0.63 ± 0.07[g]
Toris *et al.*, 2000	Cynomolgus	22.2 ± 2.4	1.5 ± 0.6	0.16 ± 0.14[b]	0.14 ± 1.20[f]
					1.05 ± 0.58[e]
Ocular hypertensive monkeys					
Nilsson *et al.*, 2006	Cynomolgus	30 ± 4	2.1 ± 0.3		
Toris *et al.*, 2006b	Cynomolgus	30.9 ± 3.3	2.0 ± 0.4	0.08 ± 0.04[b]	1.02 ± 0.28[f]
Toris *et al.*, 2005	Cynomolgus	35.1 ± 13.6	1.8 ± 0.6	0.09 ± 0.08[b]	0.37 ± 1.00[f]
Toris *et al.*, 2003	Cynomolgus	30.9 ± 12.1	2.1 ± 1.2	0.12 ± 0.13[b]	0.46 ± 1.24[f]
Toris *et al.*, 2000	Cynomolgus	33.8 ± 8.0	1.9 ± 0.9	0.06 ± 0.04[b]	0.92 ± 0.65[f]

Values are means ± standard error or standard deviation as provided in the original papers. The listed articles are those published within the past decade that reported one or more values of aqueous humor dynamics in untreated or vehicle-treated ocular normotensive or hypertensive monkeys.

C, outflow facility measured by [a]two-level constant pressure perfusion method, [b]fluorophotometry, [c]flow-to-blood perfusion method, [d]tonography; F_a, aqueous flow measured by fluorophotometry; F_u, uveoscleral outflow measured by [e]intracameral tracer method, [f]mathematical calculation, [g]indirect isotope intracameral tracer method; IOP, intraocular pressure measured by tonometry.

1. Tonometry

Tonometry is the measurement of the IOP by a deformation of the globe related to the force responsible for the deformation. Classes of tonometers used in research animals include indentation, applanation, and rebound varieties. The Shiotz tonometer is a simple handheld indentation tonometer developed decades ago. The instrument is used to measure IOP and to perform tonography. In a clinical setting, the Goldmann applanation tonometer has been considered to be the gold standard against which all other tonometers are compared. This instrument is difficult to use in animals because of the need to place the subject in front of a slit-lamp with eyes facing forward. Functional portable varieties of the Goldmann tonometer are the handheld Perkins tonometer (Haag-Streit USA, Inc., Mason, OH) and the Tono-Pen (Reichert, Dewep, NY). These instruments can measure the IOPs of subjects in a variety of positions including seated and supine but not prone. Another type of applanation tonometer is the pneumatic tonometer that uses air pressure to press a probe on the cornea causing corneal deformation. An air pressure sensor measures the IOP as the corneal deformation is transferred to surrounding structures. Pneumatonometers include the Model 30 Classic (Reichert, Dewep, NY) and ocular blood flow tonometer (Silver and Farrell, 1994). The ocular blood flow tonometer also exploits the pulsatile nature of the IOP to estimate ocular blood flow. Pneumatonometers are often used to measure the IOP in animals whose eyes are similar in size to humans. These tonometers were designed originally for human use and some of the assumptions inherent in the measurement may not apply to animal eyes. A new class of tonometer is the rebound tonometer, a device touted to provide IOP readings independent of cornea thickness, a factor that may affect the measurement. In this category are the Pascal Dynamic Contour tonometer (Zeimer Ophthalmic, Port, Switzerland) designed predominantly for human use and the Tonolab (Tiolat Oy, Helsinki, Finland) designed for rats and mice.

2. Manometry

Manometry provides a direct measurement of the IOP. In the anesthetized animal, a small needle is placed through the cornea into the anterior chamber or through the pars plana into the vitreous cavity. The latter approach avoids trauma to the tissues of the anterior chamber. This needle is connected by saline-filled tubing to a pressure transducer that detects the spontaneous pressure. The measurement is not affected by cornea thickness, scleral rigidity, or other factors that plague tonometry. The disadvantages of this method are that it is invasive and does require anesthesia, both factors that disturb the IOP.

3. Telemetry

Continuous measurement of the IOP is obtainable in research animals by surgically implanting a sensor catheter into the anterior chamber or midvitreous of an eye and a telemetry transmitter into a subcutaneous pocket in the cheek or back. The transmitter sends the IOP information to a receiver attached to the animal's cage. The receiver sends the digital signal to a computer-based data-acquisition system. Each animal's IOP is measured continuously at around 100 Hz for several seconds in cycle runs of a few minutes. Pressure measurements occur in this manner until the battery fails, usually after weeks to months of continuous use. This method was first reported in rabbits (McLaren *et al.*, 1996; Schnell *et al.*, 1996) and is currently being developed for other laboratory animals. Success of the surgical implantation and long-term maintenance of an active signal requires great technical skill but the investment in time and effort is rewarded by continuous undisturbed circadian IOPs, data not obtainable by any other means.

B. Aqueous Humor Flow

The production rate of aqueous humor into the posterior chamber is the sum of the flow rate from the posterior chamber through the pupil into the anterior chamber (*aqueous flow*), from the posterior chamber into the vitreous cavity and across the retinal pigment epithelium (*posterior flow*), and the loss of aqueous humor by other routes such as across the cornea, which is thought to be minimal (Fig. 1). Aqueous flow is less than aqueous production but it is generally assumed that changes in aqueous flow reflect changes in aqueous production.

The formation of aqueous humor involves several steps starting with the ultrafiltration of plasma through the capillaries of the ciliary processes. This is followed by active secretion of fluid from the ciliary process core across the ciliary epithelial layers and into the posterior chamber. Sodium is pumped into the intercellular spaces of the nonpigmented epithelium by sodium–potassium ATPase. Bicarbonate and other negative ions follow the sodium. The bicarbonate is produced by the action of carbonic anhydrase, which catalyzes the formation of bicarbonate ions from carbon dioxide and water. The tight junctions at the stromal ends of adjoining cells and ion pumps along the cell walls create a concentration gradient of ions in the intercellular cleft. Water follows the ions into the cleft and then flows in the direction of the posterior chamber (Diamond and Bossert, 1967). Nutrients and other substances necessary for the survival of the lens and cornea are added to the fluid by the process of diffusion or facilitated transport.

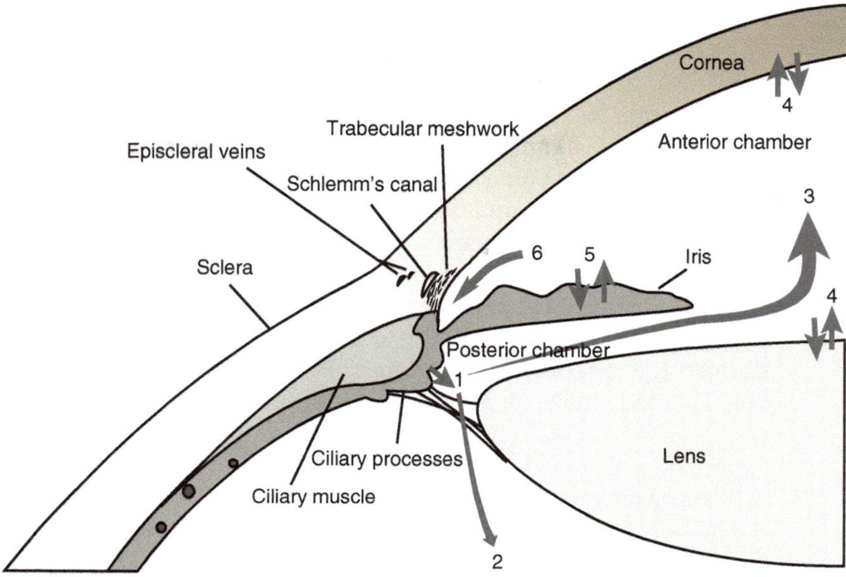

FIGURE 1 The flow of aqueous humor from the posterior chamber. Aqueous humor that is secreted into the posterior chamber (1) flows across the vitreous cavity (2) or through the pupil into the anterior chamber (3). In the anterior chamber, there is exchange of fluid with the lens and cornea (4) and iris vasculature (5). Fluid circulates around the anterior chamber and eventually drains into the anterior chamber angle (6). (See Color Insert.)

Aqueous flow has a distinctive circadian rhythm that varies among species. In humans, who generally are more active during the day than at night, the diurnal rate of aqueous flow is about twice the nocturnal rate (Reiss *et al.*, 1984; Brubaker, 1991; Koskela and Brubaker, 1991). Rabbits have a shifted circadian rhythm with higher rates at night when the animal is most active and lower rates during the day when the animal is sedentary (Smith and Gregory, 1989).

The rate of aqueous flow varies among species and is dependent somewhat on the size of the anterior chamber; the larger the chamber, the greater the metabolic needs of the eye and the faster the flow rate. Mice have an anterior chamber volume of 6 μl and an aqueous flow of 0.1–0.2 μl/min (Aihara *et al.*, 2003a). Rats have a slightly larger anterior chamber volume (15 μl) and correspondingly higher rate of aqueous flow (0.35 μl/min) (Mermoud *et al.*, 1996). Cynomolgus monkeys have anterior chamber volumes in the range of 90–110 μl and aqueous flow rates in the range of 1.5–2.1 μl/min (Table VI). Anterior chamber volume and aqueous flow in rabbits average about 200 μl and 2.5 μl/min, respectively. Although cats are similar in body size to rabbits, their anterior chamber volume is substantially larger (810 μl) (Toris *et al.*, 1995) and their aqueous flow rate is correspondingly higher (3.5–12 μl/min, Table IV). Beagles are significantly larger animals than cats but they have relatively smaller anterior chamber volumes (325–400 μl) (Ward *et al.*, 2001; Toris *et al.*, 2006a) and slower aqueous flow rates (5–7 μl/min, Table V).

In early studies of aqueous humor dynamics, aqueous flow was calculated from the Goldmann equation [Eq. (1)]. Intraocular pressure and outflow facility were measured, episcleral venous pressure was measured or estimated, and uveoscleral outflow was not considered. Current methods to assess aqueous flow measure the disappearance rate of a tracer from the anterior chamber, the assumption being that all intracameral tracer drains solely through the anterior chamber angle. First-order fluorescein disappearance kinetics applies after an initial equilibration period.

1. Fluorophotometry

The noninvasive technique of fluorophotometry was first introduced by Goldmann (1951), developed further by Maurice (1963), and perfected by Brubaker (1982), although many other people had a hand in its development along the way. Fluorescein is placed into the anterior chamber by intracameral injection, iontophoresis, or topical application. Over time, the fluorescein becomes well mixed with the aqueous humor. Periodic scans of the eye are taken with a fluorescence detector (fluorophotometer) that measures the mass of fluorescein in the anterior chamber and cornea. Plotting fluorescein mass over time yields a fluorescein decay curve. The aqueous flow is calculated from the decay slopes and the volumes of the cornea and anterior chamber. Details of this method for use in humans are found in several very comprehensive reviews (Brubaker, 1982, 1998; Brubaker *et al.*, 1990). The same instrument is used for cats, dogs, and monkeys, and a modified fluorophotometer for mice was described recently at the 2007 annual meeting of the Association for Research in Vision and Ophthalmology (Fan *et al.*, 2007).

2. Confocal Microscopy

Aqueous flow has been measured in mice by means of confocal microscopy and the kinetics of the disappearance of fluorescein from the anterior chamber fluid. Fluorescein is administered by iontophoresis, and fluorescence of the anterior chamber over time is detected with a z-scanning confocal microscope, fluorescein filter, photomultiplier detector, and long working distance objective lens. The scan through the eye is accomplished using a stepper motor controlling the microscope fine focus. Scans are taken periodically starting 2 hours after fluorescein administration and for 4 hours thereafter. The slope of a log plot of fluorescence over time provides the decay rate, which is multiplied by the anterior chamber volume to yield an aqueous flow rate (Zhang *et al.*, 2002).

3. Aqueous Humor Sampling Method

A direct sampling method to determine aqueous flow requires placing two microneedles into the anterior chamber and infusing a fluorescence tracer through one microneedle while simultaneously withdrawing, at the same rate, fluid through the second microneedle. The infused tracer is diluted by newly formed aqueous humor. The difference in fluorescence of the infused fluid and that of the withdrawn fluid over a precise interval of time provides a dilution rate of the tracer. This is multiplied by the anterior chamber volume to yield aqueous flow (Sperber and Bill, 1984).

C. *Outflow Resistance*

Fluid traverses the trabecular outflow pathway through the trabecular meshwork, juxtacanalicular connective tissue, endothelial lining of Schlemm's canal, collector channels (Fig. 2), aqueous veins, and episcleral veins. The region providing the main resistance to fluid drainage is the inner wall of Schlemm's canal, its basement membrane, and the adjacent juxtacanalicular connective tissue (Ethier, 2002; Johnson, 2006). Cells in the trabecular meshwork regulate hydraulic conductivity of the inner wall region possibly by modulating extracellular matrix turnover and by actively distorting the meshwork and changing cell shape (Lütjen-Drecoll, 1999). This route provides a hydrostatic pressure resistance that is overcome by a hydrostatic pressure gradient between the inside of the eye, the IOP, and the outside of the eye, episcleral venous pressure. The hydrodynamic effect is termed *outflow resistance.* Its reciprocal is *outflow facility.*

Outflow facility in healthy animal eyes, range from 0.15–0.45 in rabbits, 0.25–1.37 in cats, 0.09–0.24 in dogs, and 0.11–0.90 μl/min/mmHg in monkeys (Tables III–VI). It is 2- to 10-fold smaller in rats (Table II) and 100-fold smaller still in mice (Table I).

Outflow facility is reduced in aged rhesus monkey eyes compared with young healthy animals (Table VI) (Gabelt *et al.*, 1991, 2003). This reduction has been explained by the loss of movement of the ciliary muscle, causing reduced stretch and pull on the trabecular meshwork leading to accumulation of extracellular material in the meshwork and increased outflow resistance (Gabelt and Kaufman, 2005).

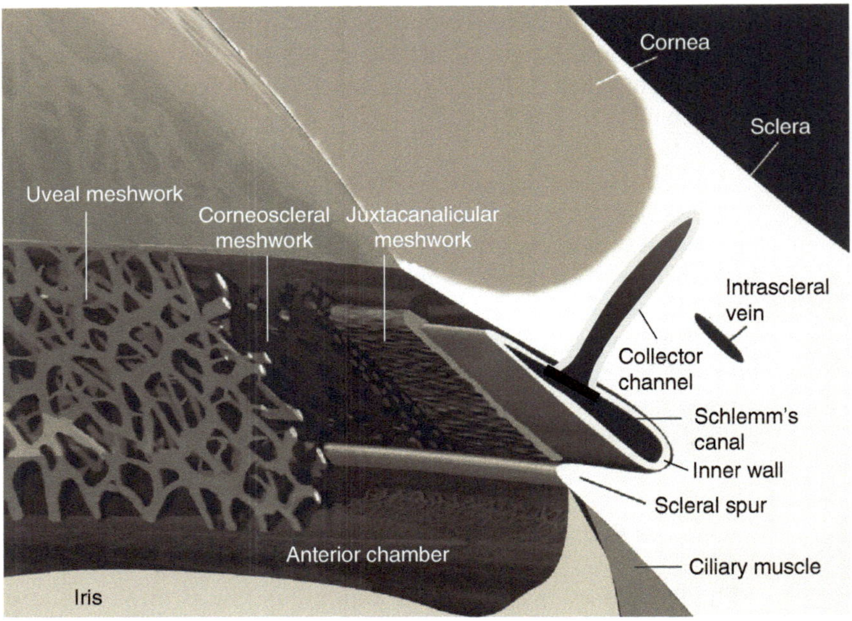

FIGURE 2 Trabecular outflow pathway. Aqueous humor drains from the anterior chamber angle via two routes, one of which is through the trabecular meshwork. Trabecular meshwork consists of uveal meshwork, corneoscleral meshwork, and juxtacanalicular meshwork. Aqueous humor traverses this meshwork, crosses the inner wall endothelium, and enters the Schlemm's canal. Fluid then drains from the canal through collector channels and then into aqueous veins and episcleral veins (not drawn) before entering the systemic circulation. (See Color Insert.)

Outflow facility also is reduced in glaucoma, both naturally occurring and experimentally induced. In beagles with naturally occurring glaucoma, outflow facility has been reported to be 0.09–0.15 μl/min/mmHg, which is significantly lower than the 0.21–0.24 μl/min/mmHg in healthy beagles (Table V). Similarly, monkeys with unilateral laser-induced glaucoma have reduced outflow facility (0.08–0.12 μl/min/mmHg) compared with contralateral normotensive eyes (0.15–0.55 μl/min/mmHg; Table VI).

All methods to determine outflow facility ascertain a change in pressure associated with a change in fluid flow. The pressure change is easily measured by tonometry or manometry but the flow change is harder to establish. It is determined by the use of standardized tables, by fluorophotometry, by measuring a dilution rate of tracer infused into the anterior chamber, or by measuring the rate of appearance of tracer in blood after intracameral administration. Outflow facility has different names depending on the method used to make the assessment. The measured value is called tonographic outflow facility (C_{ton}) when determined by tonography, fluorophotometric outflow facility (C_{fl}) when determined by fluorophotometry (Hayashi *et al.*, 1989), total outflow facility (C_{tot}) when measured by two-level constant pressure infusion (Bárány, 1964), and trabecular outflow facility (C_{trab}) when measured by intracamerally administered tracers detected in the blood (Gabelt and Kaufman, 1990).

1. Tonography

Tonography is a noninvasive measurement of outflow facility that uses either a Schiotz tonometer or the tonography setting on a pneumatonometer. Tonography once was used as a routine clinical test to aid in the diagnosis of glaucoma, but now it is used mainly for research purposes. Occasionally, it is used on research animals with eyes of similar size to humans, such as monkeys, cats, and rabbits, but it should be kept in mind that these instruments reference standardized tables that were developed in human eyes. Some of the inherent assumptions may not be valid in research animals. The procedure involves placing a weighted tonometer probe on the anesthetized cornea of the recumbent animal for 2 or 4 min. The probe is of a standard weight and when applied to the cornea, the IOP increases. When the weight is maintained on the eye, the IOP slowly decreases and aqueous humor drains through the anterior chamber angle at an increased rate. It is assumed that the change in the IOP during the measurement results from displacement of aqueous humor from the eye. The inferred volume of aqueous humor displaced from the eye by the weight is obtained in reference tables developed by Friedenwald (1948). If the displacement of fluid from the eye, ΔV, by the weight of the tonometer, is the only factor to account for the IOP decrease, then the rate of fluid outflow from the eye is the change in fluid volume over the time (t) of the test, $\Delta V/t$. Tonographic outflow facility, C_{ton}, is calculated from Grant's equation (Grant, 1950).

$$C_{ton} = \frac{\Delta V/t}{\mathrm{IOP}_t - \mathrm{IOP}_0} \qquad (5)$$

IOP_0 is the intraocular pressure before the weighted probe is applied to the eye (at $t = 0$). At this time, it is assumed that inflow and outflow of the aqueous humor are equal, and the IOP and volume are stable. IOP_t is the average IOP at the end of the test, which lasts for either 2 or 4 min ($t = 2$ or 4), at which time the rate of aqueous humor outflow from the eye is greater than inflow and the ocular volume has diminished. The greater the IOP decrease during the test, the greater the expected volume change to account for the pressure change and the larger the trabecular outflow facility. Eyes with low outflow facility, as in glaucoma or ocular hypertension, will show relatively little change in the IOP during the test.

Ocular rigidity is a confounding factor in the tonography measurement because of the indentation of the cornea caused by the weight of the unit. Tonography performed with an indentation tonometer assumes that the pressure change, as a function of time, is based on the accuracy of the ocular rigidity coefficient both at the beginning and during the measurement. Individual variations in ocular rigidity can be large and indentation tonography makes no compensation for this. Tonography by a pneumatic tonography unit is less affected by ocular rigidity than the Schiotz unit because the probe that is placed on the eye creates a relatively small indentation of the cornea. Both instruments derive a change in ocular fluid volume from standard tables.

Assumptions in addition to normal ocular rigidity need to be considered when interpreting tonography data. It is assumed that corneal curvature is normal and the rate of aqueous humor production (inflow into the anterior chamber) during tonography remains unchanged by the applied pressure. The change in pressure during the test is assumed to be caused by fluid being forced out of the eye across the trabecular meshwork only. A decrease in the rate of aqueous humor formation or the drainage of fluid by nontrabecular routes would be measured erroneously as increased outflow facility. Examples of such circumstances include a decrease in ocular blood volume or extracellular fluid volume or an increase in the facility of aqueous humor drainage through the uvea. The outflow facility measured by tonography (C_{ton}) includes pseudofacility (C_{ps}) and uveoscleral outflow facility (C_{fu}) in addition to trabecular outflow facility (C_{trab})

$$C_{\mathrm{ton}} = C_{\mathrm{trab}} + C_{\mathrm{fu}} + C_{\mathrm{ps}} \tag{6}$$

A change in outflow facility measured by tonography does not always indicate a change in true trabecular outflow facility if pseudofacility and uveoscleral outflow facility are also disturbed.

2. Fluorophotometry

As a means to avoid the problems of pseudofacility and ocular rigidity, a fluorophotometry method to assess outflow facility (Hayashi *et al.*, 1989) was developed that measures rather than assumes a change in ocular fluid flow. First, intraocular pressure (IOP_1) is measured by tonometry and aqueous flow (F_1) is determined by fluorophotometry. Next, a drug is given that reduces the IOP by reducing the aqueous flow without affecting the aqueous humor drainage pathways. The topical β-blocker, timolol, and the systemic carbonic anhydrase inhibitor, acetazolamide, are usually used for this purpose. Again intraocular pressure (IOP_2) and aqueous flow (F_2) are measured and Eq. (7) is used to calculate outflow facility (C):

$$C = \frac{F_1 - F_2}{\mathrm{IOP}_1 - \mathrm{IOP}_2} \tag{7}$$

Like all methods to estimate trabecular outflow facility, the fluorophotometric method rests on a number of assumptions. It is assumed that uveoscleral outflow varies little with changes in the IOP, and uveoscleral outflow facility is very small relative to trabecular outflow facility. This has been measured directly in animals and appears to be true under normal circumstances (Bill, 1966, 1967; Toris and Pederson, 1985). The fluorophotometry method would not be valid in some conditions that greatly affect uveoscleral outflow facility such as the presence of a cyclodialysis cleft (Suguro *et al.*, 1985). Additionally, there is a debate over the effect, if any, of topical prostaglandins on uveoscleral outflow facility (Camras, 2003). The fluorophotometric method to assess outflow facility has been used in rabbits, cats, dogs, monkeys (Tables III–VI), and humans (See Chapter 8).

3. Perfusion Methods

The two-level constant pressure perfusion technique (Bárány, 1964) is used in research animals to achieve a more accurate estimate of outflow facility than the indirect methods of tonography and fluorophotometry. This technique requires insertion of needles into the eye, and if great care is taken to avoid infection and minimize injury, multiple measurements can be made in the same animal. The needle is attached to a reservoir of mock aqueous humor and the IOP is set by the level of the reservoir above the eye. The rate of fluid flow into the anterior chamber (F_1) that is needed to maintain a stable intraocular pressure (IOP_1) is measured. The reservoir is raised to a new level, and the new intraocular pressure (IOP_2) and fluid inflow (F_2) are measured. Equation (7) is used to calculate outflow facility. A similar method, the two-level constant flow technique, measures the pressures needed to maintain a constant fluid flow into the eye at each of two different flow rates. Like tonography, this method includes ocular rigidity, pseudofacility, and uveoscleral outflow facility in the measurement.

Arguably the most accurate method to assess trabecular outflow facility is the "flow to blood" method. A radioactive isotope is infused into the anterior chamber and the radioactivity detected in the blood during a specified time interval is considered to have drained solely through the trabecular meshwork. Changes in outflow facility can be made by sampling blood at different times and infusion pressures and using Eq. (7) to calculate trabecular outflow facility.

D. Uveoscleral Outflow

The aqueous humor that enters the ciliary muscle exits the eye by multiple routes (Fig. 3). It percolates through the supraciliary space and across the anterior sclera or into the suprachoroidal space and across the posterior sclera

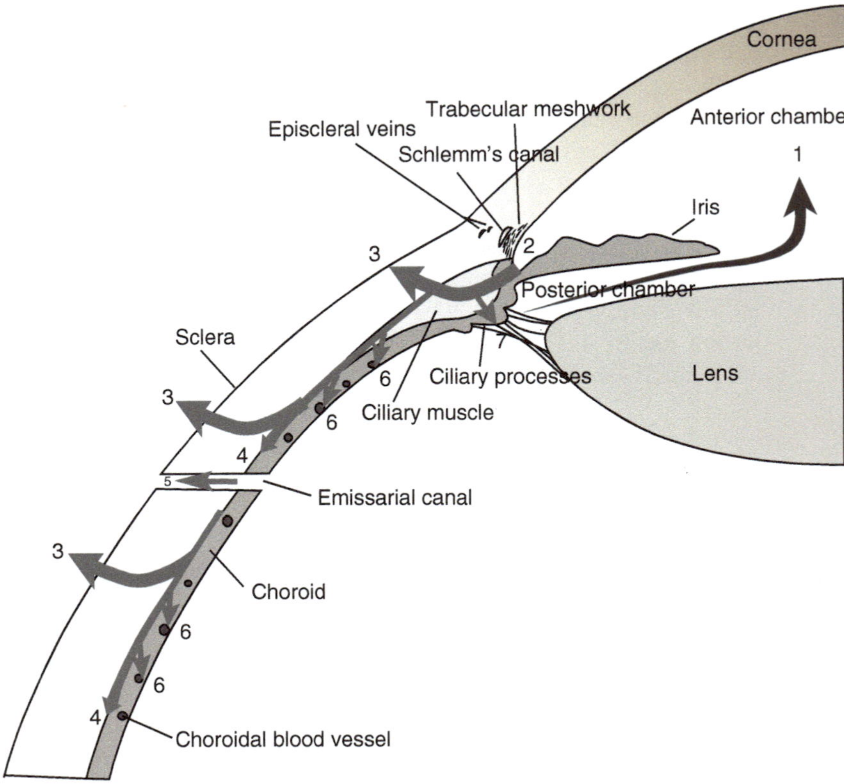

FIGURE 3 Uveoscleral outflow pathways. Most of the aqueous humor that flows anteriorly through the pupil (1) will drain into the anterior chamber angle (2). If it does not traverse the trabecular meshwork, it will enter the uveoscleral outflow pathway. This pathway starts with the ciliary muscle. From there, fluid can flow in many directions, including: across the sclera (3), within the supraciliary and suprachoroidal spaces (4), through emissarial canals (5), into choroidal vessels (6) and vortex veins (not drawn), and into ciliary processes (7) where it is secreted again. (See Color Insert.)

(uveoscleral outflow) (Bill, 1971). It flows through the emissarial canals around the vortex veins (McMaster and Macri, 1968; Green *et al.*, 1977; Krohn and Bertelsen, 1997) or traverses uveal vessels and into vortex veins (uveovortex outflow) (Pederson *et al.*, 1977). Additional evidence suggests that the fluid enters the ciliary processes from the ciliary muscle and is secreted back into the posterior chamber. Figure 4 is an unpublished micrograph of our study tracking intracameral fluorescein isothiocyanate dextran (1×10^{-4} M, 70,000 MW) into the ciliary body of ketamine/xylazine-anesthetized New Zealand white rabbits. The tracer was infused into the anterior chamber for 30 min at a pressure of 15 mmHg. The eye was quickly frozen in isopentane

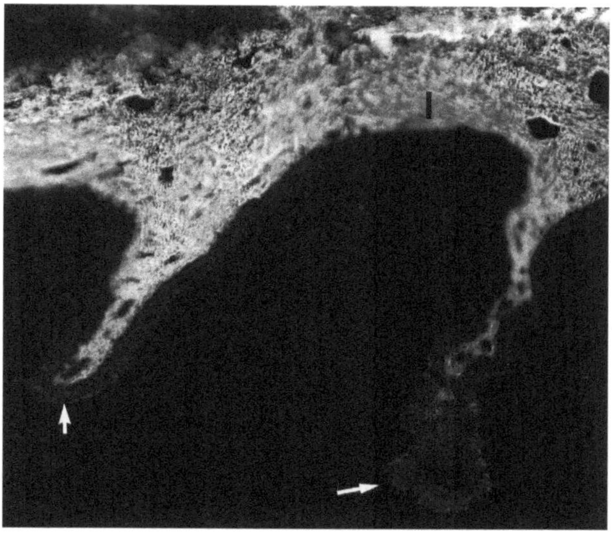

FIGURE 4 Tracer-filled ciliary processes of a rabbit. Fluorescence micrograph of two ciliary processes (arrows) of a rabbit infused into the anterior chamber with fluorescein isothiocyanate dextran (1×10–4 M, 70,000 MW) for 30 min at a pressure of 15 mmHg. The core of the left process is completely filled with tracer (light color) and the core of the right process is partially filled, leaving the tip void of tracer. This suggests that the direction of fluid movement is from the ciliary muscle into the process.

cooled in liquid nitrogen and slowly freeze dried over a period of 3 weeks to minimize postmortem tracer movement (Grayson and Laties, 1971). Cryosections were observed under a fluorescence microscope. Among other uveal tissues, the tracer was found in the core of ciliary processes. Fluid in the core is the source of material for aqueous humor. Therefore, some fluid originating from the anterior chamber is very likely secreted back into the posterior chamber.

A decrease in uveoscleral outflow has been found as a function of age in humans (Toris *et al.*, 1999, 2002) and rhesus monkeys (Gabelt *et al.*, 2003). Normal healthy monkeys 3–10 years of age drain 45–70% of their aqueous humor through the uveoscleral outflow pathway, whereas older animals 25–29 years of age drain aqueous humor at about half this rate (Gabelt *et al.*, 2003). Uveoscleral outflow amounts to more than 80% of total outflow in mice, 1–20% in rabbits, 7–25% in cats, and 15% in dogs (Tables I and III–V).

1. Mathematical Calculation

The only noninvasive method to measure uveoscleral outflow requires assessment of the IOP, aqueous flow, outflow facility, and episcleral venous pressure in the same eye on the same day and calculating uveoscleral outflow from Eq. (4). Large standard deviations are generated using this method because each variable has its own inherent variability. This approach is more useful to detect within and between group differences than it is to determine absolute values. Despite this limitation, the calculation method has been used to provide valuable information about drug efficacy and disease effects.

2. Intracameral Tracer Methods

Intracameral tracer methods are often used in research animals to determine uveoscleral outflow. The *direct tracer method* requires insertion of two needles into the eye: one connected by tubing to a pressure transducer to monitor the IOP and the other to a reservoir filled with a fluorescent or radioactive tracer. The IOP is maintained at a stable level by adjusting the height of the reservoir above the eye. At a precise time interval, usually 30–60 min, the eye is enucleated and dissected into tissues of the outflow pathway including iris, sclera, ciliary body, and choroid. The tracer is extracted from each tissue and quantitated. Uveoscleral outflow is calculated as the total volume of tracer that had accumulated in the tissue during the specified time interval. The sacrifice of the animal makes this method nonrepeatable.

The direct tracer method also has been used to measure the facility of uveoscleral outflow. Uveoscleral outflow is determined twice (F_1, F_2), once at each of two different infusion pressures (IOP_1, IOP_2), necessarily in different eyes, and uveoscleral outflow facility is calculated from Eq. (7).

The *indirect isotope method* involves infusing a radioactive tracer into the anterior chamber and monitoring the appearance rate of the tracer in the blood (an indication of trabecular outflow) and the disappearance rate of tracer from the anterior chamber (an indication of aqueous flow). Uveoscleral outflow is calculated as the difference between aqueous flow and trabecular outflow. This method does not involve sacrifice of the animal, and changes in uveoscleral outflow can be assessed over time.

E. Episcleral Venous Pressure

The aqueous humor that reaches Schlemm's canal leaves the eye through collector channels and episcleral veins. The pressure in these veins averages about 7–11 mmHg in humans (Zeimer, 1989) and 10 mmHg in monkeys (Table VI). Normal values in mice, rabbits, cats, and dogs (8–14 mmHg, Tables I and III–V) are remarkably similar to primates despite the anatomical differences of the venous pathways among the various species.

1. Episcleral Venomanometry

In clinical studies, a commercially available venomanometer (Eyetech, Morton Grove, IL) is used to obtain the pressure in episcleral veins. This device is attached to a slit-lamp and the subject is positioned such that the episcleral veins near the limbus can be seen through the binoculars. The flexible membrane of the venomanometer is placed on the surface of the eye above an episcleral vein and pressure is applied behind the membrane until the vessel collapses. This pressure is considered episcleral venous pressure. This method requires a cooperative subject and a clear conjunctiva to allow an unobstructed view of an appropriate vessel. Anesthesia is necessary in uncooperative animals to keep the eye still and the vessels in focus. The instrument was designed for human use and works best in them.

2. Force-Displacement Method

Episcleral venous pressure has been measured in dogs using an applanating Lucite cone and an isometric force-displacement transducer. The cone is pressed onto an appropriate vein on the scleral surface with enough force to collapse the vessel. The pressure at which blanching is first observed is recorded by the transducer and considered to be episcleral venous pressure (Gelatt *et al.*, 1982).

3. Direct Cannulation

A servo-null technique has been used to measure episcleral venous pressure directly. A microneedle with a diameter of 1–2 μm and filled with sodium chloride solution is inserted into an episcleral vein with the aid of a piezoelectric micromanipulator. Hydrostatic forces within the eye push blood into the microneedle tip. As the blood–salt solution interface moves within the microneedle, the servo-null device senses the change in resistance to electrical flow caused by a pressure change outside the tip of the microneedle. A signal is sent to the piezoelectric pressure pump and pressure transducer and a counterpressure is generated by the pump that is equal to the pressure outside the tip thus restoring the original resistance. This counterpressure is equal to the hydrostatic pressure within the vessel. The analog output from the servo-null device is converted to a digital signal with a data acquisition instrument. Using this method in anesthetized monkeys (Mäepea and Bill, 1989), episcleral venous pressure averaged 10.4 mmHg at spontaneous IOP and increased slightly when the IOP was increased experimentally.

4. Intracameral Microneedle Method

In the tiny mouse eye, episcleral venous pressure is measured by an intracameral microneedle method. The mouse is placed under a dissecting microscope and a glass microneedle, connected by tubing to a reservoir of physiological saline, is inserted into the anterior chamber. The IOP is set by the height of the reservoir above the eye. Schlemm's canal is visualized through the transparent tissues of the anterior chamber angle of the albino mouse. The reservoir is slowly lowered and the IOP thus decreased until erythrocytes reflux into Schlemm's canal. The pressure at which reflux occurs is considered episcleral venous pressure (Aihara *et al.*, 2003a,b).

III. AQUEOUS HUMOR DYNAMICS IN RESEARCH ANIMALS

Live animal models are used to address basic physiological, biochemical, and genetic questions related to aqueous humor dynamics. These models are also used to test potential therapies for the treatment of elevated IOP in order to establish proof of principle prior to human study. Selection of the most appropriate animal model requires an appreciation of the structural diversity of the outflow pathways and physiological distinctions among the species.

A. Mice

The mouse is recognized as a valuable research tool in the study of genetic factors related to ocular hypertension and various forms of glaucoma. The murine eye has important cellular and molecular similarities to humans and genetic manipulation of this animal is a major research asset (John *et al.*, 1998; Chang *et al.*, 1999; Gross *et al.*, 2003; Grozdanic *et al.*, 2003; Ruiz-Ederra and Verkman, 2006). The value of this animal model ultimately lies in the assessment of its IOP and aqueous humor dynamics with accuracy, reproducibility, and minimal trauma.

Intraocular pressures of many strains of mice are being published at an impressive rate. Initial experiments to measure the IOP (Avila *et al.*, 2001) used the microneedle method similar to that used to assess episcleral venous pressure. Tonometers originally designed for large eyes have been adapted for the small murine eye (Reitsamer *et al.*, 2004; Avila *et al.*, 2005) and new tonometers have been developed specifically for the mouse (Wang *et al.*, 2005; Fan *et al.*, 2006; Pease *et al.*, 2006). Intraocular pressures in the mouse are very similar to IOPs in healthy humans, ranging from 15 to 17 mmHg (Table I). Type and duration of anesthesia should be taken into consideration when interpreting the IOP data because these factors can have profound effects on murine physiology (Johnson *et al.*, 2008).

Aqueous flow and outflow facility in the mouse are less than 10% of aqueous flow and outflow facility in humans (Table I), findings that are not surprising considering the small size of the murine eye. Interestingly, the predominant outflow pathway in the mouse appears to be the uveoscleral outflow pathway, draining more than 80% of the anterior chamber aqueous humor (Aihara *et al.*, 2003a; Crowston *et al.*, 2004). No other species has shown a rate of uveoscleral outflow of this magnitude. The physiological significance of this finding warrants further investigation.

B. Rats

Rats in glaucoma research are used primarily to study ganglion cell loss from elevated IOP. In normal healthy rats, IOPs range from 15 to 18 mmHg. Aqueous flow reported in one study was 0.35 μl/min (Table II), about double that of mice and 12% of human. Outflow facility ranges from 0.03 to 0.05 μl/min/mmHg, 10 times that of mouse and 1/10 that of human. There is no report of uveoscleral outflow in rats.

The rat model of glaucoma is created by damaging tissues along the aqueous humor drainage routes. Methods yielding elevated IOP for weeks to months include hypertonic saline injection into the aqueous humor outflow pathways (Morrison *et al.*, 1997), cautery of the episcleral veins (Shareef *et al.*, 1995), or laser burns to tissues of the anterior chamber angle (WoldeMussie *et al.*, 2001). Aqueous humor dynamics in the rat model of glaucoma have not been reported.

C. Rabbits

Rabbits are often used in studies of aqueous humor dynamics because they are easy to handle, have big eyes, and are very sensitive to ocular procedures. However, rabbits have a particularly labile blood–aqueous barrier and may respond to ocular treatments in a manner quite often different from humans (Bito, 1984). There are distinctive differences in the structures of the rabbit outflow pathways when compared with primates. Rabbits do not have a true Schlemm's canal or highly developed trabecular meshwork and scleral spur and there is no functional relationship between their ciliary muscle and outflow mechanisms (Bito, 1984). Only nonhuman primates share these anatomical features with humans. Values of aqueous flow and outflow facility in rabbits are similar to primates but the relative rate of uveoscleral outflow appears to be slower (Table III).

Experimental ocular hypertension has been created in rabbits by several methods: multiple drops of glucocorticoid steroids (Levene *et al.*, 1974), single or multiple intraocular injections of α-chymotrypsin (Chee and Hamasaki, 1971; Zhu and Cai, 1992), ligation of vortex veins (Zhu and Cai, 1992), or injection of chondroitin sulfate, hydroxypropyl methylcellulose, sodium carboxymethylcellulose, or methylcellulose (Zhu and Cai, 1992). These procedures produce an elevation in IOP for a few days to months. Two studies of α-chymotrypsin-treated eyes found that outflow facility was slightly reduced compared with normotensive eyes (Table III). No other information is available regarding aqueous humor dynamics in the rabbit model of elevated IOP.

D. Cats

Cats have relatively large eyes with correspondingly large anterior chambers, and they produce aqueous humor at higher rates than other research animals. Most of their aqueous humor drains through the trabecular meshwork. Cats are used occasionally to investigate drug effects on aqueous humor dynamics (Table IV). However, interspecies differences in drug response are at times striking. For example, prostaglandin A_2 is a potent IOP-lowering drug in cats whereas prostaglandin $F_{2\alpha}$ is relatively weak. In primates, the converse is true (Bito *et al.*, 1989). Ocular hypertension has been created in cats by administration of multiple topical doses of corticosteroids (Zhan *et al.*, 1992). The IOP remains elevated as long as the topical dexamethasone is administered. There is no information on aqueous humor dynamics in this model.

E. Dogs

Aqueous humor dynamics in dogs is of interest because of the prevalence of spontaneous glaucoma in some breeds including American cocker spaniels, basset hounds, and beagles. As in humans, the glaucoma is characterized by an increase in the IOP that, if left untreated, causes irreversible optic nerve damage and blindness. Canine glaucoma with increased IOP is associated with reduced outflow facility (Gelatt *et al.*, 1977; Peiffer *et al.*, 1980) and reduced uveoscleral outflow (Barrie *et al.*, 1985). Episcleral venous pressure remains normal (Gelatt *et al.*, 1982). Aqueous flow has not been measured. In healthy beagles, aqueous flow ranges from 5.2 to 6.8 μl/min (Table V). Interestingly, aqueous flow in mongrel dogs is about 25% of the beagle rate (Table V). Type of anesthesia, measurement method, and genetic background are possible explanations for the different rates of aqueous flow between the purebred beagle and the mixed breed mongrel.

Uveoscleral outflow accounts for approximately 15% of total outflow in clinically normal beagles. This percentage drops to 3% in beagles with glaucoma (Barrie *et al.*, 1985). Outflow facility ranges from 0.21 to 0.24 μl/min/mmHg in healthy beagles and this is reduced to a range of 0.09 to 0.15 μl/min/mmHg in glaucomatous beagles. Studies of aqueous humor dynamics in dogs are summarized in Table V.

F. Nonhuman Primates

Only nonhuman primates and humans have a Schlemm's canal, scleral spur, scleral sulcus, limbal trabecular meshwork, and Schwalbe's line. Studies of aqueous humor dynamics have been conducted in healthy rhesus, cynomolgus, and vervet monkeys; rhesus monkeys with spontaneous glaucoma; or cynomolgus monkeys with laser-induced glaucoma. The monkey is preferred for testing IOP-lowering drugs, evaluating surgical procedures designed to treat glaucoma, and elucidating the physiology of aging. Monkeys are also used in studies difficult to conduct in a controlled manner in humans such as longitudinal studies of glaucoma progression.

A genetically isolated colony of rhesus monkeys in Cayo Santiago was found to have a high incidence of chronic open-angle glaucoma (Dawson *et al.*, 1993). This animal is advantageous as a model for human open-angle glaucoma because of the spontaneity of the disease, its familial inheritance, and the animal's relatively rapid rate of aging (roughly four times the human rate). A preliminary report (Toris *et al.*, 2007, presented at the 2007 annual meeting of the Association for Research in Vision and Ophthalmology) on aqueous humor dynamics in these animals found that the IOP is elevated because outflow facility and uveoscleral outflow are significantly reduced. This is similar to patients with ocular hypertension (Toris *et al.*, 2002). These animals also have reduced aqueous flow, unlike patients with ocular hypertension (Toris *et al.*, 2002) or glaucoma (Brubaker, 1991).

Monkeys can be made glaucomatous experimentally by lasering the trabecular meshwork that causes scarring and increased resistance in the aqueous humor drainage tissues (Quigley and Hohman, 1983; Lee *et al.*, 1985; Podos *et al.*, 1987). One eye is lasered and the contralateral normotensive eye serves as a control. The IOP remains elevated for years but fluctuates widely by as much as 40 mmHg (Pederson and Gaasterland, 1984) compared with a fluctuation of 15 mmHg in human chronic open-angle glaucoma. The laser-induced glaucoma model has reduced outflow facility but increased uveoscleral outflow (Table VI). Aqueous flow is reduced for a short time after lasering but returns to normal after a few months (Toris *et al.*, 2000). Monkeys are expensive and difficult to manage, but their similarities to humans make them especially valuable in the study of glaucoma and aqueous humor dynamics.

IV. SUMMARY

The use of animal models to investigate the circulation of aqueous humor in normal ocular health and chronic ocular diseases that disturb the IOP requires an appreciation for the diversity among species in the anatomical and physiological differences in the iridocorneal angle. Extrapolating animal findings to human disease requires discretion. Methods have been developed recently to measure the components of aqueous humor dynamics in the tiny murine eye, making the mouse the preferred model for genetic studies of glaucoma. Some breeds of dog suffer from spontaneous glaucoma that provides clues to the pathogenesis of glaucoma in humans. Study of these breeds has contributed to improved treatments for human as well as canine glaucoma. The monkey remains the most valuable animal model for understanding aqueous humor dynamics in human glaucoma because of its anatomical and physiological similarities to humans.

There are a myriad of methods available to assess the four main parameters of aqueous humor dynamics, aqueous flow, outflow facility, uveoscleral outflow, and episcleral venous pressure. The most accurate methods are invasive and often terminal. Methods designed for one species may not be easily adaptable for another. The noninvasive methods are indirect, highly variable, and fraught with many limitations and assumptions. Nevertheless, these methods remain useful in clinical research. Imaging techniques are evolving rapidly and may soon become sufficiently sensitive to allow simultaneous visualization and quantitation of ocular hydrodynamics in a direct noninvasive manner.

Acknowledgment

Figures were made with the assistance of Richard Tamesis, MD.

References

Aihara, M., Lindsey, J. D., and Weinreb, R. N. (2003a). Aqueous humor dynamics in mice. *Invest. Ophthalmol. Vis. Sci.* **44,** 5168–5173.

Aihara, M., Lindsey, J. D., and Weinreb, R. N. (2003b). Episcleral venous pressure of mouse eye and effect of body position. *Curr. Eye Res.* **27,** 355–362.

Artru, A. A. (1995). Rate of anterior chamber aqueous formation, trabecular outflow facility, and intraocular compliance during desflurane or halothane anesthesia in dogs. *Anesth. Analg.* **81,** 585–590.

Artru, A. A., and Momota, Y. (1999). Trabecular outflow facility and formation rate of aqueous humor during anesthesia with sevoflurane-nitrous oxide or sevoflurane-remifentanil in rabbits. *Anesth. Analg.* **88,** 781–786.

Avila, M. Y., Carré, D. A., Stone, R. A., and Civan, M. M. (2001). Reliable measurement of mouse intraocular pressure by a servo-null micropipette system. *Invest. Ophthalmol. Vis. Sci.* **42,** 1841–1846.

Avila, M. Y., Munera, A., Guzman, A., Do, C. W., Wang, Z., Stone, R. A., and Civan, M. M. (2005). Noninvasive intraocular pressure measurements in mice by pneumotonometry. *Invest. Ophthalmol. Vis. Sci.* **46,** 3274–3280.

Bárány, E. H. (1964). Simultaneous measurement of changing the intraocular pressure and outflow facility in the vervet monkey by constant pressure infusion. *Invest. Ophthalmol.* **3,** 135–143.

Barrie, K. P., Gum, G. G., Samuelson, D. A., and Gelatt, K. N. (1985). Quantitation of uveoscleral outflow in normotensive and glaucomatous beagles by ^{3}H-labeled dextran. *Am. J. Vet. Res.* **46,** 84–88.

Bill, A. (1966). Conventional and uveo-scleral drainage of aqueous humour in the cynomolgus monkey (*Macaca irus*) at normal and high intraocular pressures. *Exp. Eye Res.* **5,** 45–54.

Bill, A. (1967). Further studies on the influence of the intraocular pressure on aqueous humor dynamics in cynomolgus monkeys. *Invest. Ophthalmol.* **6,** 364–372.

Bill, A. (1971). Aqueous humor dynamics in monkeys (*Macaca irus* and *Cercopithecus ethiops*). *Exp. Eye Res.* **11,** 195–206.

Bito, L. Z. (1984). Species differences in the responses of the eye to irritation and trauma: A hypothesis of divergence in ocular defense mechanisms, and the choice of experimental animals for eye research. *Exp. Eye Res.* **39,** 807–829.

Bito, L. Z., Camras, C. B., Gum, G. G., and Resul, B. (1989). The ocular hypotensive effects and side effects of prostaglandins on the eyes of experimental animals. *In* "The Ocular Effects of Prostaglandins and Other Eicosanoids" (L. Z. Bito and J. Stjernschantz, eds.), Vol. 312, pp. 349–368. Alan R. Liss, Inc, New York, NY.

Brubaker, R. F. (1982). The flow of aqueous humor in the human eye. *Trans. Am. Ophthalmol. Soc.* **80,** 391–474.

Brubaker, R. F. (1991). Flow of aqueous humor in humans. *Invest. Ophthalmol. Vis. Sci.* **32,** 3145–3166.

Brubaker, R. F. (1998). Clinical measurements of aqueous dynamics: Implications for addressing glaucoma. *In* "The Eye's Aqueous Humor. From Secretion to Glaucoma" (M. M. Civan, ed.), Vol. 45, pp. 233–284. Academic Press, San Diego.

Brubaker, R. F., Maurice, D. M., and McLaren, J. W. (1990). Fluorometry of the anterior segment. *In* "Noninvasive Diagnostic Techniques in Ophthalmology" (B. R. Masters, ed.), pp. 248–280. Springer-Verlag, New York, NY.

Camras, C. B. (2003). Some thoughts on the pressure dependence of uveoscleral flow. *J. Glaucoma* **12,** 92–93.

Cawrse, M. A., Ward, D. A., and Hendrix, D. V. H. (2001). Effects of topical application of a 2% solution of dorzolamide on intraocular pressure and aqueous humor flow rate in clinically normal dogs. *Am. J. Vet. Res.* **62,** 859–863.

Chang, B., Smith, R. S., Hawes, N. L., Anderson, M. G., Zabaleta, A., Savinova, O., Roderick, T. H., Heckenlively, J. R., Davisson, M. T., and John, S. W. M. (1999). Interacting loci cause severe iris atrophy and glaucoma in DBA/2J mice. *Nat. Genet.* **21,** 405–409.

Chee, P., and Hamasaki, D. I. (1971). The basis for chymotrypsin-induced glaucoma. *Arch. Ophthalmol.* **85,** 103–106.

Chidlow, G., Cupido, A., Melena, J., and Osborne, N. N. (2001). Flesinoxan, a 5-HT_{1A} receptor agonist/α_1-adrenoceptor antagonist, lowers intraocular pressure in NZW rabbits. *Curr. Eye Res.* **23,** 144–153.

Chien, F. Y., Wang, R.-F., Mittag, T. W., and Podos, S. M. (2003). Effect of WIN 55212–2, a cannabinoid receptor agonist, on aqueous humor dynamics in monkeys. *Arch. Ophthalmol.* **121,** 87–90.

Chu, T.-C., and Potter, D. E. (2002). Ocular hypotension induced by electroacupuncture. *J. Ocul. Pharmacol. Ther.* **18,** 293–305.

Colasanti, B. K. (1990). A comparison of the ocular and central effects of Δ^9-tetrahydrocannabinol and cannabigerol. *J. Ocul. Pharmacol.* **6,** 259–269.

Crosson, C. E. (2001). Intraocular pressure responses to the adenosine agonist cyclohexyladenosine: Evidence for a dual mechanism of action. *Invest. Ophthalmol. Vis. Sci.* **42,** 1837–1840.

Crosson, C. E., and Petrovich, M. (1999). Contributions of adenosine receptor activation to the ocular actions of epinephrine. *Invest. Ophthalmol. Vis. Sci.* **40,** 2054–2061.

Crowston, J. G., Aihara, M., Lindsey, J. D., and Weinreb, R. N. (2004). Effect of latanoprost on outflow facility in the mouse. *Invest. Ophthalmol. Vis. Sci.* **45,** 2240–2245.

Dawson, W. W., Brooks, D. E., Hope, G. M., Samuelson, D. A., Sherwood, M. B., Engel, H. M., and Kessler, M. J. (1993). Primary open angle glaucomas in the rhesus monkey. *Br. J. Ophthalmol.* **77,** 302–310.

Diamond, J. M., and Bossert, W. H. (1967). Standing-gradient osmotic flow. A mechanism for coupling of water and solute transport in epithelia. *J. Gen. Physiol.* **50,** 2061–2083.

Ethier, C. R. (2002). The inner wall of Schlemm's canal. *Exp. Eye Res.* **74,** 161–172.

Fan, S., Johnson, T. V., and Toris, C. B. (2006). Measurement of intraocular pressure in mice by TonoLab tonometry. *Invest. Ophthalmol. Vis. Sci.* **47,** 215.

Fan, S., Johnson, T. V., Liu, H., Ishimoto, B. M., and Toris, C. B. (2007). Aqueous flow measured by fluorophotometry in the mouse. *Invest. Ophthalmol. Vis. Sci.* **48,** 3921.

Friedenwald, J. S. (1948). Some problems with the calibration of tonometers. *Am. J. Ophthalmol.* **31,** 935–944.

Gabelt, B. T., and Kaufman, P. L. (1990). The effect of prostaglandin $F_{2\alpha}$ on trabecular outflow facility in cynomolgus monkeys. *Exp. Eye Res.* **51,** 87–91.

Gabelt, B. T., and Kaufman, P. L. (2005). Changes in aqueous humor dynamics with age and glaucoma. *Prog. Retin. Eye Res.* **24,** 612–637.

Gabelt, B. T., Crawford, K., and Kaufman, P. L. (1991). Outflow facility and its response to pilocarpine decline in aging rhesus monkeys. *Arch. Ophthalmol.* **109,** 879–882.

Gabelt, B. T., Millar, C. J., Kiland, J. A., Peterson, J. A., Seeman, J. L., and Kaufman, P. L. (2001). Effects of serotonergic compounds on aqueous humor dynamics in monkeys. *Curr. Eye Res.* **23,** 120–127.

Gabelt, B. T., Gottanka, J., Lütjen-Drecoll, E., and Kaufman, P. L. (2003). Aqueous humor dynamics and trabecular meshwork and anterior ciliary muscle morphologic changes with age in rhesus monkeys. *Invest. Ophthalmol. Vis. Sci.* **44,** 2118–2125.

Gabelt, B. T., Seeman, J. L., Podos, S. M., Mittag, T. W., and Kaufman, P. L. (2004). Aqueous humor dynamics in monkeys after topical 8-iso PGE_2. *Invest. Ophthalmol. Vis. Sci.* **45,** 892–899.

Gabelt, B. T., Okka, M., Dean, T. R., and Kaufman, P. L. (2005). Aqueous humor dynamics in monkeys after topical R-DOI. *Invest. Ophthalmol. Vis. Sci.* **46,** 4691–4696.

Gelatt, K. N., Gwin, R. M., Peiffer, R. L., Jr., and Gum, G. G. (1977). Tonography in the normal and glaucomatous beagle. *Am. J. Vet. Res.* **38,** 515–520.

Gelatt, K. N., Gum, G. G., Merideth, R. E., and Bromberg, N. (1982). Episcleral venous pressure in normotensive and glaucomatous beagles. *Invest. Ophthalmol. Vis. Sci.* **23,** 131–135.

Goh, Y., Oshima, T., and Araie, M. (1994). Mechanism of intraocular pressure reduction observed after topical application of S-1033 in animals. *Jpn. J. Ophthalmol.* **38,** 228–235.

Goldmann, H. (1951). Abflussdruck, minutenvolumen und widerstand der kammerwasserströmung des menschen. *Doc. Ophthalmol.* **5–6,** 278–356.

Grant, W. M. (1950). Tonographic method for measuring the facility and rate of aqueous flow in human eyes. *Arch. Ophthalmol.* **44,** 204–214.

Grayson, M. C., and Laties, A. M. (1971). Ocular localization of sodium fluorescein. Effects of administration in rabbit and monkey. *Arch. Ophthalmol.* **85,** 600–609.

Green, K., Sherman, S. H., Laties, A. M., Pederson, J. E., Gaasterland, D. E., and MacLellan, H. M. (1977). Fate of anterior chamber tracers in the living rhesus monkey eye with evidence for uveo-vortex outflow. *Trans. Ophthalmol. Soc. UK* **97,** 731–739.

Gross, R. L., Ji, J., Chang, P., Pennesi, M. E., Yang, Z., Zhang, J., and Wu, S. M. (2003). A mouse model of elevated intraocular pressure: Retina and optic nerve findings. *Trans. Am. Ophthalmol. Soc.* **101,** 163–169.

Grozdanic, S. D., Betts, D. M., Sakaguchi, D. S., Allbaugh, R. A., Kwon, Y. H., and Kardon, R. H. (2003). Laser-induced mouse model of chronic ocular hypertension. *Invest. Ophthalmol. Vis. Sci.* **44,** 4337–4346.

Hayashi, M., Yablonski, M. E., and Bito, L. Z. (1987). Eicosanoids as a new class of ocular hypotensive agents. 2. Comparison of the apparent mechanism of the ocular hypotensive effects of A and F type prostaglandins. *Invest. Ophthalmol. Vis. Sci.* **28,** 1639–1643.

Hayashi, M., Yablonski, M. E., and Novack, G. D. (1989). Trabecular outflow facility determined by fluorophotometry in human subjects. *Exp. Eye Res.* **48,** 621–625.

Higginbotham, E. J., Lee, D. A., Bartels, S. P., Richardson, T., and Miller, M. (1988). Effects of cyclocryotherapy on aqueous humor dynamics in cats. *Arch. Ophthalmol.* **106,** 396–403.

Honjo, M., Inatani, M., Kido, N., Sawamura, T., Yue, B. Y., Honda, Y., and Tanihara, H. (2001). Effects of protein kinase inhibitor, HA1077, on intraocular pressure and outflow facility in rabbit eyes. *Arch. Ophthalmol.* **119,** 1171–1178.

Inoue, T., Yokoyoma, T., Mori, Y., Sasaki, Y., Hosokawa, T., Yanagisawa, H., and Koike, H. (2001). The effect of topical CS-088, an angiotensin AT_1 receptor antagonist, on intraocular pressure and aqueous humor dynamics in rabbits. *Curr. Eye Res.* **23,** 133–138.

John, S. W., Smith, R. S., Savinova, O. V., Hawes, N. L., Chang, B., Turnbull, D., Davisson, M., Roderick, T. H., and Heckenlively, J. R. (1998). Essential iris atrophy, pigment dispersion, and glaucoma in DBA/2J mice. *Invest. Ophthalmol. Vis. Sci.* **39,** 951–962.

Johnson, M. (2006). What controls aqueous humour outflow resistance?. *Exp. Eye Res.* **82,** 545–557.

Johnson, T. V., Fan, S., and Toris, C. B. (2008). Rebound tonometry in conscious, conditioned mice avoids the acute and profound effects of anesthesia on intraocular pressure. *J. Ocular Pharmacol. Therapeutics.* **24,** 175–185.

Kanno, M., Araie, M., Tomita, K., and Sawanobori, K. (1998). Effects of topical nipradilol, a β-blocking agent with α-blocking and nitroglycerin-like activities, on aqueous humor dynamics and fundus circulation. *Invest. Ophthalmol. Vis. Sci.* **39,** 736–743.

Kee, C., and Seo, K. (1997). The effect of interleukin-1α on outflow facility in rat eyes. *J. Glaucoma* **6,** 246–249.

Kee, C., Hong, T., and Choi, K. (1997). A sensitive ocular perfusion apparatus measuring outflow facility. *Curr. Eye Res.* **16,** 1198–1201.

Kiel, J. W., and Reitsamer, H. A. (2007). Paradoxical effect of phentolamine on aqueous flow in the rabbit. *J. Ocul. Pharmacol. Ther.* **23,** 21–26.

Kiel, J. W., Reitsamer, H. A., Walker, J. S., and Kiel, F. W. (2001). Effects of nitric oxide synthase inhibition on ciliary blood flow, aqueous production and intraocular pressure. *Exp. Eye Res.* **73,** 355–364.

Koskela, T., and Brubaker, R. F. (1991). The nocturnal suppression of aqueous humor flow in humans is not blocked by bright light. *Invest. Ophthalmol. Vis. Sci.* **32,** 2504–2506.

Kotikoski, H., Vapaatalo, H., and Oksala, O. (2003). Nitric oxide and cyclic GMP enhance aqueous humor outflow facility in rabbits. *Curr. Eye Res.* **26,** 119–123.

Krohn, J., and Bertelsen, T. (1997). Corrosion casts of the suprachoroidal space and uveoscleral drainage routes in the human eye. *Acta Ophthalmol.* **75,** 32–35.

Kurata, K., Fujimoto, H., Tsukuda, R., Suzuki, T., Ando, T., and Tokuriki, M. (1998). Aqueous humor dynamics in beagle dogs with caffeine-induced ocular hypertension. *J. Vet. Med. Sci.* **60,** 737–739.

Lee, P.-Y., Podos, S. M., Howard-Williams, J. R., Severin, C. H., Rose, A. D., and Siegel, M. J. (1985). Pharmacological testing in the laser-induced monkey glaucoma model. *Curr. Eye Res.* **4,** 775–781.

Levene, R. Z., Rothberger, M., and Rosenberg, S. (1974). Corticosteroid glaucoma in the rabbit. *Am. J. Ophthalmol.* **78,** 505–510.

Lütjen-Drecoll, E. (1999). Functional morphology of the trabecular meshwork in primate eyes. *Prog. Retin. Eye Res.* **18,** 91–119.

Mäepea, O., and Bill, A. (1989). The pressures in the episcleral veins, Schlemm's canal and the trabecular meshwork in monkeys: Effects of changes in intraocular pressure. *Exp. Eye Res.* **49,** 645–663.

Maurice, D. M. (1963). A new objective fluorophotometer. *Exp. Eye Res.* **2,** 33–38.

McLaren, J. W., Brubaker, R. F., and FitzSimon, J. S. (1996). Continuous measurement of intraocular pressure in rabbits by telemetry. *Invest. Ophthalmol. Vis. Sci.* **37,** 966–975.

McMaster, P. R. B., and Macri, F. J. (1968). Secondary aqueous humor outflow pathways in the rabbit, cat, and monkey. *Arch. Ophthalmol.* **79,** 297–303.

Melena, J., Santafé, J., Segarra-Domenech, J., and Puras, G. (1999). Aqueous humor dynamics in α-chymotrypsin-induced ocular hypertensive rabbits. *J. Ocul. Pharmacol. Ther.* **15,** 19–27.

Mermoud, A., Baerveldt, G., Minckler, D. S., Prata, J. A., Jr., and Rao, N. A. (1996). Aqueous humor dynamics in rats. *Graefes Arch. Clin. Exp. Ophthalmol.* **234,** S198–S203.

Morrison, J. C., Moore, C. G., Deppmeier, L. M. H., Gold, B. G., Meshul, C. K., and Johnson, E. C. (1997). A rat model of chronic pressure-induced optic nerve damage. *Exp. Eye Res.* **64,** 85–96.

Muta, K., Tamaki, Y., Araie, M., and Matsubara, M. (2000). Effect of ifenprodil on aqueous humor dynamics and optic nerve head circulation in rabbits. *J. Ocul. Pharmacol. Ther.* **16,** 241–250.

Nguyen, C. T. O., Bui, B. V., Sinclair, A. J., and Vingrys, A. J. (2007). Dietary omega 3 fatty acids decrease intraocular pressure with age by increasing aqueous outflow. *Invest. Ophthalmol. Vis. Sci.* **48,** 756–762.

Nilsson, S. F. E., Drecoll, E., Lütjen-Drecoll, E., Toris, C. B., Krauss, A. H. P., Kharlamb, A., Nieves, T., Guerra, T., and Woodward, D. F. (2006). The prostanoid EP_2 receptor agonist butaprost increases uveoscleral outflow in the cynomolgus monkey. *Invest. Ophthalmol. Vis. Sci.* **47,** 4042–4049.

Ogidigben, M. J., and Potter, D. E. (2001). Central imidazoline (I_1) receptors modulate aqueous hydrodynamics. *Curr. Eye Res.* **22,** 358–366.

Oka, T., Taniguchi, T., Kitazawa, Y., Sagara, T., and Nishida, T. (2006). Aqueous humor dynamics associated with the phorbol ester-induced decrease in intraocular pressure in the rabbit. *Jpn. J. Ophthalmol.* **50,** 497–503.

Okka, M., Tian, B., and Kaufman, P. L. (2004). Effects of latrunculin B on outflow facility, intraocular pressure, corneal thickness, and miotic and accommodative responses to pilocarpine in monkeys. *Trans. Am. Ophthalmol. Soc.* **102,** 251–257.

Oksala, O., and Stjernschantz, J. (1988). Increase in outflow facility of aqueous humor in cats induced by calcitonin gene-related peptide. *Exp. Eye Res.* **47,** 787–790.

Pease, M. E., Hammond, J. C., and Quigley, H. A. (2006). Manometric calibration and comparison of TonoLab and Tono-Pen tonometers in rats with experimental glaucoma and in normal mice. *J. Glaucoma* **15,** 512–519.

Pederson, J. E., and Gaasterland, D. E. (1984). Laser-induced primate glaucoma. I. Progression of cupping. *Arch. Ophthalmol.* **102,** 1689–1692.

Pederson, J. E., Gaasterland, D. E., and MacLellan, H. M. (1977). Uveoscleral aqueous outflow in the rhesus monkey: Importance of uveal reabsorption. *Invest. Ophthalmol. Vis. Sci.* **16,** 1008–1017.

Peiffer, R. L., Jr., Gelatt, K. N., and Gum, G. G. (1976). Determination of the facility of aqueous humor outflow in the dog, comparing in vivo and in vitro tonographic and constant pressure perfusion techniques. *Am. J. Vet. Res.* **37,** 1473–1477.

Peiffer, R. L., Jr., Gum, G. G., Grimson, R. C., and Gelatt, K. N. (1980). Aqueous humor outflow in beagles with inherited glaucoma: Constant pressure perfusion. *Am. J. Vet. Res.* **41,** 1808–1813.

Peterson, J. A., Tian, B., Geiger, B., and Kaufman, P. L. (2000a). Effect of latrunculin-B on outflow facility in monkeys. *Exp. Eye Res.* **70,** 307–313.

Peterson, J. A., Tian, B., McLaren, J. W., Hubbard, W. C., Geiger, B., and Kaufman, P. L. (2000b). Latrunculins' effects on intraocular pressure, aqueous humor flow, and corneal endothelium. *Invest. Ophthalmol. Vis. Sci.* **41,** 1749–1758.

Podos, S. M., Camras, C. B., Serle, J. B., and Lee, P.-Y. (1987). Pharmacologic alteration of aqueous humor dynamics in normotensive and glaucomatous monkey eyes. *In* "Glaucoma Update III" (G. K. Krieglstein, ed.), pp. 225–235. Springer-Verlag, Berlin.

Puras, G., Santafé, J., Segarra, J., Garrido, M., and Melena, J. (2002a). Effects of topical natural ergot alkaloids on intraocular pressure and aqueous humor dynamics in ocular normotensive rabbits. *J. Ocul. Pharmacol. Ther.* **18,** 41–52.

Puras, G., Santafé, J., Segarra, J., Garrido, M., and Melena, J. (2002b). The effect of topical natural ergot alkaloids on the intraocular pressure and aqueous humor dynamics in rabbits with α-chymotrypsin-induced ocular hypertension. *Graefes Arch. Clin. Exp. Ophthalmol.* **240,** 322–328.

Quigley, H. A., and Hohman, R. M. (1983). Laser energy levels for trabecular meshwork damage in the primate eye. *Invest. Ophthalmol. Vis. Sci.* **24,** 1305–1307.

Reiss, G. R., Lee, D. A., Topper, J. E., and Brubaker, R. F. (1984). Aqueous humor flow during sleep. *Invest. Ophthalmol. Vis. Sci.* **25,** 776–778.

Reitsamer, H. A., and Kiel, J. W. (2002). A rabbit model to study orbital venous pressure, intraocular pressure, and ocular hemodynamics simultaneously. *Invest. Ophthalmol. Vis. Sci.* **43,** 3728–3734.

Reitsamer, H. A., Kiel, J. W., Harrison, J. M., Ransom, N. L., and McKinnon, S. J. (2004). Tono-Pen measurement of intraocular pressure in mice. *Exp. Eye Res.* **78,** 799–804.

Rosenberg, L. F., Karalekas, D. P., Krupin, T., and Hyderi, A. (1996). Cyclocryotherpay and noncontact Nd:YAG laser cyclophotocoagulation in cats. *Invest. Ophthalmol. Vis. Sci.* **37,** 2029–2036.

Ruiz-Ederra, J., and Verkman, A. S. (2006). Mouse model of sustained elevation in intraocular pressure produced by episcleral vein occlusion. *Exp. Eye Res.* **82,** 879–884.

Russell, K. R. M., Wang, D. R., and Potter, D. E. (2000). Modulation of ocular hydrodynamics and iris function by bremazocine, a kappa opioid receptor agonist. *Exp. Eye Res.* **70,** 675–682.

Schnell, C. R., Debon, C., and Percicot, C. L. (1996). Measurement of intraocular pressure by telemetry in conscious, unrestrained rabbits. *Invest. Ophthalmol. Vis. Sci.* **37,** 958–965.

Shareef, S. R., Garcia-Valenzuela, E., Salierno, A., Walsh, J., and Sharma, S. C. (1995). Chronic ocular hypertension following episcleral venous occlusion in rats. *Exp. Eye Res.* **61,** 379–382.

Silver, D. M., and Farrell, R. A. (1994). Validity of pulsatile ocular blood flow measurements. *Surv. Ophthalmol.* **38,** S72–S80.

Skorobohach, B. J., Ward, D. A., and Hendrix, D. V. H. (2003). Effects of oral administration of methazolamide on intraocular pressure and aqueous humor flow rate in clinically normal dogs. *Am. J. Vet. Res.* **64,** 183–187.

Smith, S. D., and Gregory, D. S. (1989). A circadian rhythm of aqueous flow underlies the circadian rhythm of IOP in NZW rabbits. *Invest. Ophthalmol. Vis. Sci.* **30,** 775–778.

Sperber, G. O., and Bill, A. (1984). A method for near-continuous determination of aqueous humor flow; effects of anaesthetics, temperature and indomethacin. *Exp. Eye Res.* **39,** 435–453.

Sugiyama, T., Kida, T., Mizuno, K., Kojima, S., and Ikeda, T. (2001). Involvement of nitric oxide in the ocular hypotensive action of nipradilol. *Curr. Eye Res.* **23,** 346–351.

Suguro, K., Toris, C. B., and Pederson, J. E. (1985). Uveoscleral outflow following cyclodialysis in the monkey eye using a fluorescein tracer. *Invest. Ophthalmol. Vis. Sci.* **26,** 810–813.

Takagi, Y., Nakajima, T., Shimazaki, A., Kageyama, M., Matsugi, T., Matsumura, Y., Gabelt, B. T., Kaufman, P. L., and Hara, H. (2004). Pharmacological characteristics of AFP-168 (tafluprost), a new prostanoid FP receptor agonist, as an ocular hypotensive drug. *Exp. Eye Res.* **78,** 767–776.

Tian, B., and Kaufman, P. L. (2005). Effects of the Rho kinase inhibitor Y-27632 and the phosphatase inhibitor calyculin A on outflow facility in monkeys. *Exp. Eye Res.* **80,** 215–225.

Tian, B., Wang, R.-F., Podos, S. M., and Kaufman, P. L. (2004). Effects of topical H-7 on outflow facility, intraocular pressure, and corneal thickness in monkeys. *Arch. Ophthalmol.* **122,** 1171–1177.

Toris, C. B., and Pederson, J. E. (1985). Effect of intraocular pressure on uveoscleral outflow following cyclodialysis in the monkey eye. *Invest. Ophthalmol. Vis. Sci.* **26,** 1745–1749.

Toris, C. B., Yablonski, M. E., Wang, Y.-L., and Hayashi, M. (1995). Prostaglandin A_2 increases uveoscleral outflow and trabecular outflow facility in the cat. *Exp. Eye Res.* **61,** 649–657.

Toris, C. B., Yablonski, M. E., Wang, Y.-L., and Camras, C. B. (1999). Aqueous humor dynamics in the aging human eye. *Am. J. Ophthalmol.* **127,** 407–412.

Toris, C. B., Zhan, G.-L., Wang, Y.-L., Zhao, J., McLaughlin, M. A., Camras, C. B., and Yablonski, M. E. (2000). Aqueous humor dynamics in monkeys with laser-induced glaucoma. *J. Ocul. Pharmacol. Ther.* **16,** 19–27.

Toris, C. B., Koepsell, S. A., Yablonski, M. E., and Camras, C. B. (2002). Aqueous humor dynamics in ocular hypertensive patients. *J. Glaucoma* **11,** 253–258.

Toris, C. B., Zhan, G.-L., and McLaughlin, M. A. (2003). Effects of brinzolamide on aqueous humor dynamics in monkeys and rabbits. *J. Ocul. Pharmacol. Ther.* **19,** 397–404.

Toris, C. B., Zhan, G.-L., Camras, C. B., and McLaughlin, M. A. (2005). Effects of travoprost on aqueous humor dynamics in monkeys. *J. Glaucoma* **14,** 70–73.

Toris, C. B., Lane, J. T., Akagi, Y., Blessing, K. A., and Kador, P. F. (2006a). Aqueous flow in galactose-fed dogs. *Exp. Eye Res.* **83,** 865–870.

Toris, C. B., Zhan, G.-L., Feilmeier, M. R., Camras, C. B., and McLaughlin, M. A. (2006b). Effects of a prostaglandin DP receptor agonist, AL-6598, on aqueous humor dynamics in a nonhuman primate model of glaucoma. *J. Ocul. Pharmacol. Ther.* **22,** 86–92.

Toris, C. B., Dawson, W. W., and McLaughlin, M. A. (2007). Aqueous humor dynamics in Rhesus monkeys with naturally occurring ocular hypertension. *Invest. Ophthalmol. Vis. Sci.* **48,** 4894.

Wang, J., Song, Y., Ren, B., Sun, N., Yu, Y., and Hu, H. (2003). Different effects of topical prazosin and pilocarpine on uveoscleral outflow in rabbit eyes. *Yan Ke Xue Bao* **19,** 191–194.

Wang, W.-H., Millar, J. C., Pang, I.-H., Wax, M. B., and Clark, A. F. (2005). Noninvasive measurement of rodent intraocular pressure with a rebound tonometer. *Invest. Ophthalmol. Vis. Sci.* **46,** 4617–4621.

Wang, Y.-L., Toris, C. B., Zhan, G., and Yablonski, M. E. (1999). Effects of topical epinephrine on aqueous humor dynamics in the cat. *Exp. Eye Res.* **68,** 439–445.

Ward, D. A., Cawrse, M. A., and Hendrix, D. V. H. (2001). Fluorophotometric determination of aqueous humor flow rate in clinically normal dogs. *Am. J. Vet. Res.* **62,** 853–858.

WoldeMussie, E., Ruiz, G., Wijono, M., and Wheeler, L. A. (2001). Neuroprotection of retinal ganglion cells by brimonidine in rats with laser-induced chronic ocular hypertension. *Invest. Ophthalmol. Vis. Sci.* **42,** 2849–2855.

Zeimer, R. C. (1989). Episcleral venous pressure. *In* "The Glaucomas" (R. Ritch, M. B. Shields, and T. Krupin, eds.), pp. 249–255. The C. V. Mosby Company, St. Louis.

Zhan, G.-L., Miranda, O. C., and Bito, L. Z. (1992). Steroid glaucoma: Corticosteroid-induced ocular hypertension in cats. *Exp. Eye Res.* **54,** 211–218.

Zhan, G.-L., Toris, C. B., Camras, C. B., Wang, Y.-L., and Yablonski, M. E. (1998). Bunazosin reduces intraocular pressure in rabbits by increasing uveoscleral outflow. *J. Ocul. Pharmacol. Ther.* **14,** 217–228.

Zhan, G.-L., Lee, P.-Y., Ball, D. C., Mayberger, C. J., Tafoya, M. E., Camras, C. B., and Toris, C. B. (2002). Time dependent effects of sympathetic denervation on aqueous humor dynamics and choroidal blood flow in rabbits. *Curr. Eye Res.* **25,** 99–105.

Zhang, D., Vetrivel, L., and Verkman, A. S. (2002). Aquaporin deletion in mice reduces intraocular pressure and aqueous fluid production. *J. Gen. Physiol.* **119,** 561–569.

Zhu, M. D., and Cai, F. Y. (1992). Development of experimental chronic intraocular hypertension in the rabbit. *Aust. N. Z. J. Ophthalmol.* **20,** 225–234.

CHAPTER 8

Aqueous Humor Dynamics II

Clinical Studies

Carol B. Toris and Carl B. Camras

Department of Ophthalmology, University of Nebraska Medical Center, Omaha, Nebraska, 68198-5840

Current Topics in Membranes, Volume 62

1063-5823/08 $35.00
DOI: 10.1016/S1063-5823(08)00408-0

I. OVERVIEW

A stable rate of production and drainage of aqueous humor is essential for the health of the eye and maintenance of normal visual function. This chapter reviews the contributions of aqueous humor dynamics in normal and pathological conditions affecting intraocular pressure (IOP). IOP remains relatively stable throughout one's lifetime but subtle changes do occur in the outflow pathways that could increase IOP and the risk for glaucoma. Abnormalities in aqueous humor dynamics have been found in various clinical syndromes that affect IOP. Most of the abnormalities have been localized to the aqueous humor outflow pathways. Surprisingly, aqueous humor production remains relatively stable in all of these conditions and ranges of IOPs. Some older drugs to treat elevated IOP work by reducing aqueous humor production, theoretically placing the avascular lens and cornea at risk for damage from limited nutrients and accumulation of toxic metabolites. The recently approved drugs and the ones currently under development for future glaucoma therapy are those that target the outflow pathways. From a physiological perspective, this is a logical approach because this is the location of the pathology in glaucoma and the region in need of repair. These drugs and their effects on aqueous humor dynamics also are discussed in this chapter.

II. INTRODUCTION

Continued interest in aqueous humor dynamics in humans is fueled by the finding that the production, circulation, and drainage of ocular aqueous humor determine IOP, and excessive IOP is a major risk factor for the development of glaucoma (Gordon *et al.*, 2002). It follows then that abnormalities in aqueous humor dynamics translate into the development of glaucoma. Noninvasive methods are used in a clinical setting to measure IOP and three of the components of aqueous humor dynamics: aqueous flow, outflow facility, and episcleral venous pressure. A noninvasive method to measure uveoscleral outflow remains elusive. Currently, this component is determined indirectly by mathematical calculation using the modified Goldmann equation. Despite their limitations, these methods have provided a good estimate of aqueous humor dynamics in healthy human eyes and in eyes with various diseases and/or treatments affecting IOP. The methods are described in detail in Chapter 7 and elsewhere (Friedenwald, 1957; Brubaker, 1982; Zeimer *et al.*, 1983; Hayashi *et al.*, 1989b; Brubaker, 1991). Results of key studies using these methods are reported herein.

III. NORMAL VALUES OF AQUEOUS HUMOR DYNAMICS IN HUMANS

A. Aqueous Flow

Aqueous flow averages about 2.9 μl/min in young healthy humans and 2.2 μl/min in octogenarians. The difference between these values amounts to a reduction of 2.4% per decade (Brubaker, 1991). This decrease in aqueous flow over the lifetime of an individual is statistically significant (Becker, 1958; Brubaker *et al.*, 1981; Brubaker, 1991; Diestelhorst and Krieglstein, 1992; Diestelhorst and Krieglstein, 1994; Toris *et al.*, 1999b), but is relatively small compared to age-related changes in IOP and anterior chamber volume (Brubaker *et al.*, 1981). The age-related aqueous flow reduction may not be of much clinical significance as the small decrease has not been found to be associated with the ocular diseases plaguing elderly humans.

Aqueous flow not only varies throughout one's lifetime but has a distinctive circadian rhythm. The flow rate at night during sleep is only 43% of the rate during the morning after awakening (Brubaker, 1991). This rhythm is not eliminated by sleeping under bright lights (Koskela and Brubaker, 1991b), closing of an eye (Reiss *et al.*, 1984; Topper and Brubaker, 1985), reclining during the day (Reiss *et al.*, 1984; Topper and Brubaker, 1985), or sleep deprivation (Reiss *et al.*, 1984). The nocturnal aqueous flow suppression is more pronounced than that accomplished by treatment with any of the aqueous flow suppressants now on the market. This potent physiological effect continues to fuel interest in identifying the factors such as hormonal interactions (Brubaker, 1998) that drive the circadian rhythm of aqueous flow.

Whether the aqueous flow rate is dependent on IOP has been of interest for decades. Early studies have suggested that a negative correlation exists between aqueous flow and IOP (Kupfer and Ross, 1971; Kupfer *et al.*, 1971; Gaasterland *et al.*, 1973; Gaasterland *et al.*, 1975), but later studies did not. In one study, IOP was raised by placing a subject on a tilt table with head down but aqueous flow was not reduced under these conditions (Carlson *et al.*, 1987). Compared with healthy age-matched controls or pretreatment levels, patients with elevated IOP associated with glaucoma (Grant, 1951; Martin *et al.*, 1992a; Larsson *et al.*, 1995b) and ocular hypertension with (Brown and Brubaker, 1989; Camras *et al.*, 2003) or without pigment dispersion syndrome (Brubaker, 1991; Toris *et al.*, 2002) do not have reduced aqueous flow rates and patients with reduced IOP associated with laser trabeculoplasty (Brubaker and Liesegang, 1983; Araie *et al.*, 1984; Yablonski *et al.*, 1985a) and myotonic dystrophy (Walker *et al.*, 1982; Khan and Brubaker, 1993) do not have increased rates. These consistently negative IOP effects on aqueous flow indicate that there is no compensatory mechanism to reduce aqueous flow as IOP increases. It appears that a stable rhythm of aqueous humor production sustains the metabolic needs of the eye throughout one's lifetime.

B. Outflow Facility

Outflow facility in healthy human eyes is thought to be in the range of 0.1–0.4 μl/min/mmHg (Grant, 1951; Becker, 1958; Gaasterland *et al.*, 1978; Toris *et al.*, 1999b, 2002). When measured by tonography, outflow facility is smaller in healthy older human eyes compared with younger eyes (Becker, 1958; Gaasterland *et al.*, 1978). Studies of perfused enucleated human cadaver eyes also have reported reductions in outflow facility with aging (Gabelt and Kaufman, 2005). When measured by fluorophotometry, outflow facility was found not to be different between young and old healthy individuals (Toris *et al.*, 1999b). The tonographic method includes pseudofacility and ocular rigidity in the measurement, whereas the fluorophotometry method does not (Toris *et al.*, 1995b). Pseudofacility is the apparent but not necessarily actual change in aqueous flow caused by the tonographic measurement itself. The pressure on the eye from the weighted probe compresses the anterior chamber preventing, for a time, posterior chamber fluid from entering the anterior chamber at the unperturbed rate. This reduction in aqueous flow into the anterior chamber is termed pseudofacility because it is not a true facility and does not reflect a true change in aqueous production. The fluorophotometric method does not include pseudofacility in the measurement because a weight is not placed on the eye and aqueous flow is measured and not inferred from tables. Ocular rigidity is a measure of the resistance that the eye exerts to distending forces and, like pseudofacility, is a component of the tonography measurement. The slope of the semilogarithmic ocular pressure–volume relationship was proposed by Friedenwald (1957) to be a quantitative measure of ocular rigidity. This relationship is strongly dependent on the systemic arterial pressure that affects choroidal blood flow and volume (Kiel, 1995). Ocular rigidity when measured by Schiotz tonometry was found to increase about 25% in older versus younger humans (Armaly, 1959; Gaasterland *et al.*, 1978). The increase in ocular rigidity with aging could contribute to the apparent age-related decrease in outflow facility as measured by tonography.

C. Episcleral Venous Pressure

Aqueous humor outflow is dependent on the difference between the pressure inside the eye (IOP) and the pressure in the aqueous humor drainage vessels outside the eye (episcleral venous pressure). In the human eye, aqueous humor in the canal of Schlemm drains into the venous network that encircles the sclera near the corneal limbus. The pressure in these drainage vessels is determined by vascular and hydrostatic factors. In rabbits, episcleral venous pressure varies significantly with acute changes in arterial blood pressure and cranial venous pressure (Reitsamer and Kiel, 2002a) and appears to be under neural control (Funk *et al.*, 1996). Episcleral venous pressure is largely independent of IOP and aqueous flow yet remains a major determinant of the steady-state pressure of the eye.

All the methods to measure episcleral venous pressure in humans require identification of the aqueous veins on the surface of the eye and application of pressure to the walls of these veins until indentation or collapse of the vessels is observed. The applied pressure at which the vessel collapses is considered the episcleral venous pressure. In research studies, the pressure is applied to the vessels by one of three methods: a rigid device, a jet of air, and a flexible membrane (Brubaker, 1967). The flexible membrane venomanometer (Zeimer *et al.*, 1983) remains in use even today for research purposes. Visualization and identification of an appropriate vessel and determination of the precise pressure at which the vessel collapses are the major problems with this technique making it impractical in routine clinical practice.

Episcleral venous pressure in healthy human eyes has been reported to be in the range of 7–14 mmHg (Zeimer, 1989). It appears to be relatively stable and the magnitude of any change is relatively small. Situations in which episcleral venous pressure of rabbits is altered include inhalation of O_2 (Yablonski *et al.*, 1985b), application of cold temperature (Ortiz *et al.*, 1988), and treatment with vasoactive drugs (Reitsamer *et al.*, 2006). Because of the difficulty in obtaining an accurate measurement, often a value of 9 or 10 mmHg is used and assumed to be unchanged during the course of a study. If episcleral venous pressure were to change, erroneous conclusions could be made concerning the cause of an IOP response.

D. Uveoscleral Outflow

Aqueous humor that reaches the ciliary muscle can exit the eye by several routes including through the sclera, into choroidal vessels, in and around the vortex veins, or into the ciliary processes. This outflow is usually called "uveoscleral outflow," although this term is inadequate to describe all of the possible routes of fluid egress. All of these pathways have in common the ciliary muscle (a uveal tissue) as the initial site of outflow, but not all pathways include seepage through the sclera. The term "uveal outflow" is a more accurate descriptive term but "uveoscleral" is the more familiar term.

Therefore, "uveoscleral" outflow will be used in this chapter to describe drainage of fluid from the anterior chamber angle other than through the trabecular meshwork.

Uveoscleral outflow once was thought to be very slow in humans. This is based on a study (Bill and Phillips, 1971) of eyes to be enucleated for reasons that included posterior segment tumors. Prior to enucleation, the anterior chamber was infused with ^{131}I-labeled serum albumin. One to 7 hours later, the enucleated eyes were analyzed for radioactivity. The anterior chamber volume and the radioactivity of the ocular tissues at the end of the perfusion period were used to calculate uveoscleral outflow. In the two nonglaucomatous eyes that had received no ocular drug for 48 hours prior to the study, uveoscleral outflow was 4% and 14% of total drainage. However, these eyes each had retinal detachments involving one-third to two-thirds of the total retinal surface area. Uveoscleral outflow is known to be affected by retinal detachment (Pederson, 1986). Eyes treated with atropine had greater uveoscleral outflow rates than eyes treated with pilocarpine, suggesting that the ciliary muscle tone helps to regulate this drainage. More than 35 years later, there remains no better direct way to measure uveoscleral outflow in humans. However, with today's strict regulatory requirements dictating the conduct of clinical research, a similar study likely would not be approved by an internal review board in the United States. In fact, no subsequent study with a similar invasive design has been reported in humans.

Today uveoscleral outflow is calculated mathematically using the modified Goldmann equation (Chapter 7) and measured values of IOP, episcleral venous pressure, aqueous flow, and outflow facility. Each of these values has its own inherent variability, making the power to detect differences in uveoscleral outflow quite low unless large numbers (30 or more) of subjects are enrolled in a study. Of the needed values, episcleral venous pressure is the most difficult to measure. Calculated uveoscleral outflow can change tremendously depending on which value of episcleral venous pressure is used in the equation. Because of the distrust in methods used to measure episcleral venous pressure, some publications calculate a range of uveoscleral outflow values from the modified Goldmann equation based on assumptions of different arbitrary levels of episcleral venous pressure. An important assumption in the evaluation of uveoscleral outflow is that the condition or manipulation being assessed does not alter episcleral venous pressure during the course of the study. Despite the limitations of the method, the mathematical calculation of uveoscleral outflow has provided plausible explanations for the changes in IOP with respect to aging, clinical syndromes, pharmacological agents, and surgical treatments. Ultimately, it is the difference between groups not the absolute value of uveoscleral outflow that is of the greater clinical relevance.

Uveoscleral outflow is often described as pressure independent. This is based on some early studies in monkeys in which uveoscleral outflow changed little at IOPs from the normal to high range (11–35 mmHg) (Bill, 1966; Toris and Pederson, 1985). However, at low IOP (4 mmHg), uveoscleral outflow is low and pressure dependent (Toris and Pederson, 1985). Pressure does not affect uveoscleral outflow to the extent that pressure affects trabecular outflow.

Calculated uveoscleral outflow is 25–57% of total aqueous flow in young healthy subjects 20–30 years of age (Townsend and Brubaker, 1980; Mishima *et al.*, 1997; Toris *et al.*, 1999b). A study (Toris *et al.*, 1999b) comparing young versus old healthy volunteers found uveoscleral outflow to be significantly reduced in those 60 years of age or older (Table I). The reduction in uveoscleral outflow with aging helps to explain why age is a risk factor for ocular hypertension or glaucoma.

IV. CLINICAL SYNDROMES

A. *Ocular Hypertension*

Ocular hypertension is the condition in which the IOP is elevated above what is considered normal but no evidence of pathological optic nerve cupping or visual field defects exists. Patients with ocular hypertension

TABLE I

Effects of Aging on Aqueous Humor Dynamics

	Group 1 (20–30 years old)		Group 2 (60 years and older)		
Parameter	*n*	Mean ± SD	*n*	Mean ± SD	*p* value*
IOP (mmHg)	51	14.7±2.6	53	14.3±2.6	0.5
P_{ev} (mmHg)	51	8.8±1.6	53	8.9±1.5	0.7
F_a (μl/min)	51	2.8±0.8	53	2.4±0.6	**0.002**
F_u (μl/min)	51	1.5±0.8	53	1.1±0.8	**0.01**
C_{fl} (μl/min/mmHg)	51	0.23±0.10	53	0.27±0.13	0.06
C_{ton} (μl/min/mmHg)	34	0.25±0.11	24	0.20±0.09	0.07
AC_{vol} (μl)	51	247±39	53	160±39	**<0.00001**
CCT (μm)	51	530±38	53	537±36	0.4

AC_{vol}, anterior chamber volume; CCT, central corneal thickness; C_{fl}, fluorophotometric outflow facility; C_{ton}, tonographic outflow facility; F_a, aqueous flow; F_u, uveoscleral outflow; IOP, intraocular pressure; P_{ev}, episcleral venous pressure (Table from Toris *et al.*, 1999b)

*Comparing the two groups with unpaired, two-tailed *t*-test.

have normal aqueous flow (Brubaker, 1991; Ziai *et al.*, 1993; Toris *et al.*, 2002) but reduced outflow facility (Grant, 1951; Ziai *et al.*, 1993; Toris *et al.*, 2002) and uveoscleral outflow (Toris *et al.*, 2002; Table II). The changes in the outflow pathways account for the elevated IOP in these patients.

B. Primary Open-Angle Glaucoma

Primary open-angle glaucoma is a disease in which elevated IOP is associated with pathological optic nerve cupping and/or visual field loss. It has been well-established that the major contributing factor for the elevated IOP in most glaucomas is increased resistance through the trabecular meshwork. Tonography was the method used in the early 1950s to identify abnormal outflow in patients with glaucoma compared with healthy subjects (Grant, 1951). In a later study, tonographic outflow facility was measured in the early morning and aqueous flow by fluorophotometry during the day and night in a group of patients with chronic simple glaucoma (Larsson *et al.*, 1995b). During the day, aqueous flow rates (2.39 μl/min) in the patients were not significantly ($p = 0.11$) different from those (2.70 μl/min) in healthy control subjects. At night, the same patients had significantly ($p = 0.02$) higher aqueous flow rates (1.29 μl/min) than the control subjects (1.02 μl/min). The lower

TABLE II
Aqueous Humor Dynamics in Patients with Ocular Hypertension

	Ocular hypertension		Ocular normotension		
Parameter	n	Mean ± SD	n	Mean ± SD	p value*
IOP (mmHg)	55	21.4±4.6	55	14.9±2.5	**<0.0001**
P_{ev} (mmHg)	55	9.5±1.4	55	9.0±1.7	0.1
F_a (μl/min)	55	2.6±0.8	55	2.6±0.7	0.8
F_u (μl/min)	55	0.7±0.8	54	1.1±0.8	**0.005**
C_{fl} (μl/min/mmHg)	55	0.17±0.08	54	0.27±0.13	**<0.0001**
AC_{vol} (μl)	55	195±55	55	198±55	0.8
CCT (μm)	55	533±33	55	525±38	0.2

AC_{vol}, anterior chamber volume; CCT, corneal thickness; C_{fl}, fluorophotometric outflow facility; F_a, aqueous flow; F_u, uveoscleral outflow; IOP, intraocular pressure; P_{ev}, episcleral venous pressure (Table From Toris *et al.*, 2002).

*Comparing ocular hypertensive patients with normotensive volunteers using unpaired, two-tailed t-test.

outflow facility in the glaucoma patients compared with healthy controls (0.14 μl/min/mmHg versus 0.23 μl/min/mmHg, respectively) was the major cause of the elevated IOP. The smaller aqueous flow reduction at night likely contributed little to the ocular hypertension. Uveoscleral outflow was not assessed.

There is little known about uveoscleral outflow in patients with primary open-angle glaucoma. A small study (Yablonski *et al.*, 1985a) of 14 patients with IOPs uncontrolled on maximally tolerated medical therapy found elevated rates of uveoscleral outflow (80% of total outflow, compared with 37% in contralateral eyes with less severe glaucoma). The facility of outflow through the trabecular meshwork was very low in these patients on multiple medications (0.02 μl/min/mmHg) (Yablonski *et al.*, 1985a) compared with a separate study of healthy subjects (0.25 μl/min/mmHg) on no known prescription medication (Toris *et al.*, 1999b). The high resistance in the trabecular meshwork of the patients with glaucoma (Yablonski *et al.*, 1985a) may have caused redirection of a major portion of the aqueous humor into the uvea, a region where flow is less dependent upon IOP. Systemic and ocular medications also may have contributed to this large difference in uveoscleral outflow between studies. Interestingly, monkeys with low outflow facility (0.06 compared with 0.16 μl/min/mmHg in the contralateral control eyes) from laser burns to the trabecular meshwork also demonstrated elevated uveoscleral outflow (2.25 compared with 1.05 μl/min in the contralateral control eyes) in the absence of drugs that might alter uveoscleral drainage (Toris *et al.*, 2000). It is possible that in the initial stages of glaucoma, both uveoscleral outflow and outflow facility are below normal, similar to that reported in patients with ocular hypertension (Toris *et al.*, 2002). As the disease progresses and outflow facility continues to decline, conditions favor redirection of aqueous humor from the trabecular to the uveal pathway. Further study of aqueous humor drainage in patients with primary open-angle glaucoma is needed to provide support for this idea.

In addition to changes in aqueous humor drainage, other factors may be involved in the elevated IOP and optic neuropathy in primary open-angle glaucoma. One theory is that blood pressure decreases (Hayreh *et al.*, 1994) and IOP increases at night (Liu *et al.*, 1998, 1999a,b), a combination that places eyes at increased risk for glaucomatous damage. In one study (Liu *et al.*, 2003), circadian IOPs of patients with early glaucoma and high diurnal IOP were compared with healthy subjects during the day and at night. Patients with glaucoma had relatively smaller changes in the diurnal versus nocturnal IOPs compared with healthy subjects. Both groups had higher IOPs at night (supine) than during the day (seated). Additionally, the supine IOP pattern around the normal awakening time (5:30–7:30 AM) was different between the two groups. The clinical importance of these findings is unknown. What is clear is that the need for IOP control at night is at least as important as during the day (Weinreb and Liu, 2006).

C. Normal Tension Glaucoma

Normal tension glaucoma is defined as cupping and visual field loss characteristic of glaucoma with no evidence of elevated IOP. In a study (Larsson *et al.*, 1993) of 10 patients with normal tension glaucoma compared with 10 age-matched healthy controls, there were no differences in IOP (15 vs 14 mmHg, respectively), outflow facility (0.18 vs 0.22 μl/min/mmHg, respectively), and daytime aqueous flow (2.48 vs 2.45 μl/min, respectively). Aqueous flow at night was slightly higher for the patients than the controls (1.24 vs 0.96 μl/min, respectively) but this was not statistically significant ($p = 0.09$). Patients with normal tension glaucoma exhibit increased variability of nighttime blood pressure (Plange *et al.*, 2006) and nocturnal hypotension that may reduce the optic nerve head blood flow below a healthy level (Hayreh *et al.*, 1994). These factors may lead to fluctuations in ocular perfusion pressure causing ischemic episodes and subsequent damage at the optic nerve head. These events are not detectable on routine clinical examination and are independent of aqueous humor dynamics.

D. Pigment Dispersion Syndrome

Pigment dispersion syndrome is a condition in which the posterior surface of the iris makes contact with the anterior zonules of the lens. Friction between the two surfaces causes release of pigment and cells from the iris which then move with the flow of aqueous humor into the anterior chamber and the trabecular meshwork. These patients have deeper anterior chambers than normal that predisposes them to the condition (Brown and Brubaker, 1989; Camras *et al.*, 2003). Patients with pigment dispersion syndrome but without ocular hypertension have similar patterns of aqueous humor inflow and outflow as healthy controls (Brown and Brubaker, 1989; Camras *et al.*, 2003). In some eyes, the cellular debris in the trabecular meshwork appears to obstruct the outflow of aqueous humor sufficiently to elevate IOP. The elevated IOP is caused by reduced outflow facility (Brown and Brubaker, 1989; Camras *et al.*, 2003). No change in uveoscleral outflow was found (Camras *et al.*, 2003) that is different from ocular hypertension without pigment dispersion syndrome, a condition in which both uveoscleral outflow and outflow facility are reduced (Table II) (Toris *et al.*, 2002). Therefore, pigment dispersion syndrome with ocular hypertension is sufficiently different from ocular hypertension without pigment dispersion syndrome so that it should be considered a separate entity.

E. Exfoliation Syndrome

Patients with exfoliation syndrome have a characteristic pattern of white deposits on the anterior capsule of the lens and tissues of the ciliary body, iris, cornea, and trabecular meshwork (Hammer *et al.*, 2001). Ocular hypertension in exfoliation syndrome has been attributed in part to deposits of exfoliative material near the endothelial cells of the trabecular meshwork and Schlemm's canal with subsequent degradation of the tissues and obstruction of the aqueous humor outflow pathways. The amount of material deposited in and around the trabecular meshwork has been positively correlated with increasing IOP and the presence of glaucoma (Schlötzer-Schrehardt and Naumann, 1995).

In a study (Gharagozloo *et al.*, 1992) of 18 untreated patients with unilateral exfoliation syndrome and ocular normotension, a comparison of the affected eye with its contralateral unaffected eye found similar IOPs (14 mmHg and 12 mmHg, respectively), the same aqueous flow rates (2.4 μl/min), and similar outflow facilities (0.15 μl/min/mmHg and 0.19 μl/min/mmHg, respectively). An earlier study (Johnson and Brubaker, 1982) of 10 patients with unilateral exfoliation syndrome and ocular hypertension (mean IOP of 32 mmHg in the affected eye and 18 mmHg in the unaffected eye) reported that aqueous flow and outflow facility were significantly lower in the affected eye (2.02 μl/min and 0.07 μl/min/mmHg, respectively) than the unaffected eye (2.38 μl/min and 0.15 μl/min/mmHg, respectively). The authors concluded that the lower flow was due to damage to the ciliary epithelia from the disease process. A later explanation was that the lower flow was the result of insufficient washout of timolol that had been used to treat the affected eye (Brubaker, 1998). In a recent study (Johnson *et al.*, 2008), comparisons were made between a group of 40 patients with exfoliation syndrome with and without ocular hypertension and a group of 40 age-matched patients without exfoliation syndrome with and without ocular hypertension. There was no significant difference in aqueous flow and a significantly lower uveoscleral outflow in the exfoliation syndrome group. When these groups were further divided by IOP, patients with ocular hypertension with or without exfoliation syndrome had reduced outflow facility compared with ocular normotensive controls. The effect on outflow facility was IOP dependent and unrelated to the exfoliation syndrome. The effect on uveoscleral outflow was exfoliation syndrome dependent and unrelated to

IOP. Exfoliation syndrome has distinctive effects on aqueous humor dynamics that are different from those found in ocular hypertension and pigment dispersion syndrome.

F. Diabetes Mellitus

Diabetes mellitus is associated with problems of the general circulation including increased blood viscosity, decreased tissue oxygenation, vascular leakiness, capillary shunting, and capillary nonperfusion. Ocular effects include changes in aqueous humor dynamics, IOP, aqueous flare, permeability of the blood–ocular barrier, and retinal vasculature. These effects are dependent on the duration of the diabetes, the age of the patient, and the severity of the retinopathy. The IOP tends to be slightly lower in diabetes mellitus than normal, but agreement from one study to another is not consistent. Studies reporting reduced IOP suggest the cause to be a reduction in the rate of aqueous flow (Hayashi *et al.*, 1989a; Larsson *et al.*, 1995a). A study (Larsson *et al.*, 1995a) of 61 patients with type 1 diabetes mellitus and of 60 patients with type 2 diabetes mellitus found that IOP, outflow facility, and aqueous flow were correlated with severity of diabetic retinopathy, age of onset, duration of diabetes, and age of the patient at the time of the study. There was a significant inverse correlation between the degree of retinopathy and aqueous flow rate. Apparently, as the degree of retinopathy increased, aqueous flow decreased.

Some of the published findings of reduced aqueous flow in diabetes may have been related to the insulin treatment. Increases in insulin and glucose concentrations have been shown to increase ocular blood flow (Bursell *et al.*, 1996; Schmetterer *et al.*, 1997; Luksch *et al.*, 2001) and might affect aqueous flow. To control for the insulin treatment, aqueous flow was measured in 11 patients with type 1 diabetes and 17 age-matched healthy control subjects while maintained on a hyperinsulinemic-euglycemic glucose clamp (Lane *et al.*, 2001). Aqueous flow was determined at two different insulin concentrations. A concentration-dependent effect of insulin on aqueous flow was not found. However, despite the absence of microvascular complications (retinopathy, microalbuminuria), patients with type 1 diabetes had reduced aqueous flow. The decrease in aqueous flow occurred despite the normal IOPs. This suggests the involvement of other parameters of aqueous humor dynamics such as a decrease in uveoscleral outflow and/or trabecular outflow facility, or an increase in episcleral venous pressure. No change in tonographic outflow facility was found in patients with type 1 diabetes (Larsson *et al.*, 1995a; Lane *et al.*, 2001). Uveoscleral outflow and episcleral venous pressure have yet to be investigated.

G. Uveitis

Fuchs' uveitis syndrome or Fuchs heterochromic iridocyclitis is a chronic, unilateral iridocyclitis characterized by iris heterochromia. Development of abnormal uveal pigment is associated with chronic low-grade inflammation that causes iris atrophy and secondary glaucoma in some patients. A study (Johnson *et al.*, 1983) of 10 patients with unilateral Fuchs' uveitis syndrome and normal IOP (17 mmHg) reported no change in outflow facility but increased permeability of the blood–aqueous barrier when compared with the unaffected contralateral eye. The 7% greater clearance of fluorescein in the affected eye was explained by an increased diffusional clearance of fluorescein. This complicated the measurement of aqueous flow. The authors suggested that aqueous flow could be lower in the affected eyes. A study (Toris and Pederson, 1985) in monkeys with experimental iridocyclitis found hypotony associated with reduced aqueous flow and increased uveoscleral outflow.

H. Glaucomatocyclitic Crisis

Glaucomatocyclitic crisis is a condition with recurrent episodes of markedly elevated IOP usually ranging between 40 to 60 mmHg accompanied by anterior chamber inflammation. The condition, also called Posner–Schlossman syndrome, was described by Posner and Schlossman (1948). Most of the available evidence suggests that the cause of the elevated IOP in this syndrome is a reduction of outflow facility (Camras *et al.*, 1996). During the interval between attacks, outflow facility returns to normal or slightly increases compared to the contralateral healthy eye (Grant, 1951; Mansheim, 1953; Higgitt, 1956; Spivey and Armaly, 1963). One study found reduced outflow facility in both the affected and healthy eye in 6 of 11 patients (Kass *et al.*, 1973). One theory suggested that the inflammation may be mediated by prostaglandins, especially prostaglandin E, which has been found in higher concentration in the aqueous humor of patients during but not between attacks (Masuda *et al.*, 1975). Not all studies are in agreement with this theory. Topical prostaglandins have been shown to increase, rather than decrease, outflow facility (Table III), which would contribute to a reduction in IOP.

Another suggested contribution to the elevated IOP in glaucomatocyclitic crisis is a possible increase in aqueous production (Spivey and Armaly, 1963; Sugar, 1965). Not all studies that evaluated this parameter came to this conclusion (Grant, 1951; Mansheim, 1953; Hart and Weatherill, 1968). All of these studies were conducted decades ago when aqueous flow was not measured directly but rather was calculated using the Goldmann equation.

TABLE III

The Effects of Prostaglandin Analogues on Aqueous Humor Dynamics in Humans

Analogue	References	Diagnosis and *n*	Duration of treatment	IOP	F_a	C	P_{ev}	F_u
Bimatoprost	Brubaker *et al.*, 2001	25 ONT	QD × 2 days	↓ (day)	↑ (day)	↑ day		↑ day
			QD × 3 days	↓ (night)	↑ (night)			
	Christiansen *et al.*, 2004	29 OHT or POAG	QD × 7 days	↓	↔	↑		↑
	Lim *et al.*, 2008	30 ONT	QD × 7 days	↓	↔	↑		↑
Latanoprost	Toris *et al.*, 1993	22 ONT or OHT	BID × 7 days	↓	↔	↔	↔	↑
	Ziai *et al.*, 1993	40 ONT or OHT	One drop or BID × 5 days	↓	↔ (one drop)	↑(5 days)		
	Mishima *et al.*, 1997	13 ONT	One drop	↓ (day and night)	↔ (day) ↑ (night)	↔ (day and night)		↑ (day and night)
	Lindén and Alm, 1997a	18 ONT	One drop QD × 14 days	↓ (more than BID)	↑			
			Three drops QD × 14 days	↓ (more than BID)	↔			
			BID × 14 days	↓	↔			
	Toris *et al.*, 2001	30 OHT	QD × 7 days	↓	↔	↔	↔	↑
	Dinslage *et al.*, 2004	42 OHT or POAG	QD × 14 days	↓	↔	↑		
	Lim *et al.*, 2008	30 ONT	QD × 7 days	↓	↔	↑		↑

Travoprost	Toris *et al.*, 2007	26 OHT and POAG	QD × 17 days	↓ (day and night)	↔	↑ (day)	↔	↑ (marginally insignificant)
	Lim *et al.*, 2008	30 ONT	QD × 7 days	↓	↔	↑		↑
Unoprostone	Sakurai *et al.*, 1991	10 ONT	One drop and BID × 4 weeks	↓	↔	↔	↔	
				↓	↔			
	Tetsuka *et al.*, 1992	8 ONT	One drop	↓	↔	↔		
	Toris *et al.*, 2004b	30 OHT	BID 5 and 28 days	↓	↔	↑	↔	↔

BID, twice daily; C_{fl}, outflow facility determined by fluorophotometry; F_a, aqueous flow; F_u, uveoscleral outflow IOP, intraocular pressure; OHT, ocular hypertension; ONT, ocular normotension; P_{ev}, episcleral venous pressure; POAG, primary open-angle glaucoma; QD, once daily. Down arrow indicates a reduction, up arrow indicates an increase, horizontal arrowheads indicates no change and no arrow indicates that the measurement was not reported.

Uveoscleral outflow was not considered in the calculation and episcleral venous pressure was assumed to be normal. Later, when aqueous flow was measured by fluorophotometry (Nagataki and Mishima, 1976), it was thought to be increased during an attack. Errors in the measurement may exist due to the increase in proteins and flare in the anterior chamber (Brubaker, 1997). It is unlikely that hypersecretion contributes to the elevated IOP in glaucomatocyclitic crisis (Camras *et al.*, 1996).

I. Myotonic Dystrophy

Myotonic dystrophy is a common adult form of muscular dystrophy. It is associated with a mutation that affects a gene on chromosome 19. The muscle weakness is accompanied by myotonia. Ocular effects include hypotony. Two studies were performed to determine if the ocular hypotony could be explained by aqueous humor hyposecretion. A study (Walker *et al.*, 1982) of 26 patients with myotonic dystrophy found IOPs averaging 7.1 mmHg and aqueous flow rates of 10% lower than normal. A later study (Khan and Brubaker, 1993) of 17 patients with muscular dystrophy and 17 age-matched controls found that IOPs were significantly lower in the myotonic dystrophy group (8.4 mm Hg) compared to the controls (14.0 mm Hg). Aqueous humor flow was reduced 9% in agreement with the earlier study but this alone was insufficient to explain the ocular hypotony. The authors hypothesized that atrophy of the ciliary muscle may have resulted in increased uveoscleral outflow.

V. DRUGS AFFECTING AQUEOUS HUMOR DYNAMICS

In most countries, the initial treatment for a newly diagnosed glaucoma patient is to prescribe topical ocular hypotensive medications. All of the clinically available drugs reduce IOP by affecting aqueous humor dynamics in some manner. They can be classified according to the mechanism(s) by which they decrease IOP. This includes aqueous humor suppressants (carbonic anhydrase inhibitors, β-adrenergic antagonists), drugs that enhance uveal and/or trabecular drainage (cholinergic agonists, prostaglandin analogues), or drugs that have combined inflow and outflow effects (adrenergic agonists).

A. Carbonic Anhydrase Inhibitors

Carbonic anhydrase inhibitors are used for treatment of primary open-angle glaucoma, secondary glaucomas, and acute angle-closure glaucoma. The mechanism by which carbonic anhydrase inhibitors reduce aqueous flow involves bicarbonate ions that are produced in the ciliary body by hydration of carbon dioxide under the control of carbonic anhydrase. Under normal conditions, bicarbonate, chloride, and sodium are secreted into the posterior chamber, drawing water into the posterior chamber by osmosis. Inhibition of the carbonic anhydrase enzyme slows the rate of water movement into the posterior chamber and the IOP decreases.

The first carbonic anhydrase inhibitor for clinical use was the sulfonamide derivative acetazolamide given orally four times daily. Its ocular hypotensive effects were reported initially in three clinical studies by Becker (1954), Breinin and Gortz (1954), and Grant and Trotter (1954). When given systemically, acetazolamide had no effect on outflow facility, suggesting that a reduction in aqueous humor formation was the cause of the IOP decrease. This conclusion was confirmed in many subsequent experiments in animals and humans. Other systemic carbonic anhydrase inhibitors, methazolamide, ethoxzolamide, and dichlorphenamide, reduce IOP by a similar mechanism (Stein *et al.*, 1983; Bar-Ilan *et al.*, 1984; Kalina *et al.*, 1988; Vogh *et al.*, 1989; Brechue and Maren, 1993; Skorobohach *et al.*, 2003).

The fluorophotometric method was used to study aqueous flow in healthy subjects treated once during the day with acetazolamide (Dailey *et al.*, 1982). Compared to a placebo pill, acetazolamide reduced aqueous flow by 27%. A later study of healthy subjects (McCannel *et al.*, 1992) found that acetazolamide lowered aqueous flow during the day by 21% and at night by an additional 24% below the normal nocturnal reduction of aqueous flow of 59%. This nocturnal effect of acetazolamide on aqueous flow was not a consistent finding (Topper and Brubaker, 1985).

Systemic carbonic anhydrase inhibitors may cause some unpleasant and potentially serious side effects in some patients. In an effort to reduce the unwanted systemic effects and to increase tolerability and compliance, topical carbonic anhydrase inhibitors have been developed. Several decades of research were required to produce a topical drop that was nontoxic, could readily penetrate the cornea, and had a very high affinity to ocular carbonic anhydrase. To be effective, inhibition of 99% of the activity of ocular carbonic anhydrase was needed (Maren *et al.*, 1977). Dorzolamide hydrochloride, a heterocyclic water-soluble sulfonamide and a potent inhibitor of the carbonic anhydrase isoenzyme II, was the first to be developed into a successful topical formulation. Its ocular hypotensive mechanism of action was compared against acetazolamide in several clinical studies. In one study (Maus *et al.*, 1997), acetazolamide more effectively suppressed the daytime flow of aqueous humor (30%) than dorzolamide (17%). Acetazolamide, given orally, was additive to dorzolamide, given topically, but dorzolamide was not additive to acetazolamide in healthy individuals (Maus *et al.*, 1997), ocular hypertensive subjects (Toris *et al.*, 2004a), or glaucoma patients (Rosenberg *et al.*, 1998). Apparently, topical dorzolamide is only half as effective as acetazolamide at slowing the production rate of aqueous humor. Lack of sufficient penetration of topical dorzolamide to target tissues or some extraocular effects of systemically administered acetazolamide are a couple of possible explanations for these findings (Brubaker, 1998; Toris *et al.*, 2004a).

B. β-Adrenergic Antagonists

The first report of a systemically administered β-adrenergic antagonist (propranolol) that lowered IOP appeared in 1967 (Phillips *et al.*, 1967). The following year, propranolol was reported to be effective when given topically (Bucci *et al.*, 1968) but this was associated with severe side effects. Eight years later, timolol was shown to reduce IOP and to be well tolerated in healthy subjects (Katz *et al.*, 1976; Zimmerman and Kaufman, 1977). It was approved for clinical use shortly thereafter, and for a time, it became the most widely prescribed drug for glaucoma treatment despite the reports of some severe systemic side effects including asthma exacerbation, worsening congestive heart failure, heart block, and even sudden death (Nelson *et al.*, 1986). For years, timolol was the "gold-standard" against which newer drugs were compared.

The mechanism by which timolol lowers IOP is by reducing aqueous flow. The aqueous flow suppression ranges from 28% to 50% in many clinical studies (Coakes and Brubaker, 1978; Yablonski *et al.*, 1978; Schenker *et al.*, 1981; Brubaker, 1982; Larsson *et al.*, 1993, 1995b; Wayman *et al.*, 1997; Larsson, 2001; Toris *et al.*, 2004a). Other β-adrenergic antagonists also reduced aqueous flow in a similar manner. These drugs include betaxolol (Reiss and Brubaker, 1983; Gaul *et al.*, 1989; Coulangeon *et al.*, 1990), carteolol (Coulangeon *et al.*, 1990), and levobunolol (Yablonski *et al.*, 1987; Gaul *et al.*, 1989). None of these drugs appear to affect aqueous humor outflow. When measured at night, timolol has no effect on aqueous flow (Topper and Brubaker, 1985; McCannel *et al.*, 1992). Apparently, it is unable to reduce aqueous flow further than the normally low nocturnal rate (Reiss *et al.*, 1984).

When given chronically, the initial effect of timolol on aqueous flow fades so that half the effect is gone after one year of treatment (Brubaker *et al.*, 1982). After discontinuation of timolol, recovery of the aqueous flow rate occurs slowly but completely to pretreatment levels over several weeks to months (Schlecht and Brubaker, 1988). Similarly, there is complete recovery of aqueous flow after treatment with betaxolol and levobunolol (Gaul *et al.*, 1989).

There is additivity of the aqueous flow reduction when timolol and carbonic anhydrase inhibitors are combined (Dailey *et al.*, 1982; Wayman *et al.*, 1997, 1998; Brubaker *et al.*, 2000; Toris *et al.*, 2004a). Additivity of these two aqueous flow suppressants can be explained by the different mechanisms whereby each of these different classes of drug reduces aqueous flow. Timolol blocks β-adrenoceptors located in the ciliary processes (Bartels *et al.*, 1980) and carbonic anhydrase inhibitors inactivate carbonic anhydrase enzymes also located in the ciliary processes (Maren, 1987; Brechue and Maren, 1993).

C. Adrenergic Agonists

Adrenergic agonists are drugs that stimulate α_1, α_2, and/or β-adrenergic receptors. The first drug of this class to be used as an antiglaucoma drug is epinephrine, a direct adrenergic agonist of all three receptors. It was first used in 1900 (Darier, 1900) and it is still used occasionally today. In the last 50 years, a large number of studies [summarized elsewhere (Townsend and Brubaker, 1980; Wang *et al.*, 2002)] have reported the effects of epinephrine on aqueous humor dynamics in humans. The reason for the large number of reports is the lack of agreement between studies and the inability to conclude with confidence the precise mechanism for the IOP reduction.

In clinical studies, epinephrine was found to increase aqueous flow (Higgins and Brubaker, 1980; Townsend and Brubaker, 1980; Schenker *et al.*, 1981; Wentworth and Brubaker, 1981; Lee *et al.*, 1983; Topper and Brubaker, 1985; Kacere *et al.*, 1992), to have no effect (Kronfeld, 1963; Galin *et al.*, 1966; Nagataki and Brubaker, 1981; Schneider and Brubaker, 1991; Mori *et al.*, 1992), and even to reduce aqueous flow (Weekers *et al.*, 1954a; Garner *et al.*, 1959; Nagataki, 1977; Araie and Takase, 1981; Wang *et al.*, 2002). When epinephrine was found to increase aqueous flow, the increase was as much as 32% (Wentworth and Brubaker, 1981). Epinephrine may have a different pharmacodynamic effect on aqueous humor production depending on which receptor action is predominant at the time of the measurement. The final aqueous flow effect is likely the sum of the activities of epinephrine at all adrenergic receptors.

The effect of epinephrine on outflow is as conflicting as it is on inflow. Increases in the facility of trabecular outflow have been reported in some studies (Ballintine and Garner, 1961; Becker *et al.*, 1961; Kronfeld, 1963; Becker and Morton, 1966; Townsend and Brubaker, 1980; Schenker *et al.*, 1981; Wang *et al.*, 2002) but not in others (Weekers *et al.*, 1954a,b; Garner *et al.*, 1959; Galin *et al.*, 1966; Wentworth and Brubaker, 1981). Calculations of uveoscleral outflow have found that epinephrine increases drainage through the uvea in two studies (Townsend and Brubaker, 1980; Schenker *et al.*, 1981) but not in a third (Wang *et al.*, 2002).

It is clear that epinephrine has complex and dynamic effects on the production and drainage of aqueous humor likely because of its differential activity at α_1-, α_2-, and β-receptors. Relatively recently, selective agonists at the α-receptors have been developed into efficacious IOP-lowering drugs. Studies of aqueous humor dynamics following treatment with these drugs have helped to clarify the mechanism for the IOP reduction of adrenergic agonists as a class.

Clonidine is a direct-acting adrenergic agonist prescribed historically as an antihypertensive agent. In addition to blood pressure reduction, it was found also to substantially lower IOP. However, the systemic side effects, including sedation, precludes its use as a chronic IOP-lowering treatment. Apraclonidine, an α_2-adrenergic agonist that also has α_1-adrenergic activity, was developed for the topical treatment of ocular hypertension with fewer side effects than clonidine. Apraclonidine was approved in 1997 to lower IOP, but it is rarely prescribed for chronic therapy because of the high incidence of allergic reaction with prolonged use. Occasionally, apraclonidine is used to prevent IOP spikes after certain anterior segment laser procedures including laser trabeculoplasties, laser iridotomies, and Nd:YAG laser posterior capsulotomies. Apraclonidine was found to reduce IOP in humans by reducing aqueous flow from 30% to 34% (Gharagozloo *et al.*, 1988; Koskela and Brubaker, 1991a; Toris *et al.*, 1995b) and by increasing trabecular outflow facility about 50% when measured by fluorophotometry (Toris *et al.*, 1995b).

Brimonidine, a more selective α_2-agonist with some α_1-activity, was approved for clinical use 4 years after apraclonidine. As with apraclonidine, brimonidine reduces IOP initially by decreasing aqueous flow. The aqueous flow effect is rapid, occurring within 1 hour of administration (Toris *et al.*, 1999a). This effect persists during 2 days to 1 week of twice-daily dosing in healthy subjects (Schadlu *et al.*, 1998; Maus *et al.*, 1999; Larsson, 2001; Tsukamoto and Larsson, 2004) and ocular hypertensive patients (Toris *et al.*, 1995a, 1999a). However, with continued dosing for a month, the aqueous flow effect fades and an increase in uveoscleral outflow becomes more significant (Toris *et al.*, 1999a). During this transition, the IOP remains significantly below pretreatment levels.

The substantial aqueous flow drop with one dose of brimonidine may be the result of strong vasoconstriction in blood vessels of the uvea causing reduced blood flow and volume of the ciliary body (Reitsamer *et al.*, 2006). With vasoconstriction, the tissue volume would be replaced by posterior chamber fluid volume and the rate of aqueous humor flow from the posterior chamber into the anterior chamber would be slowed down. Aqueous flow would not equal the rate of aqueous humor production until a new steady state is reached, which should be accomplished within a half hour. Reduced aqueous flow lasting for hours to days (Toris *et al.*, 1997) may be explained with more satisfaction by insufficient delivery of oxygen and/or other essential energy sources to the ciliary processes attributable to the reduction in blood flow. Slowed aqueous production may persist until the vasoconstrictive effect fades and blood flow returns to its normal rate.

An increase in uveoscleral outflow is the reason that the IOP remains significantly reduced with continued brimonidine treatment in humans (Toris *et al.*, 1995a, 1999a) and rabbits (Lee *et al.*, 1992). Locally enhanced production and release of endogenous prostaglandins by brimonidine is one possible cause of the uveoscleral outflow effect. Topical prostaglandins have been found to increase uveoscleral outflow in most studies (Table III). Cyclooxygenase inhibitors in humans suppress the ocular hypotensive effects of adrenergic agents including epinephrine, brimonidine, and apraclonidine (Camras and Podos, 1989; Sponsel *et al.*, 2002). Adrenergic agonists stimulate the release of prostaglandins into the anterior chamber *in vivo* and the synthesis of prostaglandins in ocular tissues *in vitro* (Camras and Podos, 1989). It is also possible that brimonidine increases uveoscleral outflow in a manner independent of endogenous prostaglandin release. There is evidence that relaxation of the ciliary muscle can be accomplished by α_2-adrenergic agonists (Kubo and Suzuki, 1992) in a manner similar to the prostaglandin $F_{2\alpha}$-analogue, latanoprost (Nilsson *et al.*, 1989).

The different mechanisms of action of brimonidine and apraclonidine may be related to their receptor selectivity. Possible receptors targeted differentially by these drugs include two imidazoline subtypes (Michel and Ernsberger, 1992) and four α_2-adrenergic subtypes (Bylund, 1988). There is functional evidence for at least two of the four known subtypes of α_2-agonists in the anterior segment of the eye (Potter *et al.*, 1990). The ocular hypotensive effect of brimonidine in the primate may be mediated through stimulation of an imidazoline receptor rather than an α_2-receptor (Burke *et al.*, 1995). The effect of apraclonidine may be via stimulation of a combination of α_1- and α_2-receptors.

The mechanism by which apraclonidine and brimonidine reduce aqueous flow appears to be somewhat different from the mechanism by which timolol affects aqueous flow. Apraclonidine reduces aqueous flow at night (Koskela and Brubaker, 1991a), whereas timolol does not (Topper and Brubaker, 1985). Additionally, although apraclonidine and timolol were not additive when one drop of each was given to 20 healthy subjects, apraclonidine did reduce aqueous flow further when one drop was given to 17 glaucoma patients on long-term timolol treatment (Gharagozloo and Brubaker, 1991). The authors suggested that apraclonidine reduced aqueous flow in eyes that had adapted to chronic timolol use to the level seen before adaptation occurred (Gharagozloo and Brubaker, 1991). A study (Larsson, 2001) of brimonidine and timolol additivity in 20 healthy subjects found that treatment with both drugs twice-daily for three treatments caused a further reduction in aqueous humor flow and IOP than each drug alone. A third study (Maus *et al.*, 1999) compared the additivity of brimonidine and apraclonidine with timolol in the same 19 healthy subjects dosed once in the evening and once on the day of the study. Both α_2-agonists reduce IOP in the timolol-treated eye, primarily by further suppressing aqueous flow. These studies suggest that the aqueous flow reduction by α_2-adrenergic agonists and β-adrenergic antagonists are not mediated entirely by the same pathway.

D. Prostaglandin Analogues

It was a long road from the discovery in 1955 of prostaglandins in rabbit ocular tissues (Ambache, 1957) to the finding in 1981 that topical prostaglandins could lower IOP effectively in monkeys (Camras and Bito, 1981; Stjernschantz, 2001), to the approval in 1997 of the first prostaglandin analogue for the treatment of glaucoma. Currently, four prostaglandin $F_{2\alpha}$-analogues are used to reduce IOP: latanoprost, travoprost, bimatoprost, and unoprostone. Both latanoprost and travoprost are ester compounds and are pharmacologically classified as prostaglandin analogues. Bimatoprost is an amide prodrug of a prostaglandin analogue described as a "prostamide" (Woodward *et al.*, 2003). Unoprostone is an analogue of a pulmonary metabolite of prostaglandin $F_{2\alpha}$ and is sometimes labeled as a "docosanoid" (Haria and Spencer, 1996). All four drugs are very similar in structure and will be considered compounds of the same class for the purposes of this review.

Latanoprost, travoprost, and bimatoprost appear to have similar ocular hypotensive efficacy and work by similar mechanisms, mainly by increasing uveoscleral outflow, and trabecular outflow facility. Unoprostone is the least efficacious and works mainly by increasing trabecular outflow facility (Toris *et al.*, 2004b). The many supporting clinical studies involving aqueous humor dynamics of the four prostaglandin $F_{2\alpha}$-analogues are summarized in Table III.

Prostaglandin $F_{2\alpha}$ and its various analogues appear to increase uveoscleral outflow by remodeling the extracellular matrix of the ciliary muscle (Weinreb *et al.*, 1997; Ocklind, 1998; Sagara *et al.*, 1999), by widening the connective tissue-filled spaces among the ciliary muscle bundles (Lütjen-Drecoll and Tamm, 1988; Tamm *et al.*, 1990), and by relaxing the ciliary muscle, which may contribute to widening of the intermuscular spaces (Van Alphen *et al.*, 1977; Poyer *et al.*, 1995). There also is evidence for a change in the shape of ciliary muscle cells, with alterations in the actin and vinculin localization within the cells (Stjernschantz *et al.*, 1998). Thus, it appears that prostaglandin analogues have complex effects on the ciliary muscle, the net effect being increased flow of aqueous humor through this tissue.

It has been suggested that prostaglandins may increase the pressure sensitivity of uveoscleral outflow, a flow that is usually considered to be relatively pressure insensitive. Uveoscleral outflow facility is 10% of trabecular outflow facility in young healthy monkeys (Bill, 1966, 1967b; Toris and Pederson, 1985).

The prostaglandin effect on uveoscleral outflow facility results from the morphological and biochemical changes to tissues lining the uveal pathways (Weinreb *et al.*, 2002). Providing indirect evidence for this is a study in monkeys treated with prostaglandin $F_{2\alpha}$, in which trabecular outflow facility remains unchanged despite a concurrent 60% increase in total outflow facility (Crawford *et al.*, 1987). However, using a more direct tracer method for the assessment, uveoscleral outflow facility in cats treated with prostaglandin A_2 was unchanged despite an increase in uveoscleral outflow (Toris *et al.*, 1995c). That many studies reported no effect of prostaglandins on tonographic outflow facility despite an increase in uveoscleral outflow also provides indirect evidence that uveoscleral outflow facility is not affected by prostaglandins. Species differences, measurement technique, and/or type, duration, and dose of prostaglandin all have been explanations for the differing findings among studies. The definitive experiment to answer this question should be done in nonhuman primates in which intracameral tracer is infused at different IOPs and uveoscleral outflow is determined at each pressure.

Effects of prostaglandin analogues on aqueous flow are mixed. Many studies report no effect of prostaglandin analogues on aqueous flow whereas a few studies report an increase (Table III). That IOP is significantly reduced by these drugs indicates that the strong increased outflow effect more than compensates for a small increased inflow effect.

E. Cholinergic Agonists

For more than a century, pilocarpine has been used to treat elevated IOP. This drug is a muscarinic receptor agonist obtained from the leaves of tropical American shrubs from the genus *Pilocarpus*. Pilocarpine reduces IOP by increasing outflow facility through the trabecular meshwork (Bill and Wålinder, 1966). This is accomplished by stimulating postsynaptic muscarinic receptors in the ciliary muscle causing it to contract and pull on the scleral spur that enlarges the fluid channels and reduces the resistance in the trabecular meshwork (Kaufman and Gabelt, 1997). With high doses of pilocarpine, the contraction of the ciliary muscle decreases uveoscleral outflow in monkeys (Bill, 1967a). However, at clinical doses, pilocarpine increases outflow facility without diminishing uveoscleral outflow in humans (Toris *et al.*, 2001). A small increase in aqueous flow of 14% was found in one study of pilocarpine (Nagataki and Brubaker, 1982) but not in others (Araie and Takase, 1981; Toris *et al.*, 2001). Similar studies of other cholinergic agonists are lacking.

Detailed assessment of the ocular hypotensive mechanism of action of glaucoma medications has helped to predict the benefit or detriment of combined therapy. Knowing that prostaglandin $F_{2\alpha}$ and its analogues relax the ciliary muscle, modify its extracellular matrix, and increase uveoscleral outflow (Weinreb *et al.*, 2002) and that pilocarpine in monkeys contracts the ciliary muscle and reduces uveoscleral outflow (Bill, 1967a), it was predicted that the two drugs would not have an additive IOP effect. This prediction proved to be true when pilocarpine and prostaglandin $F_{2\alpha}$ were used in combination therapy in monkeys (Crawford and Kaufman, 1987; Camras *et al.*, 1990). However, latanoprost was additive in humans treated with a variety of cholinergic agonists, including pilocarpine, carbachol, phospholine iodide (Villumsen and Alm, 1992; Friström and Nilsson, 1993; Patelska *et al.*, 1997; Shin *et al.*, 1999; Kent *et al.*, 1999), or physostigmine (Lindén and Alm, 1997b). In a clinical study (Toris *et al.*, 2001) to explain the IOP additivity of latanoprost and pilocarpine, it was found that latanoprost when used alone increased uveoscleral outflow and pilocarpine when used alone increased outflow facility. Additivity of latanoprost and pilocarpine was achieved because pilocarpine did not block the uveoscleral outflow increase induced by latanoprost, and latanoprost appeared to potentiate the trabecular outflow facility increase induced by pilocarpine.

Reports of aqueous flow effects of pilocarpine are mixed. One study found no effect of pilocarpine on aqueous flow (Toris *et al.*, 2001) when given to ocular hypertensive patients four times daily for a week and another study found that pilocarpine (Nagataki and Brubaker, 1982) increases aqueous flow in healthy volunteers. The effect was small and of little clinical significance.

F. Experimental Drugs

New classes of drugs are under investigation as potential therapies to increase IOP and treat hypotony or decrease IOP and treat glaucoma. Discussed below are studies of some promising new classes of drugs affecting aqueous humor dynamics.

1. Dopaminergic Agonists and Antagonists

The six types of dopamine receptors (D1, D2a, D2b, D3, D4, and D5) are metabotropic G–protein-coupled receptors with dopamine as their endogenous ligand. D1 and D5 typically are excitatory and the rest are inhibitory. Growing evidence suggests a role for dopamine receptors in the modulation of IOP. D1-receptor blockade [SDZ PSD-958 (Prünte *et al.*, 1997)], D2-receptor stimulation [quinpirole (Prünte *et al.*, 1997)], and D3-receptor stimulation [PD128,907 (Chu *et al.*, 2004) and 7-hydroxy-2-dipropylaminotetralin (Chu *et al.*, 2000)] contribute to IOP lowering in rabbits. A possible mechanism for this IOP decrease is a reduction in aqueous flow (Chu *et al.*, 2000; Reitsamer and Kiel, 2002b). In rabbits, high doses of dopamine decrease aqueous flow (Reitsamer and Kiel, 2002b) in a manner similar to brimonidine, that is, by reducing ciliary blood flow (Reitsamer *et al.*, 2006). Interestingly, low doses of dopamine increase rather than decrease aqueous flow (Reitsamer and Kiel, 2002b). Evidence in rabbits suggest that D1 receptors are not involved in the outflow of aqueous humor. Tonographic outflow facility was unchanged in rabbits treated with a mixed D1-receptor antagonist and D2-receptor agonist (Prünte *et al.*, 1997). The D1-receptor antagonist, SCH 23390, binds to sites within the epithelium of the ciliary processes, but not in the iridocorneal angle (Mancino *et al.*, 1992).

Clinical studies have found that the selective D1-receptor agonist, fenoldopam, increases IOP in healthy humans (Piltz *et al.*, 1998) and fenoldopam, along with another selective D1-receptor agonist, 3B90, increase IOP in patients with primary open-angle glaucoma (Virno *et al.*, 1992). Additionally, D1 receptors have been found in human uvea and sclera (Cavallotti *et al.*, 1999), suggesting an outflow effect, unlike in rabbits. Obviously, further research is needed to elucidate the role of dopamine in the regulation of aqueous humor dynamics.

Ibopamine, a prodrug of *N*-methyldopamine with α-adrenergic properties, is under development for the purpose of increasing IOP to treat hypotony. Ibopamine apparently increases IOP by stimulating aqueous humor production. Two studies (Virno *et al.*, 1997; Azevedo *et al.*, 2003) compared the IOP and aqueous flow effect of multiple drops of topical ibopamine in healthy volunteers and patients with glaucoma. In both studies, a significant increase in aqueous humor production was found in both groups of subjects but only the glaucoma group had the expected increase in IOP. These puzzling findings were explained in a third study (McLaren *et al.*, 2003). It was observed that ibopamine caused mydriasis that can interfere with the fluorophotometric method to measure aqueous flow. The effect of mydriasis on aqueous flow was evaluated in 24 healthy subjects treated with either tropicamide or ibopamine to create mydriasis, dapiprazole to block the mydriasis, or placebo. It was found that a rapid transient doubling of aqueous flow was an artifact of increased fluorescein clearance from the anterior chamber back through the dilated pupil. Ibopamine coupled with dapiprazole did increase aqueous flow compared with tropicamide coupled with dapiprazole, but the increase was slight and would not have been detected as an increase in IOP in the healthy subjects. The increased IOP in the glaucoma patients treated with ibopamine was due to the poor outflow facility in these patients.

2. Angiotensin AT_1-Receptor Antagonists

The renin–angiotensin system is a hormone system that helps to regulate blood pressure and extracellular volume in the body. Angiotensin I is formed by the action of renin on angiotensinogen and then is converted to angiotensin II by angiotensin-converting enzyme. The action of angiotensin II is targeted by angiotensin II-receptor antagonists, which directly block angiotensin II AT_1 receptors, one of two receptor subtypes able to bind angiotensin II. Blockade of AT_1 receptors causes vasodilation, and reduces the secretion of vasopressin and aldosterone, the combined effect of which is a decrease in blood pressure. Components of the renin–angiotensin system are thought to be involved in the regulation IOP as well as blood pressure. In support of this is the localization of AT_1 receptors in ocular tissues of rabbits (Ramirez *et al.*, 1996) and humans (Savaskan *et al.*, 2004). Two effects mediated by the AT_1 receptor include vasoconstriction and extracellular matrix formation, two factors that can affect aqueous humor dynamics.

AT_1-receptor antagonists (sartans) have been shown to have effects on IOP and aqueous humor outflow. When given topically to monkeys with unilateral ocular hypertension, IOP was reduced in a dose-dependent manner (Wang *et al.*, 2005). CS-088 (olmesartan) (Inoue *et al.*, 2001b) at 1% and 2% significantly lowered the IOP in ocular hypertensive rabbits with a maximum IOP reduction of 10 mmHg. Aqueous flow and outflow facility were unchanged and a slight increase in uveoscleral outflow (17%) was observed, which was insufficient to explain the IOP reduction. Further study (Inoue *et al.*, 2001a) in rabbits found that ocular angiotension II attenuates uveoscleral outflow via AT_1 receptors, but again this physiological effect was insufficient to explain the IOP effect. This leads to the intriguing possibility that unidentified factors independent of those in the modified Goldmann equation may contribute to the effect.

3. Regulators of the Actin Cytoskeleton

The trabecular meshwork cytoskeleton currently is a major focus of research interest because this is the region that apparently is most altered in glaucomas associated with elevated IOP, and this is the target region for some new therapeutic approaches to treat glaucoma (Tian *et al.*, 2000). The trabecular meshwork consists of multiple arrays of collagen beams covered by endothelium-like cells, with extracellular matrix filling the spaces between the beams. Adjacent to this is the Schlemm's canal, which is a continuous endothelium-lined channel through which aqueous humor drains into the aqueous veins and general venous circulation. As one ages, resistance to fluid flow increases, the number of trabecular meshwork cells decreases, and structural alterations of the extracellular matrix in the juxtacanalicular region occur (Kaufman and Gabelt, 1993; Gabelt *et al.*, 2003). Eyes diagnosed with glaucoma have abnormal-appearing juxtacanalicular extracellular matrix and abnormal number of trabecular meshwork cells compared with age-matched healthy individuals (Joseph and Grierson, 1994; Sihota *et al.*, 2001; Read *et al.*, 2007). These findings suggest that cells and extracellular matrix in the juxtacanalicular region may be key components of resistance regulation. Cell volume, shape, contractility, adhesion to neighboring cells and the surrounding extracellular matrix, and the amount and composition of the extracellular matrix could affect resistance to fluid drainage by altering the dimensions or direction of flow pathways. Several cytoskeletal drugs are now under investigation as potential therapeutic agents for the treatment of glaucoma.

a. Cytochalasins. Cytochalasins are fungal metabolites that can bind to actin filaments and block polymerization and elongation of actin. Cytochalasin affects the kinetics of actin-filament polymerization at both the barbed and pointed ends. Inhibition of actin polymerization causes cells to change shape and structure, stop cellular processes such as cell division, and undergo apoptosis (Haidle and Myers, 2004).

Cytochalasin-B and -D increase outflow facility in living monkey and organ-cultured human eyes (Kaufman and Bárány, 1977; Kaufman and Erickson, 1982; Johnson, 1997), accompanied by separation of endothelial cells of the beams, the juxtacanalicular region, and the inner wall from their neighboring cells and surrounding extracellular matrix. The accompanying meshwork distension and inner wall ruptures improve fluid flow leading to washout of extracellular matrix (Svedbergh *et al.*, 1978; Johnstone *et al.*, 1980). Cytochalasins are no longer under investigation as potential antiglaucoma drugs as better candidates have evolved from this research. One of these includes the latrunculins.

b. Latrunculins. The latrunculins are a family of natural products and toxins produced by certain sponges, including the genus *Latrunculia*. The mode of action of latrunculins is less complex than that of cytochalasins and the compound is more potent. For these reasons, latrunculins currently are of greater scientific interest than cytochalasins. By sequestering actin monomers, latrunculins prevent the monomers from repolymerizing into filaments. The result is disassembly of actin filaments and reduction in cell–cell and cell–matrix adhesion (Ethier *et al.*, 2006) and expansion of the juxtacanalicular space (Sabanay *et al.*, 2006). Latrunculin-B increases outflow facility in postmortem human eyes (Ethier *et al.*, 2006) and latrunculin-A and -B have similar effects in living monkey eyes (Peterson *et al.*, 1999, 2000). The mechanism for the facility increase is likely due to loss of mechanical integrity of the trabecular meshwork (Ethier *et al.*, 2006). Aqueous humor flow in monkeys is increased by 87% at 3 hours after treatment with latrunculin-A and unchanged after treatment with latrunculin-B (Peterson *et al.*, 2000). Some cornea toxicity was noted with one drop of 0.02% latrunculin-B (Peterson *et al.*, 2000) but not with 5 days of twice-daily dosing of a fourfold lower concentration (Okka *et al.*, 2004).

Other cytoskeletal drugs used in preclinical studies are described elsewhere (Tian *et al.*, 2000). None of these drugs has reached clinical trials.

c. Rho-Kinase Inhibitors. Rho kinase has been identified as one of the effectors of the small GTP-binding protein Rho. The Rho/Rho-kinase pathway appears to play an important role in many cellular functions including vascular smooth muscle cell contraction, cell motility and adhesion, actin cytoskeleton organization, and cytokinesis. Rho-kinase inhibitors have been reported to inhibit agonist-induced smooth muscle contraction both *in vitro* and *in vivo* and to prevent the stress fiber development and focal adhesion formation induced by Rho-kinase in cultured cells (Uehata *et al.*, 1997). In the eye, trabecular meshwork and ciliary muscle cells express morphological and electrophysiological properties that are typical of smooth muscle cells elsewhere in the body (Wiederholt *et al.*, 2000). The Rho-associated protein kinase inhibitor, Y-27632, has been shown to alter the contractility of the trabecular meshwork and ciliary muscle, and to modulate the behavior of trabecular meshwork cells. Human trabecular meshwork and Schlemm's canal cells treated with Y-27632 cause cellular relaxation and loss of cell–substratum adhesions (Rao *et al.*, 2001). Bovine trabecular meshwork strips express major contractility-regulating proteins, which are involved in tissue function. Inhibition of the signaling pathways by Y-27632, which leads to myosin phosphorylation, causes inhibition of contractile force in trabecular meshwork (Rosenthal *et al.*, 2005). These changes in trabecular meshwork cells appear to alter outflow facility. Topical treatment with Y-27632 produced a significant IOP decrease that was associated with increased outflow facility and uveoscleral outflow in rabbits (Honjo *et al.*, 2001; Waki *et al.*, 2001) and monkeys (McLaughlin *et al.*, 2004). Modulation of the contractile or relaxant components of the trabecular meshwork and ciliary muscle appears to play a direct role in the regulation of aqueous humor outflow (Wiederholt *et al.*, 1995; Tian *et al.*, 2000). The outflow facility effect might be explained by increased paracellular fluid flow across Schlemm's canal or altered flow through the juxtacanalicular tissue.

4. Serotonin Agonists

Serotonin (5-hydroxy tryptamine, 5-HT) is an important endogenous neurotransmitter in the mammalian central nervous system. It appears to play a role in human ocular physiology because serotonin (Martin *et al.*, 1992b) and serotonergic nerves (Tobin *et al.*, 1988) have been found in the iris-ciliary body and $5HT_{2A}$- and $5HT_{2B}$-receptor mRNAs have been localized in trabecular meshwork cells (Sharif *et al.*, 2006). Additionally, serotonin has been identified in the human aqueous humor (Martin *et al.*, 1988; Martin *et al.*, 1992b; Veglio *et al.*, 1998). $5HT_{2A}$ agonists lower IOP in ocular normotensive (Gabelt *et al.*, 2005) and hypertensive monkeys (Sharif *et al.*, 2006). The IOP effect of the 5-HT_2 agonist, R-DOI, in ocular normotensive monkeys (Gabelt *et al.*, 2005) was explained by an increase in uveoscleral outflow. Unrelated to the IOP reduction was an increase in aqueous flow and no change in outflow facility.

VI. SUMMARY

Intraocular pressure remains stable throughout the lifetime of the healthy human eye. Small changes in aqueous flow are counterbalanced by small changes in outflow. Elevated IOP in various clinical syndromes is associated with abnormalities in the aqueous humor outflow pathways. Aqueous flow is surprisingly stable under all of these conditions and ranges of IOP. Some earlier drugs used to treat ocular hypertension were those that had direct effects on the ciliary processes to reduce the production rate of aqueous humor. Drugs currently under development to be the next generation of glaucoma therapy are those that target the outflow pathways. From a physiological perspective, this is the logical approach because in glaucoma this is the location of the pathology and the region in need of repair.

The study of aqueous humor outflow in humans is limited by the methods currently available to make each assessment. Some methods are indirect and many assumptions apply that may not be valid under all conditions. Development of improved noninvasive methods to measure aqueous humor outflow is a major challenge in the study of the aqueous humor circulatory system. Visual imaging systems are showing promise in this regard. Not only do these methods provide detailed morphology but quantitation of fluid dynamics soon may be possible.

References

Ambache, N. (1957). Properties of irin, a physiological constituent of the rabbit iris. *J. Physiol.* **135,** 114–132.

Araie, M., and Takase, M. (1981). Effects of various drugs on aqueous humor dynamics in man. *Jpn. J. Ophthalmol.* **25,** 91–111.

Araie, M., Yamamoto, T., Shirato, S., and Kitazawa, Y. (1984). Effects of laser trabeculoplasty on the human aqueous humor dynamics: A fluorophotometric study. *Ann. Ophthalmol.* **16,** 540–544.

Armaly, M. F. (1959). The consistency of the 1955 calibration for various tonometer weights. *Am. J. Ophthalmol.* **48,** 602–611.

Azevedo, H., Ciarniello, M. G., Rosignoli, M. T., Dionisio, P., and Cunha-Vaz, J. (2003). Effects of ibopamine eye drops on intraocular pressure and aqueous humor flow in healthy volunteers and patients with open-angle glaucoma. *Eur. J. Ophthalmol.* **13,** 370–376.

Ballintine, E. J., and Garner, L. L. (1961). Improvement of the coefficient of outflow in glaucomatous eyes. *Arch. Ophthalmol.* **66,** 314–317.

Bar-Ilan, A., Pessah, N. I., and Maren, T. H. (1984). The effects of carbonic anhydrase inhibitors on aqueous humor chemistry and dynamics. *Invest. Ophthalmol. Vis. Sci.* **25,** 1198–1205.
Bartels, S. P., Roth, H. O., Jumblatt, M. M., and Neufeld, A. H. (1980). Pharmacological effects of topical timolol in the rabbit eye. *Invest. Ophthalmol. Vis. Sci.* **19,** 1189–1197.
Becker, B. (1954). Decrease in intraocular pressure in man by a carbonic anhydrase inhibitor, diamox. A preliminary report. *Am. J. Ophthalmol.* **37,** 13–15.
Becker, B. (1958). The decline in aqueous secretion and outflow facility with age. *Am. J. Ophthalmol.* **46,** 731–736.
Becker, B., and Morton, W. R. (1966). Topical epinephrine in glaucoma suspects. *Am. J. Ophthalmol.* **62,** 272–277.
Becker, B., Pettit, T. H., and Gay, A. J. (1961). Topical epinephrine therapy of open-angle glaucoma. *Arch. Ophthalmol.* **66,** 219–225.
Bill, A. (1966). Conventional and uveo-scleral drainage of aqueous humour in the cynomolgus monkey (*Macaca irus*) at normal and high intraocular pressures. *Exp. Eye Res.* **5,** 45–54.
Bill, A. (1967a). Effects of atropine and pilocarpine on aqueous humour dynamics in cynomolgus monkeys (*Macaca irus*). *Exp. Eye Res.* **6,** 120–125.
Bill, A. (1967b). Further studies on the influence of the intraocular pressure on aqueous humor dynamics in cynomolgus monkeys. *Invest. Ophthalmol.* **6,** 364–372.
Bill, A., and Phillips, C. I. (1971). Uveoscleral drainage of aqueous humour in human eyes. *Exp. Eye Res.* **12,** 275–281.
Bill, A., and Wålinder, P.-E. (1966). The effects of pilocarpine on the dynamics of aqueous humor in a primate *(Macaca irus)*. *Invest. Ophthalmol.* **5,** 170–175.
Brechue, W. F., and Maren, T. H. (1993). A comparison between the effect of topical and systemic carbonic anhydrase inhibitors on aqueous humor secretion. *Exp. Eye Res.* **57,** 67–78.
Breinin, G. M., and Gortz, H. (1954). Carbonic anhydrase inhibitor acetazolamide (Diamox): A new approach to the therapy of glaucoma. *AMA Arch. Ophthalmol.* **52,** 333–348.
Brown, J. D., and Brubaker, R. F. (1989). A study of the relation between intraocular pressure and aqueous humor flow in the pigment dispersion syndrome. *Ophthalmology* **96,** 1468–1470.
Brubaker, R. F. (1967). Determination of episcleral venous pressure in the eye. A comparison of three methods. *Arch. Ophthalmol.* **77,** 110–114.
Brubaker, R. F. (1982). The flow of aqueous humor in the human eye. *Trans. Am. Ophthalmol. Soc.* **80,** 391–474.
Brubaker, R. F. (1991). Flow of aqueous humor in humans. *Invest. Ophthalmol. Vis. Sci.* **32,** 3145–3166.
Brubaker, R. F. (1997). Clinical evaluation of the circulation of aqueous humor. *In* "Duane's Clinical Ophthalmology" (W. Tasman and E. A. Jaeger, eds.), Vol. 3, pp. 1–11. Lippincott-Raven Publishers, Philadelphia.
Brubaker, R. F. (1998). Clinical measurements of aqueous dynamics: Implications for addressing glaucoma. *In* "The Eye's Aqueous Humor. From Secretion to Glaucoma" (M. M. Civan, ed.), Vol. 45, pp. 233–284. Academic Press, San Diego.
Brubaker, R. F., and Liesegang, T. J. (1983). Effect of trabecular photocoagulation on the aqueous humor dynamics of the human eye. *Am. J. Ophthalmol.* **96,** 139–147.
Brubaker, R. F., Nagataki, S., Townsend, D. J., Burns, R. R., Higgins, R. G., and Wentworth, W. (1981). The effect of age on aqueous humor formation in man. *Ophthalmology.* **88,** 283–287.
Brubaker, R. F., Nagataki, S., and Bourne, W. M. (1982). Effect of chronically administered timolol on aqueous humor flow in patients with glaucoma. *Ophthalmology.* **89,** 280–283.
Brubaker, R. F., Ingram, C. J., Schoff, E. O., and Nau, C. B. (2000). Comparison of the efficacy of betaxolol-brinzolamide and timolol-dorzolamide as suppressors of aqueous humor flow in human subjects. *Ophthalmology* **107,** 283–287.

Brubaker, R. F., Schoff, E. O., Nau, C. B., Carpenter, S. P., Chen, K., and VanDenburgh, A. M. (2001). Effects of AGN 192024, a new ocular hypotensive agent, on aqueous dynamics. *Am. J. Ophthalmol.* **131,** 19–24.

Bucci, M. G., Missiroli, A., Pecori, G. J., and Virno, M. (1968). The local administration of propranolo in the therapy of glaucoma. *Boll. Ocul.* **47,** 51–60.

Burke, J., Kharlamb, A., Shan, T., Runde, E., Padillo, E., Manlapaz, C., and Wheeler, L. (1995). Adrenergic and imidazoline receptor-mediated responses to UK-14,304–18 (brimonidine) in rabbits and monkeys. A species difference. *In* "The Imidazoline Receptor: Pharmacology, Functions, Ligands, and Relevance to Biology and Medicine" (D. J. Reis, P. Bousquet, and A. Parini, eds.), Vol. 763, pp. 78–95. The New York Academy of Sciences, New York.

Bursell, S.-E., Clermont, A. C., Kinsley, B. T., Simonson, D. C., Aiello, L. M., and Wolpert, H. A. (1996). Retinal blood flow changes in patients with insulin-dependent diabetes mellitus and no diabetic retinopathy. A video fluorescein angiography study. *Invest. Ophthalmol. Vis. Sci.* **37,** 886–897.

Bylund, D. B. (1988). Subtypes of α_2-adrenoceptors: Pharmacological and molecular biological evidence converge. *Trends Pharmacol. Sci.* **9,** 356–361.

Camras, C. B., and Bito, L. Z. (1981). Reduction of intraocular pressure in normal and glaucomatous primate *(Aotus trivirgatus)* eyes by topically applied prostaglandin $F_{2\alpha}$. *Curr. Eye Res.* **1,** 205–209.

Camras, C. B., and Podos, S. M. (1989). The role of endogenous prostaglandin in clinically-used and investigational glaucoma therapy. *In* "The Ocular Effects of Prostaglandins and Other Eicosanoids" (L. Z. Bito and J. Stjernschantz, eds.), Vol. 312, pp. 459–475. Alan R. Liss, Inc, New York, USA.

Camras, C. B., Wang, R.-F., and Podos, S. M. (1990). Effect of pilocarpine applied before or after prostaglandin $F_{2\alpha}$ on IOP in glaucomatous monkey eyes. *Invest. Ophthalmol. Vis. Sci.* **31,** 150. Abstract.

Camras, C. B., Chacko, D. M., Schlossman, A., and Posner, A. (1996). Posner-Schlossman syndrome. *In* "Ocular Infection & Immunity" (J. S. Prepose, G. N. Holland, and K. R. Wilhelmus, eds.), pp. 529–537. Mosby-Year Book, Inc., St. Louis.

Camras, C. B., Haecker, N. R., Zhan, G., and Toris, C. B. (2003). Aqueous humor dynamics in patients with pigment dispersion syndrome. *Invest. Ophthalmol. Vis. Sci.* **44,** 2207.

Carlson, K. H., McLaren, J. W., Topper, J. E., and Brubaker, R. F. (1987). Effect of body position on intraocular pressure and aqueous flow. *Invest. Ophthalmol. Vis. Sci.* **28,** 1346–1352.

Cavallotti, C., Pescosolido, N., Pescosolido, V., and Iannetti, G. (1999). Determination of dopamine D_1 receptors in the human uveo scleral tissue by light microscope autoradiography. *Int. Ophthalmol.* **23,** 171–179.

Christiansen, G. A., Nau, C. B., McLaren, J. W., and Johnson, D. H. (2004). Mechanism of ocular hypotensive action of bimatoprost (Lumigan) in patients with ocular hypertension or glaucoma. *Ophthalmology* **111,** 1658–1662.

Chu, E., Chu, T.-C., and Potter, D. E. (2000). Mechanisms and sites of ocular action of 7-hydroxy-2-dipropylaminotetralin: A dopamine$_3$ receptor agonist. *J. Pharmacol. Exp. Ther.* **293,** 710–716.

Chu, E., Socci, R., and Chu, T.-C. (2004). PD128,907 induces ocular hypotension in rabbits: involvement of D_2/D_3 dopamine receptors and brain natriuretic peptide. *J. Ocul. Pharmacol. Ther.* **20,** 15–23.

Coakes, R. L., and Brubaker, R. F. (1978). The mechanism of timolol in lowering intraocular pressure. In the normal eye. *Arch. Ophthalmol.* **96,** 2045–2048.

Coulangeon, L. M., Sole, M., Menerath, J. M., and Sole, P. (1990). Aqueous humor flow measured by fluorophotometry. A comparative study of the effect of various beta-blocker eyedrops in patients with ocular hypertension. *Ophtalmologie* **4,** 156–161.

Crawford, K. S., and Kaufman, P. L. (1987). Pilocarpine antagonizes prostaglandin F_2-induced ocular hypotension in monkeys. Evidence for enhancement of uveoscleral outflow by prostaglandin $F_{2\alpha}$. *Arch. Ophthalmol.* **105,** 1112–1116.

Crawford, K., Kaufman, P. L., and Gabelt, B. T. (1987). Effects of topical $PGF_{2\alpha}$ on aqueous humor dynamics in cynomolgus monkeys. *Curr. Eye Res.* **6,** 1035–1044.

Dailey, B. A., Brubaker, R. F., and Bourne, W. M. (1982). The effects of timolol maleate and acetazolamide on the rate of aqueous formation in normal human subjects. *Am. J. Ophthalmol.* **93,** 232–237.

Darier, A. (1900). De l'extrait de capsule surrénales en thérapeutique oculaire. *La Clinique Ophthalmologique.* **6,** 141–143.

Diestelhorst, M., and Krieglstein, G. K. (1992). Does aqueous humor secretion decrease with age? *Int. Ophthalmol.* **16,** 305–309.

Diestelhorst, M., and Krieglstein, G. K. (1994). Physiologic aging in aqueous humor minute volume of the human eye. *Ophthalmologe* **91,** 575–577.

Dinslage, S., Hueber, A., Diestelhorst, M., and Krieglstein, G. K. (2004). The influence of latanoprost 0.005% on aqueous humor flow and outflow facility in glaucoma patients: a double-masked placebo-controlled clinical study. *Graefes Arch. Clin. Exp. Ophthalmol.* **242,** 654–660.

Ethier, C. R., Read, A. T., and Chan, D. W.-H. (2006). Effects of latrunculin-B on outflow facility and trabecular meshwork structure in human eyes. *Invest. Ophthalmol. Vis. Sci.* **47,** 1991–1998.

Friedenwald, J. S. (1957). Tonometer calibration. An attempt to remove discrepancies found in the 1954 calibration scale for Schiotz tonometers. *Trans. Am. Acad. Ophthalmol. Otolaryngol.* **61,** 108–123.

Friström, B., and Nilsson, S. E. G. (1993). Interaction of PhXA41, a new prostaglandin analogue, with pilocarpine. A study on patients with elevated intraocular pressure. *Arch. Ophthalmol.* **111,** 662–665.

Funk, R. H. W., Gehr, J., and Rohen, J. W. (1996). Short-term hemodynamic changes in episcleral arteriovenous anastomoses correlate with venous pressure and IOP changes in the albino rabbit. *Curr. Eye Res.* **15,** 87–93.

Gaasterland, D., Kupfer, C., Ross, K., and Gabelnick, H. L. (1973). Studies of aqueous humor dynamics in man. III. Measurements in young normal subjects using norepinephrine and isoproterenol. *Invest. Ophthalmol.* **12,** 267–279.

Gaasterland, D., Kupfer, C., and Ross, K. (1975). Studies of aqueous humor dynamics in man. IV. Effects of pilocarpine upon measurements in young normal volunteers. *Invest. Ophthalmol.* **14,** 848–853.

Gaasterland, D., Kupfer, C., Milton, R., Ross, K., McCain, L., and MacLellan, H. (1978). Studies of aqueous humour dynamics in man. VI. Effect of age upon parameters of intraocular pressure in normal human eyes. *Exp. Eye Res.* **26,** 651–656.

Gabelt, B. T., and Kaufman, P. L. (2005). Changes in aqueous humor dynamics with age and glaucoma. *Prog. Retin. Eye Res.* **24,** 612–637.

Gabelt, B. T., Gottanka, J., Lütjen-Drecoll, E., and Kaufman, P. L. (2003). Aqueous humor dynamics and trabecular meshwork and anterior ciliary muscle morphologic changes with age in rhesus monkeys. *Invest. Ophthalmol. Vis. Sci.* **44,** 2118–2125.

Gabelt, B. T., Okka, M., Dean, T. R., and Kaufman, P. L. (2005). Aqueous humor dynamics in monkeys after topical R-DOI. *Invest. Ophthalmol. Vis. Sci.* **46,** 4691–4696.

Galin, M. A., Baras, I., and Glenn, J. (1966). L-epinephrine and intraocular pressure. *Invest. Ophthalmol.* **5,** 120–124.

Garner, L. L., Johnstone, W. W., Ballintine, E. J., and Carroll, M. E. (1959). Effect of 2% levo-rotary epinephrine on the intraocular pressure of the glaucomatous eye. *AMA Arch. Ophthalmol.* **62,** 230–238.

Gaul, G. R., Will, N. J., and Brubaker, R. F. (1989). Comparison of a noncardioselective beta-adrenoceptor blocker and a cardioselective blocker in reducing aqueous flow in humans. *Arch. Ophthalmol.* **107,** 1308–1311.
Gharagozloo, N. Z., and Brubaker, R. F. (1991). Effect of apraclonidine in long-term timolol users. *Ophthalmology* **98,** 1543–1546.
Gharagozloo, N. Z., Relf, S. J., and Brubaker, R. F. (1988). Aqueous flow is reduced by the alpha-adrenergic agonist, apraclonidine hydrochloride (ALO 2145). *Ophthalmology* **95,** 1217–1220.
Gharagozloo, N. Z., Baker, R. H., and Brubaker, R. F. (1992). Aqueous dynamics in exfoliation syndrome. *Am. J. Ophthalmol.* **114,** 473–478.
Gordon, M. O., Beiser, J. A., Brandt, J. D., Heuer, D. K., Higginbotham, E. J., Johnson, C. A., Keltner, J. P., Miller, J. P., Parrish, R. K. I., Wilson, M. R., Kass, M. A., and for the Ocular Hypertension Treatment Study Group (2002). The Ocular Hypertension Treatment Study: Baseline factors that predict the onset of primary open-angle glaucoma. *Arch. Ophthalmol.* **120,** 714–720.
Grant, W. M. (1951). Clinical measurements of aqueous outflow. *AMA Arch. Ophthalmol.* **46,** 113–131.
Grant, W. M., and Trotter, R. R. (1954). Diamox (acetazolamide) in treatment of glaucoma. *A. M. A. Arch. Ophthalmol.* **51,** 735–739.
Haidle, A. M., and Myers, A. G. (2004). An enantioselective, modular, and general route to the cytochalasins: Synthesis of L-696,474 and cytochalasin B. *Proc. Natl. Acad. Sci. USA* **101,** 12048–12053.
Hammer, T., Schlötzer-Schrehardt, U., and Naumann, G. O. H. (2001). Unilateral or asymmetric pseudoexfoliation syndrome? An ultrastructural study. *Arch. Ophthalmol.* **119,** 1023–1031.
Haria, M., and Spencer, C. M. (1996). Unoprostone (isopropyl unoprostone). *Drugs Aging* **9,** 213–218.
Hart, C. T., and Weatherill, J. R. (1968). Gonioscopy and tonography in glaucomatocyclitic crises. *Br. J. Ophthalmol.* **52,** 682–687.
Hayashi, M., Yablonski, M. E., Boxrud, C., Fong, N., Berger, C., and Jovanovic, L. J. (1989a). Decreased formation of aqueous humour in insulin-dependent diabetic patients. *Br. J. Ophthalmol.* **73,** 621–623.
Hayashi, M., Yablonski, M. E., and Novack, G. D. (1989b). Trabecular outflow facility determined by fluorophotometry in human subjects. *Exp. Eye Res.* **48,** 621–625.
Hayreh, S. S., Zimmerman, M. B., Podhajsky, P., and Alward, W. L. M. (1994). Nocturnal arterial hypotension and its role in optic nerve head and ocular ischemic disorders. *Am. J. Ophthalmol.* **117,** 603–624.
Higgins, R. G., and Brubaker, R. F. (1980). Acute effect of epinephrine on aqueous humor formation in the timolol-treated normal eye as measured by fluorophotometry. *Invest. Ophthalmol. Vis. Sci.* **19,** 420–423.
Higgitt, A. C. (1956). Secondary glaucoma. *Trans. Opthal. Soc. UK* **76,** 73–82.
Honjo, M., Tanihara, H., Inatani, M., Kido, N., Sawamura, T., Yue, B. Y. J., Narumiya, S., and Honda, Y. (2001). Effects of rho-associated protein kinase inhibitor Y-27632 on intraocular pressure and outflow facility. *Invest. Ophthalmol. Vis. Sci.* **42,** 137–144.
Inoue, T., Yokoyoma, T., and Koike, H. (2001a). The effect of angiotensin II on uveoscleral outflow in rabbits. *Curr. Eye Res.* **23,** 139–143.
Inoue, T., Yokoyoma, T., Mori, Y., Sasaki, Y., Hosokawa, T., Yanagisawa, H., and Koike, H. (2001b). The effect of topical CS-088, an angiotensin AT_1 receptor antagonist, on intraocular pressure and aqueous humor dynamics in rabbits. *Curr. Eye Res.* **23,** 133–138.
Johnson, D., Liesegang, T. J., and Brubaker, R. F. (1983). Aqueous humor dynamics in Fuchs' uveitis syndrome. *Am. J. Ophthalmol.* **95,** 783–787.

Johnson, D. H. (1997). The effect of cytochalasin D on outflow facility and the trabecular meshwork of the human eye in perfusion organ culture. *Invest. Ophthalmol. Vis. Sci.* **38,** 2790–2799.

Johnson, D. H., and Brubaker, R. F. (1982). Dynamics of aqueous humor in the syndrome of exfoliation with glaucoma. *Am. J. Ophthalmol.* **93,** 629–634.

Johnson, T. V., Fan, S., Camras, C. B., and Toris, C. B. (2008). Aqueous humor dynamics in exfoliation syndrome. *Arch. Ophthalmol.* **126,** 1–7.

Johnstone, M., Tanner, D., Chau, B., and Kopecky, K. (1980). Concentration-dependent morphologic effects of cytochalasin B in the aqueous outflow system. *Invest. Ophthalmol. Vis. Sci.* **19,** 835–841.

Joseph, J., and Grierson, I. (1994). Anterior seqment changes in glaucoma. *In* "Pathobiology of Ocular Disease; a Dynamic Approach" (A. Garner and G. K. Klintworth, eds.), pp. 433–468. Marcel Dekker, New York.

Kacere, R. D., Dolan, J. W., and Brubaker, R. F. (1992). Intravenous epinephrine stimulates aqueous formation in the human eye. *Invest. Ophthalmol. Vis. Sci.* **33,** 2861–2865.

Kalina, P. H., Shetlar, D. J., Lewis, R. A., Kullerstrand, L. J., and Brubaker, R. F. (1988). 6-Amino-2-benzothiazole-sulfonamide. The effect of a topical carbonic anhydrase inhibitor on aqueous humor formation in the normal human eye. *Ophthalmology* **95,** 772–777.

Kass, M. A., Becker, B., and Kolker, A. E. (1973). Glaucomatocyclitic crisis and primary open-angle glaucoma. *Am. J. Ophthalmol.* **75,** 668–673.

Katz, I. M., Hubbard, W. A., Getson, A. J., and Gould, A. L. (1976). Intraocular pressure decrease in normal volunteers following timolol ophthalmic solution. *Invest. Ophthalmol.* **15,** 489–492.

Kaufman, P. L., and Bárány, E. H. (1977). Cytochalasin B reversibly increases outflow facility in the eye of the cynomolgus monkey. *Invest. Ophthalmol. Vis. Sci.* **16,** 47–53.

Kaufman, P. L., and Erickson, K. A. (1982). Cytochalasin B and D dose-outflow facility response relationships in the cynomolgus monkey. *Invest. Ophthalmol. Vis. Sci.* **23,** 646–650.

Kaufman, P. L., and Gabelt, B. T. (1993). Aging, accommodation and outflow facility. *In* "Basic Aspects of Glaucoma Research III" (E. Lütjen-Drecoll, ed.), pp. 257–274. Schattauer, Stuttgart.

Kaufman, P. L., and Gabelt, B. T. (1997). Direct, indirect, and dual-action parasympathetic drugs. *In* "Textbook of Ocular Pharmacology" (T. J. Zimmerman, K. S. Kooner, M. Sharir, and R. D. Fechtner, eds.), pp. 221–238. Lippincott-Raven, Philadelphia.

Kent, A. R., Vroman, D. T., Thomas, T. J., Herbert, R. L., and Crosson, C. E. (1999). Interaction of pilocarpine with latanoprost in patients with glaucoma and ocular hypertension. *J. Glaucoma* **8,** 257–262.

Khan, A. R., and Brubaker, R. F. (1993). Aqueous humor flow and flare in patients with myotonic dystrophy. *Invest. Ophthalmol. Vis. Sci.* **34,** 3131–3139.

Kiel, J. W. (1995). The effect of arterial pressure on the ocular pressure-volume relationship in the rabbit. *Exp. Eye Res.* **60,** 267–278.

Koskela, T., and Brubaker, R. F. (1991a). Apraclonidine and timolol. Combined effects in previously untreated normal subjects. *Arch. Ophthalmol.* **109,** 804–806.

Koskela, T., and Brubaker, R. F. (1991b). The nocturnal suppression of aqueous humor flow in humans is not blocked by bright light. *Invest. Ophthalmol. Vis. Sci.* **32,** 2504–2506.

Kronfeld, P. C. (1963). Society Proceedings, Glaucoma Research Conference. Del Monte, CA (D.J. Lyle ed.). *Am.J.Ophthalmol.* **55,** 821–832.

Kubo, C., and Suzuki, R. (1992). Involvement of prejunctional alpha 2-adrenoceptor in bovine ciliary muscle movement. *J. Ocul. Pharmacol.* **8,** 225–231.

Kupfer, C., and Ross, K. (1971). Studies of aqueous humor dynamics in man. I. Measurements in young normal subjects. *Invest. Ophthalmol.* **10,** 518–522.

Kupfer, C., Gaasterland, D., and Ross, K. (1971). Studies of aqueous humor dynamics in man. II. Measurements in young normal subjects using acetazolamide and L-epinephrine. *Invest. Ophthalmol.* **10,** 523–533.

Lane, J. T., Toris, C. B., Nakhle, S. N., Chacko, D. M., Wang, Y.-L., and Yablonski, M. E. (2001). Acute effects of insulin on aqueous humor flow in patients with type 1 diabetes. *Am. J. Ophthalmol.* **132,** 321–327.

Larsson, L.-I. (2001). Aqueous humor flow in normal human eyes treated with brimonidine and timolol, alone and in combination. *Arch. Ophthalmol.* **119,** 492–495.

Larsson, L.-I., Rettig, E. S., Sheridan, P. T., and Brubaker, R. F. (1993). Aqueous humor dynamics in low-tension glaucoma. *Am. J. Ophthalmol.* **116,** 590–593.

Larsson, L.-I., Pach, J. M., and Brubaker, R. F. (1995a). Aqueous humor dynamics in patients with diabetes mellitus. *Am. J. Ophthalmol.* **120,** 362–367.

Larsson, L.-I., Rettig, E. S., and Brubaker, R. F. (1995b). Aqueous flow in open-angle glaucoma. *Arch. Ophthalmol.* **113,** 283–286.

Lee, D. A., Brubaker, R. F., and Nagataki, S. (1983). Acute effect of thymoxamine on aqueous humor formation in the epinephrine-treated normal eye as measured by fluorophotometry. *Invest. Ophthalmol. Vis. Sci.* **24,** 165–168.

Lee, P.-Y., Serle, J. B., Podos, S. M., and Severin, C. (1992). Time course of the effect of UK 14304–18 (brimonidine tartrate) on rabbit uveoscleral outflow. *Invest. Ophthalmol. Vis. Sci.* **33,** 1118. Abstract.

Lim, K. S., Nau, C. B., O'Byrne, M. M., Hodge, D., Toris, C. B., McLaren, J. W., and Johnson, D. H. (2008). Comparative study on the mechanism of action of bimatoprost, latanoprost and travoprost in healthy subjects. A double-masked, placebo-controlled, randomized, paired comparison, 4-session crossover study. *Ophthalmology* **115,** 790–795.

Lindén, C., and Alm, A. (1997a). Effects on intraocular pressure and aqueous flow of various dose regimens of latanoprost in human eyes. *Acta Ophthalmol.* **75,** 412–415.

Lindén, C., and Alm, A. (1997b). Physostigmine increases aqueous humor production in human eyes. *Curr. Eye Res.* **16,** 1166–1170.

Liu, J. H. K., Kripke, D. F., Hoffman, R. E., Twa, M. D., Loving, R. T., Rex, K. M., Gupta, N., and Weinreb, R. N. (1998). Nocturnal elevation of intraocular pressure in young adults. *Invest. Ophthalmol. Vis. Sci.* **39,** 2707–2712.

Liu, J. H. K., Kripke, D. F., Hoffman, R. E., Twa, M. D., Loving, R. T., Rex, K. M., Lee, B. L., Mansberger, S. L., and Weinreb, R. N. (1999a). Elevation of human intraocular pressure at night under moderate illumination. *Invest. Ophthalmol. Vis. Sci.* **40,** 2439–2442.

Liu, J. H. K., Kripke, D. F., Twa, M. D., Hoffman, R. E., Mansberger, S. L., Rex, K. M., Girkin, C. A., and Weinreb, R. N. (1999b). Twenty-four-hour pattern of intraocular pressure in the aging population. *Invest. Ophthalmol. Vis. Sci.* **40,** 2912–2917.

Liu, J. H. K., Zhang, X., Kripke, D. F., and Weinreb, R. N. (2003). Twenty-four-hour intraocular pressure pattern associated with early glaucomatous changes. *Invest. Ophthalmol. Vis. Sci.* **44,** 1586–1590.

Luksch, A., Polak, K., Matulla, B., Dallinger, S., Kapiotis, S., Rainer, G., Wolzt, M., and Schmetterer, L. (2001). Glucose and insulin exert additive ocular and renal vasodilator effects on healthy humans. *Diabetologia* **44,** 95–103.

Lütjen-Drecoll, E., and Tamm, E. (1988). Morphological study of the anterior segment of cynomolgus monkey eyes following treatment with prostaglandin $F_2\alpha$. *Exp. Eye Res.* **47,** 761–769.

Mancino, R., Cerulli, L., Ricci, A., and Amenta, F. (1992). Direct demonstration of dopamine D_1-like receptor sites in the ciliary body of the rabbit eye by light microscope autoradiography. *Naunyn Schmiedebergs Arch. Pharmacol.* **346,** 644–648.

Mansheim, B. J. (1953). Aqueous outflow measurements by continuous tonometry in some unusual forms of glaucoma. *AMA Arch. Ophthalmol.* **50,** 580–587.

Maren, T. H. (1987). Carbonic anhydrase: General perspectives and advances in glaucoma research. *Drug Dev. Res.* **10,** 255–276.

Maren, T. H., Haywood, J. R., Chapman, S. K., and Zimmerman, T. J. (1977). The pharmacology of methazolamide in relation to the treatment of glaucoma. *Invest. Ophthalmol. Vis. Sci.* **16,** 730–742.

Martin, X. D., Brennan, M. C., and Lichter, P. R. (1988). Serotonin in human aqueous humor. *Ophthalmology* **95,** 1221–1226.

Martin, P. B., Vila, P. C. F., Martinez, T. M. P., and Peréz, D. A. (1992a). A fluorophotometric study on the aqueous humor dynamics in primary open angle glaucoma. *Int. Ophthalmol.* **16,** 311–314.

Martin, X. D., Malina, H. Z., Brennan, M. C., Hendrickson, P. H., and Lichter, P. R. (1992b). The ciliary body—the third organ found to synthesize indoleamines in humans. *Eur. J. Ophthalmol.* **2,** 67–72.

Masuda, K., Izawa, Y., and Mishima, S. (1975). Prostaglandins and glaucomato-cyclitic crisis. *Jpn. J. Ophthalmol.* **19,** 368–375.

Maus, T. L., Larsson, L.-I., McLaren, J. W., and Brubaker, R. F. (1997). Comparison of dorzolamide and acetazolamide as suppressors of aqueous humor flow in humans. *Arch. Ophthalmol.* **115,** 45–49.

Maus, T. L., Nau, C., and Brubaker, R. F. (1999). Comparison of the early effects of brimonidine and apraclonidine as topical ocular hypotensive agents. *Arch. Ophthalmol.* **117,** 586–591.

McCannel, C. A., Heinrich, S. R., and Brubaker, R. F. (1992). Acetazolamide but not timolol lowers aqueous humor flow in sleeping humans. *Graefes Arch. Clin. Exp. Ophthalmol.* **230,** 518–520.

McLaren, J. W., Herman, D. C., Brubaker, R. F., Nau, C. B., Wayman, L. L., Ciarniello, M. G., Rosignoli, M. T., and Dionisio, P. (2003). Effect of ibopamine on aqueous humor production in normotensive humans. *Invest. Ophthalmol. Vis. Sci.* **44,** 4853–4858.

McLaughlin, M. A., Toris, C. B., Brooks, D. E., Scott, D. A., Cage, T., Fan, S., Zhan, G.-L., Ollivier, F. J., Camras, C. B., and Holt, W. F. (2004). Effects of rho kinase inhibitors on intraocular pressure (IOP) and aqueous humor dynamics in rabbits and monkeys. *Invest. Ophthalmol. Vis. Sci.* **45,** 3534.

Michel, M. C., and Ernsberger, P. (1992). Keeping an eye on the I site: Imidazoline-preferring receptors. *Trends Pharmacol. Sci.* **13,** 369–370.

Mishima, H. K., Kiuchi, Y., Takamatsu, M., Rácz, P., and Bito, L. Z. (1997). Circadian intraocular pressure management with latanoprost: diurnal and nocturnal intraocular pressure reduction and increased uveoscleral outflow. *Surv. Ophthalmol.* **41,** S139–S144.

Mori, M., Araie, M., Sakurai, M., and Oshika, T. (1992). Effects of pilocarpine and tropicamide on blood-aqueous barrier permeability in man. *Invest. Ophthalmol. Vis. Sci.* **33,** 416–423.

Nagataki, S. (1977). Effects of adrenergic drugs on aqueous humor dynamics in man. *J. Jpn. Ophthalmol. Soc.* **81,** 1795–1800.

Nagataki, S., and Brubaker, R. F. (1981). Early effect of epinephrine on aqueous formation in the normal human eye. *Ophthalmology* **88,** 278–282.

Nagataki, S., and Brubaker, R. F. (1982). Effect of pilocarpine on aqueous humor formation in human beings. *Arch. Ophthalmol.* **100,** 818–821.

Nagataki, S., and Mishima, S. (1976). Aqueous humor dynamics in glaucomato-cyclitic crisis. *Invest. Ophthalmol.* **15,** 365–370.

Nelson, W. L., Fraunfelder, F. T., Sills, J. M., Arrowsmith, J. B., and Kuritsky, J. N. (1986). Adverse respiratory and cardiovascular events attributed to timolol ophthalmic solution, 1987–1985. *Am. J. Ophthalmol.* **102,** 606–611.

Nilsson, S. F. E., Sperber, G. O., and Bill, A. (1989). The effect of prostaglandin $F_{2\alpha}$-1-isopropyl ester ($PGF_{2\alpha}$-IE) on uveoscleral outflow. *In* "The Ocular Effects of Prostaglandins and Other Eicosanoids" (L. Z. Bito and J. Stjernschantz, eds.), Vol. 312, pp. 429–436. Alan R. Liss, Inc, New York, USA.

Ocklind, A. (1998). Effect of latanoprost on the extracellular matrix of the ciliary muscle. A study on cultured cells and tissue sections. *Exp. Eye Res.* **67,** 179–191.

Okka, M., Tian, B., and Kaufman, P. L. (2004). Effects of latrunculin B on outflow facility, intraocular pressure, corneal thickness, and miotic and accommodative responses to pilocarpine in monkeys. *Trans. Am. Ophthalmol. Soc.* **102,** 251–257.

Ortiz, G. J., Cook, D. J., Yablonski, M. E., Masonson, H., and Harmon, G. (1988). Effect of cold air on aqueous humor dynamics in humans. *Invest. Ophthalmol. Vis. Sci.* **29,** 138–140.

Patelska, B., Greenfield, D. S., Liebmann, J. M., Wand, M., Kushnick, H., and Ritch, R. (1997). Latanoprost for uncontrolled glaucoma in a compassionate case protocol. *Am. J. Ophthalmol.* **124,** 279–286.

Pederson, J. E. (1986). Ocular hypotony. *Trans. Ophthalmol. Soc. UK* **105,** 220–226.

Peterson, J. A., Tian, B., Bershadsky, A. D., Volberg, T., Gangnon, R. E., Spector, I., Geiger, B., and Kaufman, P. L. (1999). Latrunculin-A increases outflow facility in the monkey. *Invest. Ophthalmol. Vis. Sci.* **40,** 931–941.

Peterson, J. A., Tian, B., McLaren, J. W., Hubbard, W. C., Geiger, B., and Kaufman, P. L. (2000). Latrunculins' effects on intraocular pressure, aqueous humor flow, and corneal endothelium. *Invest. Ophthalmol. Vis. Sci.* **41,** 1749–1758.

Phillips, C. I., Howitt, G., and Rowlands, D. J. (1967). Propranolol as ocular hypotensive agent. *Br. J. Ophthalmol.* **51,** 222–226.

Piltz, J. R., Stone, R. A., Boike, S., Everitt, D. E., Shusterman, N. H., Audet, P., Zariffa, N., and Jorkasky, D. K. (1998). Fenoldopam, a selective dopamine-1 receptor agonist, raises intraocular pressure in males with normal intraocular pressure. *J. Ocul. Pharmacol. Ther.* **14,** 203–216.

Plange, N., Kaup, M., Daneljan, L., Predel, H. G., Remky, A., and Arend, O. (2006). 24-h blood pressure monitoring in normal tension glaucoma: Night-time blood pressure variability. *J. Hum. Hypertens* **20,** 137–142.

Posner, A., and Schlossman, A. (1948). Syndrome of unilateral recurrent attacks of glaucoma with cyclitic symptoms. *Arch. Ophthalmol.* **39,** 517–535.

Potter, D. E., Crosson, C. E., Heath, A. R., and Ogidigben, M. (1990). Functional evidence for heterogeneity of ocular α_2-adrenoceptors. *Proc. West. Pharmacol. Soc.* **33,** 257–260.

Poyer, J. F., Millar, C., and Kaufman, P. L. (1995). Prostaglandin $F_{2\alpha}$ effects on isolated rhesus monkey ciliary muscle. *Invest. Ophthalmol. Vis. Sci.* **36,** 2461–2465.

Prünte, C., Nuttli, I., Markstein, R., and Kohler, C. (1997). Effects of dopamine D-1 and D-2 receptors on intraocular pressure in conscious rabbits. *J. Neural Transm.* **104,** 111–123.

Ramirez, M., Davidson, E. A., Luttenauer, L., Elena, P.-P., Cumin, F., Mathis, G. A., and De Gasparo, M. (1996). The renin-angiotensin system in the rabbit eye. *J. Ocul. Pharmacol. Ther.* **12,** 299–312.

Rao, P. V., Deng, P.-F., Kumar, J., and Epstein, D. L. (2001). Modulation of aqueous humor outflow facility by the Rho kinase-specific inhibitor Y-27632. *Invest. Ophthalmol. Vis. Sci.* **42,** 1029–1037.

Read, A. T., Chan, D. W.-H., and Ethier, C. R. (2007). Actin structure in the outflow tract of normal and glaucomatous eyes. *Exp. Eye Res.* **84,** 214–226.

Reiss, G. R., and Brubaker, R. F. (1983). The mechanism of betaxolol, a new ocular hypotensive agent. *Ophthalmology* **90,** 1369–1372.

Reiss, G. R., Lee, D. A., Topper, J. E., and Brubaker, R. F. (1984). Aqueous humor flow during sleep. *Invest. Ophthalmol. Vis. Sci.* **25,** 776–778.

Reitsamer, H. A., and Kiel, J. W. (2002a). A rabbit model to study orbital venous pressure, intraocular pressure, and ocular hemodynamics simultaneously. *Invest. Ophthalmol. Vis. Sci.* **43,** 3728–3734.

Reitsamer, H. A., and Kiel, J. W. (2002b). Effects of dopamine on ciliary blood flow, aqueous production, and intraocular pressure in rabbits. *Invest. Ophthalmol. Vis. Sci.* **43,** 2697–2703.

Reitsamer, H. A., Posey, M., and Kiel, J. W. (2006). Effects of a topical $\alpha 2$ adrenergic agonist on ciliary blood flow and aqueous production in rabbits. *Exp. Eye Res.* **82,** 405–415.

Rosenberg, L. F., Krupin, T., Tang, L.-Q., Hong, P. H., and Ruderman, J. M. (1998). Combination of systemic acetazolamide and topical dorzolamide in reducing intraocular pressure and aqueous humor formation. *Ophthalmology* **105,** 88–93.

Rosenthal, R., Choritz, L., Schlott, S., Bechrakis, N. E., Jaroszewski, J., Wiederholt, M., and Thieme, H. (2005). Effects of ML-7 and Y-27632 on carbachol- and endothelin-1-induced contraction of bovine trabecular meshwork. *Exp. Eye Res.* **80,** 837–845.

Sabanay, I., Tian, B., Gabelt, B. T., Geiger, B., and Kaufman, P. L. (2006). Latrunculin B effects on trabecular meshwork and corneal endothelial morphology in monkeys. *Exp. Eye Res.* **82,** 236–246.

Sagara, T., Gaton, D. D., Lindsey, J. D., Gabelt, B. T., Kaufman, P. L., and Weinreb, R. N. (1999). Topical prostaglandin $F_{2\alpha}$ treatment reduces collagen types I, III, and IV in the monkey uveoscleral outflow pathway. *Arch. Ophthalmol.* **117,** 794–801.

Sakurai, M., Araie, M., Oshika, T., Mori, M., Masuda, K., Ueno, R., and Takase, M. (1991). Effects of topical application of UF-021, a novel prostaglandin derivative, on aqueous humor dynamics in normal human eyes. *Jpn. J. Ophthalmol.* **35,** 156–165.

Savaskan, E., Löffler, K. U., Meier, F., Müller-Spahn, F., Flammer, J., and Meyer, P. (2004). Immunohistochemical localization of angiotensin-converting enzyme, angiotensin II and AT1 receptor in human ocular tissues. *Ophthalmic Res.* **36,** 312–320.

Schadlu, R., Maus, T. L., Nau, C. B., and Brubaker, R. F. (1998). Comparison of the efficacy of apraclonidine and brimonidine as aqueous suppressants in humans. *Arch. Ophthalmol.* **116,** 1441–1444.

Schenker, H. I., Yablonski, M. E., Podos, S. M., and Linder, L. (1981). Fluorophotometric study of epinephrine and timolol in human subjects. *Arch. Ophthalmol.* **99,** 1212–1216.

Schlecht, L. P., and Brubaker, R. F. (1988). The effects of withdrawal of timolol in chronically treated glaucoma patients. *Ophthalmology* **95,** 1212–1216.

Schlötzer-Schrehardt, U., and Naumann, G. O. H. (1995). Trabecular meshwork in pseudoexfoliation syndrome with and without open-angle glaucoma. A morphometric, ultrastructural study. *Invest. Ophthalmol. Vis. Sci.* **36,** 1750–1764.

Schmetterer, L., Müller, M., Fasching, P., Diepolder, C., Gallenkamp, A., Zanaschka, G., Findl, K., Strenn, K., Mensik, C., Tschernko, E., Eichler, H.-G., and Wolzt, M. (1997). Renal and ocular hemodynamic effects of insulin. *Diabetes* **46,** 1868–1874.

Schneider, T. L., and Brubaker, R. F. (1991). Effect of chronic epinephrine on aqueous humor flow during the day and during sleep in normal healthy subjects. *Invest. Ophthalmol. Vis. Sci.* **32,** 2507–2510.

Sharif, N. A., Kelly, C. R., and McLaughlin, M. (2006). Human trabecular meshwork cells express functional serotonin-2A ($5HT_{2A}$) receptors: Role in IOP reduction. *Invest. Ophthalmol. Vis. Sci.* **47,** 4001–4010.

Shin, D. H., McCracken, M. S., Bendel, R. E., Pearlman, R., Juzych, M. S., Hughes, B. A., Schulz, C., Kim, C., and Baek, N. H. (1999). The additive effect of latanoprost to maximum-tolerated medications with low-dose, high-dose and no pilocarpine therapy. *Ophthalmology* **106,** 386–390.

Sihota, R., Lakshmaiah, N. C., Walia, K. B., Sharma, S., Pailoor, J., and Agarwal, H. C. (2001). The trabecular meshwork in acute and chronic angle closure glaucoma. *Indian J. Ophthalmol.* **49,** 255–259.

Skorobohach, B. J., Ward, D. A., and Hendrix, D. V. H. (2003). Effects of oral administration of methazolamide on intraocular pressure and aqueous humor flow rate in clinically normal dogs. *Am. J. Vet. Res.* **64,** 183–187.

Spivey, B. E., and Armaly, M. F. (1963). Tonographic findings in glaucomatocyclitic crises. *Am. J. Ophthalmol.* **55,** 47–51.

Sponsel, W. E., Paris, G., Trigo, Y., Pena, M., Weber, A., Sanford, K., and McKinnon, S. (2002). Latanoprost and brimonidine: therapeutic and physiologic assessment before and after oral nonsteroidal anti-inflammatory therapy. *Am. J. Ophthalmol.* **133,** 11–18.

Stein, A., Pinke, R., Krupin, T., Glabb, E., Podos, S. M., Serle, J., and Maren, T. H. (1983). The effect of topically administered carbonic anhydrase inhibitors on aqueous humor dynamics in rabbits. *Am. J. Ophthalmol.* **95,** 222–228.

Stjernschantz, J. W. (2001). From $PGF_{2\alpha}$-isopropyl ester to latanoprost: A review of the development of xalatan: The Proctor Lecture. *Invest. Ophthalmol. Vis. Sci.* **42,** 1134–1145.

Stjernschantz, J., Selén, G., Ocklind, A., and Resul, B. (1998). Effects of latanoprost and related prostaglandin analogues. *In* "Uveoscleral Outflow. Biology and Clinical Aspects" (A. Alm and R. N. Weinreb, eds.), pp. 57–72. Mosby International Limited, London, UK.

Sugar, H. S. (1965). Heterochromia iridis with special consideration of its relation to cyclitic disease. *Am. J. Ophthalmol.* **60,** 1–18.

Svedbergh, B., Lütjen-Drecoll, E., Ober, M., and Kaufman, P. L. (1978). Cytochalasin B-induced structural changes in the anterior ocular segment of the cynomolgus monkey. *Invest. Ophthalmol. Vis. Sci.* **17,** 718–734.

Tamm, E., Rittig, M., and Lütjen-Drecoll, E. (1990). Elektronenmikroskopische und immunohistochemische untersuchungen zur augendrucksenkenden wirkung von prostaglandin $F_{2\alpha}$. [Electron microscopy and immunohistochemical studies of the intraocular pressure lowering effect of prostaglandin $F_{2\alpha}$. *Fortschr. Ophthalmol.* **87,** 623–629.

Tetsuka, H., Tsuchisaka, H., Kin, K., Takahashi, Y., and Takase, M. (1992). A mechanism for reducing intraocular pressure in normal volunteers using UF-021, a prostaglandin-related compound. *J. Jpn. Ophthalmol. Soc.* **96,** 496–500.

Tian, B., Geiger, B., Epstein, D. L., and Kaufman, P. L. (2000). Cytoskeletal involvement in the regulation of aqueous humor outflow. *Invest. Ophthalmol. Vis. Sci.* **41,** 619–623.

Tobin, A. B., Unger, W., and Osborne, N. N. (1988). Evidence for the presence of serotonergic nerves and receptors in the iris-ciliary body complex of the rabbit. *J. Neurosci.* **8,** 3713–3721.

Topper, J. E., and Brubaker, R. F. (1985). Effects of timolol, epinephrine, and acetazolamide on aqueous flow during sleep. *Invest. Ophthalmol. Vis. Sci.* **26,** 1315–1319.

Toris, C. B., and Pederson, J. E. (1985). Effect of intraocular pressure on uveoscleral outflow following cyclodialysis in the monkey eye. *Invest. Ophthalmol. Vis. Sci.* **26,** 1745–1749.

Toris, C. B., Camras, C. B., and Yablonski, M. E. (1993). Effects of PhXA41, a new prostaglandin $F_{2\alpha}$ analog, on aqueous humor dynamics in human eyes. *Ophthalmology* **100,** 1297–1304.

Toris, C. B., Gleason, M. L., Camras, C. B., and Yablonski, M. E. (1995a). Effects of brimonidine on aqueous humor dynamics in human eyes. *Arch. Ophthalmol.* **113,** 1514–1517.

Toris, C. B., Tafoya, M. E., Camras, C. B., and Yablonski, M. E. (1995b). Effects of apraclonidine on aqueous humor dynamics in human eyes. *Ophthalmology* **102,** 456–461.

Toris, C. B., Yablonski, M. E., Wang, Y.-L., and Hayashi, M. (1995c). Prostaglandin A_2 increases uveoscleral outflow and trabecular outflow facility in the cat. *Exp. Eye Res.* **61,** 649–657.

Toris, C. B., Camras, C. B., and Yablonski, M. E. (1999a). Acute versus chronic effects of brimonidine on aqueous humor dynamics in ocular hypertensive patients. *Am. J. Ophthalmol.* **128,** 8–14.

Toris, C. B., Yablonski, M. E., Wang, Y.-L., and Camras, C. B. (1999b). Aqueous humor dynamics in the aging human eye. *Am. J. Ophthalmol.* **127,** 407–412.

Toris, C. B., Zhan, G.-L., Wang, Y.-L., Zhao, J., McLaughlin, M. A., Camras, C. B., and Yablonski, M. E. (2000). Aqueous humor dynamics in monkeys with laser-induced glaucoma. *J. Ocul. Pharmacol. Ther.* **16,** 19–27.

Toris, C. B., Zhan, G.-L., Zhao, J., Camras, C. B., and Yablonski, M. E. (2001). Potential mechanism for the additivity of pilocarpine and latanoprost. *Am. J. Ophthalmol.* **131,** 722–728.

Toris, C. B., Koepsell, S. A., Yablonski, M. E., and Camras, C. B. (2002). Aqueous humor dynamics in ocular hypertensive patients. *J. Glaucoma* **11,** 253–258.

Toris, C. B., Zhan, G.-L., Yablonski, M. E., and Camras, C. B. (2004a). Effects on aqueous flow of dorzolamide combined with either timolol or acetazolamide. *J. Glaucoma* **13,** 210–215.

Toris, C. B., Zhan, G., and Camras, C. B. (2004b). Increase in outflow facility with unoprostone treatment in ocular hypertensive patients. *Arch. Ophthalmol.* **122,** 1782–1787.

Toris, C. B., Zhan, G., Fan, S., Dickerson, J. E., Landry, T. A., Bergamini, M. V. W., and Camras, C. B. (2007). Effects of travoprost on aqueous humor dynamics in patients with elevated intraocular pressure. *J. Glaucoma* **16,** 189–195.

Townsend, D. J., and Brubaker, R. F. (1980). Immediate effect of epinephrine on aqueous formation in the normal human eye as measured by fluorophotometry. *Invest. Ophthalmol. Vis. Sci.* **19,** 256–266.

Tsukamoto, H., and Larsson, L.-I. (2004). Aqueous humor flow in normal human eyes treated with brimonidine and dorzolamide, alone and in combination. *Arch. Ophthalmol.* **122,** 190–193.

Uehata, M., Ishizaki, T., Satoh, H., Ono, T., Kawahara, T., Morishita, T., Tamakawa, H., Yamagami, J., Inui, J., Maekawa, M., and Narumiya, S. (1997). Calcium sensitization of smooth muscle mediated by a Rho-associated protein kinase in hypertension. *Nature* **389,** 990–994.

Van Alphen, G. W. H. M., Wilhelm, P. B., and Elsenfeld, P. W. (1977). The effect of prostaglandins on the isolated internal muscles of the mammalian eye, including man. *Doc. Ophthalmol.* **42,** 397–415.

Veglio, F., De Sanctis, U., Schiavone, D., Cavallone, S., Mulatero, P., Grignolo, F. M., and Chiandussi, L. (1998). Evaluation of serotonin levels in human aqueous humor. *Ophthalmologica* **212,** 160–163.

Villumsen, J., and Alm, A. (1992). Effect of the prostaglandin $F_{2\alpha}$ analogue PhXA41 in eyes treated with pilocarpine and timolol. *Invest. Ophthalmol. Vis. Sci.* **33,** 1248. Abstract.

Virno, M., Gazzaniga, A., Taverniti, L., Pecori, G. J., and De Gregorio, F. (1992). Dopamine, dopaminergic drugs and ocular hypertension. *Int. Ophthalmol.* **16,** 349–353.

Virno, M., Taverniti, L., De Gregorio, F., Sedran, L., and Longo, F. (1997). Increase in aqueous humor production following D1 receptors activation by means of ibopamine. *Int. Ophthalmol.* **20,** 141–146.

Vogh, B. P., Godman, D. R., and Maren, T. H. (1989). Measurement of aqueous humor flow following scleral injection of sulfacetamide as marker: effect of methazolamide, timolol, and pilocarpine. *J. Ocul. Pharmacol.* **5,** 293–302.

Waki, M., Yoshida, Y., Oka, T., and Azuma, M. (2001). Reduction of intraocular pressure by topical administration of an inhibitor of the Rho-associated protein kinase. *Curr. Eye Res.* **22,** 470–474.

Walker, S. D., Brubaker, R. F., and Nagataki, S. (1982). Hypotony and aqueous humor dynamics in myotonic dystrophy. *Invest. Ophthalmol. Vis. Sci.* **22,** 744–751.

Wang, Y.-L., Hayashi, M., Yablonski, M. E., and Toris, C. B. (2002). Effects of multiple dosing of epinephrine on aqueous humor dynamics in human eyes. *J. Ocul. Pharmacol. Ther.* **18,** 53–63.

Wang, R.-F., Podos, S. M., Mittag, T. W., and Yokoyoma, T. (2005). Effect of CS-088, an angiotensin AT_1 receptor antagonist, on intraocular pressure in glaucomatous monkey eyes. *Exp. Eye Res.* **80,** 629–632.

Wayman, L., Larsson, L.-I., Maus, T., Alm, A., and Brubaker, R. F. (1997). Comparison of dorzolamide and timolol as suppressors of aqueous humor flow in humans. *Arch. Ophthalmol.* **115,** 1368–1371.

Wayman, L. L., Larsson, L.-I., Maus, T. L., and Brubaker, R. F. (1998). Additive effect of dorzolamide on aqueous humor flow in patients receiving long-term treatment with timolol. *Arch. Ophthalmol.* **116,** 1438–1440.

Weekers, R., Prijot, E., and Gustin, J. (1954a). Mesure de la résistance à l'écoulement de l'humeur aqueuse au moyen du tonemètre électronique. 6e partie: Mode d'action de l'adrénaline dans le glaucome chronique. *Ophthalmologica* **128,** 213–217.

Weekers, R., Prijot, E., and Gustin, J. (1954b). Recent advances and future prospects in the medical treatment of ocular hypertension. *Br. J. Ophthalmol.* **38,** 742–746.

Weinreb, R. N., and Liu, J. H. (2006). Nocturnal rhythms of intraocular pressure. *Arch. Ophthalmol.* **124,** 269–270.

Weinreb, R. N., Kashiwagi, K., Kashiwagi, F., Tsukahara, S., and Lindsey, J. D. (1997). Prostaglandins increase matrix metalloproteinase release from human ciliary smooth muscle cells. *Invest. Ophthalmol. Vis. Sci.* **38,** 2772–2780.

Weinreb, R. N., Toris, C. B., Gabelt, B. T., Lindsey, J. D., and Kaufman, P. L. (2002). Effects of prostaglandins on the aqueous humor outflow pathways. *Surv. Ophthalmol.* **47,** S53–S64.

Wentworth, W. O., and Brubaker, R. F. (1981). Aqueous humor dynamics in a series of patients with third neuron Horner's syndrome. *Am. J. Ophthalmol.* **92,** 407–415.

Wiederholt, M., Bielka, S., Schweig, F., Lütjen-Drecoll, E., and Lepple-Wienhues, A. (1995). Regulation of outflow rate and resistance in the perfused anterior segment of the bovine eye. *Exp. Eye Res.* **61,** 223–234.

Wiederholt, M., Thieme, H., and Stumpff, F. (2000). The regulation of trabecular meshwork and ciliary muscle contractility. *Prog. Retin. Eye Res.* **19,** 271–295.

Woodward, D. F., Krauss, A. H. P., Chen, J., Liang, Y., Li, C., Protzman, C. E., Bogardus, A., Chen R., Kedzie, K. M., Krauss, H. A., Gil, D. W., Kharlamb, A., *et al.* (2003). Pharmacological characterization of a novel antiglaucoma agent, bimatoprost (AGN 192024). *J. Pharmacol. Exp. Ther.* **305,** 772–785.

Yablonski, M. E., Cook, D. J., and Gray, J. (1985a). A fluorophotometric study of the effect of argon laser trabeculoplasty on aqueous humor dynamics. *Am. J. Ophthalmol.* **99,** 579–582.

Yablonski, M. E., Gallin, P., and Shapiro, D. (1985b). Effect of oxygen on aqueous humor dynamics in rabbits. *Invest. Ophthalmol. Vis. Sci.* **26,** 1781–1784.

Yablonski, M. E., Zimmerman, T. J., Waltman, S. R., and Becker, B. (1978). A fluorophotometric study of the effect of topical timolol on aqueous humor dynamics. *Exp. Eye Res.* **27,** 135–142.

Yablonski, M. E., Novack, G. D., Burke, P. J., Cook, D. J., and Harmon, G. (1987). The effect of levobunolol on aqueous humor dynamics. *Exp. Eye Res.* **44,** 49–54.

Zeimer, R. C. (1989). Episcleral venous pressure. *In* "The Glaucomas" (R. Ritch, M. B. Shields, and T. Krupin, eds.), pp. 249–255. The C. V. Mosby Company, St. Louis.

Zeimer, R. C., Gieser, D. K., Wilensky, J. T., Noth, J. M., Mori, M. M., and Odunukwe, E. E. (1983). A practical venomanometer. Measurement of episcleral venous pressure and assessment of the normal range. *Arch. Ophthalmol.* **101,** 1447–1449.

Ziai, N., Dolan, J. W., Kacere, R. D., and Brubaker, R. F. (1993). The effects on aqueous dynamics of PhXA41, a new prostaglandin $F_{2\alpha}$ analogue, after topical application in normal and ocular hypertensive human eyes. *Arch. Ophthalmol.* **111,** 1351–1358.

Zimmerman, T. J., and Kaufman, H. E. (1977). Timolol, dose response and duration of action. *Arch. Ophthalmol.* **95,** 605–607.

CHAPTER 9

Effects of Circulatory Events on Aqueous Humor Inflow and Intraocular Pressure

Herbert A. Reitsamer* and Jeffrey W. Kiel†

*Department of Ophthalmology, Paracelsus Medical University, Salzburg, Austria
†Department of Ophthalmology, University of Texas Health Science Center at San Antonio, San Antonio, Texas 78229

I. OVERVIEW

This chapter will review the role of the ocular circulations in intraocular pressure (IOP) homeostasis. Historically, glaucoma was considered an ischemic disease caused by elevated IOP; however, it is now evident that ocular hypertension is not a prerequisite for glaucoma and that the progressive death of retinal ganglion cells can arise from multiple etiologies. Nonetheless, high IOP is a primary risk factor for glaucoma and lowering IOP is its primary treatment (Heijl *et al.*, 2002). Since either situation has the potential to affect the ocular circulations, we will begin with a brief description of IOP

Current Topics in Membranes, Volume 62 1063-5823/08 $35.00

DOI: 10.1016/S1063-5823(08)00409-2

effects on ocular blood flow. From there, we will address the more complex and less well-understood topic of ocular blood flow effects on IOP, both in terms of transient IOP responses to changes in ocular blood volume and the roles of the ciliary and episcleral circulations in aqueous formation and outflow which set steady-state IOP.

II. IOP EFFECTS ON OCULAR BLOOD FLOW

The arteriovenous pressure gradient provides the energy that moves blood through the network of vessels in any tissue. As in all tissues, the cardiac output and the total peripheral resistance of the systemic circulation set the arterial pressure for the ocular circulations. Because the arteries supplying the eye are relatively inaccessible, the arterial pressure is usually measured at a site remote from the eye and the ocular arterial pressure is estimated by correcting for the hydrostatic column effect; for example in a sitting human subject, ocular arterial pressure is roughly 67% of that measured at the brachial artery. By contrast, the venous pressure for the ocular circulations is more complex than in most tissues. Because veins are thin-walled and deform easily, their caliber is primarily determined by the transmural pressure gradient, that is, the distending pressure inside the vein minus the compressing pressure outside the vein. In most tissues, the pressure outside the veins is negligible; in the eye, the pressure outside the veins is the IOP. Consequently, the ocular veins behave as Starling resistors (Patterson and Starling, 1914) such that the pressure inside the veins just before they exit the eye slightly exceeds the IOP, so long as the IOP is less than the arterial pressure. Otherwise, the veins would collapse and flow cease (Duke-Elder, 1926; Bill, 1962, 1963; Moses, 1963). In practical terms, this means that the IOP is the effective venous pressure in the eye, and the arteriovenous pressure gradient (commonly called the ocular perfusion pressure, PP) is the mean arterial pressure (MAP, at eye level) minus the IOP. Thus, if the IOP is raised while holding the MAP constant at different levels, blood flow will decrease and go to zero when IOP equals MAP. This behavior is shown for one ocular circulation in Fig. 1; the flow behavior in the other ocular circulations is qualitatively similar (Kiel and van Heuven, 1995). Autoregulatory mechanisms that act to maintain blood flow constant despite modest changes in PP have been demonstrated in all ocular circulations (Alm and Bill, 1972a, 1973; Kiel and Shepherd, 1992; Kiel *et al.*, 2001; Weigert *et al.*, 2005). The important point is that ocular hypertension has the potential to cause ischemia in any ocular circulation, particularly if coupled with systemic hypotension.

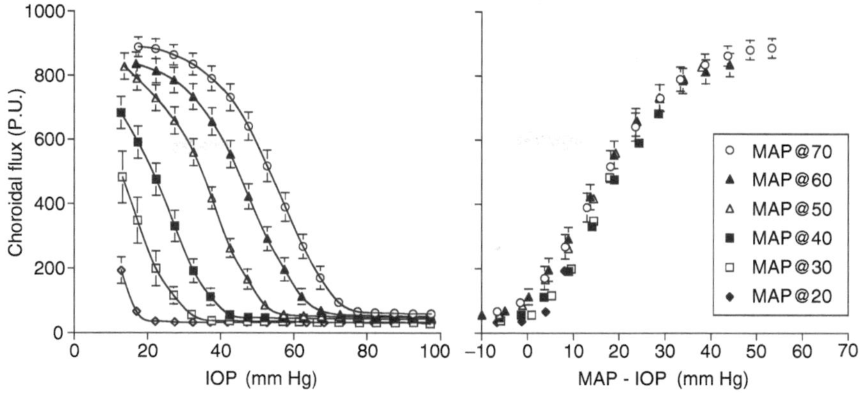

FIGURE 1 Relationship between choroidal blood flow (choroidal flux), intraocular pressure (IOP) and perfusion pressure (MAP – IOP). The left panel shows the response of choroidal blood flow to increasing intraocular IOP at different fixed values of MAP (right panel). Choroidal blood flow declines as IOP approaches the MAP and stops when IOP exceeds the MAP. The right panel shows the same dataset with choroidal blood flow plotted against perfusion pressure; the curves superimpose demonstrating that MAP – IOP is the effective ocular perfusion pressure. [Modified from Kiel and Van Heuven (1995)].

III. OCULAR BLOOD FLOW EFFECTS ON IOP

Historically, two conceptual models have served as frameworks for understanding IOP. One model views the IOP in terms of the ocular pressure–volume relationship, which is an exponential function of the total volume of the ocular contents and the elastic properties of the ocular coats. This approach provides the theoretical basis for indentation tonometry and tonography (Friedenwald, 1937; Grant, 1950). The other model considers the steady-state IOP in hydraulic terms as an ohmic function of aqueous flow and outflow resistance (Barany, 1963). This approach provides a theoretical basis for understanding ocular hypertension and hypotony as well as the current pharmacological and surgical manipulations of IOP. Each model is useful for understanding different aspects of the IOP (Figure 2).

If the elasticity of the eye is constant, any change in IOP must involve a change in ocular volume. Under normal conditions, the primary contributors to the total ocular volume are the vitreous, lens, aqueous, and blood. The vitreous and lens volumes are relatively stable over time and seldom have an acute influence on the IOP. By contrast, the volumes of aqueous and blood are more labile and account for most variations in the IOP. Changes in aqueous volume result from transient imbalances in aqueous production and outflow. Similarly, changes in ocular blood volume occur during transient imbalances in the flow of blood into and out of the eye. Most of the

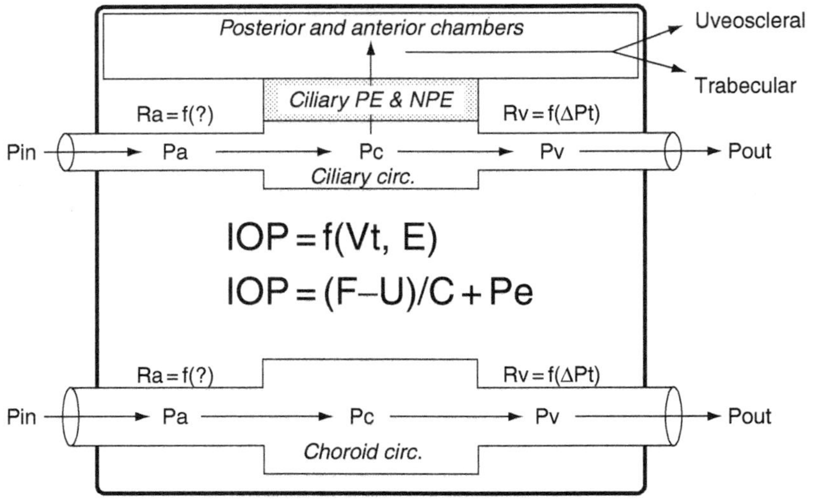

FIGURE 2 Schematic illustration of the factors generating the IOP. (P_{in}, extraocular arterial pressure; P_a, intraocular arterial pressure; P_c, intraocular capillary pressure; P_v, intraocular venous pressure; P_{out}, extraocular venous pressure; R_a, arterial resistance; R_v, venous resistance; ΔP_t, transmural pressure gradient; PE, pigmented epithelium; NPE, nonpigmented epithelium; V_t, total ocular volume; E, elastance or "rigidity" of the ocular coats; F, aqueous flow; C, outflow conductance or "facility"; P_e, episcleral venous pressure) (Kiel, 1998). R_a in the choroid and ciliary circulations is designated with a question mark to reflect the ill-defined nature of the local and neurohumoral inputs involved.

ocular blood volume is in the choroid, the highly vascularized tissue between the retina and sclera supplied by the short posterior ciliary arteries and drained by the vortex veins.

Because most IOP measurement techniques are discontinuous, the effect of blood flow (or more specifically, blood volume) on IOP generally goes unnoticed. However, with continuous IOP measurement it is readily apparent that ocular blood volume contributes to IOP. For example, blood pressure synchronous changes in IOP are evident in the movement of the mires during applanation tonometry and clearly detected manometrically (Fig. 3, top). The IOP pulse is caused by pulsatile arterial inflow and steady venous outflow giving rise to fluctuations in ocular blood volume (Fig. 3, bottom). The IOP pulse is used to estimate the pulsatile component of ocular blood flow (Silver and Farrell, 1994).

Another demonstration of the ocular blood volume contribution to IOP is seen when the heart stops (Fig. 4). When the heart is stopped abruptly, the arterial pressure falls toward the mean circulatory filling pressure (Guyton *et al.*, 1954), blood flow into the eye ceases, and the resident blood volume drains from the eye resulting in a rapid net decrease in blood volume and an equally rapid fall in IOP. In the authors' experience, for anesthetized rabbits

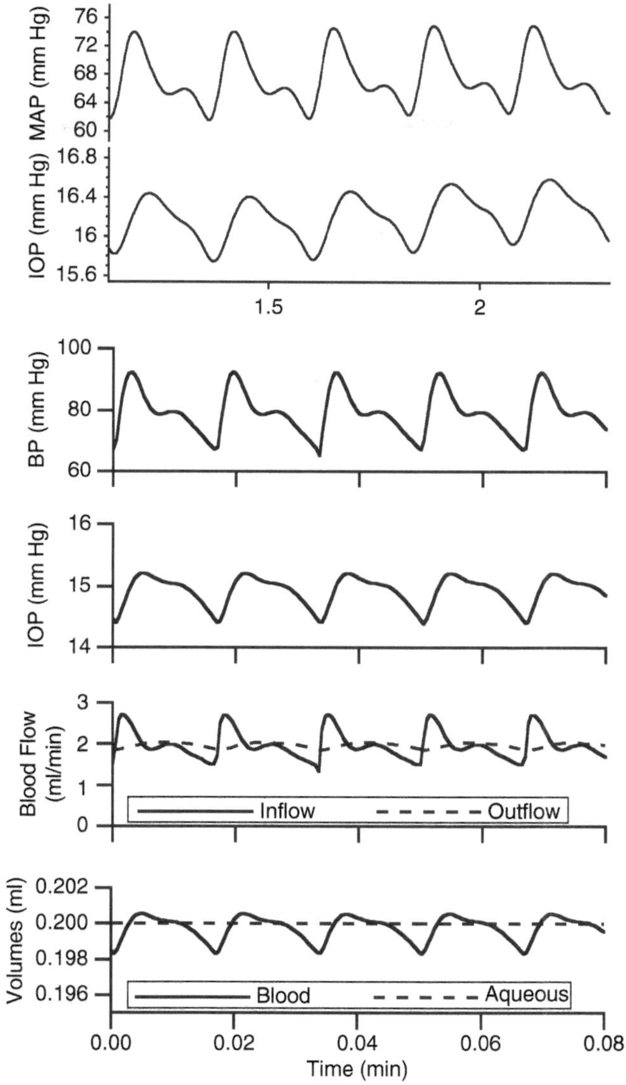

FIGURE 3 (Top) Blood pressure and IOP recorded by direct cannulation of the central ear artery and vitreous compartment in an anesthetized rabbit. (Bottom) Computer simulation showing cardiac synchronous IOP pulsations due to fluctuations in blood volume caused by pulsatile inflow of blood and steady outflow (Kiel, 1998).

under control conditions, the IOP immediately before and after death are typically 15 and 7 mm Hg, respectively. This occurs in seconds, which is too fast for aqueous dynamics to play a role in the IOP decrease.

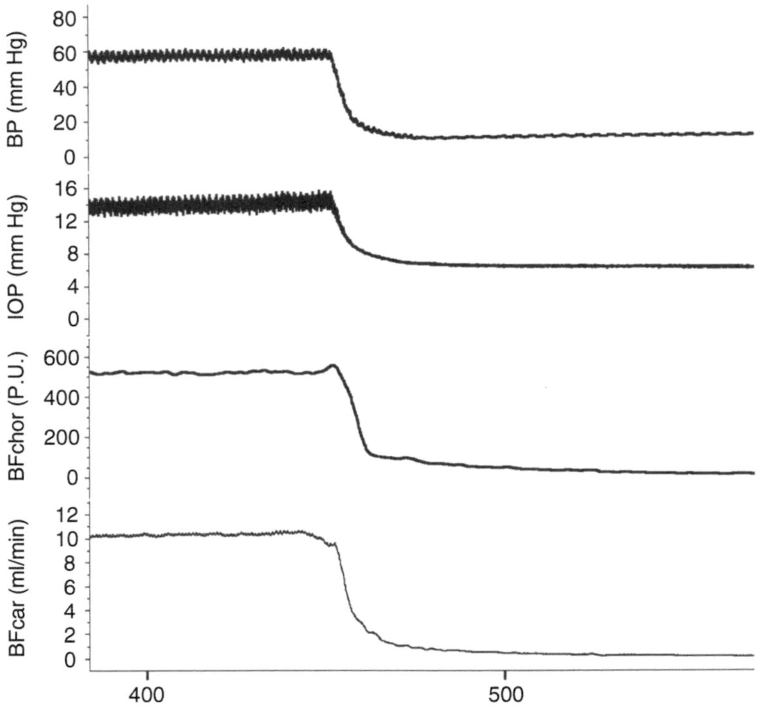

FIGURE 4 IOP falls immediately when the heart is stopped with an overdose of pentobarbital (100 mg/kg, iv) in a deeply anesthetized rabbit. Continued venous outflow without corresponding arterial inflow results in the net loss of blood volume responsible for the fall in IOP. (Time in seconds.)

It should be noted that ocular blood volume is relatively well regulated under normal conditions, at least in response to changes in arterial pressure (Kiel, 1994). As shown in Fig. 5, an acute increase in arterial pressure induced mechanically by occluding the descending aorta elicits only a small increase in IOP due to choroidal vasoconstriction under control conditions. Although both autoregulatory myogenic (Kiel, 1994) and autonomic neural mechanisms (Bill *et al.*, 1977) have been proposed to explain this response, the regulatory mechanism responsible for this choroidal vasoconstriction is beyond the scope of this chapter. What is important to note, however, is that when this choroidal regulation is abolished with a systemic vasodilator, a similar acute elevation of arterial pressure can elicit a significantly larger increase in IOP. It is unclear whether choroidal regulation is impaired in glaucoma (Weigert *et al.*, 2005), but if such large spikes in IOP occur during normal variations in arterial pressure they could contribute to optic neuropathy.

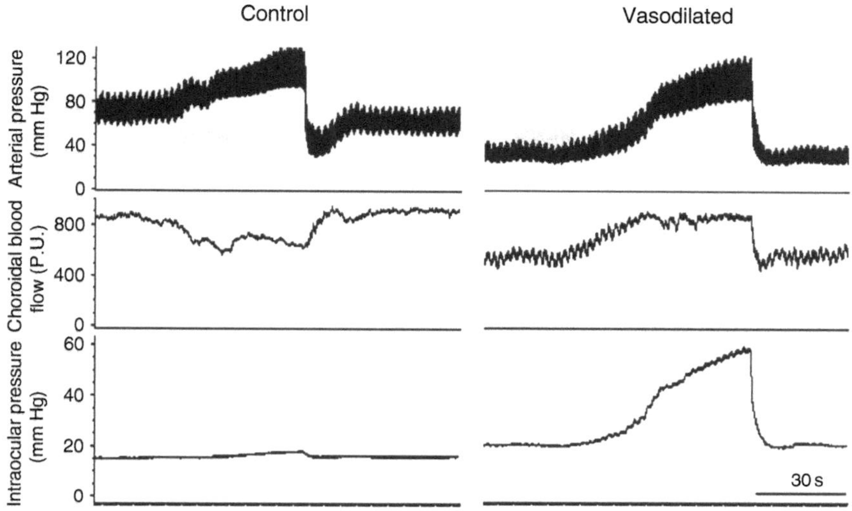

FIGURE 5 IOP responses to acute mechanically induced increases in arterial pressure in an anesthetized rabbit. Raising arterial pressure to 110 mm Hg elicits a modest increase in IOP under control conditions (left) and a much larger increase in IOP when choroidal regulation is impaired with a systemic vasodilator (right). The vasodilated IOP response is the largest we have recorded (Kiel, 1994).

As in tonography, a sustained pressure-induced increase in ocular blood volume does not produce a sustained increase in IOP as shown in Fig. 6. Instead, the elevated IOP increases the pressure gradient for aqueous outflow, which causes a compensatory decrease of aqueous volume and IOP gradually returns to baseline (Kiel, 1994). If the increase in blood volume is small, the compensation is relatively quick, whereas compensation for a larger increase in blood volume, as occurs when choroidal regulation is impaired, takes longer. In either situation, IOP falls below baseline when the arterial pressure-induced distention of the vasculature is abruptly ended, reflecting the compensatory loss of aqueous volume, which is gradually restored by continued aqueous production, which returns IOP to baseline. Such a compensatory volume shift was noted almost 100 years ago by Duke-Elder who observed a marked shallowing of the anterior chamber and rise of IOP to 80–90 mm Hg upon ligation of the vortex veins in anesthetized dogs, the most extreme method to cause choroidal engorgement.

Tonometry and tonography are based on the ocular pressure–volume (P–V) relationship. Here too, the effect of blood volume on IOP is evident. In the anesthetized rabbit, step increases in volume when arterial pressure is held at different levels give different P–V relationships, which are in turn different from that obtained in postmortem eyes (Fig. 7; Kiel, 1995). The

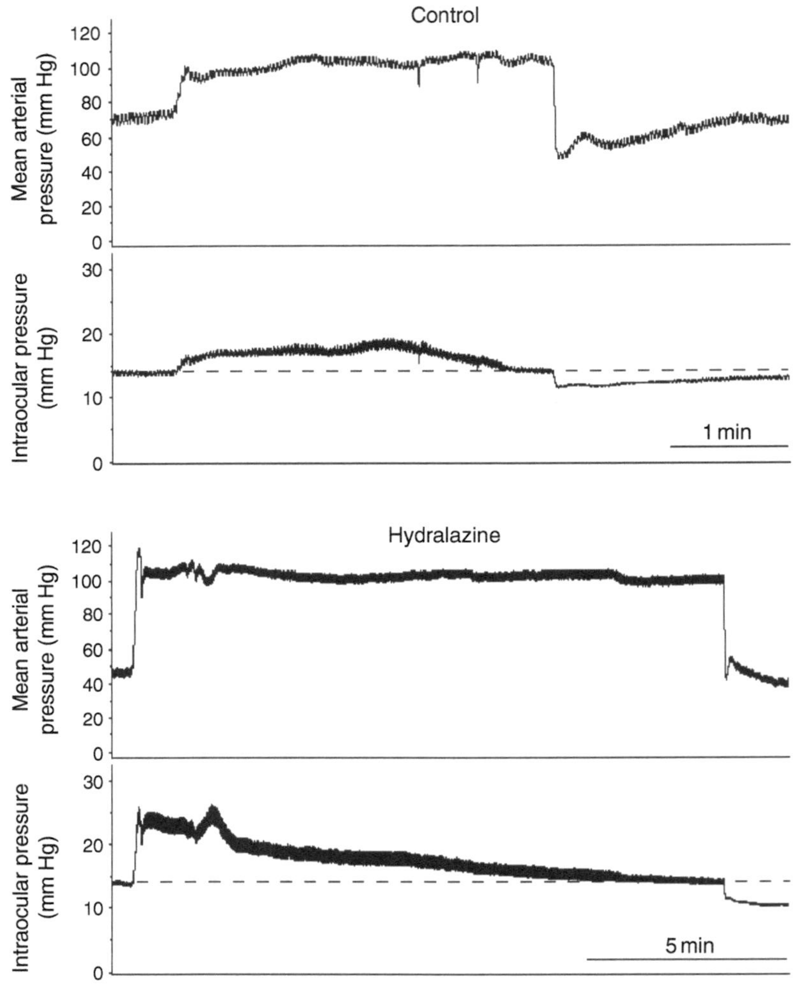

FIGURE 6 IOP responses to sustained increases in arterial pressure under control and vasodilated conditions in anesthetized rabbits. Raising arterial pressure mechanically to 110 mm Hg under control conditions elicits a relatively small increase in IOP that returns to baseline relatively quickly; raising arterial pressure to the same level under vasodilated conditions results in a larger increase in IOP that takes longer to return to baseline. Restoration of baseline arterial pressure ends vascular engorgement and the undershoot of IOP reveals the compensatory lose of aqueous volume that returned IOP to baseline during the arterial pressure elevation (Kiel, 1994).

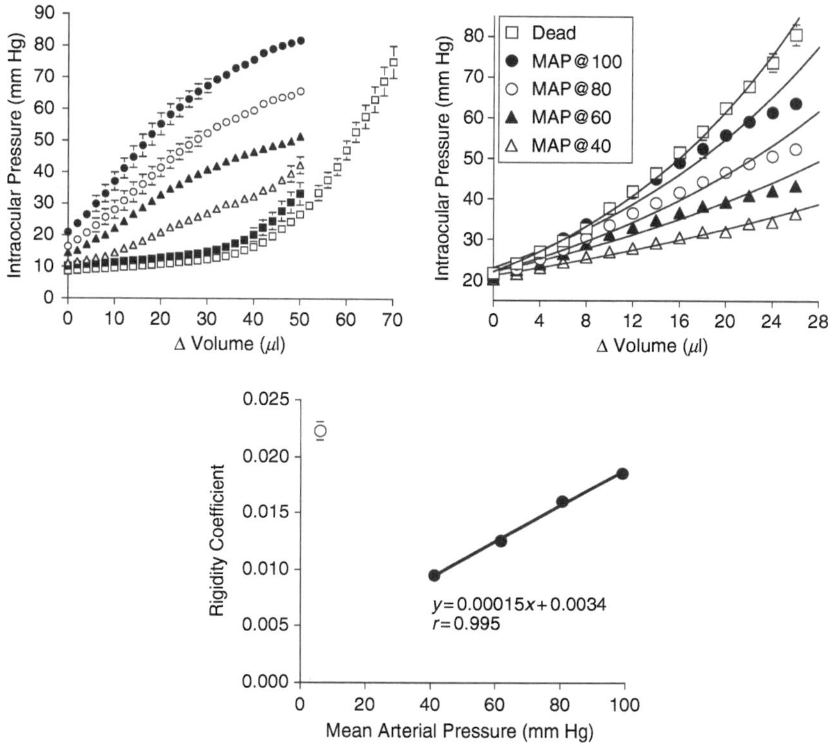

FIGURE 7 Effect of arterial pressure on the ocular pressure–volume relationship and ocular rigidity in anesthetized rabbits (Kiel, 1995).

original Friedenwald tables used in tonometry and tonography were based on enucleated eyes refilled with saline to achieve a normal IOP. This procedure precluded any blood volume buffering of the IOP response to subsequent saline injections. However, as Fig. 7 shows, the P–V relationship and the ocular rigidity coefficient are significantly dependent on the arterial pressure distending the ocular circulations.

IV. CILIARY BLOOD FLOW AND AQUEOUS PRODUCTION

A prevalent assumption in glaucoma pharmacology is that drugs that reduce aqueous production act directly on the ciliary epithelium, either by interfering with neural stimulation (e.g., prejunctional inhibition of norepinephrine release by α2-adrenergic agonists), antagonism of neurohumoral stimulation (e.g., adrenergic receptor binding by β-adrenergic antagonists), or by altering key intracellular enzymes underlying the ionic transport mechanisms that drive fluid flux across the bilayer (e.g., carbonic anhydrase inhibitors). Not often considered is the possibility of an indirect vascular mechanism, although many of these drugs are known to have vasoactive effects on ciliary blood flow (Van Buskirk *et al.*, 1990).

Because ciliary blood flow is difficult to measure, this is perhaps not surprising. Moreover, the literature suggests that alterations in ciliary blood flow have little effect on aqueous production under normal conditions. Much of that evidence comes from studies of pseudofacility, an index of the ability of elevated IOP to decrease aqueous production that was once considered a potentially significant source of error in tonographic measurements of outflow facility (Barany, 1963; Bill and Barany, 1966). Initial estimates of pseudofacility suggested that aqueous production was quite sensitive to increased IOP (e.g., 0.13μl/min/mm Hg) (Bill and Barany, 1966), but later estimates were revised downward (e.g., 0.02–0.06μl/min/mm Hg) (Bill, 1971; Kupfer, 1971; Carlson *et al.*, 1987), with some authors arguing that it was insignificant or perhaps an artifact (Moses *et al.*, 1985). Since raising IOP decreases the ocular PP and potentially decreases ciliary blood flow, it was reasonable to assume that aqueous production was insensitive to changes in ciliary blood flow. However, this assumption was rarely checked, and the few studies that measured the ciliary blood flow response to raised IOP found ciliary blood flow was autoregulated (Alm and Bill, 1972b; Kiel *et al.*, 2001). Figure 8 shows examples of ciliary autoregulation in the anesthetized rabbit during stepwise increases in IOP and during acute ramp increases and decreases in arterial pressure.

Pseudofacility was premised on the idea that raising the IOP diminished the pressure gradient for the ultrafiltration component of aqueous production. Thus, as it became clear that pseudofacility was negligible, it bolstered the view that passive ultrafiltration plays at most a permissive role in aqueous production, and that active ionic transport is primarily responsible for the fluid flux across the ciliary epithelial bilayer (Bill, 1973; Brubaker, 1991a). Little changed since then is the current view that aqueous starts as an ultrafiltrate of plasma into the ciliary stroma driven by a favorable Starling equilibrium (i.e., the balance of hydrostatic and oncotic pressures across the blood vessel wall), followed by fluid flux across the bilayer driven by an osmotic gradient established by active ionic transport (Gabelt *et al.*, 2006). As noted by Bill (1973), epithelial active transport overcomes the normal shifts in the stromal Starling equilibrium, and so ultrafiltration is not particularly sensitive to changes in ciliary blood flow. However, what remain unresolved are the epithelial metabolic requirements to sustain active transport and how those requirements are provided by ciliary blood flow. In other words, what is the minimum ciliary blood flow needed to deliver the oxygen and nutrients to meet the metabolic demands of aqueous production? To address this issue it is necessary to measure ciliary blood flow, which is a difficult challenge.

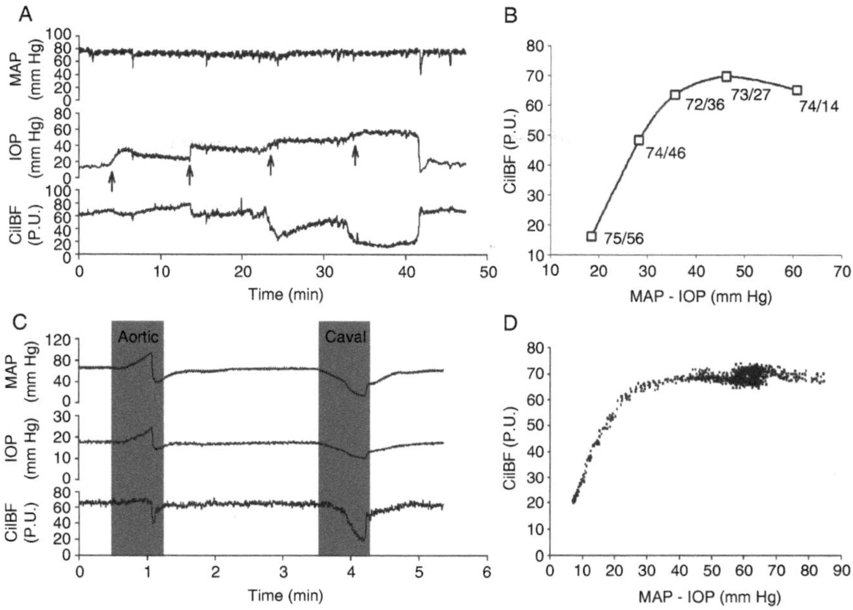

FIGURE 8 Ciliary blood flow (CilBF) responses to changes in perfusion pressure. (A) Perfusion pressure was changed by step increases (arrows) in IOP, while mean arterial pressure (MAP) was left unchanged. (B) Pressureflow relationship calculated from the tracings shown in Fig. 7A. At each point MAP and IOP are shown (as MAP/IOP). (C) Manipulation of arterial pressure by occlusions of the descending thoracic aorta (aortic) and the inferior vena cava (caval). (D) Pressureflow relationship for the ascending and descending ramp of the aortic and caval occlusions shown in Fig. 7C. Figure 7B and D show evidence of ciliary blood flow autoregulation at perfusion pressures above 30–35 mm Hg.

A. Ciliary Body Blood Supply

The ciliary body is supplied by branches off the major arterial circle of the iris which is fed by the long posterior ciliary arteries and, in some species, from the anterior ciliary arteries. The vessels of the ciliary processes divide into zones (Morrison *et al.*, 1987a; Funk and Rohen, 1990; Lutjen-Drecoll and Rohen, 1994). One zone is at the anterior base of the processes and consists of arterioles and capillaries that drain into a venular system separate from the other zones. This zone is the boundary between the nonfenestrated capillaries of the iris and the fenestrated capillaries of the ciliary processes. A second zone originates at the anterior base but extends more anteriorly into the processes and then drains into marginal venules running along the inner edge of the processes. A third zone supplies the posterior portion of the major processes and the minor processes. Most of the venous effluent from the ciliary body travels posteriorly through the *pars plana* and into the vortex venous system.

B. Measurement of Ciliary Blood Flow

Because of its location and complex vascular organization, ciliary blood flow and its regulation are poorly understood. However, the plasma clearance of ascorbate provides a rough estimate of 73 μl/min for ciliary plasma flow in humans, or a blood flow of 133 μl/min assuming a normal hematocrit (Linner, 1950, 1952). Measurements by microsphere entrapment in anesthetized monkeys give a somewhat lower ciliary blood flow at 89 μl/min (Alm and Bill, 1973). While both methods have the advantage of giving volumetric flow measurements (i.e., in μl/min), both are discontinuous (i.e., "snapshots") and provide a limited number of measurements, which make it hard to know if ciliary blood flow is at steady state when the measurement is taken, or to follow responses to pressure perturbations or drugs. An alternative continuous technique is fiber optic-based laser Doppler flowmetry (LDF), which can be used for transscleral ciliary blood flow measurements. LDF was developed in the early 1980s and has been validated and used in a variety of tissues, but it was not used to measure ciliary blood flow until recently. A thorough description of the technique can be found in the monograph by Shepherd and Oberg (1990). However, since what follows depends on whether LDF can measure ciliary blood flow, a brief summary of the evidence supporting this unique application of LDF is appropriate.

C. Sampling Depth

Fiber optic LDF uses one fiber to convey photons from a laser light source to the tissue and one or more fibers to convey photons collected from the tissue to a photodetector for processing. In unperfused tissue, photons are scattered by static tissue elements, which generate no Doppler shifts. In perfused tissue, photons deflected by moving red blood cells (RBCs) undergo Doppler shifts causing a broadening of the frequency spectrum detected by the photodetector, which is then analyzed to determine the mean velocity and number of RBCs used to calculate the RBC flux (i.e., the tissue blood flow). Most commercial LDF instruments use infrared lasers (780 nm) with fiber optic probe separations of 250 μm, a configuration predicted to give a sampling volume of 1 mm^3 (Shepherd and Öberg, 1990). Figure 9 confirms that this sampling depth is sufficient to measure through the sclera and detect blood flow in underlying intraocular tissue.

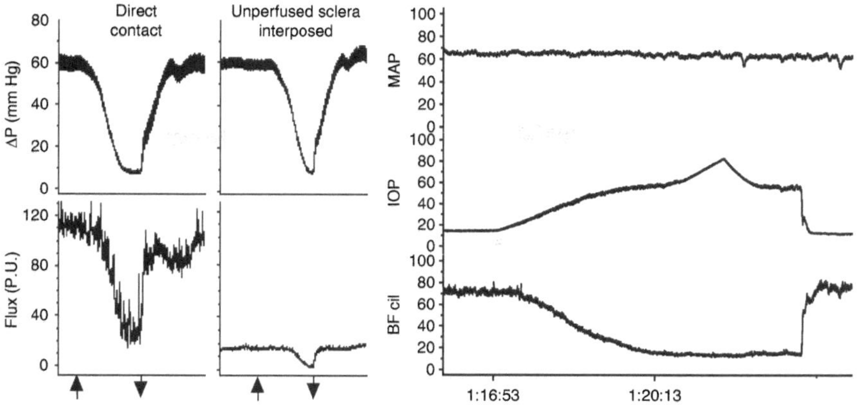

FIGURE 9 LDF measurement depth is sufficient for transscleral ciliary blood flow measurement. (Left) rabbit ciliary blood flow (Flux) response to caval occlusion at the same site overlying the ciliary body with and without a layer of unperfused sclera harvested from the fellow eye (~300-μm thick) interposed between the probe and the sclera. Flow response detected through the interposed tissue shows LDF sampling depth is sufficient to measure through sclera. (ΔP=MAP−IOP; up and down arrows mark beginning and end of caval occlusion; adapted from Kiel *et al.*, 2001.) (Right) Transscleral LDF measurement as IOP is increased by saline infusion (on, off) and fluid withdrawal (ΔV) to restore normal IOP while mean arterial pressure (MAP) remained constant. Detection of blood flow changes in response to IOP indicates LDF detects intraocular blood flow (authors' unpublished observation).

D. Measurement Site

Transscleral LDF measurements have a distinct anterior-to-posterior profile corresponding to the underlying tissues as shown in Fig. 10. The blood flow (flux) recorded between the visible limbal vessels and the signal nadir over the *pars plana* originates primarily from the ciliary body. The ciliary muscle in the rabbit is poorly developed, so most of the flow signal is from the ciliary processes (Prince and Eglitis, 1964). However, the vascular casts by Morrison *et al.* (1987a,b) suggest the anterior processes are more specialized for secretion. Whether the LDF measurement includes both the anterior and posterior processes is unclear, but Fig. 11 shows that LDF probe placement over the ciliary body does not detect changes in iris blood flow.

E. Aqueous Flow Measurement

Measurements of aqueous flow are based on the clearance of fluorescein from the cornea and anterior chamber after topical application. The commercially available FM2 fluorophotometer (Ocumetrics, Mountain View,

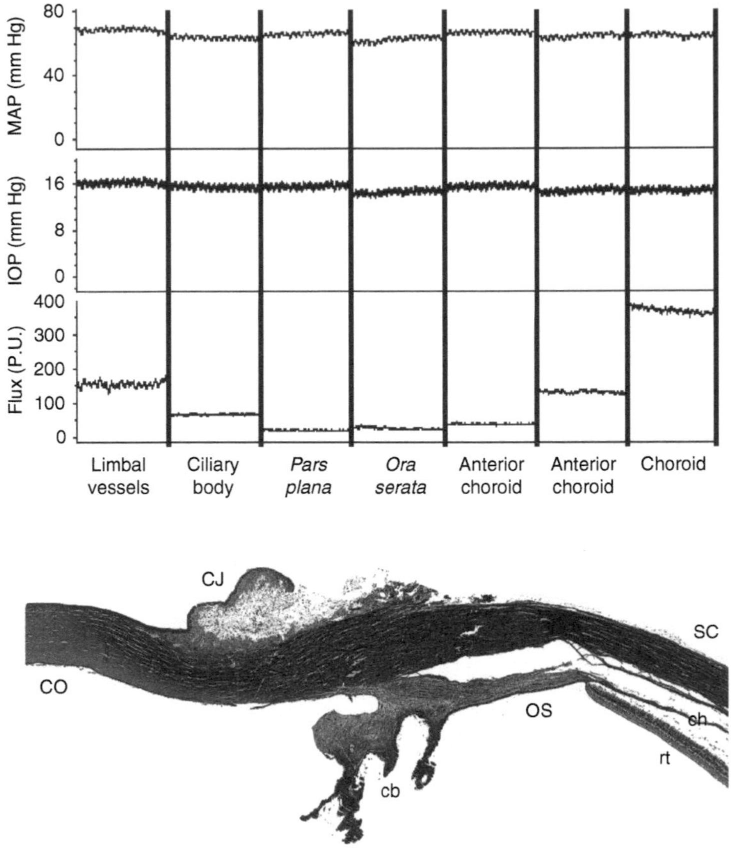

FIGURE 10 LDF transscleral ciliary blood flow measurement based on distinct blood flows corresponding to underlying tissues. Typical probe placement is ~1 mm posterior to the ring of superficial limbal vessels. (Top) Blood flow profile as laser Doppler probe is moved posteriorly from the limbus in ~1 mm steps. (Bottom) Corresponding rabbit anatomy. co, cornea; cj, conjunctiva; cb, ciliary body; os, *ora serata*; rt, retina; ch, choroid; sc, sclera. (Adapted from Kiel *et al.*, 2001.)

CA) is typically used to measure the changes in fluorescein concentration over time. Detailed descriptions of the technique can be found elsewhere (Brubaker, 1991a,b). However, in anesthetized animals held in a stereotaxic head holder, the standard procedure can be modified to take advantage of the fact that the fluorophotometer can measure the same optical path throughout the experiment, which minimizes scan variability and permits flow calculations over shorter time intervals. This is important when ciliary blood flow is varied by mechanically manipulating MAP, which limits the

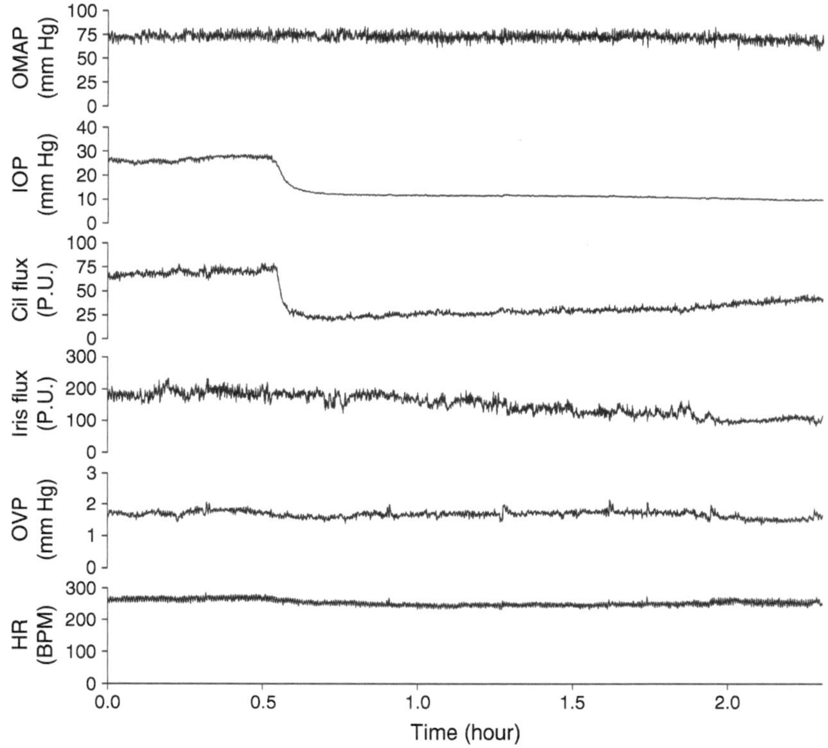

FIGURE 11 Evidence that the LDF sampling volume in the rabbit is deep enough to reach the ciliary body, yet small enough to avoid the iris. Trace shows responses to topical brimonidine (0.15%, 40 μl) given at 0.5 hours. Ocular mean arterial pressure (OMAP), orbital venous pressure (OVP), and heart rate (HR) were unaffected. LDF measurement through the sclera over the ciliary body indicates a rapid decrease in blood flow that was not detected by a second LDF probe directed through the cornea at the iris (~3 mm from the iris root). The different flow responses at neighboring sites indicate the LDF sampling volume is adequately selective for the ciliary body (Reitsamer *et al.*, 2006).

duration of the experiments. Figure 12 shows representative triplicate scans at 15-min intervals illustrating the minimal scan variability and an example of the rapid changes in fluorescein clearance that can be detected with more frequent scans.

It should be noted that fluorophotometry measures the flow of aqueous through the anterior chamber, which is assumed to be ~90% of total aqueous production; the remaining 10% is thought to flow posteriorly through the vitreous and retina to the choroid where it is reabsorbed (Brubaker, 1991a).

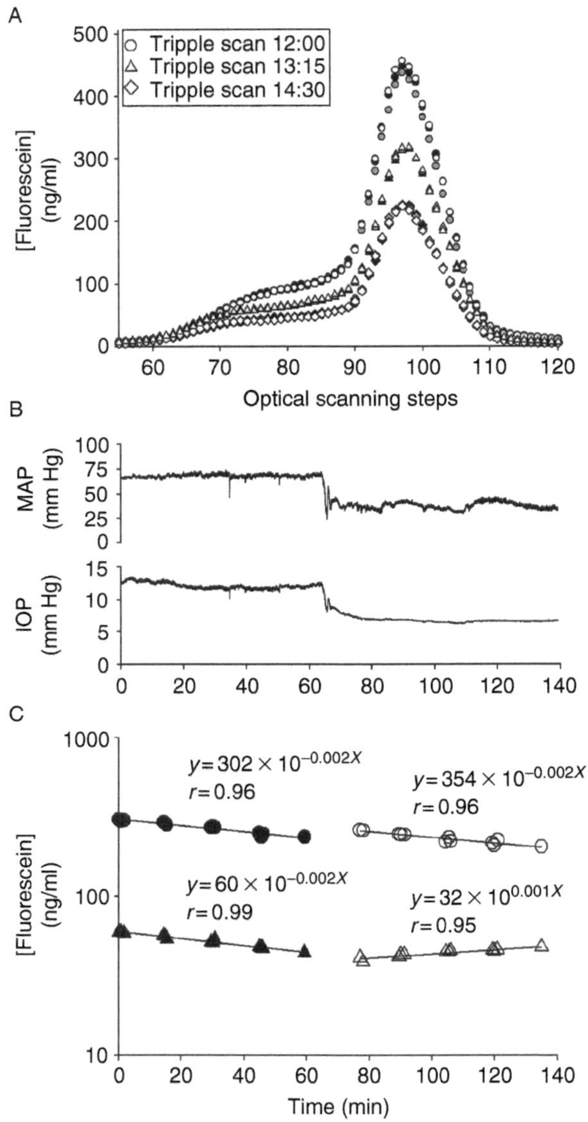

FIGURE 12 Measurement of aqueous flow. Three triplicate scans taken over 150 min show a steady decrease of fluorescein concentration (A). During experiments, scans can be made for 60–90 min before and after drug administration or perturbations of ocular perfusion pressure (B) to obtain the fluorescein concentration decay curves (C) used to calculate the aqueous flow rate. Panels B and C are from the same animal and show a step decrease of blood pressure causing the decrease of fluorescein removal from the anterior chamber, consistent with the calculated 50% reduction in aqueous flow. (Panels B and C from Reitsamer and Kiel, 2003.)

F. Relationship Between Ciliary Blood Flow and Aqueous Flow

In order to determine the relationship between ciliary blood flow and aqueous production, it is necessary to vary ciliary blood flow and measure the resulting effect on aqueous production. Ideally, only ciliary blood flow should be varied (i.e., neurohumoral input to the eye, the ocular PP, and blood flow through the other ocular circulations should all be constant), but this is not technically possible. What has been done instead is to manipulate ciliary blood flow by mechanically holding the arterial pressure at different levels (80, 70, 55, and 40 mm Hg) in anesthetized rabbits. The results of the initial study are shown in Fig. 13 (left); the relationship between ciliary blood flow and aqueous production based on additional experiments are also shown (Fig. 13, right).

The graphs in Fig. 13 show that when ciliary blood flow is increased above its normal level, aqueous production is relatively unchanged. This is also true for reductions in ciliary blood flow of ~25%, but further reductions result in proportional decreases in aqueous production. In order words, there is a critical level of ciliary blood flow below which aqueous production is blood flow-dependent. Above that critical level, aqueous production is independent of ciliary blood flow. A similar relationship is found in the gastric mucosa, where acid secretion is blood flow-independent until mucosal blood flow is

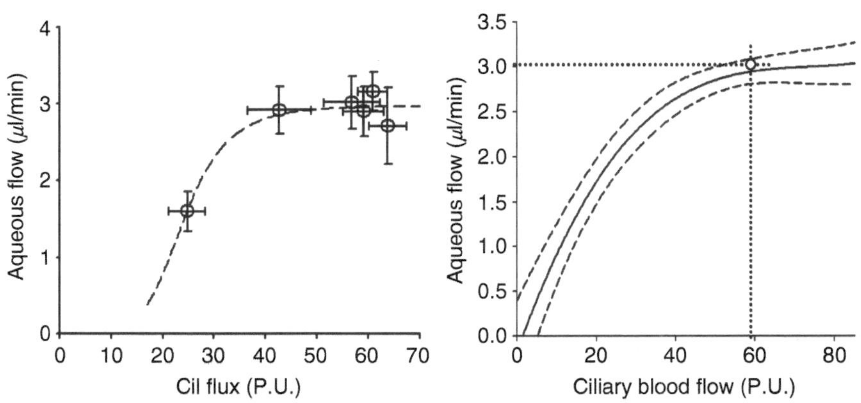

FIGURE 13 Relationship between ciliary blood flow and aqueous humor production in anesthetized rabbits. (Left) Original relationship based on 43 experiments. (Right) Revised relationship based on additional experiments. The open circle is the average baseline value at spontaneous ciliary blood flow and aqueous flow. The curve shows a plateau, where aqueous flow is relatively independent of ciliary blood flow; however, if ciliary blood flow decreases below roughly 75% of baseline, aqueous flow becomes dependent on ciliary blood flow and decreases with further reductions in ciliary blood flow (Reitsamer and Kiel, 2003).

reduced below a critical level, whereupon it becomes blood flow-dependent. In addition, work with secretory stimulants in the stomach shift the relationship upward, whereas secretory inhibitors shift the relationship downward (Perry *et al.*, 1983; Holm and Perry, 1988). If ciliary secretion behaves similarly, the following hypotheses seem reasonable: (1) aqueous production requires an adequate supply of oxygen and nutrients delivered by the ciliary circulation; (2) the neurohumoral milieu of the ciliary processes determines the level of secretory stimulation so that the rate of aqueous production at a given PP and ciliary blood flow can be stimulated or inhibited pharmacologically; and (3) the autoregulatory mechanisms governing ciliary blood flow are modulated by endogenous neurohumoral factors and that ciliary autoregulation can be overridden by the administration of exogenous vasoactive compounds or large changes in PP can exceed the autoregulatory ability of the ciliary circulation.

The dynamics of these hypotheses are more readily apparent in Fig. 14. The figure has three curves generated by a mathematical model (Kiel, 2000) of aqueous production plotted as a function of ciliary blood flow: a control curve for a normal level of secretory stimulation (Ctrl), a curve for a state of

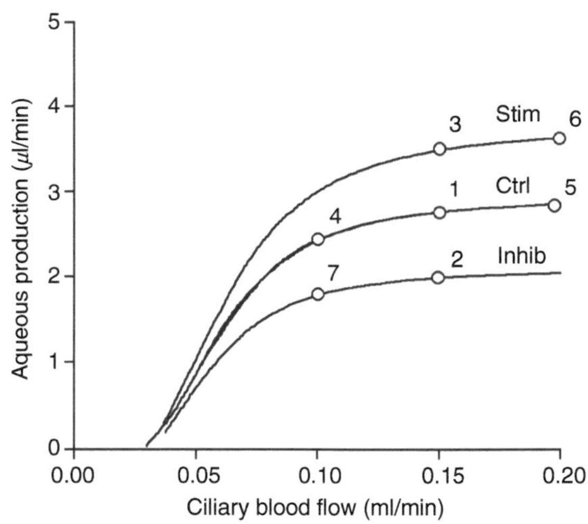

FIGURE 14 Hypothetical curves for aqueous production versus ciliary blood flow under stimulated, control and inhibited conditions. Points show: (1) control production and blood flow; (2) inhibited production without change in blood flow; (3) stimulated production without change in blood flow; (4) ciliary vasoconstriction with flow-dependent decrease in production; (5) ciliary vasodilation with small flow-dependent increase in production; (6) stimulated production with metabolic-dependent vasodilation; and (7) inhibited production with metabolic-dependent vasoconstriction.

heightened secretory stimulation (Stim), and a curve for an inhibited secretory state (Inhib). Point 1 on the control curve occurs at a normal ciliary blood flow (150 μl/min) and aqueous production (2.75 μl/min). A drug that acts directly on the active secretion by the epithelial cells (e.g., by changing intracellular cAMP) can decrease (Point 2) or increase (Point 3) production without changing ciliary blood flow. In this scenario, imposing large changes in PP sufficient to overcome ciliary autoregulation will define the stimulated and inhibited curves. Alternatively, a drug that causes ciliary vasoconstriction (Point 4) or vasodilation (Point 5) without altering the stimulus for secretion can nonetheless decrease or slightly increase production. In this case, imposing large changes in PP will generate the normal curve and reveal that the drug had no direct effect on secretion. A third scenario is that ciliary autoregulation is linked to metabolism so that a drug that stimulates secretion will cause a concomitant vasodilation (Point 6), while a drug that inhibits secretion will cause a vasoconstriction (Point 7). In this case, varying the PP will reveal that the system has shifted to a new level of secretory stimulation rather than the effect being due simply to a change in blood flow. Some drugs will affect both secretion and vascular tone, and in this case varying the PP will help to discern the relative contributions to the overall effect on production.

Pharmacological studies support this model of the role of ciliary blood flow in aqueous production and provide insight into the sometimes-ambiguous mechanisms of action of glaucoma drugs that lower IOP by reducing aqueous production. For example, three mechanisms have been proposed to explain the suppression of aqueous production by adrenergic α2-agonists: (1) activation of prejunctional α2 receptors causing the inhibition of norepinephrine release, thereby reducing a stimulus for aqueous production; (2) activation of postjunctional vascular α2 receptors, causing ciliary vasoconstriction and decreased ciliary blood flow; and (3) activation of postjunctional epithelial α2 receptors inhibiting adenylate cyclase (Reitsamer *et al.*, 2006). Although the prejunctional mechanism may be involved, it does not appear to be the primary mechanism since brimonidine, an α2 agonist, decreases aqueous production in the absence of sympathetic tone (Gabelt *et al.*, 1994). This leaves postjunctional vasoconstriction or inhibition of adenylate cyclase as the likely candidates. In anesthetized rabbits, acute topical brimonidine decreases ciliary blood flow and aqueous production. As Fig. 15 shows, the brimonidine response falls on the control curve, consistent with the decrease in ciliary blood flow being responsible for the decrease in aqueous production. Brimonidine may also inhibit adenylate cyclase, but the ciliary vasoconstriction appears to account for much of the decrease in aqueous production. In contrast, carbonic anhydrase inhibitors like dorzolamide are not vasoconstrictors and are thought to inhibit aqueous production directly by interfering with epithelial ionic transport (Gabelt *et al.*, 2006). Consistent with a

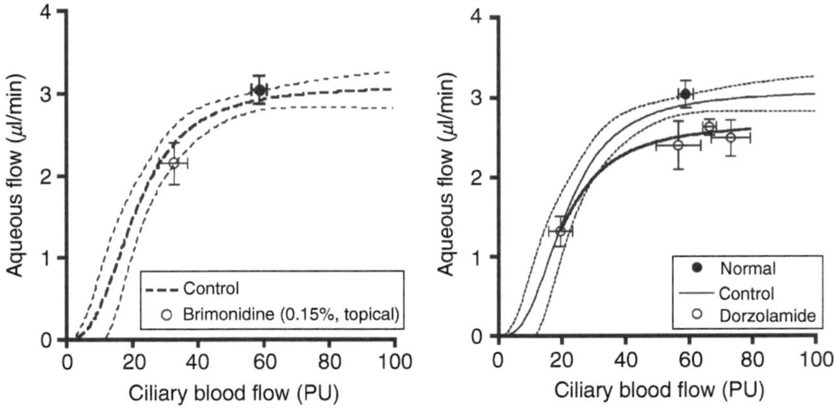

FIGURE 15 (Left) Topical brimonidine reduces ciliary blood flow and the attendant fall in aqueous production is similar to that found when ciliary blood flow is reduced to the same extent by decreasing MAP so that the datum falls on the control curve (Reitsamer *et al.*, 2006). (Right) Topical dorzolamide shifts the relationship downward consistent with a nonvascular, direct inhibition of aqueous production (Kiel and Reitsamer, 2006). (Note that the dorzolamide curve was obtained by varying ciliary blood flow with mechanical MAP manipulation after topical dorzolamide application.)

direct mechanism of action, topical dorzolamide in the rabbit causes a downward shift in the ciliary blood flow—aqueous flow relationship as shown in Fig. 15.

Figure 16 shows an example of a seemingly paradoxical drug response. In the rabbit, intravenous infusions of low and high doses of dopamine have opposite effects on aqueous production—the low dose causes an increase in production, and the high dose causes a decrease (Reitsamer and Kiel, 2002a). One explanation for these disparate responses is the binding characteristics of the different dopamine receptor subtypes in the epithelium. However, a simpler explanation might be that the high dose causes a significant decrease in ciliary blood flow, and as Fig. 16 shows, the decrease in ciliary blood flow places the high-dose datum on the control curve, suggesting that the vascular mechanism is involved. The low dose, on the other hand, has little effect on ciliary blood flow so that the datum is above the control curve, consistent with the low dose having a direct stimulatory effect.

A key question for the role of ciliary blood flow in aqueous production is what is provided by the ciliary blood flow. Ciliary blood flow provides oxygen and nutrients to the ciliary epithelia and removes metabolic waste. It is unclear which of these is critical; however, oxygen is a reasonable candidate because the ionic transport systems responsible for the osmotic

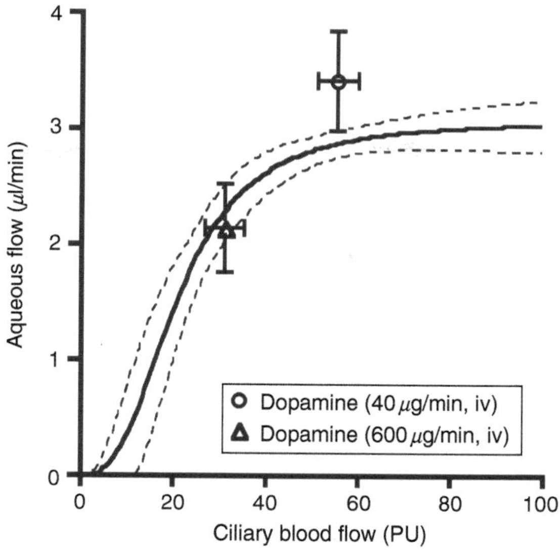

FIGURE 16 Paradoxical response to dopamine. Intravenous infusion of a low dose of dopamine increases aqueous flow while an infusion of a high dose decreases aqueous production. Both doses likely stimulate epithelial ionic transport, but the high dose also causes ciliary vasoconstriction (presumably by activating vascular alpha adrenergic receptors), which deprives the epithelium of the blood flow needed to sustain aqueous production. (Adapted from Reitsamer and Kiel, 2002a.)

gradient that produces aqueous utilize oxidative metabolism for energy and they largely stop functioning when insufficient oxygen is provided, despite the availability of other substrates (Burstein *et al.*, 1984; Krupin *et al.*, 1984). If ciliary blood flow is raised above the critical level, oxygen extraction decreases and the excess simply passes through in the venous effluent. However, if blood flow is reduced below the critical level, even maximum oxygen extraction cannot provide sufficient high-energy substrates (e.g., adenosine triphosphate) to drive ionic transport, and so aqueous production decreases. In support of oxygen as the critical factor, the change in the partial pressure of oxygen (PO2) adjacent to the ciliary body in the rabbit falls in response to topical brimonidine (Reitsamer *et al.*, 2006). As Fig. 17 shows, the brimonidine-induced decrease in ciliary blood flow is associated with a decrease in ciliary PO2. The decrease in PO2 indicates increased oxygen extraction rather than the decreased oxygen consumption that would be consistent with secretory inhibition, but whether the PO2 in the ciliary epithelia falls below the level needed to sustain ciliary metabolism and accounts completely for the brimonidine suppression of aqueous production requires further study.

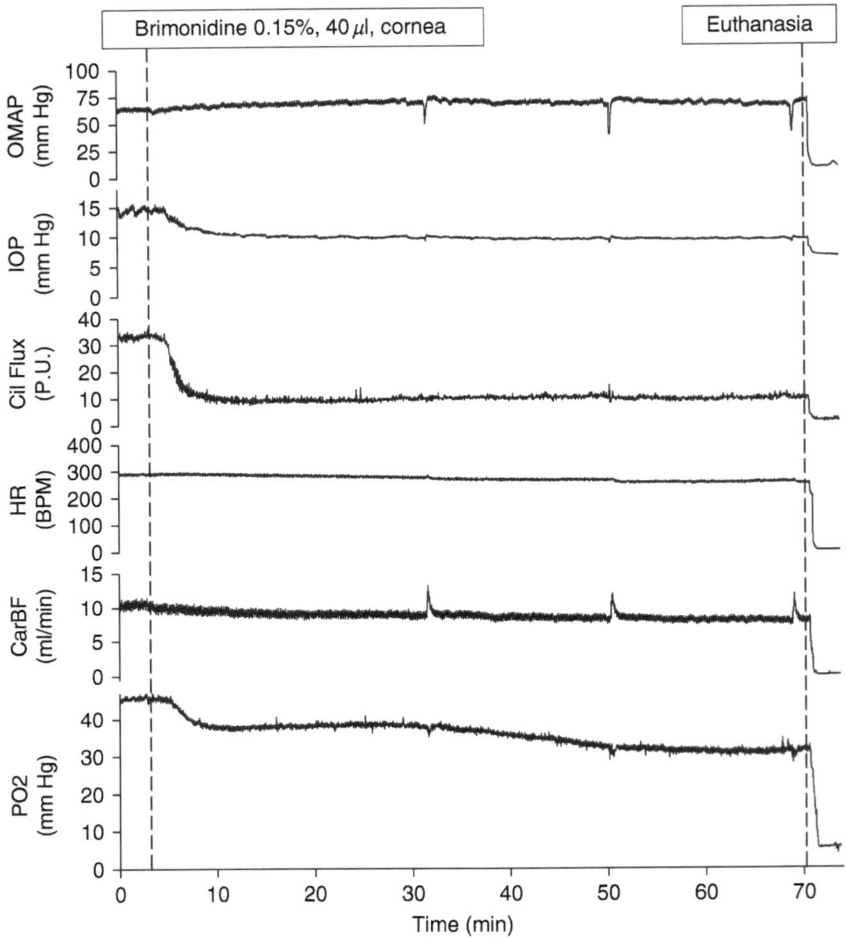

FIGURE 17 Effect of brimonidine-induced decrease in ciliary blood flow on ciliary PO2. Representative trace showing ciliary PO2 response to topical brimonidine and rapid decline in ciliary PO2 upon cessation of ciliary perfusion at death. Decrease in ciliary PO2 indicates increased oxygen extraction rather than decreased oxygen consumption.

Additional evidence that oxygen is the critical factor delivered in ciliary blood flow is shown in Fig. 18. In this experiment, anesthetized rabbits (n=12) were respired with a mixture of room air and nitrogen sufficient to lower their percent hemoglobin saturation with oxygen to 75%. Ciliary blood flow increased slightly but not significantly, yet aqueous production decreased by 22%. In this situation, the delivery of other nutrients and removal of metabolic waste remained constant and only oxygen delivery was reduced. Thus, it appears that oxygen is indeed a critical factor delivered in ciliary blood flow.

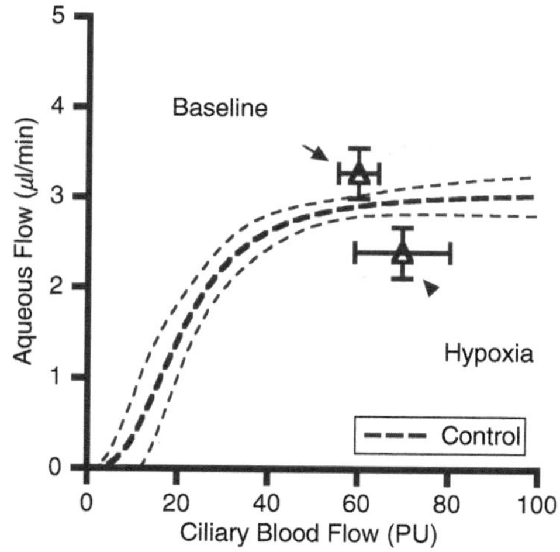

FIGURE 18 Effect of hypoxia on ciliary blood flow and aqueous production. Lowering the percent saturation of hemoglobin with oxygen from 95% to 75% decreases aqueous production without changing ciliary blood flow or the delivery of other nutrients and washout of metabolites (authors' unpublished results).

V. EPISCLERAL VENOUS PRESSURE AND IOP

The final circulatory parameter to be mentioned in this chapter is episcleral venous pressure (EVP). Of all the variables involved in IOP homeostasis, EVP is the least understood, and yet it accounts for more than 50% of the normal IOP (15mm Hg) in the Goldman equation (IOP=aqueous flow divided by outflow facility plus EVP) given current EVP estimates (8–10mm Hg) (Zeimer, 1989). Whether EVP plays a role in glaucoma or can be lowered with drugs or surgery as a treatment for ocular hypertension is unclear, but these are research questions clearly worthy of pursuit (Funk *et al.*, 1996; Toris *et al.*, 2002; Selbach *et al.*, 2005). A potential problem regarding EVP is that it is often assumed to be constant in calculations of tonographic outflow facility and uveoscleral outflow. This assumption is unlikely to be correct in many circumstances, but unfortunately it is difficult to prove one way or the other because EVP is difficult to measure.

For example, Fig. 19 shows the IOP, EVP, and orbital venous pressure (OVP) responses to changes in MAP (Reitsamer and Kiel, 2002b). The EVP response to MAP demonstrates that EVP is not constant. Another noteworthy aspect of these results is the pressure gradient from the episcleral veins to the orbital venous sinus. The source of the resistance responsible for

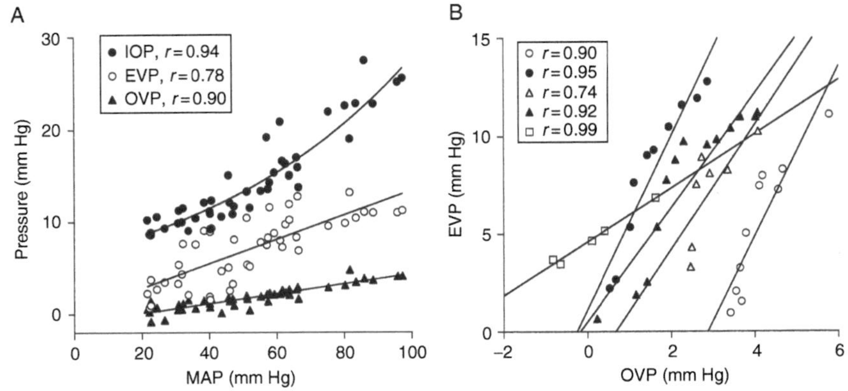

FIGURE 19 Anesthetized rabbit responses to step-changes in MAP for IOP. MAP and OVP were measured by direct cannulation. EVP was measured by inserting a fine-tipped glass pipette (2–3 μm tip) connected to a servo-null instrument into episcleral veins with a visible plume of aqueous joining the stream of red blood cells. IOP varied exponentially with MAP, whereas EVP and OVP varied linearly with MAP (A). EVP varied linearly with OVP (B). The regression lines and correlation coefficients (r) are shown for the pooled data in A and for individual experiment in B.

this pressure gradient is unknown, but the relatively long length of the small caliber episcleral veins and their course through the extraocular muscles *en route* to the orbital sinus are likely contributors.

VI. CONCLUSION

As noted at the beginning of this chapter, the ocular circulations play diverse roles in IOP homeostasis, and IOP in turn can have profound effects on the ocular circulations. How and when the ocular circulations contribute to the etiology and treatment of ocular hypertension and glaucoma are subjects worthy of the intense study they receive, which continues to provide new insights into this devastating disease.

Acknowledgments

This work was supported by NIH grant EY09702 (J.W.K.), a Research to Prevent Blindness Lew R Wasserman Merit Award (J.W.K.), Austrian FWF J1866-MED (H.A.R.), San Antonio Lions, Lions International, and an unrestricted grant from Research to Prevent Blindness Inc.

References

Alm, A., and Bill, A. (1972a). The oxygen supply to the retina, I. effects of changes in intaocular and arterial blood pressures, and in arterial Pco2 and Pco2 on the oxygen tension in the vitreous body of the cat. *Acta Physiol. Scand.* **84,** 261–274.

Alm, A., and Bill, A. (1972b). The oxygen supply to the retina, II. effects of high intraocular pressure of increased arterial carbon dioxide tension on uveal & retinal blood flow in cats. *Acta Physiol. Scand.* **84,** 306–319.

Alm, A., and Bill, A. (1973). Ocular and optic nerve blood flow at normal and increased intraocular pressures in monkeys (macaca irus): A study with radioactively labeled microspheres including flow determinations in brain and some other tissues. *Exp. Eye. Res.* **15,** 15–29.

Barany, E. H. (1963). A mathematical formulation of intraocular pressure as dependent on secretion, ultrafiltration, bulk outflow, and osmotic reabsorption of fluid. *Invest. Ophthalmol.* **2,** 584–590.

Bill, A. (1962). Aspects of regulation of the uveal venous pressure in rabbits. *Exp. Eye Res.* **1,** 193–199.

Bill, A. (1963). The uveal venous pressure. *Arch. Ophthalmol.* **69,** 780–782.

Bill, A. (1971). Effects of longstanding stepwise increments in eye pressure on the rate of aqueous humor formation in a primate (cercopithecus ethiops). *Exp. Eye Res.* **12,** 184–193.

Bill, A. (1973). The role of ciliary blood flow and ultrafiltration in aqueous humor formation. *Exp. Eye Res.* **16,** 287–298.

Bill, A., and Barany, E. H. (1966). Gross facility, facility of conventional routes, and pseudofacility of aqueous humor outflow in the cynomolgus monkey. *Arch. Ophthalmol* **75,** 665–673.

Bill, A., Linder, M., and Linder, J. (1977). The protective role of ocular sympathetic vasomotor nerves in acute arterial hypertension. *Bibl. Anat.* **16,** 30–35.

Brubaker, R. F. (1991a). Clinical evaluation of the circulation of aqueous humor. *In* "Duane's Clinical Ophthalmology" (W. Tasman and E. A. Jaeger, eds.), pp. 1–11. J.B. Lippincott Cof, Philadelphia, PA.

Brubaker, R. F. (1991b). Flow of aqueous humor in humans. *Invest. Ophthalmol. Vis. Sci.* **32,** 3145–3166.

Burstein, N. L., Fischbarg, J., Liebovitch, L., and Cole, D. F. (1984). Electrical potential, resistance, and fluid secretion across isolated ciliary body. *Exp. Eye Res.* **39,** 771–779.

Carlson, K. H., McLaren, J. W., Topper, J. E., and Brubaker, R. F. (1987). Effect of body position on intraocular pressure and aqueous flow. *Invest. Ophthalmol. Vis. Sci.* **28,** 1346–1350.

Duke-Elder, S. (1926). The venous pressure of the eye and its relation to the intra-ocular pressure. *J. Physiol.* **61,** 409–418.

Friedenwald, J. S. (1937). Contribution to the theory and practice of tonometry. *Am. J. Ophthalmol.* **20,** 985–1024.

Funk, R., and Rohen, J. W. (1990). Scanning electron microscopic study on the vasculature of the human anterior eye segment, especially with respect to the ciliary processes. *Exp. Eye Res.* **51,** 651–661.

Funk, R. H. W., Gehr, J., and Rohen, J. W. (1996). Short-term hemodynamic changes in episcleral arteriovenous anastomoses correlate with venous pressure and IOP changes in the albino rabbit. *Curr. Eye Res.* **15,** 87–93.

Gabelt, B. T., Kiland, J. A., Tian, B., and Kaufman, P. L. (2006). Aqueous humor: Secretion and dynamics. *In* "Duane's Clinical Ophthalmology" (W. Tasman and E. A. Jaeger, eds.). Lippincott Williams & Wilkins, Philadelphia.

Gabelt, B. T., Robinson, J. C., Hubbard, W. C., Peeterson, C. M., Debink, N., Wadhwa, A., and Kaufman, P. L. (1994). Apraclonidine and brimonidine effects on anterior ocular and cardiovascular physiology in normal and sympathectomized monkeys. *Exp. Eye Res.* **59,** 633–644.

Grant, W. M. (1950). Tonographic method for measuring the facility and rate of aqueous flow in human eyes. *Arch. Ophthalmol.* **44,** 204–214.

Guyton, A. C., Polizo, D., and Armstrong, G. G. (1954). Mean circulatory filling pressure measured immediately after cessation of heart pumping. *Am. J. Physiol.* **179,** 261–267.

Heijl, A., Leske, M. C., Bengtsson, B., *et al.* (2002). Reduction of intraocular pressure and glaucoma progression: Results from the early manifest glaucoma trial. *Arch Ophthalmol.* **120,** 1268–1279.

Holm, L., and Perry, M. A. (1988). Role of blood flow in gastric acid secretion. *Am. J. Physiol.* **254,** G281–G293.

Kiel, J. W. (1994). Choroidal myogenic autoregulation and intraocular pressure. *Exp. Eye Res.* **58,** 529–544.

Kiel, J. W. (1995). The effect of arterial pressure on the ocular pressure-volume relationship in the rabbit. *Exp. Eye Res.* **60,** 267–278.

Kiel, J. W. (1998). Physiology of the intraocular pressure. *In* "Pathophysiology of the Eye: Glaucoma" (J. Feher, ed.), pp. 109–144. Akademiai Kiado, Budapest.

Kiel, J. W. (2000). A computer-based, mathematical model for teaching ocular hydrodynamics. *FASEB J.* **14,** A306.

Kiel, J. W., and Reitsamer, H. A. (2006). Relationship between ciliary blood flow and aqueous production: Does it play a role in glaucoma therapy? *J. Glaucoma* **15,** 172–181.

Kiel, J. W., and Shepherd, A. P. (1992). Autoregulation of choroidal blood flow in the rabbit. *Invest. Ophthalmol. Vis. Sci.* **33,** 2399–2410.

Kiel, J. W., and van Heuven, W. A. J. (1995). Ocular perfusion pressure and choroidal blood flow in the rabbit. *Invest Ophthalmol Vis Sci.* **36,** 579–585.

Kiel, J. W., Reitsamer, H. A., Walker, J. S., and Kiel, F. W. (2001). Effects of nitric oxide synthase inhibition on ciliary blood flow, aqueous production and intraocular pressure. *Exp. Eye Res.* **73,** 355–364.

Krupin, T., Reinach, P. S., Candia, O. A., and Podos, S. M. (1984). Transepithelial electrical measurements on the isolated rabbit iris-ciliary body. *Exp. Eye Res.* **38,** 115–123.

Kupfer, C. (1971). Pseudofacility in the human eye. *Trans. Am. Ophthalmol. Soc.* **69,** 383–396.

Linner, E. (1950). A method for determining the rate of plasma flow through the secretory part of the ciliary body. *Acta Physiol. Scand.* **22,** 83–86.

Linner, E. (1952). Ascorbic acid as a test substance for measuring relative changes in the rate of plasma flow through the ciliary processes. *Acta Physiol. Scand.* **26,** 57–85.

Lutjen-Drecoll, E., and Rohen, J. W. (1994). Anatomy of aqueous humor formation and drainage. *In* "Textbook of Ophthalmology" (P. L. Kaufman and T. W. Mittag, eds.), pp. 1.1–1.16. Mosby, London.

Morrison, J. C., DeFrank, M. P., and Van Buskirk, E. M. (1987a). Comparative microvascular anatomy of mammalian ciliary processes. *Invest. Ophthalmol. Vis Sci.* **28,** 1325–1340.

Morrison, J. C., DeFrank, M. P., and Van Buskirk, E. M. (1987b). Regional microvascular anatomy of the rabbit ciliary body. *Invest. Ophthalmol. Vis Sci.* **28,** 1314–1324.

Moses, R. A. (1963). Hydrodynamic model eye. *Ophthalmologica* **146,** 137–142.

Moses, R. A., Grodzki, W. J., and Carras, P. L. (1985). Pseudofacility: Where did it go? *Arch Ophthalmol.* **103,** 1653–1655.

Patterson, S. W., and Starling, E. H. (1914). On the mechanical factors which determine the output of the ventricles. *J. Physiol.* **48,** 357–379.

Perry, M. A., Haedicke, G. J., Bulkley, G. B., Kvietys, P. R., and Granger, D. N. (1983). Relationship between acid secretion and blood flow in the canine stomach: Role of oxygen consumption. *Gastroenterology* **85,** 529–534.

Prince, J. H., and Eglitis, I. (1964). The uvea (choroid, ciliary body, and iris). *In* "The Rabbit in Eye Research" (J. Prince, ed.), pp. 140–171. Charles C Thomas Publisher, Springfield, IL.

Reitsamer, H. A., and Kiel, J. W. (2002a). Effects of dopamine on ciliary blood flow, aqueous production, and intraocular pressure in rabbits. *Invest. Ophthalmol. Vis. Sci.* **43,** 2697–2703.

Reitsamer, H. A., and Kiel, J. W. (2002b). A rabbit model to study orbital venous pressure, intraocular pressure, and ocular hemodynamics simultaneously. *Invest. Ophthalmol. Vis. Sci.* **43,** 3728–3734.

Reitsamer, H. A., and Kiel, J. W. (2003). Relationship between ciliary blood flow and aqueous production in rabbits. *Invest Ophthalmol. Vis. Sci.* **44,** 3967–3971.

Reitsamer, H. A., Posey, M., and Kiel, J. W. (2006). Effects of a topical alpha2 adrenergic agonist on ciliary blood flow and aqueous production in rabbits. *Exp. Eye Res.* **82,** 405–415.

Selbach, J. M., Posielek, K., Steuhl, K. P., and Kremmer, S. (2005). Episcleral venous pressure in untreated primary open-angle and normal-tension glaucoma. *Ophthalmologica* **219,** 357–361.

Shepherd, A. P., and Öberg, P. A. (1990). *In* "Laser-Doppler Blood Flowmetry." Kluwer Academic Publishers, Norwell, MA.

Silver, D. M., and Farrell, R. A. (1994). Validity of pulsatile ocular blood flow measurements. *Surv. Ophthalmol* **38,** S72–S80.

Toris, C. B., Koepsell, S. A., Yablonski, M. E., and Camras, C. B. (2002). Aqueous humor dynamics in ocular hypertensive patients. *J. Glaucoma* **11,** 253–258.

Van Buskirk, E. M., Bacon, D. R., and Fahrenbach, W. H. (1990). Ciliary vasoconstriction after topical adrenergic drugs. *Am. J. Ophthalmol.* **109,** 511–517.

Weigert, G., Findl, O., Luksch, A., *et al.* (2005). Effects of moderate changes in intraocular pressure on ocular hemodynamics in patients with primary open-angle glaucoma and healthy controls. *Ophthalmology* **112,** 1337–1342.

Zeimer, R. C. (1989). Episcleral venous pressure. *In* "The Glaucomas" (R. Ritch, M. Shields, and T. Krupin, eds.), pp. 249–255. C.V. Mosby Company, St. Louis.

CHAPTER 10

Retinal Ganglion Cells and Glaucoma: Traditional Patterns and New Possibilities

Claire H. Mitchell and Wennan Lu
Department of Physiology, University of Pennsylvania, Philadelphia, Pennsylvania 19104

I. OVERVIEW

A mismatch between the rate of aqueous humor production and drainage can reduce the well being of retinal ganglion cells. The mechanisms linking elevated IOP and ganglion cell distress vary with the magnitude of the pressure increase, the duration of the insult, and a variety of endogenous and environmental influences. Some suspects have been acknowledged for decades, such as a distortion of the lamina cribrosa, changes in the vascular supply, and altered levels of neurochemicals. The role of extracellular purines places a novel spin on the theory of neurochemical imbalance, with

Current Topics in Membranes, Volume 62

1063-5823/08 $35.00
DOI: 10.1016/S1063-5823(08)00410-9

extracellular ATP released through pressure-sensitive pathways either stimulating lethal $P2X_7$ receptors on ganglion cells, or being converted to adenosine and protecting them. Additional theories propose a role for reactive astrocytes, compressive forces and even increased age itself in weakening ganglion cells to the point where they eventually die. Many of the basic patterns may not be restricted to the posterior of the eye, but may hold lessons for the study on the anterior chamber.

II. INTRODUCTION

At first glance, a chapter on retinal ganglion cells may seem out of place in a book about aqueous humor. Ganglion cells have an unknown influence on the composition, production, or drainage of the humor, while only 5% of the aqueous humor flows towards the posterior of the eye, limiting even unidirectional communication (Maurice, 1987). However, a mismatch between rates of aqueous humor secretion and drainage is of clinical interest primarily because the resulting increase in intraocular pressure (IOP) is a predominant risk factor for glaucomatous optic neuropathy (Quigley, 1996; Gordon *et al.*, 2002; Sigal *et al.*, 2005). A key goal in balancing the inflow and outflow of aqueous humor in glaucoma is to maintain or restore ganglion health. As such, a basic understanding of ganglion cells and how they are injured by elevated IOP is beneficial.

This report will first summarize the general characteristics of ganglion cell injury in glaucoma, detailing the types of ganglion cells lost, the influence of pressure, the time course of their loss, and the loss of ganglion cell function that precedes cells death. The second section deals with selected theories to explain how increased IOP can lead to ganglion cell loss. It is becoming increasingly evident that glaucoma is a multifactorial disease with both genetic and environmental influences. It is also evident that the pathogenesis of glaucoma generates considerable controversy. This chapter will not attempt to address all the putative contributions to ganglion cell death in glaucoma; many of these have been detailed in extensive reviews and readers are encouraged to pursue them for more information (e.g., Hernandez, 2000; Morgan, 2000; Osborne *et al.*, 2001, 2003; Levin and Gordon, 2002; Wax and Tezel, 2002; Neufeld and Liu, 2003; Votruba, 2004; Morrison *et al.*, 2005; Tezel, 2006; Gupta and Yucel, 2007). An abridged discussion does provide a necessary structure to understand recent developments, and in some cases provides evidence for opposing viewpoints. Particular attention will be given to the emerging role of purines as a link between elevated ocular pressure, ganglion cell death, and neuroprotection.

III. INFLUENCES ON GLAUCOMATOUS DAMAGE TO GANGLION CELLS

A. Ganglion cell Type

There are 1.2–1.5 million ganglion cells in the human retina. They receive visual information from photoreceptors via the bipolar and amacrine cells, and deliver the visual signal through the axons of the optic nerve to the superior colliculus (SC), and the lateral geniculate nucleus (LGN). Morphologic criteria are used to classify ganglion cells into basic groups (Kolb *et al.*, 1992). The midget cells (P-cells) have relatively small cell bodies and dendritic trees, and project to parvocellular layers of LGN (Dacey, 1993). Parasol cells (M-cells) have larger cell bodies and dendritic fields and project to magnocellular layers of the LGN. Bistratified retinal ganglion cells have the smallest cell bodies and project to the koniocellular layers of the LGN. While all cell types are present across the retina, larger cells are concentrated in the periphery, while the central retina contains a higher percentage of cells with smaller somata (Dacey, 1994).

The distribution of ganglion cell types to particular retinal regions has apparent relevance to the identification of susceptible populations, and it is thought that a correlation could provide insight into the causes of cell death. Although ganglion cells throughout the retina are lost in glaucoma (Desatnik *et al.*, 1996), peripheral ganglion cells typically die at a higher rate (Laquis *et al.*, 1998; Sawada and Neufeld, 1999). This agrees with clinical findings where peripheral vision is affected first. The nasal field is usually compromised during early stages of glaucoma, with an arcuate pattern of loss surrounding the fovea leading to enhanced axonal loss in the superior and inferior regions (Quigley and Green, 1979).

Some psychophysical evidence indicates parasol cells may be particularly susceptible to glaucomatous damage (Anderson and O'Brien, 1997). However, evaluation of patients in the early stages of the disease found no preferential loss of sensitivity from the magnocellular pathway (Ansari *et al.*, 2002). A detailed morphologic analysis of labeled ganglion cells in primates with ocular hypertension found no significant difference between the loss of parasol and of midget ganglion cells (Morgan *et al.*, 2000). Other research has approached the problem through cell size, attempting to correlate likelihood of loss with soma or axon dimensions. The mean diameter of axons in the optic nerve of primates with experimental glaucoma was significantly smaller than in control eyes (Quigley *et al.*, 1987). Likewise, there were fewer ganglion cells with larger cell bodies in the retina of glaucomatous primate eyes (Glovinsky *et al.*, 1991). While larger cells do seem more susceptible to damage changes in cell size with the progression of glaucoma undermine this form of analysis. For example, the dendritic field, soma and axon of ganglion cells from glaucomatous primates were reduced before cell death occurred (Weber *et al.*, 1998). The extent of shrinkage correlated with evidence of optic nerve atrophy. This shrinkage may also contribute to reduced visual function, as discussed below.

B. Elevated IOP

The normal range of IOP in adult humans has traditionally been defined as between 10 and 21 mmHg, with pressure above this considered a major risk factor for the establishment of glaucoma. Multiple studies have demonstrated a correlation between elevated IOP and cell death. Pharmacological treatment that decreases IOP reduces the likelihood of disease progression (AGIS, 2000). Even in so-called "normal-tension" glaucoma, a decrease in IOP is beneficial (CNTG, 1998), suggesting that the set point for damage could be lower in some patients. Interestingly, an improvement in the patterned electroretinogram (PERG) response in patients with normotensive glaucoma accompanied a decrease in IOP, implying that reduced cellular function can be reversed (Ventura and Porciatti, 2005). The correlation between elevated pressure and ganglion cell loss is far from perfect however. A substantial proportion of patients with primary open angle glaucoma (POAG) experience an increase in optic disc cupping even after pressure is reduced below 17 mmHg (Tezel *et al.*, 2001). POAG patients whose loss of visual field progressed could not be distinguished from those whose fields remained the same on the basis of pressure alone (Martinez-Bello *et al.*, 2000).

Large diurnal variations in pressure present an increased risk of glaucomatous damage even in patients whose IOP is normal during examination (Asrani *et al.*, 2000) and diurnal variation in absolute IOP can be larger in patients with untreated POAG than in controls (Sehi *et al.*, 2005). Laboratory work indicates that fluctuations in pressure may indeed lead to pathological changes. For example, cyclical stretch of glial cells from the lamina cribrosa produced a clear change in the expression of many genes that could affect the structure of the lamina cribrosa, as well as genes affecting axonal signaling (Kirwan *et al.*, 2005). These observations indicate changes in pressure, in addition to absolute pressure levels, contribute to the pathology.

Recent analysis has provided some interesting insight into the mechanical effects of increased pressure on the eye, and there is currently some debate as to whether the moderate increases in hydrostatic pressure associated with glaucoma are sufficient to induce changes on a molecular level (Knepper *et al.*, 2005; Ethier, 2006). Differences in IOP capable of inducing considerable strain on ocular tissues are thought to produce a negligible effect on molecules present on a rigid surface. This suggests that differential rates of compression between various cellular or ocular structures could transduce pressure increases into structural damage. Stress and strain on optic nerve head tissues were strongly associated with scleral thickness and stiffness of the lamina cribrosa, supporting the importance of relative compression (Sigal *et al.*, 2005). Nevertheless, the reproducible effects of hydrostatic pressure on cells in numerous *in vitro* studies suggest hydrostatic pressure can itself initiate responses (Hernandez, 2000; Knepper *et al.*, 2005).

There is typically a long delay between the initial detection of an elevated IOP and a noticeable loss of vision. This delay reflects both the slow nature of the pathological processes and the relatively insensitive tools available to detect ganglion cell loss. The death of ganglion cell axons can be detected structurally as a thinning of the retinal nerve fiber layer and changes to the optic disk, including a loss of the neuroretinal rim and an increase in the cup to disk (C/D) measurements. However, the predominant method of detection remains standardized automated field assessment. The rate of progressive field loss in glaucomatous patients can be difficult to measure (Katz *et al.*, 1997). Assessment over eight years found an average decrease of only 1.3% per year across the entire visual field (Pereira *et al.*, 2002). This small a degree of annual change is less than the variability of the test itself. Alternative techniques for detection are being developed. For example, the PERG has been known to detect changes in the glaucomatous eye for some time (Weinstein *et al.*, 1988), and recent reports indicate it can detect relatively small changes at the early stages of glaucoma (Hood *et al.*, 2005; Bach *et al.*, 2006; Porciatti *et al.*, 2007). This brings hope that a more sensitive assessment of the progression may become available in the near future.

Novel preliminary work suggests that the acceleration of loss of ganglion cells with age may reflect a general inability of aged eyes to endure the effects of pressure, rather than just the cumulative response to a chronic elevation in IOP (Cepurna *et al.*, 2006). When the IOP of 8- and 28-month-old rats was elevated following injection of hypertonic saline into the episcleral veins, the optic nerves of the older rats displayed a significantly greater degree of degeneration than that of the younger rats for a given pressure elevation. This is consistent with increased susceptibility of ganglion cells in older animals to ischemic damage (Kawai *et al.*, 2001), and implies that older tissues are either more susceptible to injury, and/or less able to repair the damage.

It is now evident that absolute level of ocular pressure is just one of many factors that influence the occurrence and progression of glaucoma. Although lowering IOP is helpful, ganglion cell loss continues in many cases, and additional treatments which directly preserve retinal ganglion cell viability offer new potential for preventing visual loss (Hartwick, 2001; Levin and

Gordon, 2002). A more detailed understanding of the multiple mechanisms that injure them may aid in determining how pressure contributes, and perhaps more importantly, how pressure interacts with other factors to damage ganglion cells.

C. Chronic Dysfunction and Secondary Death

When discussing glaucomatous damage to ganglion cells, it is important to understand that with moderate increases in IOP, cells can perform at reduced levels for long intervals before finally dying. The characteristics of ganglion cell dysfunction, along with the spread of injury, may provide insight into the causes of eventual cell loss.

Psychophysical analysis indicates ganglion cell function is lost before significant thinning of the nerve fiber layer is detected, consistent with a defective transmission of the visual signal occurring independently from cell death (Ventura *et al.*, 2006). Microelectrode recordings found the activity of ganglion cells was modified by short-term changes in pressure, with even a moderate increased in IOP affecting the flicker-evoked responses (Grehn *et al.*, 1984). The ability to restore function by reducing IOP also strengthens the theory that functional loss is distinct from death (Ventura and Porciatti, 2005). As mentioned, a shrinkage of the dendritic tree and soma of ganglion cells can precede death in chronic glaucoma, consistent with a loss in the effectiveness of processing the visual message as a stage in disease progression (Morgan, 2002).

The susceptibility of ganglion cells to a secondary death indicates that signals emanating from injured cells are themselves detrimental. For example, partial transection of the optic nerve leads to the death of ganglion cells in quadrants corresponding to the severed axons. However, loss also extended beyond the regions of cut axons to encompass other cells not originally affected (Levkovitch-Verbin *et al.*, 2001). The mechanisms involved in this secondary death, possibly including immunological or neurochemical signals, may involve processes shared with chronically injured ganglion cells. Secondary degeneration may also explain the continued ganglion cell loss after adequate IOP control is obtained.

Together, these observations suggest ganglion cells can be injured in glaucoma long before they die, and that injured cells affect their neighbors. At least some of the functional changes in ganglion cells triggered by increased IOP are reversible, resulting in impairment that need not necessarily lead to death. Although the theories linking pressure and glaucoma described below were primarily based on assessment of cell mortality, the ability of these mechanisms to compromise function as well as survival should be considered. The development of assays to monitor sick cells as well as dead ones will aid our understanding of both disease progression and neuroprotective approaches.

IV. MECHANISMS OF GANGLION CELL DEATH

The ability to preserve ganglion cells in glaucoma is hampered by our inability to fully explain why elevated ocular pressure leads to cell loss in the first place. As discussed above, the disease is multifactorial, with several mechanisms contributing to death, and it is likely that none are as mutually exclusive as their main proponents would like to believe. A compression of the lamina cribrosa, decreased vascular supply, reduction in availability of neurotropic factors, autoimmune and neurotransmitter imbalances, and parallels to other neurodegenerative all contribute. This survey has not attempted to cover all the possible factors that may damage ganglion cells, and readers are directed to the previously mentioned reviews for more comprehensive information. Instead, the focus here is on traditional themes and the novel role of purines. The emerging ability of purines to integrate increased pressure with neurochemical imbalances may have general relevance for both neurotoxic explanations and neuroprotective strategies.

A. Distention of the Lamina Cribrosa

The lamina cribrosa lies even with the sclera and serves as a scaffold to support axons of the optic nerve as they exit the eye. Structurally it is formed by a series of beams composed of extracellular matrix material and covered with cellular material, with glial cells of particular relevance to the pressurized eye (Morgan, 2000). Beams are arranged around pores through which the unmyelinated axons of ganglion cells pass. Myelination occurs posterior to the lamina cribrosa.

The lamina cribrosa is a major site of glaucomatous damage. The inferior/superior pattern of ganglion cell loss in the retina correlates well with the topography of the lamina cribrosa, with the injured axons more likely to pass through laminar regions containing larger pores made of fewer beams (Quigley and Addicks, 1981). The lamina cribrosa of glaucomatous eyes shows sign of compression, with posterior bowing evident (Quigley *et al.*, 1983). Swelling and an accumulation of organelles are found in axons around the lamina cribrosa in glaucomatous eyes, suggesting impaired axonal transport (Quigley *et al.*, 1981).

Decreased axonal transport likely contributes to ganglion cell dysfunction and death. The retrograde transport from the LGN to the retina is sensitive to pressure (Minckler *et al.*, 1977), and the decreased transport of neurotrophic factors from the brain to the retina may contribute ganglion cell malaise (Pease *et al.*, 2000). The transport of neurotrophic factors from the brain to the ganglion cell bodies in the retina is disrupted in eyes with increased IOP, with factors accumulating at the level of the lamina cribrosa (Quigley *et al.*, 1980, 2000). The protective role of neurotrophic factors is indicated by the delayed loss of ganglion cells following transaction of the optic nerve in eyes given neurotrophic factors (Mey and Thanos, 1993). Bypassing the neurotrophic factor receptors by genetic upregulation of the effector extracellular signal-regulated kinase 1/2 (Erk1/2) is also effective and increases neuronal survival in rats with ocular hypertension (Zhou *et al.*, 2005).

Although neurotrophic factors can protect ganglion cells, their impaired transport may not be a primary cause of cell injury or death. Regions of maximal transport disruption do not correlate with areas of maximal nerve damage (Ogden *et al.*, 1988). Particularly convincing was a careful study of the chronology of glaucomatous changes which found that ganglion cell death preceded the depletion of neurotrophic factors in the retina (Johnson *et al.*, 2000). This implies that the death of retinal ganglion cells may involve additional processes that are exacerbated by the reduction in protective neurotrophic factors.

B. Vascular Compromise

It is likely that elevated IOP places a metabolic strain on ganglion cells in the optic nerve head (Osborne *et al.*, 2001). These unmyelinated axons passing through the lamina cribrosa contain a high density of the mitochondrial enzymes cytochrome *c* oxidase and succinate dehydrogenase, consistent with a high energetic demand (Andrews *et al.*, 1999). As such, the region could be particularly susceptible to a reduction in the efficiency of vascular supply to the optic nerve head. Under conditions where vascular delivery is less than optimal, energetically compromised cells may be less able to deal with environmental insults. It has been proposed that the reduction in energy production could compromise the function of the Na^+–K^+ ATPase pump and depolarize the membrane (Osborne *et al.*, 2001). Transmission of this depolarization along the axon to the cell body could convey a metabolic strain initialized at the optic nerve head to the retina.

The theory that a general metabolic disorder leads to energetically challenged ganglion cells less able to withstand insults is supported experimentally. Rats with preexisting glaucoma lost many more ganglion cells than control animals after both were exposed to ischemia (Kawai *et al.*, 2001). Whether this is because glaucomatous eyes have a higher metabolic need, or because the glaucomatous cells were already on the edge of survival is not clear. It is also not certain that a reduced vascular supply is directly caused by an increased IOP or reflects a secondary disorder. The pattern of ganglion cell loss accompanying occlusion of the carotid arteries differs from that produced by elevating IOP (Osborne *et al.*, 1999a), suggesting pressure may itself initiate additional pathologies independent of its effects on blood flow.

Astrocytes make a major contribution to ganglion cell injury under conditions of vascular compromise, as well as to distortions of the laminar cribrosa discussed above. Hypoxic challenge elevates levels of intracellular calcium in astrocytes (Peers *et al.*, 2006) and reduces their ability to remove glutamate from the extracellular space (Swanson *et al.*, 1995). Astrocytes also become reactive after ischemia, triggering a number of pathological responses (Neufeld and Liu, 2003). Astrocytes in the glaucomatous optic nerve head show morphological changes and altered expression of certain proteins (Varela and Hernandez, 1997). In this respect an elevated pressure leads to increased secretion of elastin and to remodeling of the lamina cribrosa that may contribute to progressive optic atrophy (Hernandez, 2000). Anatomical connections imply astrocytes could also convey pathological signals from the optic nerve head to retinal regions, although such spreading remains to be demonstrated directly.

C. Neurochemical Imbalances

Altered levels of extracellular neurotransmitters lead to the death of cortical neurons in chronic neurodegenerative diseases, and can likewise disturb retinal ganglion cells in glaucoma. The distribution of excitatory and inhibitory receptors present on a particular ganglion cell is likely to affect health and survival; ganglion cells with increased membrane expression of excitatory receptors capable of elevating intracellular calcium would be more vulnerable, while those cells with increased expression of inhibitory receptors that lower calcium levels would be relatively protected (Osborne *et al.*, 1999b).

Although a general theory that altered neurochemical balance can alter ganglion cell function and survival has considerable merit, the precise identity of the receptors involved remains to be determined. For example, ganglion cells responsive to the inhibitory transmitter GABA had enhanced sensitivity to the excitatory transmitter NMDA as compared to cells without a GABA response (Sun *et al.*, 2003). This argues against a simple protective effect of GABA receptors alone. Expression of NMDA receptors did not correlate with susceptibility of neurons to glaucomatous cell death (Hof *et al.*, 1998). This suggests that additional neurochemicals may alter the balance and impact survival. The following sections briefly review evidence for the involvement three neurotransmitters with the potential to damage ganglion cells in glaucoma. Although a contribution from other molecules is likely, the questions addressed are of general relevance. As discussed in the final section on purines, the balance of inhibitory and excitatory responses may be modulated by both receptor expression and the availability of specific transmitters.

1. Nitric Oxide

Nitric oxide (NO) can kill motor neurons (Estevez *et al.*, 1998) and it may also contribute to the loss of ganglion cells in glaucoma (Neufeld, 2004). The enzyme responsible for the production of NO, nitric oxide synthase (NOS), is altered in both animal and human forms of the disease. Increased levels of the NOS-2 isoform were detected in reactive astrocytes from the optic nerve head of humans with glaucoma as compared to controls (Liu and Neufeld, 2000). *In vitro* experiments with astrocytes obtained from the optic nerve head of humans found that an increase in hydrostatic pressure led to an elevation of protein and mRNA for NOS-2 (Liu and Neufeld, 2001). Rats with experimental glaucoma treated with the NOS-2 inhibitor aminoguanidine had reduced rates of ganglion cell death (Neufeld *et al.*, 1999). Activation of epidermal growth factor receptor (EGFR) in astrocytes of the optic nerve head may be an early step in the astrocyte response to stress. Attenuating this activation with a tyrosine kinase inhibitor reduced ganglion cell death (Liu *et al.*, 2006). Ganglion cell loss accompanying retinal ischemia was also decreased by the NOS inhibitors aminoguanidine and methyl ester N^{ω}-nitro-L-arginine methyl ester (L-NAME), suggesting that NOS acted in a pathway common to both stresses (Geyer *et al.*, 1995; Adachi *et al.*, 1998).

As with many areas of glaucoma research, there is some inconsistency surrounding the role of NOS in glaucoma. A study on a rat model of chronic glaucoma found no evidence for an increase in either NOS-2 protein or mRNA in the ganglion cell layer or optic nerve head (Pang *et al.*, 2005). This study also failed to find an increase in immunoreactivity for NOS-2 in humans with POAG. The discovery that L-NAME can actually lower IOP in rabbits independent of any protective effects (Giuffrida *et al.*, 2003) urges caution when interpreting evidence for involvement of NOS, and for neurochemical changes in general, in glaucoma. It is of course possible that differences found between model systems actually reflect the variety of pathological disorders clustered under the heading "glaucoma," stressing another parameter to be considered.

2. Glutamate

Perhaps no topic concerning the fate of ganglion cells in glaucoma has aroused as much debate as the role of the excitatory amino acid glutamate. Evidence exists both for and against it having a major role. Finding a way to explain these discrepancies will ultimately benefit the field.

Exogenously added glutamate agonists can kill retinal ganglion cells (Lei *et al.*, 1992; Manabe and Lipton, 2003). Over-stimulation of the ionotropic NMDA glutamate receptor can lead to an excess elevation of intracellular calcium (Sucher *et al.*, 1990; Lei *et al.*, 1992) and apoptotic cell death (Lam *et al.*, 1999), consistent with a downstream activation of endonucleases and proteases typically observed in calcium-mediated apoptosis (Choi, 1988). The central role of calcium elevation in ganglion cell death triggered by NMDA is supported by the observation that inhibition of L-type calcium channels with dihydropyridine reduced cell loss (Sucher *et al.*, 1991). The ability of the NMDA receptor antagonist memantine to prevent pressure-triggered ganglion cell death in rats is consistent with the hypothesis that glutamate might indeed play a role in the endogenous pathophysiology of glaucoma (WoldeMussie *et al.*, 2002), although the effectiveness of extending this protection to patients remains to be determined.

An excess of glutamate has been associated with a secondary susceptibility of ganglion cells under conditions of ischemic challenge (Osborne *et al.*, 1999b) and optic nerve crush (Yoles and Schwartz, 1998). However, a direct link between elevated IOP, increased glutamate and stimulation of the NMDA receptor remains elusive, and numerous inconsistencies complicate the relationship. For example, NMDA preferentially kills small and medium diameter ganglion cells (Vorwerk *et al.*, 1999) while large diameter cells are more susceptible in glaucoma (Glovinsky *et al.*, 1991). The distribution of NMDA receptors was unrelated to the patterns of ganglion cell loss in primates with experimental glaucoma (Hof *et al.*, 1998). The relationship between vitreal glutamate levels and elevated IOP is at best inconsistent (Dreyer *et al.*, 1996; Levkovitch-Verbin *et al.*, 2001; Carter-Dawson *et al.*, 2002; Honkanen *et al.*, 2003). While differences in the interval between pressure increase and vitreal sampling may explain some of the discrepancies, even the ability of NMDA to kill ganglion cells is debated, with some studies suggesting they are relatively insensitive (Ullian *et al.*, 2004). It is likely that multiple factors, particularly the membrane potential and the voltage-sensitive block of the NMDA channel by Mg^{2+}, can influence the effect of NMDA.

It is also difficult to explain how elevated IOP leads to elevated levels extracellular glutamate. Initial reports suggested that retinal glutamate transporters are decreased in glaucoma (Naskar *et al.*, 2000; Martin *et al.*, 2002). However, more recent evidence suggests the opposite may occur, with levels of certain transporters elevated in response to increased IOP (WoldeMussie *et al.*, 2004; Hartwick *et al.*, 2005; Sullivan *et al.*, 2006).

Although these inconsistencies leave many issues about the role of glutamate in glaucoma unresolved, these observations do suggest the elevation of intracellular calcium by glutamate acting on NMDA receptors may be detrimental to retinal ganglion cells under some conditions. This response may contribute to glaucomatous loss of retinal ganglion cells, but it is unclear if and how elevated pressure is linked to this process.

3. Purines

Recent work from our laboratory has suggested that purines may connect increased IOP with changes in ganglion cell health and eventually even death. This hypothesis provides for a source of excess extracellular ATP, involvement of a toxic receptor, and a negative feedback system that may preserve ganglion cell health. Preliminary findings may also clarify a role of glutamate in the process. The basic hypothesis is modeled in Fig. 1.

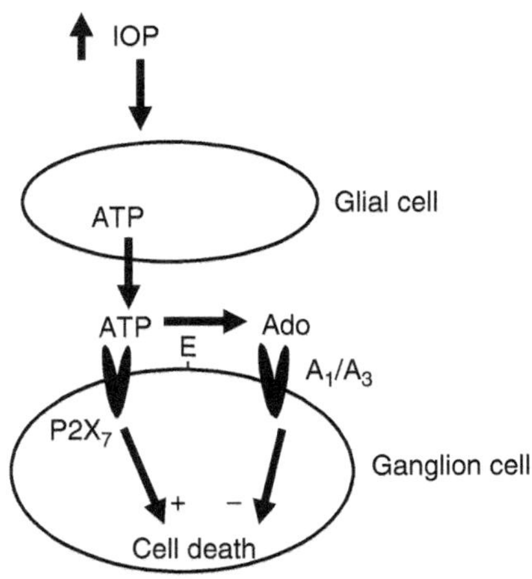

FIGURE 1 Hypothesis of effect of purines on ganglion cell health in glaucoma. Elevated pressure leads to a release of ATP. Although astrocytes are indicated, release from other sources such as Müller cells or endothelial cells is also possible. This excess extracellular ATP can stimulate $P2X_7$ receptors on retinal ganglion cells, leading to elevation of calcium, injury and eventually perhaps cell death. Alternatively, the excess ATP can be converted into to adenosine by ecto-nucleotidases (E) and prevent calcium overload and cellular damage.

a. Pressure and ATP release. Although originally known as a cellular energy source, ATP has been recognized as an extracellular transmitter for several decades (Burnstock *et al.*, 1970). It is now accepted that the extracellular signaling role of ATP has a negligible impact on cell energetics under most circumstances. ATP can be released from both neuronal and non-neuronal cells. A particularly effective trigger for ATP release from non-neural cells involves mechanical distention in the form of pressure, flow or stretch (Burnstock, 1999). We have found that ocular tissues also release ATP in response to mechanical perturbation (Mitchell *et al.*, 1998; Fleischhauer *et al.*, 2001; Mitchell, 2001). This led to the hypothesis that excess ATP might also be released in response to the changing IOP found in glaucoma.

Several preliminary lines of evidence suggest excess ATP may be released into the extracellular space in glaucoma. ATP was increased in bovine eyecups exposed to elevated hydrostatic pressure. This effect that was not due to a change in oxygen partial pressure or cell lysis (Zhang *et al.*, 2006c). Whether the observed release reflected actual hydrostatic pressure or a relative movement of the retina with respect to the sclera remains to be determined, but the response was robust and proportional to the change in pressure. The ectoATPase NTPDase1 acts as a marker for excess extracellular ATP in retinal cells (Lu *et al.*, 2007), and initial trials indicate that NTPDase1 levels are higher in retina from primate eyes with experimental glaucoma than in contralateral controls (Lu *et al.*, 2007). ATP levels are elevated fivefold in the aqueous humor of patients with acute angle closure glaucoma (Zhang *et al.*, 2007); this excess ATP could act as a precursor for the increased adenosine found in the anterior chamber of glaucomatous eyes (Daines *et al.*, 2003). The source of excess extracellular ATP in the glaucomatous posterior eye is unknown, but mechanical stimulation of astrocytes on the outer retinal layer has led to release of ATP from the Müller cells in areas adjacent to the ganglion cells, implicating glial cells in the response (Newman, 2001). The propensity of astrocytes to release ATP throughout the nervous system also suggests they could contribute to excess extracellular ATP in the retina or optic nerve head as well (Joseph *et al.*, 2003).

b. $P2X_7$ Receptors, NMDA Receptors, and Ganglion Cell Death. The effects of any excess extracellular ATP will be primarily determined by the particular receptors expressed on adjacent membranes. Several studies have demonstrated that $P2X_7$ receptors are expressed in adult retinal ganglion cells (Brandle *et al.*, 1998; Ishii *et al.*, 2003; Puthussery and Fletcher, 2004). The $P2X_7$ receptor is distinguished from $P2X_{1-6}$ receptors by its elongated C-terminus and its tendency to kill peripheral cells (Surprenant *et al.*, 1996). This finding suggested that stimulation of the $P2X_7$ receptor on ganglion cells might also be lethal. Activation of the $P2X_7$ receptor leads to an elevation of intracellular calcium that shows little inactivation, and to the death of ganglion cells (Zhang *et al.*, 2005). Although the agonist BzATP can stimulate other P2 receptors in addition to the $P2X_7$ receptor, the involvement of the $P2X_7$ was confirmed with multiple assays including inhibition by 100 nM of antagonist Brilliant Blue G. This death involved activation of caspase-3, but was not associated with the increased permeability to large fluorescent dyes associated with the activation of the $P2X_7$ receptor in peripheral cells (Surprenant *et al.*, 1996; Zhang *et al.*, 2005). Several preliminary observations suggest that the NMDA receptor may be involved in the death of retinal ganglion cells accompanying $P2X_7$ receptor stimulation (Mitchell *et al.*, 2006; Mitchell, 2008).

c. Neuroprotection by Adenosine. Extracellular nucleotidases rapidly dephosphorylate released ATP into adenosine (Robson *et al.*, 2006). This close relationship between levels of extracellular ATP and adenosine, combined with the ability of these agonists to stimulate distinct receptors, makes the study of purinergic transmission particularly complex. These contrasting effects are especially acute in retinal ganglion cells, as activation of $P2X_7$ receptors leads to cell death, while stimulation of certain adenosine receptors is neuroprotective.

In keeping with these opposing principals, a brief application of ATP clearly raised calcium levels, while prolonged exposure to ATP did not kill cells and in some cases was actually protective (Zhang *et al.*, 2006b). These protective actions of ATP were not mimicked by analogue ATPγS; levels of cell loss were similar with ATPγS and BzATP. As the bond of the terminal phosphate group in ATPγS makes it much more resistant to hydrolysis, this implied that the hydrolysis product could confer protection. The most likely candidate was adenosine, and adenosine itself was indeed able to prevent the calcium rise and death triggered by BzATP in retinal ganglion cells, consistent with the protection seen with ATP but not ATPγS (Zhang *et al.*, 2006b). This is also consistent with the general recognition of adenosine as a neuroprotective agent. Adenosine makes a major contribution to the response to retinal ischemia (Ghiardi *et al.*, 1999), while levels of adenosine rise in the ischemic retina and limit the neuronal damage (Roth *et al.*, 1997a,b).

The protective actions of adenosine following $P2X_7$ receptor activation in ganglion cells is response mediated, at least in part, by the A_3 adenosine receptor. The A_3 receptor antagonist MRS1191 prevented the ability of adenosine to block the calcium rise triggered by BzATP (Zhang *et al.*, 2006b). Cl-IB-MECA, a relatively specific agonist for the A_3 receptor, mimicked the ability of adenosine to inhibit the calcium rise triggered by BzATP. Cl-IB-MECA and another A_3 receptor agonist IB-MECA also reduced the cell death triggered by BzATP.

Although these pharmacological experiments provided clear functional evidence that the A_3 receptor is neuroprotective, molecular identification of the receptor was required as a previous *in situ* hybridization study was unable to find any message for the A_3 receptor in the rat eye (Kvanta *et al.*, 1997). However, traditional and quantitative PCR did identify the A_3 receptor in material from the ganglion cell layer of the rat retina using laser capture microdissection (Zhang *et al.*, 2006a). To ensure the message did not come from other cell types also present in the ganglion cell layer, analysis was repeated on ganglion cells isolated using the immunopanning technique. The A_3 receptor identified in ganglion cells was cloned and found to be >99% identical to that found elsewhere.

It is likely that the A_3 receptor acts with the A_1 receptor to protect ganglion cells. Agonists for the A_1 adenosine receptor are known to protect retinal ganglion cells against ischemic challenge (Larsen and Osborne, 1996), and are also known to block calcium channels on ganglion cells of amphibians (Sun *et al.*, 2002) and rats (Hartwick *et al.*, 2004). The hyperpolarization of ganglion cells by adenosine was largely inhibited by an A_1 antagonist (Newman, 2003). In other systems, the A_1 adenosine receptor inhibits voltage-dependent calcium channels, following direct block of the $Ca_{V\alpha}$ subunit by $G_{\beta\gamma}$ coupled to the A_1 receptor (Clapham, 1994; Dolphin, 2003). Although the mechanisms linking the A_3 receptor with neuroprotection on ganglion cells have not yet been fully resolved, it is likely that both A_1 and A_3 receptors share a pathway as both receptors activate PTX-sensitive Gi/Go proteins and can activate $G_{\beta\gamma}$ (Schulte and Fredholm, 2002).

The propensity of ATP to transduce mechanical stimuli into neurochemical signals, combined with the lethal effects of the $P2X_7$ receptor and the protective effects of adenosine, suggest that ATP and the $P2X_7$ receptor could provide a link between elevated IOP and the death of retinal ganglion cells. These observations also suggest that conversion of extracellular ATP into adenosine could simultaneously remove a toxic agent and produce a protective one. Upregulation of nucleotidases in response to released ATP may thus be a key adaptation that breaks the link between elevated IOP and cell death. It may also provide an entry point for intervention in the treatment of glaucoma.

V. CONCLUSION

Multiple factors are likely to compromise the function of retinal ganglion cells in glaucoma and eventually lead to their death. In so far as IOP contributes to this pathology, the regulation of aqueous humor dynamics provides the major approach currently used to protect retinal ganglion cells.

However, an awareness of the mechanisms more directly detrimental to ganglion cell survival can help tailor pharmacological approaches for successful therapies in the future.

Acknowledgements

This work was supported by NIH grants EY013434 and EY015537 to CHM. The authors thank Alan M. Laties for useful discussions.

References

Adachi, K., Fujita, Y., Morizane, C., Akaike, A., Ueda, M., Satoh, M., Masai, H., Kashii, S., and Honda, Y. (1998). Inhibition of NMDA receptors and nitric oxide synthase reduces ischemic injury of the retina. *Eur. J. Pharmacol.* **350,** 53–57.

AGIS. (2000). The Advanced Glaucoma Intervention Study (AGIS): 7. The relationship between control of intraocular pressure and visual field deterioration. The AGIS Investigators. *Am. J. Ophthalmol.* **130,** 429–440.

Anderson, R. S., and O'Brien, C. (1997). Psychophysical evidence for a selective loss of M ganglion cells in glaucoma. *Vis. Res.* **37,** 1079–1083.

Andrews, R. M., Griffiths, P. G., Johnson, M. A., and Turnbull, D. M. (1999). Histochemical localisation of mitochondrial enzyme activity in human optic nerve and retina. *Br. J. Ophthalmol.* **83,** 231–235.

Ansari, E. A., Morgan, J. E., and Snowden, R. J. (2002). Psychophysical characterisation of early functional loss in glaucoma and ocular hypertension. *Br. J. Ophthalmol.* **86,** 1131–1135.

Asrani, S., Zeimer, R., Wilensky, J., Gieser, D., Vitale, S., and Lindenmuth, K. (2000). Large diurnal fluctuations in intraocular pressure are an independent risk factor in patients with glaucoma. *J. Glaucoma* **9,** 134–142.

Bach, M., Unsoeld, A. S., Philippin, H., Staubach, F., Maier, P., Walter, H. S., Bomer, T. G., and Funk, J. (2006). Pattern ERG as an early glaucoma indicator in ocular hypertension: A long-term, prospective study. *Invest. Ophthalmol. Vis. Sci.* **47,** 4881–4887.

Brandle, U., Kohler, K., and Wheeler-Schilling, T. H. (1998). Expression of the P2X7-receptor subunit in neurons of the rat retina. *Brain Res. Mol. Brain Res.* **62,** 106–109.

Burnstock, G. (1999). Release of vasoactive substances from endothelial cells by shear stress and purinergic mechanosensory transduction. *J. Anat.* **194,** 335–342.

Burnstock, G., Campbell, G., Satchell, D., and Smythe, A. (1970). Evidence that adenosine triphosphate or a related nucleotide is the transmitter substance released by non-adrenergic inhibitory nerves in the gut. *Br. J. Pharmcol.* **40,** 668–688.

Carter-Dawson, L., Crawford, M. L., Harwerth, R. S., Smith, E. L., 3rd, Feldman, R., Shen, F. F., Mitchell, C. K., and Whitetree, A. (2002). Vitreal glutamate concentration in monkeys with experimental glaucoma. *Invest. Ophthalmol. Vis. Sci.* **43,** 2633–2637.

Cepurna, W. O., Jia, L., Johnson, E. C., and Morrison, J. C. (2006). Aging increases susceptibility to intraocular pressure (IOP) and alters retinal gene expression responses in the rat. *Invest. Ophthalmol. Vis. Sci.* **47,** 1243. (abs).

Choi, D. W. (1988). Calcium-mediated neurotoxicity: Relationship to specific channel types and role in ischemic damage. *Trends Neurosci.* **11,** 465–469.

Clapham, D. E. (1994). Direct G protein activation of ion channels? *Ann. Rev. Neurosci.* **17,** 441–464.

CNTG. (1998). Comparison of glaucomatous progression between untreated patients with normal-tension glaucoma and patients with therapeutically reduced intraocular pressures. *Am. J. Ophthalmol.* **126,** 487–497.

Dacey, D. M. (1993). The mosaic of midget ganglion cells in the human retina. *J. Neurosci.* **13,** 5334–5355.

Dacey, D. M. (1994). Physiology, morphology and spatial densities of identified ganglion cell types in primate retina. *Ciba Found. Symp.* **184,** 12–28.

Daines, B. S., Kent, A. R., McAleer, M. S., and Crosson, C. E. (2003). Intraocular adenosine levels in normal and ocular-hypertensive patients. *J. Ocul. Pharmacol. Ther.* **19,** 113–119.

Desatnik, H., Quigley, H. A., and Glovinsky, Y. (1996). Study of central retinal ganglion cell loss in experimental glaucoma in monkey eyes. *J. Glaucoma* **5,** 46–53.

Dolphin, A. C. (2003). G protein modulation of voltage-gated calcium channels. *Pharmacol. Rev.* **55,** 607–627.

Dreyer, E. B., Zurakowski, D., Schumer, R. A., Podos, S. M., and Lipton, S. A. (1996). Elevated glutamate levels in the vitreous body of humans and monkeys with glaucoma. *Arch. Ophthalmol.* **114,** 299–305.

Estevez, A. G., Spear, N., Manuel, S. M., Barbeito, L., Radi, R., and Beckman, J. S. (1998). Role of endogenous nitric oxide and peroxynitrite formation in the survival and death of motor neurons in culture. *Prog. Brain Res.* **118,** 269–280.

Ethier, C. R. (2006). Hydrostatic pressure is not a surrogate for IOP in glaucoma. *Invest. Ophthalmol. Vis. Sci.* E-letterhttp://www.iovs.org/cgi/eletters/46/48/2829.

Fleischhauer, J. C., Mitchell, C. H., Peterson-Yantorno, K., Coca-Prados, M., and Civan, M. M. (2001). PGE(2), Ca(2+), and cAMP mediate ATP activation of Cl(−) channels in pigmented ciliary epithelial cells. *Am. J. Physiol. Cell Physiol.* **281,** C1614–C1623.

Geyer, O., Almog, J., Lupu-Meiri, M., Lazar, M., and Oron, Y. (1995). Nitric oxide synthase inhibitors protect rat retina against ischemic injury. *FEBS Lett.* **374,** 399–402.

Ghiardi, G. J., Gidday, J. M., and Roth, S. (1999). The purine nucleoside adenosine in retinal ischemia-reperfusion injury. *Vis. Res.* **39,** 2519–2535.

Giuffrida, S., Bucolo, C., and Drago, F. (2003). Topical application of a nitric oxide synthase inhibitor reduces intraocular pressure in rabbits with experimental glaucoma. *J. Ocul. Pharmacol. Ther.* **19,** 527–534.

Glovinsky, Y., Quigley, H. A., and Dunkelberger, G. R. (1991). Retinal ganglion cell loss is size dependent in experimental glaucoma. *Invest. Ophthalmol. Vis. Sci.* **32,** 484–491.

Gordon, M. O., Beiser, J. A., Brandt, J. D., Heuer, D. K., Higginbotham, E. J., Johnson, C. A., Keltner, J. L., Miller, J. P., Parrish, R. K., 2nd, Wilson, M. R., and Kass, M. A. (2002). The Ocular Hypertension Treatment Study: Baseline factors that predict the onset of primary open-angle glaucoma. *Arch. Ophthalmol.* **120,** 714–720.

Grehn, F., Grusser, O. J., and Stange, D. (1984). Effect of short-term intraocular pressure increase on cat retinal ganglion cell activity. *Behav. Brain Res.* **14,** 109–121.

Gupta, N., and Yucel, Y. H. (2007). Glaucoma as a neurodegenerative disease. *Curr. Opin. Ophthalmol.* **18,** 110–114.

Hartwick, A. T. (2001). Beyond intraocular pressure: Neuroprotective strategies for future glaucoma therapy. *Optom. Vis. Sci.* **78,** 85–94.

Hartwick, A. T., Lalonde, M. R., Barnes, S., and Baldridge, W. H. (2004). Adenosine A(1)-receptor modulation of glutamate-induced calcium influx in rat retinal ganglion cells. *Invest. Ophthalmol. Vis. Sci.* **45,** 3740–3748.

Hartwick, A. T., Zhang, X., Chauhan, B. C., and Baldridge, W. H. (2005). Functional assessment of glutamate clearance mechanisms in a chronic rat glaucoma model using retinal ganglion cell calcium imaging. *J. Neurochem.* **94,** 794–807.

Hernandez, M. R. (2000). The optic nerve head in glaucoma: Role of astrocytes in tissue remodeling. *Prog. Retin Eye Res.* **19,** 297–321.

Hof, P. R., Lee, P. Y., Yeung, G., Wang, R. F., Podos, S. M., and Morrison, J. H. (1998). Glutamate receptor subunit GluR2 and NMDAR1 immunoreactivity in the retina of macaque monkeys with experimental glaucoma does not identify vulnerable neurons. *Exp. Neurol.* **153,** 234–241.

Honkanen, R. A., Baruah, S., Zimmerman, M. B., Khanna, C. L., Weaver, Y. K., Narkiewicz, J., Waziri, R., Gehrs, K. M., Weingeist, T. A., Boldt, H. C., Folk, J. C., Russell, S. R., *et al.* (2003). Vitreous amino acid concentrations in patients with glaucoma undergoing vitrectomy. *Arch. Ophthalmol.* **121,** 183–188.

Hood, D. C., Xu, L., Thienprasiddhi, P., Greenstein, V. C., Odel, J. G., Grippo, T. M., Liebmann, J. M., and Ritch, R. (2005). The pattern electroretinogram in glaucoma patients with confirmed visual field deficits. *Invest. Ophthalmol. Vis. Sci.* **46,** 2411–2418.

Ishii, K., Kaneda, M., Li, H., Rockland, K. S., and Hashikawa, T. (2003). Neuron-specific distribution of P2X7 purinergic receptors in the monkey retina. *J. Comp. Neurol.* **459,** 267–277.

Johnson, E. C., Deppmeier, L. M., Wentzien, S. K., Hsu, I., and Morrison, J. C. (2000). Chronology of optic nerve head and retinal responses to elevated intraocular pressure. *Invest. Ophthalmol. Vis. Sci.* **41,** 431–442.

Joseph, S. M., Buchakjian, M. R., and Dubyak, G. R. (2003). Colocalization of ATP release sites and ecto-ATPase activity at the extracellular surface of human astrocytes. *J. Biol. Chem.* **278,** 23331–23342.

Katz, J., Gilbert, D., Quigley, H. A., and Sommer, A. (1997). Estimating progression of visual field loss in glaucoma. *Ophthalmology* **104,** 1017–1025.

Kawai, S. I., Vora, S., Das, S., Gachie, E., Becker, B., and Neufeld, A. H. (2001). Modeling of risk factors for the degeneration of retinal ganglion cells after ischemia/reperfusion in rats: Effects of age, caloric restriction, diabetes, pigmentation, and glaucoma. *FASEB J.* **15,** 1285–1287.

Kirwan, R. P., Fenerty, C. H., Crean, J., Wordinger, R. J., Clark, A. F., and O'Brien, C. J. (2005). Influence of cyclical mechanical strain on extracellular matrix gene expression in human lamina cribrosa cells *in vitro*. *Mol. Vis.* **11,** 798–810.

Knepper, P. A., Miller, A. M., Choi, J., Wertz, R. D., Nolan, M. J., Goossens, W., Whitmer, S., Yue, B. Y., Ritch, R., Liebmann, J. M., Allingham, R. R., and Samples, J. R. (2005). Hypophosphorylation of aqueous humor sCD44 and primary open-angle glaucoma. *Invest. Ophthalmol. Vis. Sci.* **46,** 2829–2837.

Kolb, H., Linberg, K. A., and Fisher, S. K. (1992). Neurons of the human retina: A Golgi study. *J. Comp. Neurol.* **318,** 147–187.

Kvanta, A., Seregard, S., Sejersen, S., Kull, B., and Fredholm, B. B. (1997). Localization of adenosine receptor messenger RNAs in the rat eye. *Exp. Eye Res.* **65,** 595–602.

Lam, T. T., Abler, A. S., Kwong, J. M., and Tso, M. O. (1999). *N*-Methyl-D-aspartate (NMDA)-induced apoptosis in rat retina. *Invest. Ophthalmol. Vis. Sci.* **40,** 2391–2397.

Laquis, S., Chaudhary, P., and Sharma, S. C. (1998). The patterns of retinal ganglion cell death in hypertensive eyes. *Brain Res.* **784,** 100–104.

Larsen, A. K., and Osborne, N. N. (1996). Involvement of adenosine in retinal ischemia. Studies on the rat. *Invest. Ophthalmol. Vis. Sci.* **37,** 2603–2611.

Lei, S. Z., Zhang, D., Abele, A. E., and Lipton, S. A. (1992). Blockade of NMDA receptor-mediated mobilization of intracellular Ca2+ prevents neurotoxicity. *Brain Res.* **598,** 196–202.

Levin, L., and Gordon, L. (2002). Retinal ganglion cell disorders: Types and treatments. *Prog. Retin Eye Res.* **21,** 465–484.

Levkovitch-Verbin, H., Quigley, H. A., Kerrigan-Baumrind, L. A., D'Anna, S. A., Kerrigan, D., and Pease, M. E. (2001). Optic nerve transection in monkeys may result in secondary degeneration of retinal ganglion cells. *Invest. Ophthalmol. Vis. Sci.* **42,** 975–982.

Liu, B., and Neufeld, A. H. (2000). Expression of nitric oxide synthase-2 (NOS-2) in reactive astrocytes of the human glaucomatous optic nerve head. *Glia* **30,** 178–186.

Liu, B., and Neufeld, A. H. (2001). Nitric oxide synthase-2 in human optic nerve head astrocytes induced by elevated pressure *in vitro*. *Arch. Ophthalmol.* **119,** 240–245.

Liu, B., Chen, H., Johns, T. G., and Neufeld, A. H. (2006). Epidermal growth factor receptor activation: An upstream signal for transition of quiescent astrocytes into reactive astrocytes after neural injury. *J. Neurosci.* **26,** 7532–7540.

Lu, W., Reigada, D., Sevigny, J. and Mitchell, C.H. (2007). Stimulation of the P2Y1 receptor upregulates NTPDase1 in human retinal pigment epithelial cells. *J. Pharmacol. Exp. Therapeut* **323,** 157–164.

Lu, W., Rasmussen, C., Gabelt, B., Hennes, B., Kaufman, P., Laties, A. M., and Mitchell, C. H. (2007). Upregulation of NTPDase 1 in an experimental monkey glaucoma model. *Invest. Ophthalmol. Vis. Sci.* **48,** 4804. (abs).

Manabe, S., and Lipton, S. A. (2003). Divergent NMDA signals leading to proapoptotic and antiapoptotic pathways in the rat retina. *Invest. Ophthalmol. Vis. Sci.* **44,** 385–392.

Martin, K. R., Levkovitch-Verbin, H., Valenta, D., Baumrind, L., Pease, M. E., and Quigley, H. A. (2002). Retinal glutamate transporter changes in experimental glaucoma and after optic nerve transection in the rat. *Invest. Ophthalmol. Vis. Sci.* **43,** 2236–2243.

Martinez-Bello, C., Chauhan, B. C., Nicolela, M. T., McCormick, T. A., and LeBlanc, R. P. (2000). Intraocular pressure and progression of glaucomatous visual field loss. *Am. J. Ophthalmol.* **129,** 302–308.

Maurice, D. M. (1987). Flow of water between aqueous and vitreous compartments in the rabbit eye. *Am. J. Physiol.* **252,** F104–F108.

Mey, J., and Thanos, S. (1993). Intravitreal injections of neurotrophic factors support the survival of axotomized retinal ganglion cells in adult rats *in vivo*. *Brain Res.* **602,** 304–317.

Minckler, D. S., Bunt, A. H., and Johanson, G. W. (1977). Orthograde and retrograde axoplasmic transport during acute ocular hypertension in the monkey. *Invest. Ophthalmol. Vis. Sci.* **16,** 426–441.

Mitchell, C. H. (2001). Release of ATP by a human retinal pigment epithelial cell line: Potential for autocrine stimulation through subretinal space. *J. Physiol.* **534,** 193–202.

Mitchell, C. H. (2008). P2X7 receptor kills retinal ganglion cells with glutamate release and excitotoxic NMDA receptor activation. *Purinergic Signalling* **2,** Supp 1.

Mitchell, C. H., Carre, D. A., McGlinn, A. M., Stone, R. A., and Civan, M. M. (1998). A release mechanism for stored ATP in ocular ciliary epithelial cells. *Proc. Natl. Acad. Sci. USA* **95,** 7174–7178.

Mitchell, C. H., Zhang, M., Zhang, X., Lu, W., Reigada, D., and Laties, A. M. (2006). Neuronal death evoked by the P2X7 receptor mediated by the NMDA receptor. *Invest. Ophthalmol. Vis. Sci.* **47,** 2589. (abs).

Morgan, J. E. (2000). Optic nerve head structure in glaucoma: Astrocytes as mediators of axonal damage. *Eye* **14,** 437–444.

Morgan, J. E. (2002). Retinal ganglion cell shrinkage in glaucoma. *J. Glaucoma* **11,** 365–370.

Morgan, J. E., Uchida, H., and Caprioli, J. (2000). Retinal ganglion cell death in experimental glaucoma. *Br. J. Ophthalmol.* **84,** 303–310.

Morrison, J. C., Johnson, E. C., Cepurna, W., and Jia, L. (2005). Understanding mechanisms of pressure-induced optic nerve damage. *Prog. Retin Eye Res.* **24,** 217–240.

Naskar, R., Vorwerk, C. K., and Dreyer, E. B. (2000). Concurrent downregulation of a glutamate transporter and receptor in glaucoma. *Invest. Ophthalmol. Vis. Sci.* **41,** 1940–1944.

Neufeld, A. H. (2004). Pharmacologic neuroprotection with an inhibitor of nitric oxide synthase for the treatment of glaucoma. *Brain Res. Bull.* **62,** 455–459.

Neufeld, A. H., and Liu, B. (2003). Glaucomatous optic neuropathy: When glia misbehave. *Neuroscientist* **9,** 485–495.

Neufeld, A. H., Sawada, A., and Becker, B. (1999). Inhibition of nitric-oxide synthase 2 by aminoguanidine provides neuroprotection of retinal ganglion cells in a rat model of chronic glaucoma. *Proc. Natl. Acad. Sci. USA* **96,** 9944–9948.

Newman, E. A. (2001). Propagation of intercellular calcium waves in retinal astrocytes and Muller cells. *J. Neurosci.* **21,** 2215–2223.

Newman, E. A. (2003). Glial cell inhibition of neurons by release of ATP. *J. Neurosci.* **23,** 1659–1666.

Ogden, T. E., Duggan, J., Danley, K., Wilcox, M., and Minckler, D. S. (1988). Morphometry of nerve fiber bundle pores in the optic nerve head of the human. *Exp. Eye Res.* **46,** 559–568.

Osborne, N. N., Ugarte, M., Chao, M., Chidlow, G., Bae, J. H., Wood, J. P., and Nash, M. S. (1999a). Neuroprotection in relation to retinal ischemia and relevance to glaucoma. *Surv. Ophthalmol.* **43,** S102–S128.

Osborne, N. N., Wood, J. P., Chidlow, G., Bae, J. H., Melena, J., and Nash, M. S. (1999b). Ganglion cell death in glaucoma: What do we really know? *Br. J. Ophthalmol.* **83,** 980–986.

Osborne, N. N., Melena, J., Chidlow, G., and Wood, J. P. (2001). A hypothesis to explain ganglion cell death caused by vascular insults at the optic nerve head: Possible implication for the treatment of glaucoma. *Br. J. Ophthalmol.* **85,** 1252–1259.

Osborne, N. N., Chidlow, G., Wood, J., and Casson, R. (2003). Some current ideas on the pathogenesis and the role of neuroprotection in glaucomatous optic neuropathy. *Eur. J. Ophthalmol.* **13,** S19–S26.

Pang, I. H., Johnson, E. C., Jia, L., Cepurna, W. O., Shepard, A. R., Hellberg, M. R., Clark, A. F., and Morrison, J. C. (2005). Evaluation of inducible nitric oxide synthase in glaucomatous optic neuropathy and pressure-induced optic nerve damage. *Invest. Ophthalmol. Vis. Sci.* **46,** 1313–1321.

Pease, M. E., McKinnon, S. J., Quigley, H. A., Kerrigan-Baumrind, L. A., and Zack, D. J. (2000). Obstructed axonal transport of BDNF and its receptor TrkB in experimental glaucoma. *Invest. Ophthalmol. Vis. Sci.* **41,** 764–774.

Peers, C., Kang, P., Boyle, J. P., Porter, K. E., Pearson, H. A., Smith, I. F., and Kemp, P. J. (2006). Hypoxic regulation of Ca2+ signalling in astrocytes and endothelial cells. *Novartis Found. Symp.* **272,** 119–127.

Pereira, M. L., Kim, C. S., Zimmerman, M. B., Alward, W. L., Hayreh, S. S., and Kwon, Y. H. (2002). Rate and pattern of visual field decline in primary open-angle glaucoma. *Ophthalmology* **109,** 2232–2240.

Porciatti, V., Saleh, M., and Nagaraju, M. (2007). The pattern electroretinogram as a tool to monitor progressive retinal ganglion cell dysfunction in the DBA/2J mouse model of glaucoma. *Invest. Ophthalmol. Vis. Sci.* **48,** 745–751.

Puthussery, T., and Fletcher, E. L. (2004). Synaptic localization of P2X7 receptors in the rat retina. *J. Comp. Neurol.* **472,** 13–23.

Quigley, H. A. (1996). Number of people with glaucoma worldwide. *Br. J. Ophthalmol.* **80,** 389–393.

Quigley, H. A., and Addicks, E. M. (1981). Regional differences in the structure of the lamina cribrosa and their relation to glaucomatous optic nerve damage. *Arch. Ophthalmol.* **99,** 137–143.

Quigley, H. A., and Green, W. R. (1979). The histology of human glaucoma cupping and optic nerve damage: Clinicopathologic correlation in 21 eyes. *Ophthalmology* **86,** 1803–1830.

Quigley, H. A., Flower, R. W., Addicks, E. M., and McLeod, D. S. (1980). The mechanism of optic nerve damage in experimental acute intraocular pressure elevation. *Invest. Ophthalmol. Vis. Sci.* **19,** 505–517.

Quigley, H. A., Addicks, E. M., Green, W. R., and Maumenee, A. E. (1981). Optic nerve damage in human glaucoma. II. The site of injury and susceptibility to damage. *Arch. Ophthalmol.* **99,** 635–649.

Quigley, H. A., Hohman, R. M., Addicks, E. M., Massof, R. W., and Green, W. R. (1983). Morphologic changes in the lamina cribrosa correlated with neural loss in open-angle glaucoma. *Am. J. Ophthalmol.* **95,** 673–691.

Quigley, H. A., Sanchez, R. M., Dunkelberger, G. R., L'Hernault, N. L., and Baginski, T. A. (1987). Chronic glaucoma selectively damages large optic nerve fibers. *Invest. Ophthalmol. Vis. Sci.* **28,** 913–920.

Quigley, H. A., McKinnon, S. J., Zack, D. J., Pease, M. E., Kerrigan-Baumrind, L. A., Kerrigan, D. F., and Mitchell, R. S. (2000). Retrograde axonal transport of BDNF in retinal ganglion cells is blocked by acute IOP elevation in rats. *Invest. Ophthalmol. Vis. Sci.* **41,** 3460–3466.

Robson, S. C., Sévigny, J., and Zimmermann, H. (2006). The E-NTPDase family of ectonucleotidases: Structure, function, relationships and pathophysiological significance. *Purinergic Signal.* **2,** 409–430.

Roth, S., Park, S. S., Sikorski, C. W., Osinski, J., Chan, R., and Loomis, K. (1997a). Concentrations of adenosine and its metabolites in the rat retina/choroid during reperfusion after ischemia. *Curr. Eye Res.* **16,** 875–885.

Roth, S., Rosenbaum, P. S., Osinski, J., Park, S. S., Toledano, A. Y., Li, B., and Moshfeghi, A. A. (1997b). Ischemia induces significant changes in purine nucleoside concentration in the retina-choroid in rats. *Exp. Eye Res.* **65,** 771–779.

Sawada, A., and Neufeld, A. H. (1999). Confirmation of the rat model of chronic, moderately elevated intraocular pressure. *Exp. Eye Res.* **69,** 525–531.

Schulte, G., and Fredholm, B. B. (2002). Signaling pathway from the human adenosine A(3) receptor expressed in Chinese hamster ovary cells to the extracellular signal-regulated kinase 1/2. *Mol. Pharmacol.* **62,** 1137–1146.

Sehi, M., Flanagan, J. G., Zeng, L., Cook, R. J., and Trope, G. E. (2005). Relative change in diurnal mean ocular perfusion pressure: A risk factor for the diagnosis of primary open-angle glaucoma. *Invest. Ophthalmol. Vis. Sci.* **46,** 561–567.

Sigal, I. A., Flanagan, J. G., and Ethier, C. R. (2005). Factors influencing optic nerve head biomechanics. *Invest. Ophthalmol. Vis. Sci.* **46,** 4189–4199.

Sucher, N. J., Wong, L. A., and Lipton, S. A. (1990). Redox modulation of NMDA receptor-mediated Ca2+ flux in mammalian central neurons. *Neuroreport* **1,** 29–32.

Sucher, N. J., Lei, S. Z., and Lipton, S. A. (1991). Calcium channel antagonists attenuate NMDA receptor-mediated neurotoxicity of retinal ganglion cells in culture. *Br. Res.* **551,** 297–302.

Sullivan, R. K., Woldemussie, E., Macnab, L., Ruiz, G., and Pow, D. V. (2006). Evoked expression of the glutamate transporter GLT-1c in retinal ganglion cells in human glaucoma and in a rat model. *Invest. Ophthalmol. Vis. Sci.* **47,** 3853–3859.

Sun, X., Barnes, S., and Baldridge, W. H. (2002). Adenosine inhibits calcium channel currents via A1 receptors on salamander retinal ganglion cells in a mini-slice preparation. *J. Neurochem.* **81,** 550–556.

Sun, D., Rait, J. L., and Kalloniatis, M. (2003). Inner retinal neurons display differential responses to N-methyl-D-aspartate receptor activation. *J. Comp. Neurol.* **465,** 38–56.

Surprenant, A., Rassendren, F., Kawashima, E., North, R. A., and Buell, G. (1996). The cytolytic P2Z receptor for extracellular ATP identified as a P2X receptor (P2X7). *Science* **272,** 735–738.

Swanson, R. A., Farrell, K., and Simon, R. P. (1995). Acidosis causes failure of astrocyte glutamate uptake during hypoxia. *J. Cereb. Blood Flow Metab.* **15,** 417–424.

Tezel, G. (2006). Oxidative stress in glaucomatous neurodegeneration: Mechanisms and consequences. *Prog. Retin Eye Res.* **25,** 490–513.

Tezel, G., Siegmund, K. D., Trinkaus, K., Wax, M. B., Kass, M. A., and Kolker, A. E. (2001). Clinical factors associated with progression of glaucomatous optic disc damage in treated patients. *Arch. Ophthalmol.* **119,** 813–818.

Ullian, E. M., Barkis, W. B., Chen, S., Diamond, J. S., and Barres, B. A. (2004). Invulnerability of retinal ganglion cells to NMDA excitotoxicity. *Mol. Cell Neurosci.* **26,** 544–557.

Varela, H. J., and Hernandez, M. R. (1997). Astrocyte responses in human optic nerve head with primary open-angle glaucoma. *J. Glaucoma* **6,** 303–313.

Ventura, L. M., and Porciatti, V. (2005). Restoration of retinal ganglion cell function in early glaucoma after intraocular pressure reduction: A pilot study. *Ophthalmology* **112,** 20–27.

Ventura, L. M., Sorokac, N., De Los Santos, R., Feuer, W. J., and Porciatti, V. (2006). The relationship between retinal ganglion cell function and retinal nerve fiber thickness in early glaucoma. *Invest. Ophthalmol. Vis. Sci.* **47,** 3904–3911.

Vorwerk, C. K., Kreutz, M. R., Bockers, T. M., Brosz, M., Dreyer, E. B., and Sabel, B. A. (1999). Susceptibility of retinal ganglion cells to excitotoxicity depends on soma size and retinal eccentricity. *Curr. Eye Res.* **19,** 59–65.

Votruba, M. (2004). Molecular genetic basis of primary inherited optic neuropathies. *Eye* **18,** 1126–1132.

Wax, M. B., and Tezel, G. (2002). Neurobiology of glaucomatous optic neuropathy: Diverse cellular events in neurodegeneration and neuroprotection. *Mol. Neurobiol.* **26,** 45–55.

Weber, A. J., Kaufman, P. L., and Hubbard, W. C. (1998). Morphology of single ganglion cells in the glaucomatous primate retina. *Invest. Ophthalmol. Vis. Sci.* **39,** 2304–2320.

Weinstein, G. W., Arden, G. B., Hitchings, R. A., Ryan, S., Calthorpe, C. M., and Odom, J. V. (1988). The pattern electroretinogram (PERG) in ocular hypertension and glaucoma. *Arch. Ophthalmol.* **106,** 923–928.

WoldeMussie, E., Yoles, E., Schwartz, M., Ruiz, G., and Wheeler, L. A. (2002). Neuroprotective effect of memantine in different retinal injury models in rats. *J. Glaucoma* **11,** 474–480.

WoldeMussie, E., Wijono, M., and Ruiz, G. (2004). Muller cell response to laser-induced increase in intraocular pressure in rats. *Glia* **47,** 109–119.

Yoles, E., and Schwartz, M. (1998). Potential neuroprotective therapy for glaucomatous optic neuropathy. *Surv. Ophthalmol.* **42,** 367–372.

Zhang, X., Zhang, M., Laties, A. M., and Mitchell, C. H. (2005). Stimulation of $P2X_7$ receptors elevates Ca^{2+} and kills retinal ganglion cells. *Invest. Ophthalmol. Vis. Sci.* **46,** 2183–2191.

Zhang, M., Budak, M. T., Lu, W., Khurana, T. S., Zhang, X., Laties, A. M., and Mitchell, C. H. (2006a). Identification of the A3 adenosine receptor in rat retinal ganglion cells. *Mol. Vis.* **12,** 937–948.

Zhang, X., Zhang, M., Laties, A. M., and Mitchell, C. H. (2006b). Balance of purines may determine life or death as A3 adenosine receptors prevent loss of retinal ganglion cells following P2X7 receptor stimulation. *J. Neurochem.* **98,** 566–575.

Zhang, X., Reigada, D., Zhang, M., Laties, A. M., and Mitchell, C. H. (2006c). Increased ocular pressure increases vitreal levels of ATP. *Invest. Ophthalmol. Vis. Sci.* **47,** 426. (abs).

Zhang, X., Li, A., Ge, J., Reigada, D., Laties, A. M., and Mitchell, C. H. (2007). Acute increase of intraocular pressure releases ATP into the anterior chamber. *Exp. Eye Res.* **85,** 637–643.

Zhou, Y., Pernet, V., Hauswirth, W. W., and Di Polo, A. (2005). Activation of the extracellular signal-regulated kinase 1/2 pathway by AAV gene transfer protects retinal ganglion cells in glaucoma. *Mol. Ther. J. Am. Soc. Gene Ther.* **12,** 402–412.

CHAPTER 11

What is Functional Genomics Teaching us about Intraocular Pressure Regulation and Glaucoma?

Teresa Borrás

Department of Ophthalmology, University of North Carolina School of Medicine, Chapel Hill, North Carolina

Current Topics in Membranes, Volume 62 1063-5823/08 $35.00

DOI: 10.1016/S1063-5823(08)00411-0

I. OVERVIEW

Glaucoma is a complex disease characterized by the degeneration of the optic nerve and subsequent loss of vision. The disease affects a large number of people, especially those among the older population. The causes leading to the development of glaucoma can be the result of the dysfunction of several tissues. The major risk for glaucoma is elevated intraocular pressure which is the consequence of an improper regulation of aqueous humor outflow in the anterior segment of the eye (ciliary body and trabecular meshwork). Elevated pressure in turn, affects the posterior segment by exerting mechanical strain on the optic nerve and provoking deformation, distortion of axonal flow and death of the nerve cells (lamina cribrosa and retinal ganglion cells). The function of any given tissue is governed by the regulation of expression of its genes. Functional genomics offers an invaluable tool to identify the molecular differences affecting the cells under conditions associated with glaucoma. In this chapter, we have reviewed such changes. Because of the wide scope of the project, we have briefly reviewed studies in all tissues and then concentrated in those conducted on the trabecular meshwork, the keeper of outflow resistance. We have analyzed and correlated our internal data with that of other laboratories, and built a molecular signature for trabecular meshwork pathophysiology. We identified and present a set of 40 genes. These include new genes involved in known trabecular meshwork physiology as well as genes representative of new processes and mechanisms. As the functional genomics studies on RNA, proteins, protein modifications and activities continues this list will be further edited. As of today, this signature provides the opening of a small window into the global mechanisms governing the development of glaucoma.

II. INTRODUCTION

In the beginning, it was just the gene. The gene was embedded into this vast amount of DNA which needed to be deciphered, sequenced, cataloged and arranged in bits of expressed tiny fractions (expressed sequence tags, EST). The task appeared insurmountable but it was accomplished sooner than previously scheduled (Pennisi, 2001). After defining the structure of the gene, came functional genomics. Functional genomics comprises the elucidation of the genes' function, its transcription into mRNA, its translation into protein and the subsequent protein modifications which continue to occur during the life of proteins in any cell. All these regulations of gene expression, each different level, will determine the function of the tissue during its physiological and pathological conditions. The functional genomics task is thus specifically targeted to a given tissue or disease and it will take much longer to complete than the "simple" sequence of the genome.

Glaucoma is a blinding disease affecting 70 million people worldwide (Quigley, 1993, 1996; Kupfer *et al.*, 1994). Glaucoma is diagnosed clinically by a reduction of the visual field and by a change in the structure of the optic nerve, which acquires a characteristic "cupping" shape. The landmark of the development of glaucoma is the death of the retinal ganglion cells (RGC). It is generally accepted that the major risk for the development of glaucoma and thus RGC depletion is elevated intraocular pressure (IOP) (Anderson, 1989; Sommer *et al.*, 1991; Gordon *et al.*, 2002; Kass *et al.*, 2002; Higginbotham *et al.*, 2004). Nevertheless, an increasing number of reports show development of normal tension glaucoma, or individuals that experience glaucoma symptoms within an IOP considered to be inside the normal range (Werner, 1996). Whether these so called "normal" pressures are normal for the affected individual, or whether they reflect a difference pressure threshold of each individual is not yet known. In addition to pressure, other factors, such glucocorticoids have been undoubtedly associated in the clinic with the development of elevated IOP and glaucoma (Armaldy, 1963a,b; Becker and Mills, 1963; Grant, 1963b; Becker and Hank, 1964; Armaldy, 1965; Armaldy and Becker, 1965; Becker, 1965; Armaldy, 1966; Francois, 1977; Weinreb *et al.*, 1985; Jones and Rhee, 2006). Others, such an increased presence of the TGFβ2 growth factor has been observed in the aqueous of patients with the disease (Cousins *et al.*, 1991; Tripathi *et al.*, 1994b; Picht *et al.*, 2001). Still others, such the genetically glaucoma-linked myocilin have not yet yielded a clear mechanistic correlation (Borrás *et al.*, 2002, 2006; Tamm, 2002).

In addition to considering general factors affecting the development of glaucoma, it is important to take into account the number of different cell types and tissues which can contribute to the physiological maintenance of IOP and thus can be considered as contributors to the development of glaucoma. The ciliary epithelium secretes the flowing aqueous humor and has the ability to release molecules that will reach and signal the trabecular meshwork. The trabecular meshwork maintains the pressure through a number of cellular and molecular mechanisms which affect the physical properties of elasticity, tension and softness. The death of the RGC is affected not only by pressure but by other factors, like blood flow and ischemia. And the RGC axons forming the optic nerve are supported at their exit from the eye by the glial cells in the lamina cribrosa, whose health would affect the optic nerve at what is called the optic nerve head (ONH), site of the cupping in glaucoma.

In this chapter we review the impact of functional genomics on the understanding of the function of tissues which play a role in the development of the disease as well as on the elucidation of their response to glaucoma associated agents. We will briefly review the number of microarray studies on all tissues involved, and extend and focus on the genes and function of the trabecular meshwork, field of our expertise.

III. FUNCTIONAL GENOMICS: MICROARRAYS, PROTEOMICS AND PROTEIN MODIFICATION

When encountered with a given disease or insult, the cell of a certain tissue reacts by regulating the expression of its genes. This regulation occurs at several different levels, which overall can be reduced to three: transcription of the gene's DNA into mRNA, translation of the mRNA into protein, and modification of the protein molecule by a number of chemical reactions which would determine its folding, targeting for degradation and all its biochemical and physiological properties. The end point of all these levels of regulation is function. While transcription and translation of a gene could be normal, a hostile cell environment could promote generation of free radicals or favor deamination of an amino acid residue at a key location and consequently alter its normal physiological role. Analysis of gene regulation at each of the three steps is necessary to elucidate the disease mechanism and to be able to design rational drugs. The technology to investigate each of these regulatory levels has advanced considerably and is providing us with a wealth of information. If anything, we are now at the crossroads of trying to figure out the relationship between the observed changes on known genes, the new mechanisms revealed by the presence of genes previously unknown to the tissue, and what does it all mean for the overall physiology.

A. Genechips/Microarrays

The GeneChip/microarray system allows us to investigate the mRNA population of any cell (*transcriptome*) and analyze transcriptional differences between different conditions at the level of the entire genome. GeneChip microarrays determine whether a gene is transcribed or not, and whether the entire transcriptome is modulated or altered by the condition (s) under study. Most commonly Genechip/microarrays are used to analyze differences between a control and experimental condition, however, they are increasingly being used to compare changes across multiple experimental conditions/temporal changes. This GeneChip system consists of a collection of small oligonucleotide DNA fragments which have been embedded in a glass chip by optical lithography technology. The oligonucleotide sequences correspond to a collection of genes extracted from the human or animal's specific databases. In the chip, each gene of the database is represented by a number of different spots (probe set) and genes corresponding to internal controls are also included. The sequences of all spots in the chip are selected in such a way that they will undergo hybridization under similar annealing conditions to the whole population of a cell's RNA. Total RNAs extracted from a cell or tissue, affected or not by a given condition, are fluorescently labeled and hybridized (usually at Universities' core facilities) to a set of identical chips. Sophisticated software analysis packages allow determination of the degree of differential expression of each of the genes between both conditions. Although most investigators tend to confirm the data of a few genes of interest using real-time PCR, the continuous improving quality of the chips together with new and upgraded versions of analysis programs may soon preclude that step.

A number of commercial entities are offering GeneChips. These chips contain from thousands of genes representing close to the entire human genome, to a few hundred genes which are linked to a particular function or characteristic (pathway-focused microarrays). Mechanisms such as osteogenesis, signal transduction, or extracellular matrix can be independently studied. Chips with DNA representing the whole genome from most common experimental organisms such as rat, mouse or drosophila are also available. The pathway- or disease-focused pathways arrays contain less than 400 genes and use nylon membranes rather than glass slides. The membranes have a high DNA capacity and are produced by binding specifically-designed 60-mer oligonucleotides using a non-contact printing technology that prevents membrane deformations. Amounts, as small as 100 ng of total RNA are biotin-labeled, hybridized to the membrane, treated with a conjugated streptavidin molecule (usually alkaline phosphatase) and developed using chemiluminescence substrates. Special software analyzes the different spot intensities of the membrane. In addition to being focused, this technology is less costly and can be performed with standard laboratory equipment. A later, upgraded procedure involves the use of microarray 96-well plates containing in each well SYBR Green-labeled primer sets for the relevant genes of the pathway-disease (http://www.superarray.com/PCRArrayPlate.php). Addition of the reverse transcriptase reactions from the treated and untreated samples to two identical plates can then be run in a standard real-time PCR instrument, rather than in a special core facility. In addition, this latest procedure allows the use of RNA amounts as small as 5 nanograms, an important consideration when dealing with individual small tissue samples.

B. Proteomics and Protein Arrays

Proteomics refers to the identification of the protein population (*proteome*) of a cell at a given time and/or under a sought after condition. There are a few strategies to identify a protein population. The original, conventional procedure used two dimensional (2D) gel electrophoresis. This procedure involved the separation of proteins by two of their main characteristics, molecular weight and isoelectric point (pI). The separation of these properties (mass and charge) is conducted by gel electrophoresis in perpendicular directions. Proper staining at the end of the runs results in a number of spots spread along the surface of the gel. Because it is quite unlikely that two proteins would have two identical characteristics, the 2D separation between proteins is greatly improved from the one direction gel method. Proteins are subsequently identified by extracting them from the gel spots and subject them to peptide cleaving and sequencing by mass spectrometry. The cataloging of the proteins at any of the cell stages is of utmost importance to understand molecular mechanisms of a disease.

A later technology uses protein arrays. In a similar manner than the arrays used for nucleic acids, proteins are immobilized on solid surfaces that include glass, membranes, microtiter wells, mass spectrometer plates or beads. While the nucleic acid arrays are based on molecular hybridization and PCR, protein arrays are based on protein–protein interactions, which could be protein–antibody, protein–protein, protein–ligand or protein–drug and enzyme–substrate bindings. These arrays are however more complicated than those containing nucleic acids, especially because the complexity of the human proteome far exceeds that of the genome. Taken into account alternative gene splicing and post-translational modifications, the number of different protein molecules species in humans could be at least one order of magnitude higher than the number of genes, that is about 500,000 (http://www.functionalgenomics.org.uk/sections/resources/protein_arrays.htm#research). The feasibility and extent of protein expression profiles will then depend on the number of capture reagents available, such as antibodies, for the proteins of interest.

C. Protein Modifications

In addition to alternative splicing, where a single gene can give rise to proteins containing different domains, an extensive number of post-translational modifications are responsible for the diversity of the human proteome. These post-translational modifications are often key for maintenance of physiological conditions and/or development of diseases (Krueger and Srivastava, 2006). Some of the post-translational modifications associated with signaling and diseases include phosphorylation, glycosylation, ubiquitinization, prenylation, oxidation, and citrullination. To date, technology to detect protein modifications is limited to study one modification at a time. For example, in what is called phosphoproteomics, a number of antibodies developed to specific phosphorylated sites are incorporated into protein arrays. These arrays become very useful in cases where there is a knowledge of an altered particular pathway which include phosphorylated proteins associated with a given disease such as cancer, (Sheehan *et al.*, 2005). Another example, albeit less targeted, is the use of a glycomic profiling approach by isolating the total population of N-linked oligosaccharides proteins and comparing the glycan profiles between the serum of patients having the disease with normal controls (Moniaux *et al.*, 2004). Still another is the use of an activity-base protein profile (ABPP) which measures the activity of a particular class of enzymes. This approach uses active-site directed probes which detect the functional state of the enzymes (Liu *et al.*, 1999; Jessani and Cravatt, 2004; Speers and Cravatt, 2004). Binding of the probes to the arrays will label and identify active enzymes, but not their inactive precursor (Kidd *et al.*, 2001; Saghatelian *et al.*, 2004), or their inhibitor-bound forms (Greenbaum *et al.*, 2002; Jessani *et al.*, 2002; Saghatelian *et al.*, 2004). Thus these arrays will serve to detect disease associated changes which may occur in the absence of their transcriptional and/or translational abundance (Joyce *et al.*, 2004; Jessani *et al.*, 2005). A representative case could be seen in the study by Sieber *et al.* (2006) who created a library of chemical probes directed to metalloprotease's activities in a biological system. Such a library of metalloprotease-directed labeled probes identified the different activity of more than twenty metalloproteases in human cancer cell lines from invasive and noninvasive carcinoma (Sieber *et al.*, 2006). Innovative approaches similar to the one described to detect active sites of family of proteins are expected to be developed in the coming years.

IV. TISSUES INVOLVED IN THE DEVELOPMENT OF GLAUCOMA. SURVEY OF MICROARRAY STUDIES

Glaucoma is a complex disease and, as mentioned above, development of this optic neuropathy can occur as a result of the physiological dysfunction of more than one tissue. The final outcome of the disease, vision loss, could be triggered by a pathological secretion of aqueous humor (ciliary body), an altered resistance to aqueous humor outflow (trabecular meshwork), a pressure independent insult for the RGC or, by a loss of support to the optic nerve (ONH astrocytes).

The past few years have witnessed a number of functional genomics studies comparing either normal cells-tissue versus glaucoma or, normal cell-tissues versus treatments with glaucomatous associated insults. Most approaches have involved determination of differences at the transcriptional level. They have used high density oligonucleotides microarrays to find whether the global expression of thousands of genes has changed. In contrast, very few proteomic analysis and no post-translational modifications or activity profiles are yet available for any of the glaucoma tissues.

A. The Ciliary Body

The ciliary body comprises several structures which play an important role in the physiology of the eye. Two of them are relevant to the regulation of aqueous humor outflow. The first is the ciliary muscle which runs longitudinally and is directly inserted into the trabecular meshwork. Contraction of this muscle pulls on the trabecular meshwork affecting its shape and as a consequence, influencing the free passing of the aqueous humor through the tissue. The ciliary muscle is also the site of the uveoscleral outflow which constitutes the alternative route for the exit of aqueous humor of the eye and an important drainage pathway in some animal species (Bill, 1965; Bill and Hellsing, 1965). The second structure of the ciliary body is the ciliary epithelium which is responsible for the secretion of aqueous humor. This epithelium consists of two well differentiated layers, pigmented and nonpigmented, which have the same embryological origin than the pigmented epithelium and neuroretina layers of the retina.

There are numerous studies addressing individual gene expression, both directly on the microdissected ciliary processes tissue and on transformed cell lines originated from the two layers of the ciliary epithelium. These cell lines were carefully established by the separation of the pigmented and nonpigmented layers and have proven to be an excellent source for investigating potential new signaling mechanisms of this tissue (Coca-Prados and Escribano, 2007). However, to date, there are very few microarray studies available. A review of the current literature found two microarray reports, one carried out on whole human ciliary body tissue (dissected together with the iris and compared to other eye compartments) (Diehn *et al.*, 2005) and the second one conducted on human ciliary muscle cells exposed to analogs of glaucoma drugs (Zhao *et al.*, 2003).

B. Trabecular Meshwork

The trabecular meshwork is an avascular tissue whose main function is the maintenance of IOP. It is located at the angle formed by the iris and the cornea and it exhibits a distinctive architecture. There are three morphological and functionally different regions in the trabecular meshwork which are lined by a monolayer of cells at the contact with the Schlemm's canal. The uveal region, closer to the anterior chamber, is followed by the corneoscleral region, where cells are attached to trabecular beams, and by the juxtacanalicular region, whose cells lay relatively free but can connect to each other and to the cells lining the Schlemm's canal. All trabecular meshwork cells are embedded in an extracellular matrix (ECM) of distinct properties, which gives the tissue a unique spongiform three-dimensional architecture. The cells of the trabecular meshwork derive from the mesenchymal cells of the neural crest (Johnston *et al.*, 1979; Matsuo *et al.*, 1993; Baulmann *et al.*, 2002; Cvekl and Tamm, 2004) while those lining the Schlemm's canal are of vascular origin (Hamanaka *et al.*, 1992; Krohn, 1999). In humans, the trabecular meshwork is the main route for the outflow of the aqueous humor (Bill and Phillips, 1971).

To study functional genomics of the trabecular meshwork scientists have relied mainly on cultures of primary cells and perfused organ cultures. A few studies from *in vivo* animals are also available. The human cell cultures are generated from the dissected trabecular meshwork of post-mortem donors (Polansky *et al.*, 1984; Polansky and Alvarado, 1994; Stamer *et al.*, 1995; Vittitow *et al.*, 2002). More recently, the tissue is obtained from discarded surgical rims left over after a cornea transplant, which provide an excellent source of healthy cells (Rhee *et al.*, 2003). These primary cultures contain cells from both trabecular meshwork and inner wall of the Schlemm's canal. A few viral transformed cell lines from single normal and glaucomatous individuals have also been generated (Pang *et al.*, 1994). Although these cells lack some of the intrinsic characteristics of the primary cultures, such as myocilin induction by Dexamethasone (DEX), they are very useful for a number of different purposes.

The perfused organ cultures, originally developed for physiological measurements (Johnson and Tschumper, 1987; Johnson, 1997), have been adapted by our laboratory for functional genomics studies (Borrás *et al.*, 1998, 2002; Gonzalez *et al.*, 2000a,b,c; Vittitow and Borrás, 2004; Comes *et al.*, 2006). The culture procedure involves perfusion of human anterior segments from paired eyes of a single donor. The whole globes procured by the eye banks are dissected, set in culture by 24–40 hour post-mortem and perfused with serum-free medium at the physiological aqueous humor flow rate. This method revives the trabecular meshwork tissue and makes it amenable to gene expression studies. Because these cultures conserve the original architecture of the trabecular meshwork, they provide an opportunity for studying gene's responses to mechanical strain under conditions closer to those occurring *in vivo*. A unique characteristic of the outflow pathway is the existence of a pressure decrease across the inner wall of the Schlemm's canal, that is, the pressure in the trabecular meshwork proximal to the canal is higher than the pressure in the canal's lumen (Johnstone and Grant, 1973; Grierson and Lee, 1975; Johnstone, 1979; Epstein, 1997). The mechanical forces exerted by the flow of aqueous humor *in vivo* are not present in traditional cell cultures, where the bottom of the dish precludes the presence of the pressure drop. In addition, these cultures address the individual variability concern present in all comparative human studies. Since each experiment involves the use of eye pairs from the same individual, gene expression in the treated eye can be directly compared with that of its untreated contralateral eye and thus not be confounded by genetic differences. An increasing number of studies using both cells and perfused organ cultures are appearing in the literature and providing a first glance to relevant genes and mechanisms of the trabecular meshwork.

C. The Retinal Ganglion Cells

Retinal ganglion cells are the only retinal neurons that project their axons (through the optic nerve) to the brain. Death of the RGCs occurs mostly by apoptosis (Garcia-Valenzuela *et al.*, 1995; McKinnon, 2003) and it is the major hallmark of loss of vision in glaucoma. Elevated IOP plays a key role in the death of the RGCs, most likely by disrupting the axonal flow which results from the damage exerted on the optic nerve by mechanical compression forces (Pease *et al.*, 2000; Osborne *et al.*, 2001; Whitmore *et al.*, 2005). However, the apoptotic death of the RGCs and degeneration of their axons can occur via different mechanisms including deprivation of neurotrophic factors (Finn *et al.*, 2000) and immune state of individuals (Tezel and Wax, 2004). A better knowledge of functional genomics of the RGC under normal and glaucoma conditions has the potential for providing a unique understanding on the genes and mechanisms leading to their death.

To study gene expression profiles and global differential expression of RGC under glaucomatous insults, scientists have conducted microarray studies using a variety of starting material. Because the RGC layer comprises only about 1% of the cells of the entire retina, a first gene sequencing study utilized rat RGC cDNA purified by immunopanning with Thy1 antibodies (Farkas *et al.*, 2004). One study utilized the whole retina and compared an ischemia insult obtained by 1 h of elevated pressure in living rats (Yoshimura *et al.*, 2003) while other used cynomolgus monkeys and laser induced glaucoma (Miyahara *et al.*, 2003). Another compared the rat glaucoma model induced by unilateral saline injection in the rat episcleral veins (Ahmed *et al.*, 2004a). Diehn *et al.* (2005) separated the macula from the peripheral retina and compared their expression profiles with those of the different tissues of the human eye (Diehn *et al.*, 2005). Steele *et al.* (2006) used the mouse glaucoma model DBA/2J mouse and compared retinas between periods of 3 and 8 months (normal and elevated IOP) (Steele *et al.*, 2006). Piri *et al.* (2006) conducted the study after inducing RGC degeneration by optic nerve transection (ONT) (Piri *et al.*, 2006). Levkovitch-Verbin *et al.* (2006) also compared rats retinas after ONT but used instead a more focused gene array containing 18 signal transduction pathways (Levkovitch-Verbin *et al.*, 2006). Ivanov *et al.* (2006) compared expression of immunopanned-isolated rat RGC with that isolated from the whole retina (Ivanov *et al.*, 2006). Naskar and Thanos (2006) compared the whole retina of the Royal College of Surgeons (RCS) rat model, which spontaneously develops elevated IOP, with aged matched controls (Naskar and Thanos, 2006). In a human study, Kim *et al.* (2006) used laser capture microdissection to compare the retinal ganglion cell layer with other inner and outer layers of the retina (Kim *et al.*, 2006). At the time of this writing, the last report included a comparison of a transformed RGC-5 cell line under serum-free conditions to induce apoptosis (Khalyfa *et al.*, 2007). The summary of these studies is presented as part of Fig. 1.

D. Lamina Cribrosa: The Optic Nerve Supporting Tissue

There are two glial cell types characterized in the lamina cribrosa, the ONH astrocyte and the lamina cribrosa (LC) cells (Hernandez *et al.*, 1988). Astrocytes are the non neuronal brain cells which provide nutrition and structural support to other cells of the central nervous system (CNS). They constitute the most abundant cell type of the CNS and express glial fibrillary acidic protein (GFAP), a marker that distinguishes them from the neurons. The LC cells do not express GFAP. At the nonmyelinated optic nerve head, where RGC axons leave the ocular globe, astrocytes provide support to the axons and physically separate them from the surrounding capillary bed. Astrocytes form an integral part of the lamina cribrosa, a sieve-like structure in the posterior part of the sclera that allows the passage of the RGC axons and central retinal vessels leaving the ocular globe. The lamina cribrosa helps maintain the pressure gradient between the intraocular and extraocular space. Astrocytes contribute to the maintenance of the extracellular matrix of the lamina cribrosa as well as to the ion balance and extracellular pH of the intercellular space.

Under stress and injury conditions, quiescent normal astrocytes become "reactive" and stop supporting axon survival. This reactive hallmark is accompanied by the altered expression of a number of genes which influence among others ECM organization and secretion of cytokines and survival growth factors (Yang *et al.*, 2004). Reactive astrocytes have been shown to play major roles in the pathogenesis of neurodegenerative diseases, including

FIGURE 1 Summary of Functional Genomic Studies Published in Glaucoma Associated Tissues (Minus Trabecular Meshwork)

Treatment	Arrays	Sample origin	Model	Reference
Direct sequencing				
4791 clones		Living rats P21	Retina tissue and Immunopanning Thy1	Farkas 2004
Ciliary body and ciliary muscle				
Human eye compartments	Stanford FunGen Facil (43,198 probes)	Human / single adults	Human tissue 24 h pst-mort whole retina	Diehn 2005
Retinal Ganglion cells				
Elevated IOP	Affymetrix RATU34A (8,800 probes)	Rats	Ischemia1h / 12h reperfusion whole retina	Yoshimura 2003
Laser induced glaucoma	Incyte Genomics UniGEM HumanVv.2.0 (9812 probes)	Cynomolgus monkeys	Laser photocoagul / 30 days whole retina	Miyahara 2003
Elevated IOP	Affymetrix RATU34A (8,800 probes)	Rats	Episcleral vein inj saline whole retina	Ahmed 2004a
Human eye compartments	Stanford FunGen Facil (43,198 probes)	Human / single adults	Human tissue 24 h pst-mort whole retina	Diehn 2005
Elevated IOP	Affy Mouse U430.2 (39,000 probes)	Mouse / pooled retinas	DBA/2J mouse, 3&8 months whole retina	Steele 2006
RGC degeneration after ONT	Agilent Rat G4130A (22,575 probes)	Rats	Optic Nerve Transec (ONT) whole retina	Piri 2006
RGC degeneration after ONT	Superarray Mouse focused sign.transd. (96 marker genes)	Rats	Optic Nerve Transec (ONT) whole retina	Levkovitch-Verbin 2006
RGC normal	Agilent Rat G4130A (22,575 probes)	Rats	Retina tissue and Immunopanning Thy1	Ivanov 2006
Elevated IOP	MWG Biotech Rat custom cDNA (9715 probes)	Rats RCS-rdy	Spontaneous high IOP whole retina	Naskar 2006
RGC normal	Affymetrix U133 Plus2 (54,476 probes)	Human / single adults	Laser capture microdisect GLC vs inner&outer layers	Kim 2006
RGC-5 apoptosis	Agilent Rat G4130A (22,575 probes)	Rats	Transformed cell line serum depravation	Khalyfa 2007
Lamina cribrosa				
Normal /glaucomatous	Affymetrix U95Av2 (12,627 probes)	Human / single adults	Primary ONH astrocytes	Hernandez 2002
Hydrostatic pressure	Affymetrix U95Av2 (12,627 probes)	Human / single adult	Primary ONH astrocytes 60 mmHg at 6, 24 & 48 h	Yang 2004
TGFβ1	Affymetrix U133 (22,215 probes)	Human /single adult	Primary LC cells (GFAP-)	Kirwan 2005b
Mechanical stretch	Affymetrix U133 (22,215 probes)	Human / single adults	Primary LC cells (GFAP-) 15% stretch, 24h	Kirwan 2005a
Elevated IOP & ONT	OHSU Microarry Facil SMCmou8400A&6600A (15,400 probes)	Rats	ONH tissue Episcleralvein.saline & Optic Nerve Transec (ONT)	Johnson 2007

glaucoma (Hernandez, 2000; Neufeld and Liu, 2003). During the elevated IOP insult which occurs in most glaucomas, the lamina cribrosa undergoes physical deformation (Yan *et al.*, 1994), which consequently leads to subjecting astrocytes to significant mechanical strain. Such elevated hydrostatic pressure exerted on normal astrocytes as well as on those from glaucomatous models have been studied over the years on the expression of individually selected genes (Hernandez *et al.*, 2000; Agapova *et al.*, 2003). However the development of functional genomics applications is beginning to give a new global picture on the changes occurring in the cells under glaucomatous reactive conditions (Yang *et al.*, 2004).

Figure 1 includes the summary of the array studies performed with astrocytes and LC cells. Hernandez *et al.* (2002) compared primary cell lines of astrocytes from normal and glaucomatous human eyes (Hernandez *et al.*, 2002). The same laboratory subjected the cultured astrocytes to hydrostatic elevated pressure and compared with normal ones at treated times of 6, 24 and 48 hours (Yang *et al.*, 2004). Because TGFβ1 treatment affects fibrosis in the ECM, its effect was measured in lamina cribrosa, but in LC cells rather than astrocytes (Kirwan *et al.*, 2005b). Because the mechanical force of pressure induces stretch of the tissue and cells, Kirwan *et al.* (2005a) further examined the effect of mechanical stretch using the Flexercell system and subjecting the LC cells to 15% stretch for 24 hours (Kirwan *et al.*, 2005a). A last report compared expression of the ONH tissue in two models of rat glaucoma, an elevated IOP model induced by injection of saline into the episcleral veins and the RGC degeneration model induced by transection of the optic nerve (Johnson *et al.*, 2007).

Because of the wide scope of functional genomics from all glaucomatous tissues, in this chapter, we are focusing on the functional genomics of the trabecular meshwork, which is the main area of our expertise.

V. THE TRABECULAR MESHWORK TISSUE: EXPRESSED GENES (CDNA) AND PROTEINS OBTAINED BY DIRECT SEQUENCING AND MASS SPECTROMETRY

The number of functional genomic studies on the trabecular meshwork is summarized in a Fig. 2. The studies can be divided in three groups. On the first group, random cDNA clones are selected from a trabecular meshwork library and sequenced. It is an expensive and accurate procedure whose variations arise only from the origin of the sample. The second group, the most extensive, comprises differential mRNA expression analysis (in the form of cDNA) between genes differentially expressed under conditions known to be associated with glaucoma or under normal and glaucoma disease. A last group deals with the examination of the protein population and analysis of the proteome, a step in the functional genomics field which is just beginning to emerge in the trabecular meshwork.

FIGURE 2 Summary of Functional Genomic Studies Published in the Trabecular Meshwork

Type of treatment	Arrays	Sample origin Species / Age	Model	Reference
Direct sequencing				
1000 clones		Human / single adult	Intact Trabecular Meshwork AntSegment perfused 24 h	Gonzalez 2000a
1118 clones		Human / pooled infants	Primary HTM cells	Wirtz 2002
3459 clones		Human / pooled adults	Intact Trabecular Meshwork	Tomarev 2003
5000 clones		Rat / pooled angle tissue	Intact Trabecular Meshwork	Ahmed 2004b
Gene Arrays				
Culture conditions	cDNA Macroarrays (18,376 probes)	Human / pooled adults	Primary HTM cells Intact Trabecular Meshwork	Gonzalez 1999
Elevated IOP / 6 hours	cDNA Macroarrays (18,376 probes)	Human / single adults	Intact Trabecular Meshwork AntSegment perfusion	Gonzalez 2000b
Dexamethasone	cDNA Microarrays (2400 probes)	Human / pooled adults	Primary HTM cells	Ishibashi 2002
Dexamethasone	cDNA Microarrays (2400 probes)	Human / unreported	Established HTM cells	Leung 2003
Dexamethasone TM vs OpticNerve Astrocytes	Affymetrix U95Av2 (12,627 probes)	Human / single adults	Primary HTM cells Primary ONH astrocytes	Lo 2003
Prostaglandin analogs	Affymetrix U95Av2 (12,627 probes)	Human / pooled adults	Primary HTM cells	Zhao 2003
Elevated IOP / 2-4 days	cDNA Macroarrays (8,400 probes)	Human / single adults	Intact Trabecular Meshwork AntSegment perfusion	Vittitow 2004
TGFβ1 TGFβ2	Affymetrix U133A (22,215 probes)	Human / single adults	Primary HTM cells	Zhao 2004
Dexamethasone	Affymetrix U133A (22,215 probes)	Human /sngl young adults	Primary HTM cells	Rozsa 2006
Normal /glaucomatous Culture conditions	Affymetrix U133 Plus2 (54,476 probes)	Human / single adults	Primary HTM cells	Liton 2006
Mechanical stretch	OHSU human cDNA (8700 probes)	Porcine	Primary Pig TM cells	Vital 2005
Normal /glaucomatous	Affymetrix GLYCOv22 (2001 murine&human glycosilation probes)	Human / single adults	Intact Trabecular Meshwork	Diskin 2006
TIGR/MYOC overexpression	Affymetrix U133A (22,215 probes)	Human / single adult	Primary HTM cells	Borrás 2006
Elevated IOP / 7 days	Affymetrix U133 Plus2 (54,476 probes)	Human / single adults	Intact Trabecular Meshwork AntSegment perfusion	Comes 2006
Proteomics				
80 proteins	2D gel/Mass spectrom	Human	Primary HTM cells	Zhao 2004
368 proteins	1D gel/Mass spectrom	Human	Intact Trabecular Meshwork	Battacharya 2005
87 proteins	2D gel/Mass spectrom	Human	SV40 transfomed HTM cells	Steely 2006

Both types of studies, elucidation of the transcriptome and proteome of the trabecular meshwork give a first look into the relative abundance of mRNA and proteins in the trabecular meshwork tissue at a given time. Their end-point is to let us know whether some genes/proteins are more abundant than others. However, it is important to keep in mind that the fact that a mRNA/protein is highly abundant does not necessarily mean that such molecule is the most relevant for trabecular meshwork physiology. It is though, an important protein. Cells, as nature in general, do not waste resources. Even if not clearly obvious, if a cell makes a lot of a certain molecule, it probably has a good reason for it. Hence, understanding gene and protein expression levels will contribute to a general understanding of mechanisms involved in the function of a given tissue.

A. Direct Sequencing of the Transcriptome

To date, there are three studies using isolated RNA from the human trabecular meshwork, and one study using the RNA from the corneoscleral angle of the rat (Gonzalez *et al.*, 2000a; Wirtz *et al.*, 2002; Tomarev *et al.*, 2003; Ahmed *et al.*, 2004b) (Fig. 2). The source of the trabecular meshwork cDNA from each of the human studies is different. It is important to consider that for identification of the relative abundance of each of the genes, the source of the tissue is critical. Each of the sources used has its advantages and disadvantages. The intact trabecular meshwork tissue used in the Gonzalez (2000a) and Tomarev *et al.* (2003) libraries comprises genes which are transcribed while the tissue maintains its original architecture; that is, when cells of the different regions are maintaining contacts with each other in a manner close to their natural state. However, the trabecular meshwork used by Gonzalez was perfused for 24 h before RNA extraction while the tissue used by Tomarev was processed directly after procurement. Because humans at the time of death might have been exposed to drugs (often corticosteroids) affecting the expression of its genes, the perfusion system of the Gonzalez library would allow a washout out of the medication effect. On the other hand, the direct post-mortem tissue of Tomarev would conserve the expression of cells exposed to all factors of aqueous humor, which are not present in the culture medium used in perfusion studies. Another relevant difference between the two human intact tissue libraries, is that while one represents the molecular signature of one individual (67 year old male) (Gonzalez *et al.*, 2000a), the second library (Tomarev *et al.*, 2003) represents a pooled sample from 28 donors (median age 72 years). The third library, of Wirtz *et al.*, (2002) combined the RNA from early passages of primary culture cells of six young individuals (2 weeks to 2 years of age). One advantage of the cell culture procedure is the opportunity of getting considerable amounts of high quality RNA under well controlled conditions. The disadvantage is that once in culture and, under the presence of serum, cells can dramatically down- and upregulate some of their endogenous genes. Other relevant differences among the three studies, is the fact that the Gonzalez library was obtained with linearly amplified cDNA while the other two libraries were generated from unamplified cDNA. Because of all these differences plus the fact that human samples are not from an inbred population, one would expect that the relative abundance of the genes obtained from these libraries would be quite different. Surprisingly though, the three studies hold very interesting similarities (Fig. 3) and reveal common genes whose functions had not been described before in the trabecular meshwork.

FIGURE 3 Comparison of Selected Most Abundant Genes from Three Human Trabecular Meshwork libraries Number of clones/1000

Gene Name	Single adult perfused AS	Pooled infant TM cell cuture	Pooled Adult TM tissue	Basic Function
Most abundant Library #1				
Elongation factor1α	19	15	10	Translation regulator / stress?
Glyceraldehyde 3-phos-dehyd	11	7	0.7	Cytosolic energy production
Chitinase 3	9	---	0.7	Cartilage ECM
Matrix GLA	9	4	7	Inhibitor of calcification
β2-Microglobulin	8	1	1.3	MHC complex / Immune response
TIMP1	7	9	---	Inhibitor of MMPs / ECM remodeling
Apolipoprotein D	6	---	4.5	Carrier protein / Lipid transport
Heat-shock protein 90	5	2	1.3	Stress protection
Ferritin L	5	13	---	Iron storage / detoxification
MMP3 (stromelysin)	4	---	0.3	Physiological breakdown of ECM
Most abundant Library #2				
Ferritin H	1	44	1	Iron storage / detoxification
Elongation factor 1α	19	15	10	Translation regulator / stress?
Ferritin L	5	13	---	Iron storage / detoxification
Fibronectin	---	11	0.3	ECM and cell surface / cell adhesion
TIMP-1	7	9	---	Inhibitor of MMPs / ECM remodeling
Thymosin β4	2	9	3.3	Regulator of actin polymerization
Insulin-like grow fact bind prot 7	---	8	3.3	Stimulates prostacyclin / signaling
Glyceraldehyde 3-phos-dehyd	11	7	0.7	Cytosolic energy production
Tumor protein TPT1	---	7	9	Calcium binding / microtubule stabiliz
Insulin-like grow fact bind prot 2	---	6	---	IGF binding /signalling
Most abundant Library #3				
Vimentin	2	1	9.6	Intermediate filament / Cytoskeleton
Elongation factor 1α	19	15	9.6	Translation regulator / stress?
Myocilin	---	---	9.3	Stress protein
Tumor protein TPT1	---	7	9	Calcium binding / microtubule stabiliz
Matrix GLA	9	4	7	Inhibitor of calcification
Tropomyosin	---	1	6	Actin binding protein /cell contraction
Apolipoprotein D	6	---	4.5	Carrier protein / Lipid transport
Regulator of G protein signaling	4 (RGS2)	---	4.3 (RGS5)	GTPase activators / signal transduction
Angiopoietin-like factor7 / CDT6	2	---	4.3	ECM deposition
Thymosin β4	2	9	3.3	Regulator of actin polymerization
Insulin-like grow fact bind prot 7	---	8	3.3	Stimulates prostacyclin / signaling

From data on
Library #1: Single adult perfused; Gonzalez *et al*. IOVS 41, 3678 (2000a)
Library #2: Pooled infant TM cell culture; Wirtz *et al*. IOVS 44, 3698 (2002) & personal communication
Library #3: Pooled adult TM tissue; Tomarev *et al*. IOVS 44, 2588 (2003)

The first thing we learned from all libraries was that the transcriptome of the trabecular meshwork tissue is quite diverse. Results from the Gonzalez library (Gonzalez *et al.*, 2000a) showed that 74.4% of the clones correspond to single genes, with the most abundant clone (that encoding *Elongation factor 1α*, EEF1A1) present 19 times out of 1000 (1.9% of the total mRNA). In the Wirtz library (Wirtz *et al.*, 2002) the number of clones corresponding to single genes was 44.7%. The most abundant clone was that encoding Ferritin H with 44 clones out of the 1118 (4%). The Tomarev's library (Tomarev *et al.*, 2003), containing three times as many clones, showed 50% of the clones expressed once, and the most abundant, *vimentin*, being present in 29 out of the 3026 clones (close to 1%). In addition, with the exception of Ferritin H whose abundance of 4% could be explained by the influence of tissue culture conditions of the Wirtz library, the relative abundance of the most abundant genes in the three libraries was below 1.5%. These percentages are quite below those of the normalized expressions of most genes in other tissues (http://source.stanford.edu/cgi-bin/source/sourceSearch). This observation is an indication that the trabecular meshwork, whose mission is to keep a physiological resistance to aqueous humor flow, does not seem to rely in just one mechanism to perform its function, but rather it uses a number of different ones involving distinct cell characteristics. In all libraries, it was also interesting to observe that ESTs corresponding to unknown genes were primarily single clones.

Most of the gene's encoded functions were predicted to be relevant for the physiology of the trabecular meshwork, but others were totally new. Interestingly, some of the new ones would surface again as genes differentially regulated by relevant glaucomatous conditions. Some of the genes identified across all libraries deserve special attention.

Elongation factor 1α was the most abundant gene across all three libraries (Fig. 3). This gene is a subunit of the elongation factor complex, which is responsible for the delivery of aminoacyl tRNAs to the ribosomes, therefore involved in translation. However, it was recently found that EEF1A1 was required for the stress activation of the heat-shock transcription factor HSF1 by contributing to the conversion of HSF1's inactive form to a DNA-binding active conformation (Shamovsky *et al.*, 2006). When HSF1 gets activated under stress stimuli, it activates heat shock genes. HSF1 is negative regulated by *Heat shock protein 90* (HSP90) (Zou *et al.*, 1998), which also happens to be a gene that made the list of the most abundant in the three libraries. It is then probable that the presence of these genes is an indication of the occurrence of a response to stress mechanism in the trabecular meshwork. And perhaps, EEF1A's role in the outflow tissue is more related to such stress response than to its involvement in translation activity as originally speculated. As well, HSP90 is involved in the responsiveness of glucocorticoids by influencing the nuclear transport of glucocorticoid receptor β (GR β) (Zhang *et al.*, 2006), a non binding glucocorticoid receptor who has negative activity on the GR binding receptor GRα in the trabecular meshwork (Zhang *et al.*, 2007).

Matrix Gla (*MGP*). After Elongation factor 1α, MGP was the most abundant gene (Fig. 3). Matrix Gla is a vitamin K-dependent protein found most abundantly in bone and cartilage. MGP undergoes a post-translational modification, which induces a conformational change and allows its interaction with other proteins. MGP was initially discovered in demineralization extracts of bone (Price and Williamson, 1985) but was found to be expressed in a variety of tissues including vascular smooth muscle cells (Fraser and Price, 1988; Shanahan *et al.*, 1993). MGP was found to also have a key role in cardiovascular calcification (Price *et al.*, 2000). MGP knockout mice develop extensive arterial calcification of the tunica media and cartilaginous tissues with a lethal outcome at about 6 weeks of age (Luo *et al.*, 1997). Thus, MGP functions as a calcification inhibitor. Calcification is a common occurrence in the pathophysiology of the aging human artery and is associated with several cardiovascular disease states. Calcification is universally associated with atherosclerotic plaques (Tanimura *et al.*, 1986).

What made this gene intriguing for us was that calcification had never been reported in the trabecular meshwork and that no other functions had been described for MGP other than calcification. The association of aging with the development of increased outflow resistance and glaucoma, and the involvement of MGP on calcification of soft tissues led us to investigate whether MGP might play a similar role in the human trabecular meshwork. We thought that a similar process of calcification could contribute to the physiology of the human TM and to the regulation of IOP. To asses whether aging cells exhibited some of the calcification signs known to occur in vascular smooth muscle cells, we aged both cell types in culture for 12 weeks and evaluated calcification by three different calcification markers. Alizarin red, a staining method showing a red-orange color on positive calcified cells, a direct chemical calcium extraction which measures calcium levels and a biological assay which measures the activity of alkaline phosphatase, a well established enzyme marker of osteogenic differentiation. We were surprised to find that results of these three tests indicated that the trabecular meshworks cells with age are undergoing a calcification process (Fig. 4) which was indistinguishable of that occurring on aging vascular smooth muscle cells (Proudfoot *et al.*, 1998). The presence of this calcification phenomenon is perhaps one of most clear examples of the new avenues that the technology of functional genomics is bringing to the study of the trabecular meshwork.

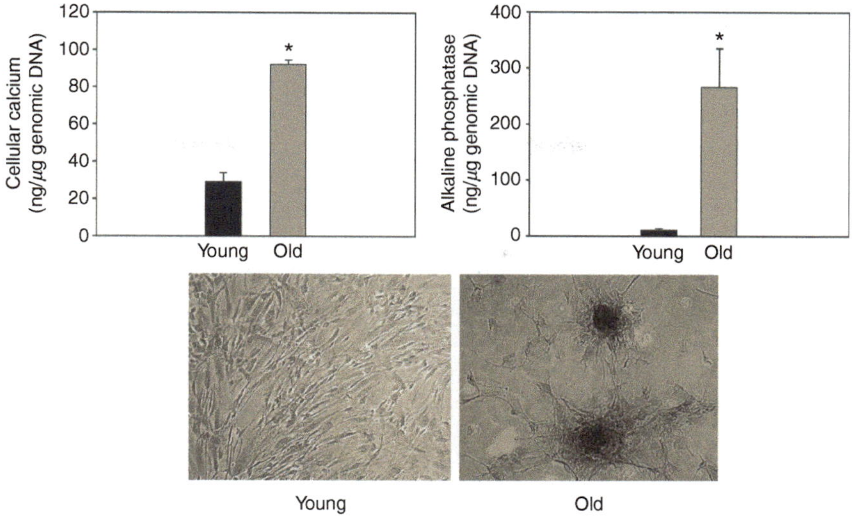

FIGURE 4 Example of a new trabecular meshwork process identified by functional genomic studies. Cells of the human trabecular meshwork undergo a calcification process as a result of ageing. The increase of calcification was evaluated on cells maintained in culture for one week (young) versus cells maintained for twelve weeks (old). *Top panel* shows the results of direct calcium measurements (left) and assay of the calcification marker alkaline phosphatase (right). *Bottom panel* shows the formation of calcification nodules in the old cells, stained with alizarin red.

The high presence of *Apolipoprotein D* (APOD) in the trabecular meshwork remains a mystery. APOD is a secreted, lipid carrier glycoprotein, member of the lipocalin family first described in 1987 (Pervaiz and Brew, 1987). It transports small hydrophobic ligands including cholesterol and sterol. APOD is widely expressed in neuronal tissues and accumulates in the cerebrospinal fluid of patients with neurological diseases such Alzheimer (Belloir *et al.*, 2001), multiple sclerosis and schizophrenia. (Helisalmi *et al.*, 2004). It has been proposed to be a robust marker for brain regions affected by particular neuropathologies (Thomas *et al.*, 2003). The speculation as to the role of APOD in trabecular meshwork is wide open. A recent publication has brought up the attention that APOD can act as a nonspecific stress protein and that most stresses causing an extended growth arrest, such as hydrogen peroxide and UV light, induce expression of APOD (Do *et al.*, 2007). This new function of APOD as an intracellular growth arrest molecule could have importance on trabecular meshwork physiology, where decreased cell number has been associated with age and development of glaucoma (Alvarado *et al.*, 1981, 1984). Curiously, APOD was very abundant in the intact tissue library but was not present in the cell culture library (Wirtz, personal communication), where cells are artificially proliferating.

β2-Microglubulin is a serum protein found in association with the major histocompatibility complex (MHC). Its presence in the cerebrospinal fluid has been used as a diagnostic for the presence of inflammation in the central nervous system (Caudie *et al.*, 2005). Thus, we have an inflammatory marker in the trabecular meshwork. Although not in the rank of the most abundant genes, other inflammation-related molecules have made their presence in the trabecular meshwork. Library #1 (Gonzalez *et al.*, 2000a) contained IL1*β*, IL-6, and IL-8, library #2 (Wirtz *et al.*, 2002) contained IL1*β*, and IL-8, and library #3 (Tomarev *et al.*, 2003) contained IL-33 (2 clones/1000) and IL-8. Perhaps their influence on vascular permeability function could be adapted by the cells of the outflow tissue as a signal to open the tight inner wall of the Schlemm's canal and lower IOP.

The *Translationally controlled tumor protein* (TCTP) encoded by the TPT1 gene was curiously very close to the top percentage abundance (1%) in the library from the pooled tissues but not present in the single individual library. TCTP is an extensively regulated protein (Thiele *et al.*, 2000; Bommer and Thiele, 2004) and it would be interesting to know whether its expression in the trabecular meshwork has an individual component. In addition to its originally assigned function as a translational factor, a number of physiological functions have been assigned to this protein, ranging from a chaperone to a histamine release factor (Bommer and Thiele, 2004). TCTP also interacts with EEF1A (Langdon *et al.*, 2004) and more recently, by using forskolin and phorbolesters it was demonstrated that TPT1 promoter is regulated by cAMP signaling (Andree *et al.*, 2006). It is worth to note that cAMP stimulation agents and consequently cAMP elevation levels have been extensively associated with regulation of outflow facility and IOP (Crawford *et al.*, 1996; Webb *et al.*, 2003). Some of these characteristics would be relevant for the physiology of the trabecular meshwork.

Cytoskeletal reorganization of the trabecular meshwork (Tian *et al.*, 2000; Gabelt *et al.*, 2006; Gonzalez *et al.*, 2006), as well as ECM remodeling (Lütjen-Drecoll *et al.*, 1986; Acott, 1992; Bradley *et al.*, 1998; Kelley *et al.*, 2007) have constituted two classic properties associated with the physiology and pathophysiology of this tissue. So the appearance of genes related to these characteristics is not unexpected. It is interesting though, to see which of the myriad of cytoskeleton or ECM-related genes seemed to have been picked up by the trabecular meshwork. These studies have revealed that perhaps *Vimentin*, *Thymosin* β, *Tropomyosin*, *Fibronectin*, *TIMP-1* and *MMP3* have a special role. Vimentin is the subunit of the intermediate filament specific to the mesenchymal tissue. But, in addition to being part of stabilizing the architecture of the cytoplasm, Vimentin was recently found to also be secreted by activated macrophages and stimulated by TNFα (Mor-Vaknin *et al.*, 2003), another known modifier of the trabecular meshwork (Bradley *et al.*, 2000; Alexander and Acott, 2001). Thymosin (Li *et al.*, 1996) and Tropomyosin, which is considered the universal actin filament regulator (Gunning *et al.*, 2005), give a fined understanding to the well established role of the actin cytoskeleton in outflow facility.

Myocilin, one of the three genes linked to primary open glaucoma (POAG) (Sarfarazi, 1997; Stone *et al.*, 1997; Rozsa *et al.*, 1998), was not present in two of the three libraries. Although the stress and corticosteroid inductions of this gene in the trabecular meshwork are well established (Polansky, 1993; Polansky *et al.*, 1997; Nguyen *et al.*, 1998; Lo *et al.*, 2003), the normalized expression level of this protein in the human trabecular meshwork is unknown. It is very likely that the relative abundance of Myocilin in the human trabecular meshwork tissue is low, individual dependent, and that, the gene is downregulated in culture (Liton *et al.*, 2006). Examination of the average intensity of the signal for Myocilin in eight Affymetrix arrays from perfused single individuals shows that its relative abundance varies between being the 2nd most abundant gene to being the 259th (our laboratory's unpublished results). Such difference in relative abundance was not correlated with age, race, gender, or perfusion time. While the tissues used in our experiments were perfused for at least 24 hours before RNA extraction, the tissues used for the generation of library #3 (Tomarev *et al.*, 2003) were not. It is possible that the high number of clones found in library #3 (Fig. 3) are due to potential steroid medication of the post-mortem tissues used to generate it. Lastly, despite the association of Myocilin only with an eye disease, its expression profile by normalized analysis of ESTs is highest in adipose tissue (18.3%) followed by the small intestine, esophagus, trachea and the eye (http://source.stanford.edu/cgi-bin/source/sourceSearch, and UniGene http://www.ncbi.nlm.nih.gov/UniGene/ESTProfileViewer.cgi?uglist=Hs.436037)

Finally, another previously unknown and potentially informative gene is *Angiopoietin-like factor 7* (ANGPTL7, alias CDT6). ANGPTL7/CDT6 is a secreted glycoprotein originally described in the cornea (Peek *et al.*, 1998) and proposed to have a role in ECM deposition and angiogenesis (Peek *et al.*, 2002). *In vitro*, expression of ANGPTL7 resulted in deposition of ECM components such as collagen type I and V (Peek *et al.*, 2002). This protein is a member of the angiopoietin family, which has been shown to be regulators of angiogenesis (Katoh and Katoh, 2006). Recently it has been reported to be present in the articular cartilage (Oike *et al.*, 2004), which as well as the trabecular meshwork and the cornea, is an avascular tissue. It is intriguing that this protein maps to 1p36, the same chromosomal locus of the GLC3B region, which has been associated with a recessive form of congenital glaucoma (Akarsu *et al.*, 1996). By comparative genomics, ANGPTL7/CDT6 has been characterized as a target gene of the WNT/β-catenin signaling pathway (Katoh and Katoh, 2006). As we will see below, the ANGPTL7/CDT6 gene is regulated by glaucomatous insults in the trabecular meshwork.

In summary, Fig. 3 shows that despite the different origin of the trabecular meshwork cells/ tissues, the majority of the most abundant genes in one library are also present in the other two. This indicates that the expression of those genes constitutes at least 0.1% of the entire transcriptome. As a summary from this figure, we have singled out the following genes as potential functional/structural candidates: Elongation factor 1α, Heat shock protein 90, Matrix Gla (MGP), Apolipoprotein D (APOD), β2-Microglubulin, Tumor protein TPT1, Vimentin, Thymosin β, Tropomyosin, Fibronectin, TIMP-1, MMP3, Myocilin and Angiopoietin-like factor 7 (alias CDT6).

B. Analysis of the Proteome and Protein Modifications

As of date, there are very few proteomic studies of the trabecular meshwork available. It is interesting to analyze them on the basis of comparing their protein findings with the mRNAs identified by the libraries or by microarray studies. Some of the data do not cross correlate between each other or with published results on individual genes, but some do, and those novel gene/proteins that appear across the board are providing a good insight as to how the trabecular meshwork functions. To follow a logistic order, I would have preferred to first examine those genes/proteins studies of normal trabecular meshwork and then those conducted on tissues affected by glaucoma or subjected to other glaucomatous insults. This was possible to do only at times, especially because no proteomics of just the normal trabecular meshwork tissue were available.

In analyzing the proteomics data it is important to bring into consideration the fact that all current trabecular meshwork proteomics studies have been conducted only on intracellular soluble proteins. Thus, in the cellular studies, media was removed and proteins were extracted from cellular pellets. Although secreted proteins are also found intracellularly, their levels inside the cells are usually lower than those of the extracellular counterpart. Hence, the relative abundance of a secreted protein versus that of the total pool analyzed might be misleading when the extracellular fraction is excluded from the experiment. Another relevant factor in interpreting proteomics results is the consideration of the chemical extraction procedure, which contributes to the kind of proteins to be subsequently analyzed by mass spectrometry. Regarding proteins from trabecular meshwork tissues, dissection procedures need also to be considered. As already indicated by Bhattacharya *et al.* (2005), the presence of contaminating surrounding tissue in some of the specimens cannot be avoided, which might lead to identification of proteins not present in the trabecular meshwork, but in neighboring tissues.

Work by Bhattacharya *et al.* (2005) analyzed proteins from 12 hour postmortem donors and from trabeculectomies. Tissues were processed individually and before mass spectrometry analysis, proteins were extracted with nonurea containing buffers, separated from their insoluble fraction and run on one-dimensional PAGE gel electrophoresis. Their work detected 368 proteins, from which 52 were present only in their glaucomatous samples.

One protein which appeared in most glaucomatous samples (five out of 6) and in none of the samples from normal individuals was *Cochlin*, a protein associated with a deafness disorder. Cochlin plays a role in maintaining the architecture of the cochlea by binding to components of the ECM (Robertson *et al.*, 1997; 1998). By immunohistochemistry it appeared to localize to the posterior region of the Schlemm's canal closest to the scleral spur (http://www.jbc.org/cgi/data/M411233200/DC1/1). Two clones of Cochlin cDNA were detected in trabecular meshwork library #3 (Tomarev *et al.*, 2003) made from the tissue of normal individuals while no clones were obtained in libraries #1&2 (Gonzalez *et al.*, 2000a; Wirtz *et al.*, 2002). This indicates that the definition of presence only in glaucomatous tissue reflects a different level of expression, rather than an absence. Cochlin was as well marked as "present" in the Affymetrix arrays of perfused untreated trabecular meshwork (our unpublished laboratory results). The findings of the association of Cochlin with only glaucomatous trabecular meshwork tissues would need further confirmation by other laboratories. In the same study, another identified protein, Myocilin, which has been linked to POAG gene appeared only in the glaucomatous pool. However, Myocilin has been detected by numerous investigators in the trabecular meshwork of normal donors (Caballero and Borrás, 2001; Tamm, 2002; Lo *et al.*, 2003; Tomarev *et al.*, 2003). Among other interesting proteins reported only in glaucomatous samples, this group identifies the two ECM binding proteins ANGPTL7/CDT6 and *Opticin* (http://www.jbc.org/cgi/data/M411233200/DC1/1). ANGPTL7/CDT6 (see also section IV.A) is highly induced by DEX in the trabecular meshwork cells (Lo *et al.*, 2003; Rozsa *et al.*, 2006) while opticin (*OPTC*) is a small leucine-rich repeat protein (Reardon *et al.*, 2000) which localizes to many tissues of the eye (Friedman *et al.*, 2002a). While ANGPTL7/CDT6 maps to a region associated with glaucoma (1p36), Opticin maps to a region in chromosome 1 associated with macular degeneration (Friedman *et al.*, 2002b).

A more recent proteomic study has used a transformed glaucomatous trabecular meshwork cell line (Steely *et al.*, 2006), urea extraction of the proteins, and high resolution two-dimensional PAGE (2D) before the mass spectrometry analysis. This study identified 87 proteins, many of them known to be relevant to trabecular meshwork function. Of them, a prominent group resolved in the acidic region of the 2D gel and contained several cytoskeletal proteins such as *Vimentin, α-Tubulin, β-Tubulin and α-Actin.* Other well known trabecular meshwork enzymes, such as *Glyceraldehyde-3-phosphate dehydrogenase, Pyruvate kinase and Enolase* dominated the basic, right side of the gel. Vimentin, an intermediate filament of the mesenchymal tissue, happened to be the most abundant cDNA in one of the TM libraries (Tomarev *et al.*, 2003) (Fig. 3) whereas Pyruvate knase and Enolase appeared downregulated by TGFβ in a different proteomic study (Zhao *et al.*, 2004). It was interesting to observe their identification of *Calreticulin, Protein disulfide isomerase, Heat shock proteins and Chaperonin containing TCP1.* These three proteins are involved in protein folding (Peterson *et al.*, 1995; Valpuesta *et al.*, 2002; Hinault *et al.*, 2006; Nagradova, 2007), and some have been found to be altered by glaucomatous insults, such TGFβ (Zhao *et al.*, 2004) and mechanical strain (Vittitow and Borrás, 2004; Vittal *et al.*, 2005). Given the potential relationship between misfolding Myocilin mutants and their link to glaucoma (Caballero *et al.*, 2000; Jacobson *et al.*, 2001; Liu and Vollrath, 2004), the presence of chaperones and folding related proteins reinforces the relevance of stress protection in the trabecular meshwork.

An earlier proteomics study involved a correlation of microarray and proteomics in primary human trabecular meshwork cells treated with TGFβ growth factors (Zhao *et al.*, 2004). The proteins for this study were urea extracted from the cellular pellets of TGFβ-treated serum-free cultures and resolved by 2D gel electrophoresis before mass spectrometry. Primary trabecular meshwork cells originated from five donors and were pooled for the analysis. The authors detected very high levels of Actin and Vimentin, which presumably obscured the resolution of Myocilin which pI and molecular weight would overlap with the Vimentin spots (Zhao *et al.*, 2004). Efforts were made by this group to identify proteins in the gels whose spot intensities were different between TGFβ1 and TGFβ2 treatments and to focus on pIs corresponding to nonmodified proteins rather than on those corresponding to proteins modified by post-translational events. Keeping these criteria, the authors gathered a list of about twenty proteins whose changes did also correlate with changes observed in the mRNA by microarrays. Among them were the *Protein disulfide isomerase, Calreticulin, Tropomyosin* and *Cu-Zn Superoxide dismutase*, whose functions were identified by the other studies.

Following the purpose of this chapter of trying to unravel trabecular meshwork physiology by the presence and behavior of their expressed genes and proteins, we made a figure with the most potentially relevant proteins identified in these studies (Fig. 5). In addition to reinforcing the importance of the contractile apparatus in the trabecular meshwork, it is interesting to observe the prevalence of functional categories representing stress protection, folding and calcium regulation. We also observed the

FIGURE 5 Selected Human Trabecular Meshwork Proteins Identified by Proteomic Analysis (See Color Insert.)

Protein name	Glaucoma Tissue[1]	Normal Tissue[1]	Glaucoma transformed cells[2]	TGFβ-treat primary cells[3]	mRNA changed by glaucom conditions*	Basic Function
Adhesion / ECM						
Angiopoiet-like factor 7 / CDT6	•	---	---	---	•	Antiangiogenesis / ECM deposition
Cochlin	•	---	---	---	---	ECM architecture (deafness)
Syndecan 1	•	•	---	---	•	Transmembrane proteoglycan / adhesion
Thrombospondin1,3&4 precur	•	•	---	---	•	Cell-cell, cell-matrix interactions
Calcification						
Mimecan (Osteoglycin)	•	•	---	---	•	Induces bone formation with TFGβ
Cytoskeletal structure						
α–Actin	•	•	•	•	---	Microfilament / cell contraction
α–Tubulin	•	•	•	•	•	Microtubule
Transgelin (smooth muscle22)	---	•	---	•	•	Actin cross-linking
Tropomyosin	•	•	---	•	•	Actin binding protein /cell contraction
Vimentin	•	•	•	•	---	Intermediate filament
Metabolism						
Enolase	•	•	•	•	---	Metabolic enzyme
GAPDH	•	•	•	---	•	Cytosolic energy production
Pyruvate kinase	•	•	•	•	---	Metabolic enzyme
Protein modification						
Calreticulin	---	---	•	•	---	Calcium homeostasis / chaperone
Chaperonin containing TCP1	---	---	•	---	•	Protein folding
Protein disulfide isomerase	---	•	•	•	•	Protein folding
Signaling / cytokines						
Calgranulin A (S100 protein)	•	•	---	---	•	Calcium-dependent signaling
Stress / Defense						
αB-crystallin	---	•	---	---	•	Stress response
Heat-shock protein 70	---	•	•	---	•	Stress protection
Myocilin	•	---	---	---	•	Stress protein
Superoxide dismutase	---	•	---	•	•	Destroys superoxide radicals
Other						
Apolipoprotein D	---	•	---	---	•	Carrier protein / Lipid transport
Cornifin	•	---	---	---	•	Membrane cross-linking

From data on
1: Bhattacharya *et al*. JBC 7, 6080 (2005)
2: Steely *et al*. Mol Vis 12, 372 (2006)
3: Zhao *et al*. IOVS 45, 4023 (2004)
*: see multiple refs throughout the text (mechanical, DEX and TGFβ);
•: present or altered; ---: not present or not altered

absence of key proteins, such as MGP (most abundant expression list, Fig. 3) and the Endothelial leukocyte adhesion molecule-1 (ELAM1, alias E-Selectin) first identified glaucoma marker (Wang *et al.*, 2001). The lack of presence of the MGP protein in the three proteomic studies is not unexpected since MGP partitions to the insoluble fraction and would not be extracted by the standard extraction procedures used by these studies. It was however surprising that ELAM1 that stained intensively the trabecular meshwork of glaucomas of different etiology (Wang *et al.*, 2001), was not present in any of the glaucomatous tissue samples.

In summary, the analysis conducted for Fig. 5 brings forward, in addition to common metabolism and structural proteins, a few selected candidates which are representative of potentially relevant mechanisms. These proteins are: ANGPTL7/CDT6, Cochlin, Mimecan, Calreticulin, Protein disulfide isomerase, Heat shock proteins, Chaperonin containing TCP1, Tropomyosin, Cu-Zn Superoxide dismutase and Myocilin.

VI. THE TRABECULAR MESHWORK TISSUE: IN SEARCH OF GENES RESPONDING TO GLAUCOMATOUS INSULTS

A. *Mechanical Insult: Intraocular Pressure and Stretch*

For the past few years our laboratory has been interested in identifying genes that are regulated by elevated IOP. This pressure inside the eye is generated by the resistance offered by the trabecular meshwork tissue to the flow of aqueous humor (Grant, 1963a; Lütjen-Drecoll, 1973; Mäepea and Bill, 1992). Because of the unique way that pressure is exerted onto the living eye, it is imperative that experiments studying gene response to IOP in the trabecular meshwork do maintain the architecture of the tissue, are performed with human specimens and use paired eyes (see expanded rationale on section III.B). Our procedure thus entails perfusing the anterior segment from the two eyes of one individual, experimentally elevating the pressure of one eye for different periods of time, and maintaining the contralateral, paired eye at normal pressure, for control. At the end of the insult, trabecular meshwork tissue is dissected, RNA extracted and expression of each gene analyzed on microarray chips.

In addition to our published studies (Gonzalez *et al.*, 2000b; Borrás *et al.*, 2002; Vittitow and Borrás, 2002; 2004; Comes *et al.*, 2005; 2006), for this chapter, we have conducted a comprehensive, comparative study of all our internal databases on genes whose expression is altered by pressure. We selected genes that had been reported as most altered in one condition and checked whether they were also changed in a different one. The results are summarized by functional categories in Fig. 6 (in alphabetical order). It is interesting to observe that many of the genes comprised in this set are surfacing in several trabecular meshwork studies from different laboratories, an indication that their relevance in the physiology of the tissue seems to be real. Among the genes which functions involve organization of the ECM and adhesion, we see again ANGPTL7/CDT6 (sections IV.A and B). The intriguing characteristic about this gene, not previously described in the trabecular meshwork, is that it is also induced by DEX (Lo *et al.*, 2003; Rozsa *et al.*, 2006), TGFβ2 (Zhao *et al.*, 2004) (see below, sections V.B and C.) and that it was reported in glaucoma tissues by proteomic analysis (Fig. 5). We and others are investigating whether its effects on the trabecular meshwork parallel those observed in other tissues (Comes *et al.*, 2006; Kuchtey *et al.*, 2007)

Several genes which have been known as osteogenesis-related genes, are gathered together here for the first time in the human trabecular meshwork. In addition to the inhibitor of calcification MGP (Fig. 3), we see the small leucine-rich repeat family of proteoglycans, *Mimecan* (alias *Osteoglycin*), *Biglycan, Osteomodulin and Fibromodulin*, which are associated with the commitment to the osteogenic phenotype in osteoblast differenciation (Balint *et al.*, 2003), and *Osteonectin* (SPARC), induced by a 7 day insult and mechanical stretch which is responsible for decreased bone density in null mice (Delany *et al.*, 2000).

Although cytoskeletal changes have been long associated with trabecular meshwork function, the potential involvement of this function through *Transgelin* gene (alias S*mooth muscle 22*) is new. Transgelin is a shape-change sensitive actin cross-linking/gelling protein which is also found in fibroblasts and smooth muscle cells. It is highly expressed in vascular smooth muscle cells and its promoter is used to direct expression of other genes to such endothelial cells (Murshed *et al.*, 2004).

The functional categories of chaperone/protein folding and signaling cytokines activity are also used by the trabecular meshwork to regulate pressure. The chaperones are represented specially by *Chaperonin containing TCP1* (also section IV.B), a part of the TCP1 ring complex which is involved in the folding of key cytoskeletal proteins such, actin and tubulin (Won *et al.*, 1998). Among the cytokines we found the neuropeptide precursors *Secretogranin, Substance P precursor and Vasoactive intestinal peptide* (VIP), all altered at more than one time period. Secretogranin is the precursor of the neuropeptide *Secretoneurin*, recently shown as a potent angiogenic in cornea *in vivo* and protector against apoptosis of HUVEC cells *in vitro* (Kirchmair *et al.*, 2004). Substance P is a neurotransmitter that interacts with smooth muscle cells and induces vasodilation. Both Substance P and VIP were reduced in a diabetic retinopathy and suggested to be involved in the lack of neovascularization in this model (Troger *et al.*, 2001). Their regulation by

FIGURE 6 Cross Comparison of Selected Human Trabecular Meshwork Genes Altered by Mechanical Insults (Fold Change) (See Color Insert.)

Function	1 hour IOP[a]	6 hours IOP[b]	2-4 days IOP[c]	7 days IOP[d]	12-48 hr Stretch[e]	Chromosome
Adhesion / ECM						
Angiopoietin-like factor7/CDT6	-	-	-	25.0	-	1p36
ICAM 4	-	-	-1.9	-	-	19p13-cen
Chitinase 3	NC	-	1.8	2.0	-	1q31-q32
Matrix remodeling asso 5 (aldican)	-	-	-	4.3	-	Xp22.33
MMP2	-	-	1.7	NC	NC	16q13-q21
MMP3	-	4	-	altered	-	11q22.3
Perlecan	NC	-	22.0	NC	-	1p36--p34
Syndecan 1	-	-	-	1.5	-	2p24.1
Syndecan 2	NC	-	-	-	1.5	8q22-q23
Tenascin C	NC	-	-	-2.1	2.0	9q32-q34
Calcification						
Biglycan	-	-	-	1.6	3.1	Xq28
Connect tiss growth factor (CTGF)	NC	-	-	1.3	1.7	6q23
Fibromodulin	NC	-	-	2.1	1.7	1q32
Matrix Gla	-	-	7.0	6.0	2.5	12p13-p12
Mimecan (Osteoglycin)	-	-	-	2.6	0.5	9q22
Osteoblast specific factor 2	NC	-	2.2	NC	0.5	13q13
Osteomodulin	NC	-	-1.9	NC	-	9q22.31
Osteonectin (SPARC)	NC	-	-	1.7	2.0	5q31-q32
Cytoskeleton / contraction						
α-Tubulin	NC	-	4.0	-	2.9	12q12-q14
Spectrin (erythrocyte)	-	-	-2.4	NC	-	1q21
Spectrin (non erythrocyte)	2.0	-	-	NC	-	9q33-q34
Transgelin (smooth muscle22)	0.7	-	4.4	-1.8	1.9	1q21-q25
Tropomodulin	NC	-	3.1	1.3	-	15q21-q21
Tropomyosin	0.7	-	-	altered	-	15q22
Protein modification						
Chaperonin containing TCP1	NC	-	1.9	altered	0.5	12q15
Protein disulfide isomerase	NC	-	-1.5	1.5	-	16p13
Proteasome 26S	-	-	2.4	NC	1.7	1q21
Signaling / cytokines						
Activin A receptor	up	-	-	NC	-	2q23-q24
Calgranulin A (S100 protein)	2.0	-	-1.2	NC	-	1q21
Chemokine (C-X-C motif) 11	4.3	-	-	4.3	-	4q21
Chemokine (C-X-C motif)* (2,3,12)		-	-	down	down	
Insul growth fact 1 (somatomedinC)	-	-	-	4.4	-	12q22-q23
Insulin growth fact bind prot*(2,3,4,7)	changed	-	2.6	-4.5	2.5	
Interleukin 6 (IL6)		4	1.8	-	-	7p21
Regulatory G-protein sign 11	NC	-	2.4	NC	-	16p13
Secretogranin II	NC	3.0	-	5.2	-	2q35-q36
Substance P precursor	NC	2.5	-	3.3	-	7q21-q22
Vasoact Intest Peptide (VIP)	NC	-	2.0	-1.5	-	6q25
VEGF	1.4	-	-	-1.7	-	6p12

FIGURE 6 (Continued)

Function	1 hour IOP[a]	6 hours IOP[b]	2-4 days IOP[c]	7 days IOP[d]	12-48 hr Stretch[e]	Chromosome
Stress / Defense						
αB-crystallin	-	4.5	1.4	-	1.6	11q22-q23
Cytochrome P450* (26A,26C,1B)	1.9	-	2.0	-	2.6	
Heat-shock protein 70k 2	-	-	-	-4.0	-	14q24
Metallothioneins*	NC	4.0	4.0	-1.3	73.0	
Myocilin	NC	-	5.1	10.3	up	1q23-q24
Superoxide dismutase 1	NC	3.0	-	NC	-	21q22
Other						
Apolipoprotein D	NC	-	13.0	2.3	-	3q26-qter
Baculo IAP rep-cont 5 (survivin)	NC	-	2.0	NC	-	17q25
Cornifin (small proline-rich B)	6.1	-	-	-3.0	-	1q21-q22
Thrombomodulin	NC	-	-	1.8	-	20p11

Criteria: selected genes include those reported as most altered in at least one condition
a: Borrás (unpublished)
b: Gonzalez *et al*. (2000b)
c: Vittitow and Borrás (2004) and unpublished laboratory data
d: Borrás (unpublished)
e: Vittal *et al.* (2005)
*: family members pooled; -: not known
NC: no change

pressure in the trabecular meshwork might reflect the mechanism used by the outflow tissue to regulate vascular permeability of the Schlemm's canal at times of mechanical insults.

Finally, stress defense proteins such as *αB-Crystallin and Myocilin* have been independently associated with trabecular meshwork and glaucoma. αB-Crystallin is induced by TGFβ2 (Siegner *et al.*, 1996; Fuchshofer *et al.*, 2006) and most importantly, it is specifically expressed in the juxtacanalicular region, which has been specifically implicated in outflow resistance (Grant, 1963a; Mäepea and Bill, 1992). Myocilin, as mentioned earlier, has been linked to POAG (Stone *et al.*, 1997). It was interesting to observe that two important protection protein families (*CytochromeP450 and Metallothioneins*) are heavily regulated by pressure. The cytochrome P450 proteins are monooxygenases which catalyze oxidative conversion of many reactions involved in drug metabolism and synthesis of cholesterol, steroids and other lipids. Metallothioneins are heavy metal binding proteins which are use to detoxify metals in all cells of the animal kingdom (Hawse *et al.*, 2006). Interestingly, the change of gene expression was not always observed in the same family member, an indication that the function, rather than the specific family member is important for pressure regulation of the trabecular meshwork. And lastly, a number of other proteins whose functions are not yet obviously relevant for outflow tissue function are identified. Such is the case of APOD, the lipid carrier protein, and *Chitinase3* (alias *cartilage GP-39*) which has been associated with conditions of inflammatory and degenerative arthritis, both showing up also on Fig. 3.

In summary, the genes altered by mechanical insults appear in Fig. 6 grouped by functional categories. It shows that the category compiling a higher number of genes is that of signaling/cytokines followed by those encoding cell adhesion/ECM and processes involved in osteogenesis. Although all functions would be relevant for trabecular meshwork physiology, perhaps those represented by the highest number of genes are preferentially selected for the regulation of IOP.

B. Dexamethasone

Dexamethasone is a potent steroid immunosuppressant agent commonly used in eye clinics for the treatment of various inflammatory conditions. It is well established that topical administration of glucocorticoids increases IOP in 30–40% of patients of the general population (Armaldy and Becker, 1965) and in 90% of patients with POAG (Armaldy, 1963a; Becker and Hank, 1964; Bartlett *et al.*, 1993). In living animals, treatment with steroids leads to an increase in outflow resistance (Knepper *et al.*, 1985). *In vitro*, DEX decreases phagocytosis in the trabecular meshwork cells (Matsumoto and Johnson, 1997), increases ECM deposition (Spaeth *et al.*, 1977; Babizhayev and Brodskaya, 1989; Johnson *et al.*, 1990, 1997; Steely *et al.*, 1992), induces the glaucoma-linked gene Myocilin (Polansky, 1993; Polansky *et al.*, 1997; Nguyen *et al.*, 1998), and decreases protease secretion and activity (Yun *et al.*, 1989; Samples *et al.*, 1993; Snyder *et al.*, 1993). For all these reasons, functional genomics studies offer a great opportunity to take a global look at the individual molecular response of trabecular meshwork cells to this agent.

With the intent of having an overall understanding of the relevant genes induced by DEX in the trabecular meshwork, we have constructed Fig. 7. We cross-checked genes identified as most altered on the four published microarray studies (Ishibashi *et al.*, 2002; Leung *et al.*, 2003; Lo *et al.*, 2003; Rozsa *et al.*, 2006) and organized them by functional categories. In this comparison, genes listed as "not reported" in one of the studies could represent not being present in the arrays used, not reported because they did not pass the authors filter criteria or simply, not being expressed because of different experimental conditions. Although all the studies used the same DEX concentration, it is relevant to point out that each of them used different starting donor material,

FIGURE 7 Selected Human Genes Altered by Dexamethasone (DEX) Treatment of HTM Primary Cells (Fold Change) (See Color Insert.)

Function	7 days[a] MicroMax 2400 (cDNA)	10 days[b] AffyU95Av2 12,627 (oligo)	10 days[c] MicroMax 2400 (cDNA)	21 days[d] AffyU133A 22,215 (oligo)	Gene Symbol	Chromosome
Adhesion / ECM						
Angiopoietin-like factor7 / CDT6	-	28.0	-	187.0	ANGPTL/CDT6	1p36
α1-Antichymotrypsin (serpin)	-	165.0	-	48.0	SERPINA3	14q32
Cadherin 2	-	-4.4	-	-	CDH2	18q11
Cadherin 4	-	-	2.4	-	CDH4	20q13
Chitinase 3	-	39.0	-	-6.1	CHI3L1	1q31-q32
Connective tissue growth factor	-	-	-	1.3	CTGF	6q23
Decorin	3.5	-	-	-1.5	DCN	12q21
Elastin	-	-66.0	-	-	ELN	7q11
Fibulin 1	2.8	7.6	-	-1.8	FBLN1	22q13
Fibronectin leucine rich transm prot 2	0.4	-	-	-2.5	FLRT2	14q24-q32
Procollagen C-proteinase enhancer	-	44.0	-	-	PCOLCE	7q22
Pigment epithelium-derived factor	24.0	46.0	-	-	PEDF	17p13
Serum amyloid A1	-	-	-	218.7	SAA1	11p15.1
Thrombospondin-2	-	-	-	-2.8	THBS2	6q27
Versican	-	-27.0	-	-	VCAN	5q14
Cytoskeleton / contraction						
γ2-actin, enteric smooth muscle	-	30.0	-	9.4	ACTG2	2p13.1
Transgelin (smooth muscle22)	-	-	1.5	3.5	TAGLN	1q21-q25
Tropomyosin 2	2.1	-	-	2.1	TPM2	9p13
Protein biosynthesis/ modification						
Aldo-keto reductase 1C1	-	126.0	-	-	AKR1C1	10p15-p14
Aldo-keto reductase 1C3	0.6	43.0	-	1.5	AKR1C3	10p15-p14
Aldo-keto reductase 1B10	3.5	175.0	-	1.3	AKR1B10	7q33
Arginigosuccinate synthetase 1	-	28.0	-	-	ASS1	9q34
Asparagine synthetase	0.3	-	down	-	ASNS	7q21
Cellular retinoic acid binding prot 1	-	-2.5	-	-	CRABP1	15q24
Cellular retinoic acid binding prot 2	0.3	-	-	-3.1	CRABP2	1q21
Cystatin C	2.8	-	-	-	CST3	20p11
24-dehydrocholesterol reductase	2.1	-	-	2.5	DHCR24	1p33-p31
Lactate dehydrogenase A	-	-	1.4	1.2	LDHA	11p15
Monoamine oxidase	2.4	-	-	13.1	MAOA	Xp11.3
Tryptophanyl-tRNA synthetase	0.5	-	0.3	-	WARS	14q32
Signaling / cytokines						
Insul-like growth fact1(somatomedinC)	-	-	-	-13.1	IGF1	12q22-q23
Insul-like growth fact binding protein 2	4.9	57.0	1.7	2.1	IGFBP2	2q33-q34
Insul-like growth fact binding protein 4	2.5	3.7	-	-1.3	IGFBP4	17q12-q21
Latent TGFβ binding protein 2	3.1	-	-	2.2	LTBP2	14q24
Gremlin	-	-39.0	-	-	GREM1	15q13-q15
Prostaglandin D2 synthase	-	38.0	-	-	PTGDS	9q34-q34
Secretogranin II	0.4	13.9	-1.6	-3.7	SCG2	2q35-q36

different times of exposure to DEX and different types of arrays. The Ishibashi report (2002) used four primary cell lines (donors age 7–28 years old) at confluent passage 4 to 5 and treated them for seven days while Leung' report (2003) utilized one established confluent cell line of unknown age exposed to

FIGURE 7 (Continued)

Function	7 days[a] MicroMax 2400 (cDNA)	10 days[b] AffyU95Av2 12,627 (oligo)	10 days[c] MicroMax 2400 (cDNA)	21 days[e] AffyU133A 22,215 (oligo)	Gene Symbol	Chromosome
Stress / Defense						
Heat shock 70kDa protein 5	-	-	0.5	-1.3	HSPA5	9q33-q34
Metallothionein 1L	-	-	2.2	-	MT1L	16q13
Metallothionein 1X	-	-	-	3.7	MT1X	16q13
Myocilin	24.0	50.0	4.3	16.7	MYOC	1q23-q24
Prostacyclin synthase	-	-35.0	-	-	PTGIS	20q13
Other						
Apolipoprotein D	-	22.0	-	5.3	APOD	3q26-qter
Citrate transport protein	-	23.0	-	-	CTP	22q11
cAMP response elem bind prot	-	-	0.6	-	ATF4	22q13
Growth arrest&DNA damage-indprot	-	-	0.8	-	DDIT3/CHOP10	12q13
Growth arrest-specific 1	-	23.0	2.4	-	GAS1	9q21-q22
Ferritin light chain	3.6	-	-	-	FTL	19q13
Mortality factor 4 like 2	2.0	-	-	1.5	MORF4L2	Xq22
Potasium voltage-gated chanel	0.5	-	-	2.4	KCNB1	20q13
BCL2/adenovirus E1B19k int prot 2	-	-2.2	-	-	BNIP2	15q22
BCL2/adenovirus E1B19k int prot 3	2.2	-	-	2.3	BNIP3	10q26
Thrombomodulin	-	-	-	3.0	THBD	20p11

Criteria: selected genes are those present in the most altered list of one report report or altered in more than one report

a: Ishibashi *et al.* (2002)

b: Lo *et al*. (2003); HTM specific (HTM-DEX vs ONH-astrocytes-DEX)

c: Leung *et al*. (2003)

d: Rozsa *et al*. (2006)

-: not reported

DEX for ten days. Both studies used and the MicroMax cDNA arrays (Perkin Elmer Life Sciences, Boston, MA), which contain 2400 genes. The Lo study (Lo *et al.*, 2003) used two primary cell lines at confluent passage 4–6 and treated them for ten days. This report used high density oligonucleotide arrays from Affymetrix (Santa Clara, CA), version U95Av2, containing 12,627 genes. Importantly, the Lo study compared DEX-induced trabecular meshwork cells with DEX-induced primary ONH astrocytes, thus selecting for genes specifically induced by DEX in the trabecular meshwork and eliminating those that were equally induced in the ONH cells. Finally, the Rozsa study comprised three confluent primary cell lines (donors 12–17 years of age) treated with DEX for 21 days. This last report also used high density oligonucleotide chip from Affymetrix, albeit a newest version, U133A, containing 22,215 genes.

Under those limited comparisons, we observed that only three genes had their expression changed in the four studies. These were *Myocilin, Secretogranin and Insulin-like growth factor binding protein2.* It is very possible that a fourth gene, ANGPTL7/*CDT6*, is also induced across all studies but that, due to the fact that is a relatively new gene, its cDNA was not present in the Micromax arrays used by Ishibasi (2002) and Leung (2003). ANGPTL7/CDT6 was the second highest DEX-altered gene in the Rozsa study (2004) and among the ten highest in Lo's report (2003). A similar reason could apply to the lack of an all across presence of *α1-Antichymotripsin* (*serpin*), *γ2-Actin*, APO*D* and *Chitinase3*, four genes preferentially selected in both Affymetrix chip studies. Other genes that appeared changed in three out of the four reports included *Fibulin*, A*ldo-keto reductase1B10* (*AKR1B10*) (alias *Aldose reductase*), *Aldo-keto reductase1C3* (*AKR1C3*) and a different member of the insulin-like growth factor binding proteins family, *IGFBP4.*

Because of the different arrays and methodologies used, special attention should be given to a number of genes that made the criteria list twice, once on the cDNA and once on the Affymetrix studies. In this case, since their appearance is not due to their absence in the hybridization arrays, their selected altered expression could be an indication of the and individual response to the glucocorticoid. Among these are *Pigment epithelium-derived factor* (PEDF), *Transgelin, Tropomyosin, and Growth arrest-specific1*, which because of their encoded functions could be relevant to the DEX response. Finally, it is intriguing to observe that the *Serum amyloid1* (SAA1) gene, the highest altered in the Rozsa study, was not reported in any of the other three. Although this gene was not present in the earlier AffyU95Av2 chip version, a close member of this gene family, *Serum amyloid 4* (SAA4), was not affected by DEX in the Lo study (our own result unpublished).

It is interesting to observe that several of the genes affected by DEX had already been brought up to our attention as associated with glaucomatous conditions in the trabecular meshwork (Figs. 3, 5 and 6). Such is the case of Myocilin, ANGPTL7/CDT6, Secretogranin, APOD, Chitinase 3, Insulin-like growth factor binding proteins, Transgelin and Tropomyosin. Most of the functions encoded by these genes comprise secreted proteins and some, such transgelin and tropomyosin are intracellular cytoskeletal modifiers.

A number of other genes were seen for the first time during the treatment of trabecular meshwork with DEX. These comprise mostly secreted proteins known to influencing ECM remodeling. SerpinA3, is a plasma protease inhibitor, member of the serine protease inhibitors and PEDF is an extracellular glycoprotein with high affinity for ECM components (Alberdi *et al.*, 1998; Kozaki *et al.*, 1998; Meyer *et al.*, 2002). Fibulin is also a secreted glycoprotein that becomes incorporated into fibrillar ECM upon binding to calcium.

The Serum amyloids are secreted proteins associated with plasma high-density lipoproteins; after inflammatory stimulus, they accumulate extracellularly and form deposits which are highly insoluble and resistant to proteolysis (Lundmark *et al.*, 2002).

Other proteins are interesting, such the Aldo-keto reductases AKR1C1 and AKR1C3, also known as 3α-hydrosteroid dehydrogenases. These are enzymes which abnormal activities result in the accumulation of 5β-dihydrocortisol, know to cause elevated IOP in rabbits (Southren *et al.*, 1994). Interestingly, AKR1C1 and AKR1C3 are also elevated in glaucomatous ONH astrocytes (Hernandez *et al.*, 2002). All together, genes listed in Fig. 7 are providing us with a practical example of the relevance of functional genomics in the elucidation of the physiology of DEX treatment of the trabecular meshwork.

Hence, the summary in Fig. 7 shows that there seems to be an individual response to DEX which could in part reflect the different steroid response observed in the clinic. Myocilin, Secretogranin, Insulin-like growth factor binding protein2 and PEDF could be among those genes that respond in an individual manner to the corticosteroid.

C. Transforming Growth Factor β2

TGFβ2 is a signaling cytokine that controls proliferation, differentiation, and other functions in many cell types. TGFβ2 is a well-established factor associated with increased trabecular meshwork resistance and glaucoma (Tripathi *et al.*, 1993; 1994a,b; Lütjen-Drecoll, 2005). Not only it is present in the aqueous humor (Cousins *et al.*, 1991) and in the trabecular meshwork (Tripathi *et al.*, 1994a) but its concentration increases in glaucoma. Fifty percent of POAG patients show significantly higher levels of this cytokine in the aqueous humor than their normal counterparts (Tripathi *et al.*, 1994b; Picht *et al.*, 2001). *In vitro*, TGFβ2 enhances ECM production and suppresses cell proliferation. In perfused human anterior segment cultures, TGFβ2 increases outflow resistance (Gottanka *et al.*, 2004). It is thus believed that TGFβ2 might be involved in the high IOP associated with glaucoma.

Surprisingly, only one gene profile study has been published addressing the effect of TGFβ2 in the human trabecular meshwork (Zhao *et al.*, 2004). The authors used primary trabecular meshwork cells from five individuals (16–76 years old) at passage 3–5 and treat them with 1 ng/ml of rhTGFβ2 for 72 hours in the absence of serum. RNA samples were pooled and compared with vehicle treated controls using Affymetrix gene chips version U133A, which contained 22,215 genes. In addition, Zhao *et al.* (2004) conducted a proteomic study on 81 excised protein spots that differed significantly between 2D gel electrophoresis of treatments and control samples (Fig. 5).

After establishing a criteria of selecting genes which levels of expression had increased or decreased greater than twofold in all their comparisons, Zhao *et al.* (2004) identified 19 upregulated and 2 downregulated genes. Many of the genes with high levels of change in the TGFβ2 samples had been previously found to be associated with other trabecular meshwork insults. Per example *ANGPTL7/CDT6* (upregulated here 5.8X) was among the most abundant in two out of the three libraries (Fig. 3), its protein was found specifically in the glaucomatous tissue (Fig. 5), it was elevated by pressure (Fig. 6), and it was highly induced by DEX (Fig. 7). *Osteoblast specific factor* (OSF2) (upregulated 2.7X), was induced by mechanical stress (Fig. 6). *Versican* (*VCAN*, *alias chondroitin sulfate proteoglycan*) (upregulated 2.7X) was among the most cell-specific downregulated genes in the trabecular meshwork by DEX (Fig. 6). A different member of the *Insulin-like growth factor binding protein* family (IGFBP3) was upregulated 3X here, stressing the relevance of this protein family in the trabecular meshwork (Figs. 3, 6 and 7. *Mimecan* (alias *osteoglycin*) (upregulated 2.6X) had been detected in the proteomics analysis (Fig. 6), and was upregulated by mechanical stress (Fig. 6). The expression of *Neuregulin* and *Thrombomodulin* (*THBD*) (upregulated 3.2X and 2.1X respectively) have been curiously shown very much increased in HTM cells overexpressing myocilin (Borrás *et al.*, 2006). Neuregulin is a regulatory signaling growth factor which interacts with the ERBB2 tyrosine kinase receptor, and Thrombomodulin is a vascular endothelial cell receptor that binds thrombin and acts as a co-factor on the activation of protein C to inhibit fibrinogen clotting. Thrombomodulin had been speculated to be involved in maintaining the fluidity of the aqueous humor (Ikeda *et al.*, 2000) and in other arrays, it was upregulated by elevated IOP and DEX (Figs. 6 and 7). These characteristics position Thrombomodulin as a potential counteracting factor to the deleterious effects of TGFβ2 increase in glaucoma.

Among the downregulated genes, the authors found *Chitinase3* (-2.1X) (alias *cartilage GP-39*), whose relationship with the trabecular meshwork has been well-established (Figs. 3, 6 and 7). Additional genes which had been individually reported as up- or downregulated by TGFβ2, such as Thrombospondin 1, Fibronectin, Transglutaminase 2, Collagen type IV and MGP were also changed in their gene chips but at levels lower than their twofold change cutoff. In contrast, Connective tissue growth factor (CTGF) which was reported induced by TGFβ2 in another glaucomatous tissue (ONH) (Fuchshofer *et al.*, 2005), showed no change in the Zhao trabecular meshwork microarray (2004). Most of the genes found in this study showed changes of expression under other trabecular meshwork related or glaucomatous conditions. Thus, Thrombospondin, a glycoprotein that mediates cell-to-cell and cell-to-matrix interactions, had been previously identified in the normal and glaucoma trabecular meshwork by proteomic analysis (http://www.jbc.org/cgi/data/M411233200/DC1/1). Fibronectin, an extracellular protein involved in cell adhesion and migration, has been reported as one of the most abundant cDNAs in one of the HTM libraries (Wirtz *et al.*, 2002) (Fig. 3). MGP, an inhibitor of calcification which showed mild downregulation in the TGFβ2 Zhao arrays (Paul Russell, personal communication), was considerably induced in the mechanical stress arrays (Fig. 6) as well as being among the most abundant TM genes (Fig. 3). CTGF, in addition of being elevated in the aqueous humor of patients with pseudoexfoliation glaucoma (Ho *et al.*, 2005), showed significant induction in the arrays using mechanical stress and DEX (Figs. 6 and 7). These results suggest that although these genes were below their criteria threshold, they are probably relevant for trabecular meshwork pathophysiology.

Finally, the authors showed that a post-translation modification isoform of Transgelin exhibits a 3X increase in the cells treated with TGFβ2. Although Transgelin mRNA was not reported as increased on the TGFβ2 arrays, its expression was increased by mechanical stress (Fig. 6) and DEX (Fig. 7). This vascular smooth muscle cell specific gene which cross links and regulates the actin cytoskeleton may be an important mediator of outflow facility.

D. Trabecular Meshwork Tissue from Glaucoma Donors

Two microarray studies have been performed comparing gene expression in trabecular meshwork tissues from POAG patients with those from individuals without a history of glaucoma (Diskin *et al.*, 2006; Liton *et al.*, 2006). In both cases, as would be expected, the original material comprised tissues from patients who had been subjected to glaucoma medications for several years. Most likely, the gene expression of the true glaucomatous conditions was masked by the effect on the trabecular meshwork cells of the medications used. A first glance of the effect of glaucoma medications on trabecular meshwork cells was reported on a microarray study exposing the cells to prostaglandins (Zhao *et al.*, 2003), an active principle used in common glaucoma drugs.

The first study, rather than investigating the differential expression of all genes, focused on the detection of differentially expressed glycogens. The microarrays used were the GLYOCOv2 (Affymetrix), which contain 2001 murine and human oligonucleotides corresponding to genes encoding carbohydrate-binding proteins and proteins involved in regulation of glycosylation. The study used ten pairs of human eyes (66–87 years old), four pairs from normal and six from glaucoma individuals. Three of the glaucoma donors were diagnosed with POAG ($n = 3$), one with low-tension glaucoma and two which were classified as glaucoma suspects. This study found that 19 genes were significantly differentially expressed with a fold-change higher than 1.4. Of these genes, 13 were upregulated and 6 downregulated. Several of these genes had been previously shown to change their expression in conditions associated with glaucoma. Among the upregulated genes the authors identified *Mimecan* (*osteoglycin*) and *Activin A*. Mimecan is upregulated by TGFβ2 (see section IV.D) (Zhao *et al.*, 2004), found in the proteomic analysis (Fig. 5) and altered by mechanical stress in two different studies (Fig. 6). The fact that the expression of this small leucine-rich repeats gene was changed under different glaucomatous conditions in five independent array studies brings forward the notion that induction of collagen fibrillogenesis and of mechanisms similar to those involved in bone formation might be associated with the pathology of the trabecular meshwork tissue. Activin A, a member of the TGFβ family and upregulated in the TGBβ2 study, plays a key role in several cellular functions including differentiation and apoptosis. Inhibition of Activin by silencing SMAD2/SMAD3 prevents keratocytes differentiation and accumulation of smooth muscle actin (You and Kruse, 2002), a relevant gene for the contraction of the trabecular meshwork cell (Lepple-Wienhues *et al.*, 1991; Tamm *et al.*, 1996). Expression of the receptor of this gene, activin A receptor (was increased after 1 hour insult of elevated pressure (Fig. 6).

The gene exhibiting the highest change in the glycogen study was the *Chemokine* (*C-C motif*) *ligand 2*, (CCL2). The CCL2 gene was very much reduced on cells treated with TGBβ1 but not affected with TGBβ2. This cytokine belongs to a family of secreted proteins involved in immunoregulatory and inflammatory processes. They are structurally related to the subfamily of cytokines with the C-X-C motifs. Expression of members of this family is altered by mechanical strain (Fig. 6). Likewise, other inflammation-related genes were altered in this study, such *RANTES*, *PECAM1* and *P-Selectin*, all known in non ocular systems to be targets of NFκB, a major mediator of inflammatory responses. The additional altered expression of *Interleukin6 receptor* reinforces the relevance of inflammation genes in the trabecular meshwork.

A second study compared tissue specimens from three control and two POAG patients (Liton *et al.*, 2006). This report used U133 Plus 2 high density microarrays from Affymetrix which have the capability of analyzing nearly 50,000 RNA transcripts. The authors found 156 genes (72 up- and 84 downregulated) in the POAG tissue with an average change value higher than 2 fold. It is surprising that, with few exceptions, such as *ELAM1* (*alias E-Selecting*)*and two members of the Cytochrome P-450* and Chemokine (C-X-C) gene families, the genes identified in this study had not been previously associated with glaucoma or glaucomatous insults in the trabecular meshwork. This could be a result of the technical upgrades associated with using a chip with such a high number of genes (approximately 50,000 versus 22,000 in the U133 and 12,000 in the U95).

The highlight of this study was the fact that *ELAM1* (*E-Selectin*) was the highest upregulated gene. ELAM1 is expressed in cytokine-stimulated endothelial cells and it is thought to be responsible for the accumulation of blood leukocytes at sites of inflammation by mediating the adhesion of cells to the vascular lining. Most important ELAM1 has been proposed to be a marker for glaucomatous trabecular meshwork (Wang *et al.*, 2001). The reason as to why this gene might not have been selected in previous trabecular meshworks arrays might have been technical. Surprised by the fact that E-Selectin did not appear upregulated in the GLYCOv2 arrays, Diskin *et al.* (2006) performed independent RQ-PCR on the same RNA sample used for the arrays. They found that E-Selectin cDNA was profoundly upregulated in the glaucomatous trabecular meshwork samples and concluded that the probe set in the array must have been defective (Diskin *et al.*, 2006). It is not then unconceivable that the same defective probe had been used by Affymetrix in earlier gene chips and that future studies using the upgraded U133A Plus 2 would be able to detect this gene.

Ceruloplasmin was another of the genes from this study whose expression had previously been reported altered in glaucoma, albeit in a different direction, and in a different tissue (retina). Ceruloplasmin is a copper-binding glycoprotein that oxidizes iron without releasing radical oxygen species. Defects in this gene are the cause of aceruloplasminemia, an autosomal recessive disorder of iron metabolism which leads to retinal degeneration, diabetes and neurological disturbances. Ceruloplasmin was shown to be downregulated here but it was upregulated in the retina of glaucomatous DBA/2 mice and in most human glaucomatous eyes. The significance of the new result is unknown.

Among other genes found altered in this study were a high proportion of cytokines associated with inflammation and acute phase response. It is worth considering that the presence of inflammatory type signals at the molecular level has been observed numerous times in the glaucomatous trabecular meshwork and in the same tissue subjected to mechanical insults (Gonzalez *et al.*, 2000b; Wang *et al.*, 2001; Liton *et al.*, 2006). It is tempting to think that perhaps the role of these cytokines in the trabecular meshwork is not related to the function typically associated with inflammatory and immune responses. Rather, the trabecular meshwork cells could be using these proteins to signal control of vascular permeability to open up the aqueous humor pathway (especially the inner wall of the Schlemm's canal) and control outflow facility. Thus the adaptation of this mechanism could represent a recruitment on the part of the trabecular meshwork of an inflammatory function for the maintenance of the physiology of the tissue.

VII. PROPOSED MOLECULAR SIGNATURE OF HUMAN GLAUCOMA

Analyzing the data from all above reports, at the end of this study we have compiled a set of 40 genes which we believe are relevant for the physiology and pathophysiology of the human trabecular meshwork. Figure 8 brings together genes whose expression has been altered on microarrays under several glaucomatous conditions, and on more than one independent study per condition. In a few cases, the change in expression of the independent study was obtained by real-time PCR instead of the arrays. We sorted out this list by the number of conditions that each gene expression had changed and color coded their functional categories for easy visualization.

It is important to consider the fact that a given gene showing up under the "one condition" category does not necessarily imply that such gene is less significant than another one listed under "four conditions". It just means the selected gene might be specifically regulated under the particular insult. On the same line of thought, the fact that a gene is not on Fig. 8 does not mean the gene is ruled out as involved on trabecular meshwork function/glaucoma. For a gene to make the cutoff of the signature presented here, it needs to be on the top of the up- or downregulated lists of at least one condition. An example of a potentially relevant gene not showing up in Fig. 8 is *Endothelin-1. However*, Endothelin-1 is a potent vasoactive peptide which is increased in the plasma of glaucoma patients (Emre *et al.*, 2005) and mediates a number of trabecular meshwork, astrocytes and retina processes associated with glaucoma (Yorio *et al.*, 2002; Prasanna *et al.*, 2003). Most likely, Endothelin-1 is a good candidate for the molecular signature of glaucoma.

Although additional genes are bound to have a role in the physiology of trabecular meshwork and glaucoma, the rationale behind the construction of Fig. 8 strongly suggests that the 40 genes listed are indeed relevant. In this molecular signature we find the presence of known mechanisms carried out by new genes, and the unearthing of new processes. For instance, the well-established contribution of adhesion/ECM to regulation of outflow facility is represented by 13 genes (32.5%); in addition of those previously identified, like CTGF, Versican and MMP3, the list brings in genes no previously correlated with trabecular meshwork function, such as ANGPTL7/CDT6, Fibulin, PEDF and Serum amyloid A1. The genes encoding signaling proteins comprised 7 of the 40 selected genes (17.5%). In this group, besides Interleukin 6 (IL-6), we had genes encoding chemokines C-X-C ligands, neuropeptides like Secretogranin and Substance P precursor, and members of the Insulin-like growth factor binding proteins. The regulation

FIGURE 8 Rational Human Trabecular Meshwork Molecular Signature of Glaucoma Based on Microarray Studies (See Color Insert.)

Gene name	Mech strain[a]	DEX[b]	TGBβ2[c]	Glaucoma Proteome[d]	Glaucoma mRNA[e]	Basic Function
Altered under four conditions						
Angiopoietin-like factor7/CDT6	●	●●	●	●	-	Antiangiogenesis/ECM deposit
Protein disulfide isomerase	●●	●	●	●	-	Protein folding
Superoxide dismutase	●	●	●	●	-	Destroys superoxide radicals
Tropomyosin*	●●	●●	●	●	-	Stabilizes actin filaments
Altered under three conditions						
Aldo-keto reductases*	●	●●●	●	-	-	Antioxidants
Calgranulin A (S100 protein)	●●	●	-	●	-	Calcium binding/cytoskelt regul
Calreticulin	-	●	●	●	-	Calcium homeostasis/chaperone
Chaperonin containing TCP1	●●●	●	-	●	-	Protein folding
Chitinase 3	●●●	●●	●	-	-	Cartilage ECM
Connective tissue growth fact	●●	●	●	-	-	Differentiation of chondrocytes
Cytochrome P450*	●●●	●	-	-	●	Detoxification
Heat-shock proteins 70 kDa*	●	●●	-	●	-	Chaperone
Insul growth fact* (1,2)	●	●	-	-	●	Signaling hormone
Insulin-like growth fact bind proteins*	●●●●	●●●●	●	-	-	Bind to IGF, increase its half life
Matrix Gla	●●●	-	●	-	●	Inhibitor of calcification
Mimecan (Osteoglycin)	●●	-	●	●	-	Induces bone formation w/TGFβ
Myocilin	●●●	●●●●	-	●	-	Stress protein
Transgelin (smooth muscle22)	●●●●	●●	●	-	-	Actin crosslinking
Thrombomodulin	●	●	●	-	-	Binds thrombin/inhibits clotting
Thrombospondin*	●	●	-	●	-	Cell-cell, cell-matrix interactions
Versican	●●	●	●	-	-	Cell-cell and cell-ECM signaling
Altered under two conditions						
Apolipoprotein D	●●	●●	-	-	-	Carrier protein/lipid transport
αB-crystallin	●●●	-	●	-	-	Chaperone
α-1-antichymotrypsin (serpin)	●	●●	-	-	-	Serine protease inhibitor
Cadherin* (2,4,10,19,24)	-	●●	-	-	●	Calcium depend cell-cell adhes
Chemokine (C-X-C motif)*	●●●	-	-	-	●	Signaling cytokine/immunity
Cochlin	●	-	-	●	-	ECM architecture (deafness)
Cornifin (small proline-rich B)	●●	-	-	●	-	Crosslink to membrane
Fibulin 1	-	●●●	-	●	-	Fibrillar ECM/calcium binding
Interleukin 6, IL6	●●	●	-	-	-	Signaling cytokine/immunity
Metallothioneins*	●●●●	●●	-	-	-	Metal binding/detoxification
Osteoblast specific factor 2	●●	-	●	-	-	Cell attach, spread, adhesion
Pigment epithelium-derived factor	●	●●	-	-	-	Inhibitor of angiogenensis
Secretogranin II	●●	●●●●	-	-	-	Neuropeptide precur/secretion
Selectin-E (ELAM-1)	-	-	-	●	●	Cell adhesion
Altered under one condition						
Biglycan	●●	-	-	-	-	Osteoblast differenciation
MMP3	●	-	-	-	-	ECM remodeling
Serum amyloid A1	-	●	-	-	-	ECM deposits/inflamm
Tropomodulin	●●	-	-	-	-	Actin depolimerization blocker
Substance P precursor	●●	-	-	-	-	Signaling neurotransmitter

●: Expression altered in one independent study; -: not reported or not altered ; functional categories color coded

☐ ECM; ☐ protein synth/modification; ☐ stress/defense; ☐ cytoskeleton; ☐ signaling/cytokines; ☐ calcification

a: From data on Figure 5.

b: From data on Figure 6

c: Zhao *et al*. (2004)

d: From data on Figure 3

e: Diskin *et al*. (2006); Liton *et al*. (2006)

*: family members pooled

of these transcripts is an indication of the presence of local stimulatory signals among the cells of the trabecular meshwork and suggests that such signaling could influence Schlemm's canal permeability and aqueous humor outflow. The functional category involving stress response and defense is represented by 6 genes (15%) and includes, in addition of the well-studied Myocilin and αB-Crystallin, detoxification enzymes encoded by Cytochrome P450 and Metallothionein genes. In these detoxifying gene families, several different members, rather than one individual gene, were found differentially expressed under three and two conditions respectively, a suggestion that the function, rather than a particular selected gene is what is actually relevant. Enzyme activities involved in protein modification and folding were represented by 4 genes (10%) including those encoding new trabecular meshwork proteins, such Protein disulfide isomerase, Calreticulin and Chaperonin containing TCP1.

Four of the genes (10%) encode proteins involved in mechanisms which have been associated with osteoblast differentiation and bone formation. This calcification process, not previously described for the trabecular meshwork, might be part of an ongoing hardening of the tissue with age and/or with the disease. To this end, we have recently showed that tissues from glaucoma patients contain higher levels of an established calcification marker and downregulate MGP (Xue *et al.*, 2007), a gene highly expressed in the trabecular meshwork (Fig. 3).

Interestingly, only three cytoskeleton modifier genes (7.5%) made the criteria for the signature list. Regulation of cytoskeleton was one of the first mechanisms associated with changes in outflow facility (Tian *et al.*, 2000). However none of the three genes, Tropomyosin, Transgelin or Tropomodulin had been identified before the arrival of functional genomics. The last three genes APOD, Cornifin and Thrombomodulin (7.5%) constitute at this time three "orphan" genes. Although very significantly regulated, their encoded functions are presently totally foreign to the trabecular meshwork.

A. Concluding Thoughts

Assessing the functional relevance of the identified genes/proteins/modified proteins would become the task of the future. Compared to other fields, there are very few functional genomic studies addressing the elucidation of glaucoma physiology. As we saw in this chapter, most of them have focused on the first step of regulation, that of transcription of the gene. Even with these few studies available, the amount of information obtained seems already too wide to be able to follow through a comprehensive evaluation of the modifications observed. The selection of which new genes, mechanisms or processes to study would need to be based on the understanding of the gene's function in other systems and on its correlation to the physiology of the glaucoma tissue. For this selection, the intuition of the investigator would be no less important.

In addition to dealing with the transcriptome, it is very important to move forward to functional genomics of the protein. In this area, we have barely scratched the surface. Protein extractions and purifications are yet more difficult than those of RNA. Proteins fraction to different compartments of the cells, some are insoluble in standard buffers and some stick badly to walls of extraction materials. Their post-translational modifications result in a very high number of molecules to analyze. But similar difficulties were overcome on the RNA field with the development of RNAse inhibitors and procedures which allowed the use of total rather than isolated polyA + mRNA. After identifying a basic proteome, additional development of specific activity libraries such as that of the metalloproteases, will add another dimension to the elucidation of what is actually responsible for the function or dysfunction of the glaucoma associated cell.

During the exponential increase of information provided by functional genomics, there is also a need to keep a balance between continuing performing microarrays and applying the information obtained from their analysis. Right now, we are still on the short side of the balance and some critical array investigations remain. The advancements on the next ten years will come in great part from the large consortiums projects handling larges amount of data. But as important as the big computer generated analyses, are the more focused and clever idea oriented contributions of small laboratories. Although at times rough and tortuous, the road ahead is very exciting. Functional genomics is offering the right tools to follow it. As technological advances continue, we will be able to accumulate more human individual gene data and discern why patients respond differently to glaucomatous insults and glaucoma drugs. Modulation of the expression of key genes by gene transfer would allow deciphering the chain reactions triggered by administration of drugs. Creating better molecular signatures will also provide glaucoma geneticists with a new pool of candidate genes for their linkage studies. And, because of the continuous discovering of additional gene functions, more unexpected and relevant mechanisms might be uncovered in the trabecular meshwork. Without a doubt, functional genomics will take us to the understanding and treatment of complex diseases such as glaucoma.

Acknowledgments

Supported by NIH grants EY11906 (TB), EY13126 (TB), EY15873 (RRA) and a Research to Prevent Blindness challenge grant to the UNC Dept. of Ophthalmology.

References

Acott, T. S. (1992). Trabecular extracellular matrix regulation. *In* "Pharmacology of Glaucoma" (S. M. Drance, E. M. Van Buskirk, and A. H. Neufeld, eds.), pp. 125–157. Williams & Wilkins, Baltimore.

Agapova, O. A., Kaufman, P. L., Lucarelli, M. J., Gabelt, B. T., and Hernandez, M. R. (2003). Differential expression of matrix metalloproteinases in monkey eyes with experimental glaucoma or optic nerve transection. *Brain Res.* **967,** 132–143.

Ahmed, F., Brown, K. M., Stephan, D. A., Morrison, J. C., Johnson, E. C., and Tomarev, S. I. (2004a). Microarray analysis of changes in mRNA levels in the rat retina after experimental elevation of intraocular pressure. *Invest. Ophthalmol. Vis. Sci.* **45,** 1247–1258.

Ahmed, F., Torrado, M., Zinovieva, R. D., Senatorov, V. V., Wistow, G., and Tomarev, S. I. (2004b). Gene expression profile of the rat eye iridocorneal angle: NEIBank expressed sequence tag analysis. *Invest. Ophthalmol. Vis. Sci.* **45,** 3081–3090.

Akarsu, A. N., Turacli, M. E., Aktan, S. G., Barsoum-Homsy, M., Chevrette, L., Sayli, B. S., and Sarfarazi, M. (1996). A second locus (GLC3B) for primary congenital glaucoma (Buphthalmos) maps to the 1p36 region. *Hum. Mol. Genet.* **5,** 1199–1203.

Alberdi, E., Hyde, C. C., and Becerra, S. P. (1998). Pigment epithelium-derived factor (PEDF) binds to glycosaminoglycans: Analysis of the binding site. *Biochemistry* **37,** 10643–10652.

Alexander, J. P., and Acott, T. S. (2001). Involvement of protein kinase C in TNFα regulation of trabecular matrix metalloproteinases and TIMPs. *Invest. Ophthalmol. Vis. Sci.* **42,** 2831–2838.

Alvarado, J., Murphy, C., Polansky, J., and Juster, R. (1981). Age-related changes in trabecular meshwork cellularity. *Invest. Ophthalmol. Vis. Sci.* **21,** 714–727.

Alvarado, J., Murphy, C., and Juster, R. (1984). Trabecular meshwork cellularity in primary open-angle glaucoma and nonglaucomatous normals. *Ophthalmology* **91,** 564–579.

Anderson, D. R. (1989). Glaucoma: The damage caused by pressure. XLVI Edward Jackson memorial lecture. *Am. J. Ophthalmol.* **108,** 485–495.

Andree, H., Thiele, H., Fahling, M., Schmidt, I., and Thiele, B. J. (2006). Expression of the human TPT1 gene coding for translationally controlled tumor protein (TCTP) is regulated by CREB transcription factors. *Gene* **380,** 95–103.

Armaldy, M. F. (1963a). Effect of corticosteroids on intraocular pressure and fluid dynamics. I. The effect of Dexamethasone in the normal eye. *Arch. Ophthalmol.* **70,** 482–491.

Armaldy, M. F. (1963b). Effect of corticosteroids on intraocular pressure and fluid dynamics. II. The effect of Dexamethasonein the glaucomatous eye. *Arch. Ophthalmol.* **70,** 492–499.

Armaldy, M. F. (1965). Statistical atributes of the steroid hypertensive response in the clinically normal eye. I. The demonstration of three levels of response. *Invest. Ophthalmol.* **26,** 187–197.

Armaldy, M. F. (1966). The heritable nature of dexamethasone-induced ocular hypertension. *Arch. Ophthalmol.* **75,** 32–35.

Armaldy, M. F., and Becker, B. (1965). Intraocular pressure response to topical corticosteroids. *Fed. Proc.* **24,** 1274–1278.

Babizhayev, M. A., and Brodskaya, M. W. (1989). Fibronectin detection in drainage outflow system of human eyes in ageing and progression of open-angle glaucoma. *Mech. Ageing Dev.* **47,** 145–157.

Balint, E., Lapointe, D., Drissi, H., van der, M. C., Young, D. W., van Wijnen, A. J., Stein, J. L., Stein, G. S., and Lian, J. B. (2003). Phenotype discovery by gene expression profiling: Mapping of biological processes linked to BMP-2-mediated osteoblast differentiation. *J. Cell. Biochem.* **89,** 401–426.

Bartlett, J. D., Woolley, T. W., and Adams, C. M. (1993). Identification of high intraocular pressure responders to topical ophthalmic corticosteroids. *J. Ocul. Pharmacol.* **9,** 35–45.

Baulmann, D. C., Ohlmann, A., Flugel-Koch, C., Goswami, S., Cvekl, A., and Tamm, E. R. (2002). Pax6 heterozygous eyes show defects in chamber angle differentiation that are associated with a wide spectrum of other anterior eye segment abnormalities. *Mech. Dev.* **118,** 3–17.

Becker, B. (1965). Intraocular pressure in response to topical corticosteroids. *Invest. Ophthalmol.* **26,** 198–205.

Becker, B., and Hank, K. A. (1964). Topical corticosteroids and hereditary in primary open-angle glaucoma. *Am. J. Ophthalmol.* **57,** 543–551.

Becker, B., and Mills, D. W. (1963). Corticosteroids and intraocular pressure. *Arch. Ophthalmol.* **70,** 500–507.

Belloir, B., Kovari, E., Surini-Demiri, M., and Savioz, A. (2001). Altered apolipoprotein D expression in the brain of patients with Alzheimer disease. *J. Neurosci. Res.* **64,** 61–69.

Bhattacharya, S. K., Rockwood, E. J., Smith, S. D., Bonilha, V. L., Crabb, J. S., Kuchtey, R. W., Robertson, N. G., Peachey, N. S., Morton, C. C., and Crabb, J. W. (2005). Proteomics reveal Cochlin deposits associated with glaucomatous trabecular meshwork. *J. Biol. Chem.* **280,** 6080–6084.

Bill, A. (1965). The aqueous humor drainage mechanism in the cynomolgus monkey (*Macaca irus*) with evidence for unconventional routes. *Invest. Ophthalmol.* **4,** 911–919.

Bill, A., and Hellsing, K. (1965). Production and drainage of aqueous humor in the cynomolgus monkey (*Macaca irus*). *Invest. Ophthalmol.* **4,** 920–926.

Bill, A., and Phillips, C. I. (1971). Uveoscleral drainage of aqueous humour in human eyes. *Exp. Eye Res.* **12,** 275–281.

Bommer, U. A., and Thiele, B. J. (2004). The translationally controlled tumour protein (TCTP). *Int. J. Biochem. Cell Biol.* **36,** 379–385.

Borrás, T., Matsumoto, Y., Epstein, D. L., and Johnson, D. H. (1998). Gene transfer to the human trabecular meshwork by anterior segment perfusion. *Invest. Ophthalmol. Vis. Sci.* **39,** 1503–1507.

Borrás, T., Rowlette, L. L., Tamm, E. R., Gottanka, J., and Epstein, D. L. (2002). Effects of elevated intraocular pressure on outflow facility and TIGR/MYOC expression in perfused human anterior segments. *Invest. Ophthalmol. Vis. Sci.* **43,** 33–40.

Borrás, T., Bryant, P. A., and Chisolm, S. S. (2006). First look at the effect of overexpression of TIGR/MYOC on the transcriptome of the human trabecular meshwork. *Exp. Eye Res.* **82,** 1002–1010.

Bradley, J. M., Vranka, J., Colvis, C. M., Conger, D. M., Alexander, J. P., Fisk, A. S., Samples, J. R., and Acott, T. S. (1998). Effect of matrix metalloproteinases activity on outflow in perfused human organ culture. *Invest. Ophthalmol. Vis. Sci.* **39,** 2649–2658.

Bradley, J. M., Anderssohn, A. M., Colvis, C. M., Parshley, D. E., Zhu, X. H., Ruddat, M. S., Samples, J. R., and Acott, T. S. (2000). Mediation of laser trabeculoplasty-induced matrix metalloproteinase expression by IL-1β and TNFα. *Invest. Ophthalmol. Vis. Sci.* **41,** 422–430.

Caballero, M., and Borrás, T. (2001). Inefficient processing of an olfactomedin-deficient myocilin mutant: Potential physiological relevance to glaucoma. *Biochem. Biophys. Res. Commun.* **282,** 662–670.

Caballero, M., Rowlette, L. L., and Borrás, T. (2000). Altered secretion of a TIGR/MYOC mutant lacking the olfactomedin domain. *Biochim. Biophys. Acta* **1502,** 447–460.

Caudie, C., Bancel, J., Dupont, M., Matanza, D., Poitevin, F., and Honnorat, J. (2005). [CSF levels and diagnostic utility of cerebrospinal fluid β2-microglobulin]. *Ann. Biol. Clin. (Paris)* **63,** 631–637.

Coca-Prados, M., and Escribano, J. (2007). New perspectives in aqueous humor secretion and in glaucoma: The ciliary body as a multifunctional neuroendocrine gland. *Prog. Retin. Eye Res.* **26,** 239–262.

Comes, N., Gasull, X., Gual, A., and Borrás, T. (2005). Differential expression of the human chloride channel genes in the trabecular meshwork under stress conditions. *Exp. Eye Res.* **80,** 801–813.

Comes, N., Xue, W., and Borrás, T. (2006). Genes upregulated after 7 d sustained elevated pressure to the trabecular meshwork reveal angiopoietin-like factor CDT6 as a candidate for outflow resistance. *Invest. Ophthalmol. Vis. Sci.* ARVO e-abstract No. 264.

Cousins, S. W., McCabe, M. M., Danielpour, D., and Streilein, J. W. (1991). Identification of transforming growth factor-β as an immunosuppressive factor in aqueous humor. *Invest. Ophthalmol. Vis. Sci.* **32,** 2201–2211.

Crawford, K. S., Gange, S. J., Gabelt, B. T., Heideman, W., Robinson, J. C., Hubbard, W. C., and Kaufman, P. L. (1996). Indomethacin and epinephrine effects on outflow facility and cyclic adenosine monophosphate formation in monkeys. *Invest. Ophthalmol. Vis. Sci.* **37,** 1348–1359.

Cvekl, A., and Tamm, E. R. (2004). Anterior eye development and ocular mesenchyme: New insights from mouse models and human diseases. *Bioessays* **26,** 374–386.

Delany, A. M., Amling, M., Priemel, M., Howe, C., Baron, R., and Canalis, E. (2000). Osteopenia and decreased bone formation in osteonectin-deficient mice. *J. Clin. Invest.* **105,** 915–923.

Diehn, J. J., Diehn, M., Marmor, M. F., and Brown, P. O. (2005). Differential gene expression in anatomical compartments of the human eye. *Genome Biol.* **6,** R74.

Diskin, S., Kumar, J., Cao, Z., Schuman, J. S., Gilmartin, T., Head, S. R., and Panjwani, N. (2006). Detection of differentially expressed glycogenes in trabecular meshwork of eyes with primary open-angle glaucoma. *Invest. Ophthalmol. Vis. Sci.* **47,** 1491–1499.

Do, C. S., Levros, L. C., Jr., and Rassart, E. (2007). Modulation of apolipoprotein D expression and translocation under specific stress conditions. *Biochim. Biophys. Acta* **1773,** 954–969.

Emre, M., Orgul, S., Haufschild, T., Shaw, S. G., and Flammer, J. (2005). Increased plasma endothelin-1 levels in patients with progressive open angle glaucoma. *Br. J. Ophthalmol.* **89,** 60–63.

Epstein, D. L. (1997). Practical aqueous humor dynamics. *In* "Chandler and Grant's Glaucoma" (D. L. Epstein, R. R. Allingham, and J. S. Schuman, eds.), pp. 18–32. Williams & Wilkins, Baltimore, Maryland, USA.

Farkas, R. H., Qian, J., Goldberg, J. L., Quigley, H. A., and Zack, D. J. (2004). Gene expression profiling of purified rat retinal ganglion cells. *Invest. Ophthalmol. Vis. Sci.* **45,** 2503–2513.

Finn, J. T., Weil, M., Archer, F., Siman, R., Srinivasan, A., and Raff, M. C. (2000). Evidence that Wallerian degeneration and localized axon degeneration induced by local neurotrophin deprivation do not involve caspases. *J. Neurosci.* **20,** 1333–1341.

Francois, J. (1977). Corticosteroid glaucoma. *Ann. Ophthalmol.* **9,** 1075–1080.

Fraser, J. D., and Price, P. A. (1988). Lung, heart, and kidney express high levels of mRNA for the vitamin K-dependent matrix Gla protein. Implications for the possible functions of matrix Gla protein and for the tissue distribution of the gamma-carboxylase. *J. Biol. Chem.* **263,** 11033–11036.

Friedman, J. S., Faucher, M., Hiscott, P., Biron, V. L., Malenfant, M., Turcotte, P., Raymond, V., and Walter, M. A. (2002a). Protein localization in the human eye and genetic screen of opticin. *Hum. Mol. Genet.* **11,** 1333–1342.

Friedman, J. S., Faucher, M., Hiscott, P., Biron, V. L., Malenfant, M., Turcotte, P., Raymond, V., and Walter, M. A. (2002b). Protein localization in the human eye and genetic screen of opticin. *Hum. Mol. Genet.* **11,** 1333–1342.

Fuchshofer, R., Birke, M., Welge-Lussen, U., Kook, D., and Lutjen-Drecoll, E. (2005). Transforming growth factor-β 2 modulated extracellular matrix component expression in cultured human optic nerve head astrocytes. *Invest. Ophthalmol. Vis. Sci.* **46,** 568–578.

Fuchshofer, R., Welge-Lussen, U., Lutjen-Drecoll, E., and Birke, M. (2006). Biochemical and morphological analysis of basement membrane component expression in corneoscleral and cribriform human trabecular meshwork cells. *Invest. Ophthalmol. Vis. Sci.* **47,** 794–801.

Gabelt, B. T., Hu, Y., Vittitow, J. L., Rasmussen, C. R., Grosheva, I., Bershadsky, A. D., Geiger, B., Borrás, T., and Kaufman, P. L. (2006). Caldesmon transgene expression disrupts focal adhesions in HTM cells and increases outflow facility in organ-cultured human and monkey anterior segments. *Exp. Eye Res.* **82,** 935–944.

Garcia-Valenzuela, E., Shareef, S., Walsh, J., and Sharma, S. C. (1995). Programmed cell death of retinal ganglion cells during experimental glaucoma. *Exp. Eye Res.* **61,** 33–44.

Gonzalez, J. M., Jr., Faralli, J. A., Peters, J. M., Newman, J. R., and Peters, D. M. (2006). Effect of heparin II domain of fibronectin on actin cytoskeleton and adherens junctions in human trabecular meshwork cells. *Invest. Ophthalmol. Vis. Sci.* **47,** 2924–2931.

Gonzalez, P., Epstein, D. L., and Borrás, T. (2000a). Characterization of gene expression in human trabecular meshwork using single-pass sequencing of 1060 clones. *Invest. Ophthalmol. Vis. Sci.* **41,** 3678–3693.

Gonzalez, P., Epstein, D. L., and Borrás, T. (2000b). Genes upregulated in the human trabecular meshwork in response to elevated intraocular pressure. *Invest. Ophthalmol. Vis. Sci.* **41,** 352–361.

Gonzalez, P., Zigler, J. S., Jr., Epstein, D. L., and Borrás, T. (1999). Identification and isolation of differentially expressed genes from very small tissue samples. *Biotechniques* **26,** 884–892.

Gordon, M. O., Beiser, J. A., Brandt, J. D., Heuer, D. K., Higginbotham, E. J., Johnson, C. A., Keltner, J. L., Miller, J. P., Parrish, R. K., Wilson, M. R., and Kass, M. A. (2002). The Ocular Hypertension Treatment Study: Baseline factors that predict the onset of primary open-angle glaucoma. *Arch. Ophthalmol.* **120,** 714–720.

Gottanka, J., Chan, D., Eichhorn, M., Lütjen-Drecoll, E., and Ethier, C. R. (2004). Effects of TGF-beta2 in perfused human eyes. *Invest. Ophthalmol. Vis. Sci.* **45,** 153–158.

Grant, W. M. (1963a). Experimental aqueous perfusion in enucleated human eyes. *Arch. Ophthalmol.* **69,** 783–801.

Grant, W. M. (1963b). Glaucoma from topical corticosteroids. *Arch. Ophthalmol.* **70,** 445–446.

Greenbaum, D. C., Arnold, W. D., Lu, F., Hayrapetian, L., Baruch, A., Krumrine, J., Toba, S., Chehade, K., Bromme, D., Kuntz, I. D., and Bogyo, M. (2002). Small molecule affinity fingerprinting. A tool for enzyme family subclassification, target identification, and inhibitor design. *Chem. Biol.* **9,** 1085–1094.

Grierson, I., and Lee, W. R. (1975). Pressure-induced changes in the ultrastructure of the endothelium lining Schlemm's canal. *Am. J. Ophthalmol.* **80,** 863–884.

Gunning, P. W., Schevzov, G., Kee, A. J., and Hardeman, E. C. (2005). Tropomyosin isoforms: divining rods for actin cytoskeleton function. *Trends Cell Biol.* **15,** 333–341.

Hamanaka, T., Bill, A., Ichinohasama, R., and Ishida, T. (1992). Aspects of the development of Schlemm's canal. *Exp. Eye Res.* **55,** 479–488.

Hawse, J. R., Padgaonkar, V. A., Leverenz, V. R., Pelliccia, S. E., Kantorow, M., and Giblin, F. J. (2006). The role of metallothionein IIa in defending lens epithelial cells against cadmium and TBHP induced oxidative stress. *Mol. Vis.* **12,** 342–349.

Helisalmi, S., Hiltunen, M., Vepsalainen, S., Iivonen, S., Corder, E. H., Lehtovirta, M., Mannermaa, A., Koivisto, A. M., and Soininen, H. (2004). Genetic variation in apolipoprotein D and Alzheimer's disease. *J. Neurol.* **251,** 951–957.

Hernandez, M. R. (2000). The optic nerve head in glaucoma: Role of astrocytes in tissue remodeling. *Prog. Retin. Eye Res.* **19,** 297–321.

Hernandez, M. R., Igoe, F., and Neufeld, A. H. (1988). Cell culture of the human lamina cribrosa. *Invest. Ophthalmol. Vis. Sci.* **29,** 78–89.

Hernandez, M. R., Pena, J. D., Selvidge, J. A., Salvador-Silva, M., and Yang, P. (2000). Hydrostatic pressure stimulates synthesis of elastin in cultured optic nerve head astrocytes. *Glia* **32,** 122–136.

Hernandez, M. R., Agapova, O. A., Yang, P., Salvador-Silva, M., Ricard, C. S., and Aoi, S. (2002). Differential gene expression in astrocytes from human normal and glaucomatous optic nerve head analyzed by cDNA microarray. *Glia* **38,** 45–64.

Higginbotham, E. J., Gordon, M. O., Beiser, J. A., Drake, M. V., Bennett, G. R., Wilson, M. R., and Kass, M. A. (2004). The Ocular Hypertension Treatment Study: Topical medication delays or prevents primary open-angle glaucoma in African American individuals. *Arch. Ophthalmol.* **122,** 813–820.

Hinault, M. P., Ben Zvi, A., and Goloubinoff, P. (2006). Chaperones and proteases: Cellular fold-controlling factors of proteins in neurodegenerative diseases and aging. *J. Mol. Neurosci.* **30,** 249–265.

Ho, S. L., Dogar, G. F., Wang, J., Crean, J., Wu, Q. D., Oliver, N., Weitz, S., Murray, A., Cleary, P. E., and O'Brien, C. (2005). Elevated aqueous humour tissue inhibitor of matrix metalloproteinase-1 and connective tissue growth factor in pseudoexfoliation syndrome. *Br. J. Ophthalmol.* **89,** 169–173.

Ikeda, T., Ishii, H., Higuchi, T., Sato, K., Hayashi, Y., Ikeda, K., and Hirabayashi, Y. (2000). Localization of thrombomodulin in the anterior segment of the human eye. *Invest. Ophthalmol. Vis. Sci.* **41,** 3383–3390.

Ishibashi, T., Takagi, Y., Mori, K., Naruse, S., Nishino, H., Yue, B. Y., and Kinoshita, S. (2002). cDNA microarray analysis of gene expression changes induced by dexamethasone in cultured human trabecular meshwork cells. *Invest. Ophthalmol. Vis. Sci.* **43,** 3691–3697.

Ivanov, D., Dvoriantchikova, G., Nathanson, L., McKinnon, S. J., and Shestopalov, V. I. (2006). Microarray analysis of gene expression in adult retinal ganglion cells. *FEBS Lett.* **580,** 331–335.

Jacobson, N., Andrews, M., Shepard, A. R., Nishimura, D., Searby, C., Fingert, J. H., Hageman, G., Mullins, R., Davidson, B. L., Kwon, Y. H., Alward, W. L., Stone, E. M., *et al.* (2001). Non-secretion of mutant proteins of the glaucoma gene myocilin in cultured trabecular meshwork cells and in aqueous humor. *Hum. Mol. Genet.* **10,** 117–125.

Jessani, N., and Cravatt, B. F. (2004). The development and application of methods for activity-based protein profiling. *Curr. Opin. Chem. Biol.* **8,** 54–59.

Jessani, N., Liu, Y., Humphrey, M., and Cravatt, B. F. (2002). Enzyme activity profiles of the secreted and membrane proteome that depict cancer cell invasiveness. *Proc. Natl. Acad. Sci. USA* **99,** 10335–10340.

Jessani, N., Niessen, S., Wei, B. Q., Nicolau, M., Humphrey, M., Ji, Y., Han, W., Noh, D. Y., Yates, J. R., III, Jeffrey, S. S., and Cravatt, B. F. (2005). A streamlined platform for high-content functional proteomics of primary human specimens. *Nat. Methods* **2,** 691–697.

Johnson, D., Gottanka, J., Flugel, C., Hoffmann, F., Futa, R., and Lutjen-Drecoll, E. (1997). Ultrastructural changes in the trabecular meshwork of human eyes treated with corticosteroids. *Arch. Ophthalmol.* **115,** 375–383.

Johnson, D. H. (1997). The effect of cytochalasin D on outflow facility and the trabecular meshwork of the human eye in perfusion organ culture. *Invest. Ophthalmol. Vis. Sci.* **38,** 2790–2799.

Johnson, D. H., and Tschumper, R. C. (1987). Human trabecular meshwork organ culture. A new method. *Invest. Ophthalmol. Vis. Sci.* **28,** 945–953.

Johnson, D. H., Bradley, J. M., and Acott, T. S. (1990). The effect of dexamethasone on glycosaminoglycans of human trabecular meshwork in perfusion organ culture. *Invest. Ophthalmol. Vis. Sci.* **31,** 2568–2571.

Johnson, E. C., Jia, L., Cepurna, W. O., Doser, T. A., and Morrison, J. C. (2007). Global changes in optic nerve head gene expression after exposure to elevated intraocular pressure in a rat glaucoma model. *Invest. Ophthalmol. Vis. Sci.* **48,** 3161–3177.

Johnston, M. C., Noden, D. M., Hazelton, R. D., Coulombre, J. L., and Coulombre, A. J. (1979). Origins of avian ocular and periocular tissues. *Exp. Eye Res.* **29,** 27–43.

Johnstone, M. A. (1979). Pressure-dependent changes in nuclei and the process origins of the endothelial cells lining Schlemm's canal. *Invest. Ophthalmol. Vis. Sci.* **18,** 44–51.

Johnstone, M. A., and Grant, W. M. (1973). Pressure-dependent changes in structures of the aqueous outflow system of human and monkey eyes. *Am. J. Ophthalmol.* **75,** 365–383.

Jones, R., III, and Rhee, D. J. (2006). Corticosteroid-induced ocular hypertension and glaucoma: A brief review and update of the literature. *Curr. Opin. Ophthalmol.* **17,** 163–167.

Joyce, J. A., Baruch, A., Chehade, K., Meyer-Morse, N., Giraudo, E., Tsai, F. Y., Greenbaum, D. C., Hager, J. H., Bogyo, M., and Hanahan, D. (2004). Cathepsin cysteine proteases are effectors of invasive growth and angiogenesis during multistage tumorigenesis. *Cancer Cell* **5,** 443–453.

Kass, M. A., Heuer, D. K., Higginbotham, E. J., Johnson, C. A., Keltner, J. L., Miller, J. P., Parrish, R. K., Wilson, M. R., and Gordon, M. O. (2002). The Ocular Hypertension Treatment Study: A randomized trial determines that topical ocular hypotensive medication delays or prevents the onset of primary open-angle glaucoma. *Arch. Ophthalmol.* **120,** 701–713.

Katoh, Y., and Katoh, M. (2006). Comparative integromics on Angiopoietin family members. *Int. J. Mol. Med.* **17,** 1145–1149.

Kelley, M. J., Rose, A., Song, K., Lystrup, B., Samples, J. W., and Acott, T. S. (2007). p38 MAP Kinase Pathway and Stromelysin Regulation in Trabecular Meshwork Cells. *Invest. Ophthalmol. Vis. Sci.* **48,** 3126–3137.

Khalyfa, A., Chlon, T., Qiang, H., Agarwal, N., and Cooper, N. G. (2007). Microarray reveals complement components are regulated in the serum-deprived rat retinal ganglion cell line. *Mol. Vis.* **13,** 293–308.

Kidd, D., Liu, Y., and Cravatt, B. F. (2001). Profiling serine hydrolase activities in complex proteomes. *Biochemistry* **40,** 4005–4015.

Kim, C. Y., Kuehn, M. H., Clark, A. F., and Kwon, Y. H. (2006). Gene expression profile of the adult human retinal ganglion cell layer. *Mol. Vis.* **12,** 1640–1648.

Kirchmair, R., Gander, R., Egger, M., Hanley, A., Silver, M., Ritsch, A., Murayama, T., Kaneider, N., Sturm, W., Kearny, M., Fischer-Colbrie, R., Kircher, B., *et al.* (2004). The neuropeptide secretoneurin acts as a direct angiogenic cytokine *in vitro* and *in vivo*. *Circulation* **109,** 777–783.

Kirwan, R. P., Fenerty, C. H., Crean, J., Wordinger, R. J., Clark, A. F., and O'Brien, C. J. (2005a). Influence of cyclical mechanical strain on extracellular matrix gene expression in human lamina cribrosa cells *in vitro*. *Mol. Vis.* **11,** 798–810.

Kirwan, R. P., Leonard, M. O., Murphy, M., Clark, A. F., and O'Brien, C. J. (2005b). Transforming growth factor-beta-regulated gene transcription and protein expression in human GFAP-negative lamina cribrosa cells. *Glia* **52,** 309–324.

Knepper, P. A., Collins, J. A., and Frederick, R. (1985). Effects of dexamethasone, progesterone, and testosterone on IOP and GAGs in the rabbit eye. *Invest. Ophthalmol. Vis. Sci.* **26,** 1093–1100.

Kozaki, K., Miyaishi, O., Koiwai, O., Yasui, Y., Kashiwai, A., Nishikawa, Y., Shimizu, S., and Saga, S. (1998). Isolation, purification, and characterization of a collagen-associated serpin, caspin, produced by murine colon adenocarcinoma cells. *J. Biol. Chem.* **273,** 15125–15130.

Krohn, J. (1999). Expression of factor VIII-related antigen in human aqueous drainage channels. *Acta Ophthalmol. Scand.* **77,** 9–12.

Krueger, K. E., and Srivastava, S. (2006). Posttranslational protein modifications: Current implications for cancer detection, prevention, and therapeutics. *Mol. Cell. Proteomics* **5,** 1799–1810.

Kuchtey, J., Curtis, M., Gelatt, K., Kallberg, M., Rinkosky, T., Komaromy, A., and Kuchtey, R. (2007). Investigations of cornea-derived transcript 6 (CDT6) as a potential glaucoma protein. *Invest. Ophthalmol. Vis. Sci.* ARVO e-abstract No. 3951.

Kupfer, C., Underwood, B., and Gillen, T. (1994). Leading causes of visual impairment worldwide. *In* "Principles and Practice of Ophthalmology" (D. M. Albert and F. A. Jakobiec, eds.), pp. 1249–1255. W.B. Saunders, Philadelphia, PA.

Langdon, J. M., Vonakis, B. M., and MacDonald, S. M. (2004). Identification of the interaction between the human recombinant histamine releasing factor/translationally controlled tumor protein and elongation factor-1 delta (also known as eElongation factor-1B beta). *Biochim. Biophys. Acta* **1688,** 232–236.

Lepple-Wienhues, A., Stahl, F., and Wiederholt, M. (1991). Differential smooth muscle-like contractile properties of trabecular meshwork and ciliary muscle. *Exp. Eye Res.* **53,** 33–38.

Leung, Y. F., Tam, P. O., Lee, W. S., Lam, D. S., Yam, H. F., Fan, B. J., Tham, C. C., Chua, J. K., and Pang, C. P. (2003). The dual role of dexamethasone on anti-inflammation and outflow resistance demonstrated in cultured human trabecular meshwork cells. *Mol. Vis.* **9,** 425–439.

Levkovitch-Verbin, H., Dardik, R., Vander, S., Nisgav, Y., Kalev-Landoy, M., and Melamed, S. (2006). Experimental glaucoma and optic nerve transection induce simultaneous upregulation of proapoptotic and prosurvival genes. *Invest. Ophthalmol. Vis. Sci.* **47,** 2491–2497.

Li, X., Zimmerman, A., Copeland, N. G., Gilbert, D. J., Jenkins, N. A., and Yin, H. L. (1996). The mouse thymosin beta 4 gene: Structure, promoter identification, and chromosome localization. *Genomics* **32,** 388–394.

Liton, P. B., Luna, C., Challa, P., Epstein, D. L., and Gonzalez, P. (2006). Genome-wide expression profile of human trabecular meshwork cultured cells, nonglaucomatous and primary open angle glaucoma tissue. *Mol. Vis.* **12,** 774–790.

Liu, Y., and Vollrath, D. (2004). Reversal of mutant myocilin non-secretion and cell killing: Implications for glaucoma. *Hum. Mol. Genet.* **13,** 1193–1204.

Liu, Y., Patricelli, M. P., and Cravatt, B. F. (1999). Activity-based protein profiling: The serine hydrolases. *Proc. Natl. Acad. Sci. USA* **96,** 14694–14699.

Lo, W. R., Rowlette, L. L., Caballero, M., Yang, P., Hernandez, M. R., and Borrás, T. (2003). Tissue differential microarray analysis of dexamethasone induction reveals potential mechanisms of steroid glaucoma. *Invest. Ophthalmol. Vis. Sci.* **44,** 473–485.

Lundmark, K., Westermark, G. T., Nystrom, S., Murphy, C. L., Solomon, A., and Westermark, P. (2002). Transmissibility of systemic amyloidosis by a prion-like mechanism. *Proc. Natl. Acad. Sci. USA* **99,** 6979–6984.

Luo, G., Ducy, P., McKee, M. D., Pinero, G. J., Loyer, E., Behringer, R. R., and Karsenty, G. (1997). Spontaneous calcification of arteries and cartilage in mice lacking matrix GLA protein. *Nature* **386,** 78–81.

Lütjen-Drecoll, E. (1973). Structural factors influencing outflow facility and its changeability under drugs. A study in Macaca arctoides. *Invest. Ophthalmol.* **12,** 280–294.

Lütjen-Drecoll, E. (2005). Morphological changes in glaucomatous eyes and the role of TGFbeta2 for the pathogenesis of the disease. *Exp. Eye Res.* **81,** 1–4.

Lütjen-Drecoll, E., Shimizu, T., Rohrbach, M., and Rohen, J. W. (1986). Quantitative analysis of 'plaque material' in the inner- and outer wall of Schlemm's canal in normal- and glaucomatous eyes. *Exp. Eye Res.* **42,** 443–455.

Mäepea, O., and Bill, A. (1992). Pressures in the juxtacanalicular tissue and Schlemm's canal in monkeys. *Exp. Eye Res.* **54,** 879–883.

Matsumoto, Y., and Johnson, D. H. (1997). Dexamethasone decreases phagocytosis by human trabecular meshwork cells *in situ*. *Invest. Ophthalmol. Vis. Sci.* **38,** 1902–1907.

Matsuo, T., Osumi-Yamashita, N., Noji, S., Ohuchi, H., Koyama, E., Myokai, F., Matsuo, N., Taniguchi, S., Doi, H., and Iseki, S. (1993). A mutation in the Pax-6 gene in rat small eye is associated with impaired migration of midbrain crest cells. *Nat. Genet.* **3,** 299–304.

McKinnon, S. J. (2003). Glaucoma: Ocular Alzheimer's disease? *Front. Biosci.* **8,** s1140–s1156.

Meyer, C., Notari, L., and Becerra, S. P. (2002). Mapping the type I collagen-binding site on pigment epithelium-derived factor – Implications for its antiangiogenic activity. *J. Biol. Chem.* **277,** 45400–45407.

Miyahara, T., Kikuchi, T., Akimoto, M., Kurokawa, T., Shibuki, H., and Yoshimura, N. (2003). Gene microarray analysis of experimental glaucomatous retina from cynomologous monkey. *Invest. Ophthalmol. Vis. Sci.* **44,** 4347–4356.

Moniaux, N., Andrianifahanana, M., Brand, R. E., and Batra, S. K. (2004). Multiple roles of mucins in pancreatic cancer, a lethal and challenging malignancy. *Br. J. Cancer* **91,** 1633–1638.

Mor-Vaknin, N., Punturieri, A., Sitwala, K., and Markovitz, D. M. (2003). Vimentin is secreted by activated macrophages. *Nat. Cell Biol.* **5,** 59–63.

Murshed, M., Schinke, T., McKee, M. D., and Karsenty, G. (2004). Extracellular matrix mineralization is regulated locally; different roles of two gla-containing proteins. *J. Cell Biol.* **165,** 625–630.

Nagradova, N. (2007). Enzymes catalyzing protein folding and their cellular functions. *Curr. Protein Pept. Sci.* **8,** 273–282.

Naskar, R., and Thanos, S. (2006). Retinal gene profiling in a hereditary rodent model of elevated intraocular pressure. *Mol. Vis.* **12,** 1199–1210.

Neufeld, A. H., and Liu, B. (2003). Glaucomatous optic neuropathy: When glia misbehave. *Neuroscientist* **9,** 485–495.

Nguyen, T. D., Chen, P., Huang, W. D., Chen, H., Johnson, D., and Polansky, J. R. (1998). Gene structure and properties of TIGR, an olfactomedin-related glycoprotein cloned from glucocorticoid-induced trabecular meshwork cells. *J. Biol. Chem.* **273,** 6341–6350.

Oike, Y., Yasunaga, K., and Suda, T. (2004). Angiopoietin-related/angiopoietin-like proteins regulate angiogenesis. *Int. J. Hematol.* **80,** 21–28.

Osborne, N. N., Melena, J., Chidlow, G., and Wood, J. P. (2001). A hypothesis to explain ganglion cell death caused by vascular insults at the optic nerve head: Possible implication for the treatment of glaucoma. *Br. J. Ophthalmol.* **85,** 1252–1259.

Pang, I. H., Shade, D. L., Clark, A. F., Steely, H. T., and DeSantis, L. (1994). Preliminary characterization of a transformed cell strain derived from human trabecular meshwork. *Curr. Eye Res.* **13,** 51–63.

Pease, M. E., McKinnon, S. J., Quigley, H. A., Kerrigan-Baumrind, L. A., and Zack, D. J. (2000). Obstructed axonal transport of BDNF and its receptor TrkB in experimental glaucoma. *Invest. Ophthalmol. Vis. Sci.* **41,** 764–774.

Peek, R., van Gelderen, B. E., Bruinenberg, M., and Kijlstra, A. (1998). Molecular cloning of a new angiopoietinlike factor from the human cornea. *Invest. Ophthalmol. Vis. Sci.* **39,** 1782–1788.

Peek, R., Kammerer, R. A., Frank, S., Otte-Holler, I., and Westphal, J. R. (2002). The angiopoietin-like factor cornea-derived transcript 6 is a putative morphogen for human cornea. *J. Biol. Chem.* **277,** 686–693.

Pennisi, E. (2001). The human genome. *Science* **291,** 1177–1180.

Pervaiz, S., and Brew, K. (1987). Homology and structure-function correlations between alpha 1-acid glycoprotein and serum retinol-binding protein and its relatives. *FASEB J.* **1,** 209–214.

Peterson, J. R., Ora, A., Van, P. N., and Helenius, A. (1995). Transient, lectin-like association of calreticulin with folding intermediates of cellular and viral glycoproteins. *Mol. Biol. Cell* **6,** 1173–1184.

Picht, G., Welge-Luessen, U., Grehn, F., and Lutjen-Drecoll, E. (2001). Transforming growth factor beta 2 levels in the aqueous humor in different types of glaucoma and the relation to filtering bleb development. *Graefes Arch. Clin. Exp. Ophthalmol.* **239,** 199–207.

Piri, N., Kwong, J. M., Song, M., Elashoff, D., and Caprioli, J. (2006). Gene expression changes in the retina following optic nerve transection. *Mol. Vis.* **12,** 1660–1673.

Polansky, J. R. (1993). HTM cell culture model for steroid effects on IOP: Overview. *In* "Basic Aspects of Glaucoma Research III" (E. Lütjen-Drecoll, ed.), pp. 307–318. Schattauer, Stuttgart, Germany.

Polansky, J. R., and Alvarado, J. A. (1994). Cellular Mechanisms Influencing the Aqueous Humor Outflow Pathway. *In* "Principles & Practice of Ophthalmology: Basic Sciences." (D. M. Albert and F. Jakobiec, eds.), pp. 226–251. Saunders WB, Philadelphia, PA.

Polansky, J. R., Wood, I. S., Maglio, M. T., and Alvarado, J. A. (1984). Trabecular meshwork cell culture in glaucoma research: Evaluation of biological activity and structural properties of human trabecular cells *in vitro*. *Ophthalmology* **91,** 580–595.

Polansky, J. R., Fauss, D. J., Chen, P., Chen, H., Lütjen-Drecoll, E., Johnson, D., Kurtz, R. M., Ma, Z. D., Bloom, E., and Nguyen, T. D. (1997). Cellular pharmacology and molecular biology of the trabecular meshwork inducible glucocorticoid response gene product. *Ophthalmologica* **211,** 126–139.

Prasanna, G., Narayan, S., Krishnamoorthy, R. R., and Yorio, T. (2003). Eyeing endothelins: A cellular perspective. *Mol. Cell. Biochem.* **253,** 71–88.

Price, P. A., and Williamson, M. K. (1985). Primary structure of bovine matrix Gla protein, a new vitamin K-dependent bone protein. *J. Biol. Chem.* **260,** 14971–14975.

Price, P. A., Faus, S. A., and Williamson, M. K. (2000). Warfarin-induced artery calcification is accelerated by growth and vitamin D. *Arterioscler. Thromb. Vasc. Biol.* **20,** 317–327.

Proudfoot, D., Skepper, J. N., Shanahan, C. M., and Weissberg, P. L. (1998). Calcification of human vascular cells *in vitro* is correlated with high levels of matrix Gla protein and low levels of osteopontin expression. *Arterioscler. Thromb. Vasc. Biol.* **18,** 379–388.

Quigley, H. A. (1993). Open-angle glaucoma. *N. Engl. J. Med.* **328,** 1097–1106.

Quigley, H. A. (1996). Number of people with glaucoma worldwide. *Br. J. Ophthalmol.* **80,** 389–393.

Reardon, A. J., Le Goff, M., Briggs, M. D., McLeod, D., Sheehan, J. K., Thornton, D. J., and Bishop, P. N. (2000). Identification in vitreous and molecular cloning of opticin, a novel member of the family of leucine-rich repeat proteins of the extracellular matrix. *J. Biol. Chem.* **275,** 2123–2129.

Rhee, D. J., Tamm, E. R., and Russell, P. (2003). Donor corneoscleral buttons: A new source of trabecular meshwork for research. *Exp. Eye Res.* **77,** 749–756.

Robertson, N. G., Skvorak, A. B., Yin, Y., Weremowicz, S., Johnson, K. R., Kovatch, K. A., Battey, J. F., Bieber, F. R., and Morton, C. C. (1997). Mapping and characterization of a novel cochlear gene in human and in mouse: A positional candidate gene for a deafness disorder, DFNA9. *Genomics* **46,** 345–354.

Robertson, N. G., Lu, L., Heller, S., Merchant, S. N., Eavey, R. D., McKenna, M., Nadol, J. B., Jr., Miyamoto, R. T., Linthicum, F. H., Jr., Lubianca Neto, J. F., Hudspeth, A. J., Seidman, C. E., *et al.* (1998). Mutations in a novel cochlear gene cause DFNA9, a human nonsyndromic deafness with vestibular dysfunction. *Nat. Genet.* **20,** 299–303.

Rozsa, F. W., Shimizu, S., Lichter, P. R., Johnson, A. T., Othman, M. I., Scott, K., Downs, C. A., Nguyen, J., Polansky, J., and Richards, J. E. (1998). GLC1A mutations point to regions of potential functional importance on the TIGR/MYOC protein. *Mol. Vis.* **4,** 20.

Rozsa, F. W., Reed, D. M., Scott, K. M., Pawar, H., Moroi, S. E., Kijek, T. G., Krafchak, C. M., Othman, M. I., Vollrath, D., Elner, V. M., and Richards, J. E. (2006). Gene expression profile of human trabecular meshwork cells in response to long-term dexamethasone exposure. *Mol. Vis.* **12,** 125–141.

Saghatelian, A., Jessani, N., Joseph, A., Humphrey, M., and Cravatt, B. F. (2004). Activity-based probes for the proteomic profiling of metalloproteases. *Proc. Natl. Acad. Sci. USA* **101,** 10000–10005.

Samples, J. R., Alexander, J. P., and Acott, T. S. (1993). Regulation of the levels of human trabecular matrix metalloproteinases and inhibitor by interleukin-1 and dexamethasone. *Invest. Ophthalmol. Vis. Sci.* **34,** 3386–3395.

Sarfarazi, M. (1997). Recent advances in molecular genetics of glaucomas. *Hum. Mol. Genet.* **6,** 1667–1677.

Shamovsky, I., Ivannikov, M., Kandel, E. S., Gershon, D., and Nudler, E. (2006). RNA-mediated response to heat shock in mammalian cells. *Nature* **440,** 556–560.

Shanahan, C. M., Weissberg, P. L., and Metcalfe, J. C. (1993). Isolation of gene markers of differentiated and proliferating vascular smooth muscle cells. *Circ. Res.* **73,** 193–204.

Sheehan, K. M., Calvert, V. S., Kay, E. W., Lu, Y., Fishman, D., Espina, V., Aquino, J., Speer, R., Araujo, R., Mills, G. B., Liotta, L. A., Petricoin, E. F., III, *et al.* (2005). Use of reverse phase protein microarrays and reference standard development for molecular network analysis of metastatic ovarian carcinoma. *Mol. Cell. Proteomics* **4,** 346–355.

Sieber, S. A., Niessen, S., Hoover, H. S., and Cravatt, B. F. (2006). Proteomic profiling of metalloprotease activities with cocktails of active-site probes. *Nat. Chem. Biol.* **2,** 274–281.

Siegner, A., May, C. A., Welge-Lussen, U. W., Bloemendal, H., and Lutjen-Drecoll, E. (1996). alpha B-crystallin in the primate ciliary muscle and trabecular meshwork. *Eur. J. Cell Biol.* **71,** 165–169.

Snyder, R. W., Stamer, W. D., Kramer, T. R., and Seftor, R. E. (1993). Corticosteroid treatment and trabecular meshwork proteases in cell and organ culture supernatants. *Exp. Eye Res.* **57,** 461–468.

Sommer, A., Tielsch, J. M., Katz, J., Quigley, H. A., Gottsch, J. D., Javitt, J., and Singh, K. (1991). Relationship between intraocular pressure and primary open angle glaucoma among white and black Americans. The Baltimore Eye Survey. *Arch. Ophthalmol.* **109,** 1090–1095.

Southren, A. L., Wandel, T., Gordon, G. G., and Weinstein, B. I. (1994). Treatment of glaucoma with 3 alpha, 5 beta-tetrahydrocortisol: A new therapeutic modality. *J. Ocul. Pharmacol.* **10,** 385–391.

Spaeth, G. L., Rodrigues, M. M., and Weinreb, S. (1977). Steroid-induced glaucoma: A. Persistent elevation of intraocular pressure B. Histopathological aspects. *Trans. Am. Ophthalmol. Soc.* **75,** 353–381.

Speers, A. E., and Cravatt, B. F. (2004). Chemical strategies for activity-based proteomics. *Chembiochem* **5,** 41–47.

Stamer, W. D., Seftor, R. E., Williams, S. K., Samaha, H. A., and Snyder, R. W. (1995). Isolation and culture of human trabecular meshwork cells by extracellular matrix digestion. *Curr. Eye Res.* **14,** 611–617.

Steele, M. R., Inman, D. M., Calkins, D. J., Horner, P. J., and Vetter, M. L. (2006). Microarray analysis of retinal gene expression in the DBA/2J model of glaucoma. *Invest. Ophthalmol. Vis. Sci.* **47,** 977–985.

Steely, H. T., Browder, S. L., Julian, M. B., Miggans, S. T., Wilson, K. L., and Clark, A. F. (1992). The effects of dexamethasone on fibronectin expression in cultured human trabecular meshwork cells. *Invest. Ophthalmol. Vis. Sci.* **33,** 2242–2250.

Steely, H. T., Dillow, G. W., Bian, L., Grundstad, J., Braun, T. A., Casavant, T. L., McCartney, M. D., and Clark, A. F. (2006). Protein expression in a transformed trabecular meshwork cell line: Proteome analysis. *Mol. Vis.* **12,** 372–383.

Stone, E. M., Fingert, J. H., Alward, W. L. M., Nguyen, T. D., Polansky, J. R., Sunden, S. L. F., Nishimura, D., Clark, A. F., Nystuen, A., Nichols, B. E., Mackey, D. A., Ritch, R., *et al.* (1997). Identification of a gene that causes primary open angle glaucoma. *Science* **275,** 668–670.

Tamm, E. R. (2002). Myocilin and glaucoma: Facts and ideas. *Prog. Retin. Eye Res.* **21,** 395–428.

Tamm, E. R., Siegner, A., Baur, A., and Lutjen-Drecoll, E. (1996). Transforming growth factor-beta 1 induces alpha-smooth muscle-actin expression in cultured human and monkey trabecular meshwork. *Exp. Eye Res.* **62,** 389–397.

Tanimura, A., McGregor, D. H., and Anderson, H. C. (1986). Calcification in atherosclerosis. I. Human studies. *J. Exp. Pathol.* **2,** 261–273.

Tezel, G., and Wax, M. B. (2004). The immune system and glaucoma. *Curr. Opin. Ophthalmol.* **15,** 80–84.

Thiele, H., Berger, M., Skalweit, A., and Thiele, B. J. (2000). Expression of the gene and processed pseudogenes encoding the human and rabbit translationally controlled tumour protein (TCTP). *Eur. J. Biochem.* **267,** 5473–5481.

Thomas, E. A., Laws, S. M., Sutcliffe, J. G., Harper, C., Dean, B., McClean, C., Masters, C., Lautenschlager, N., Gandy, S. E., and Martins, R. N. (2003). Apolipoprotein D levels are elevated in prefrontal cortex of subjects with Alzheimer's disease: No relation to apolipoprotein E expression or genotype. *Biol. Psychiatry* **54,** 136–141.

Tian, B., Geiger, B., Epstein, D. L., and Kaufman, P. L. (2000). Cytoskeletal involvement in the regulation of aqueous humor outflow. *Invest. Ophthalmol. Vis. Sci.* **41,** 619–623.

Tomarev, S. I., Wistow, G., Raymond, V., Dubois, S., and Malyukova, I. (2003). Gene expression profile of the human trabecular meshwork: NEIBank sequence tag analysis. *Invest. Ophthalmol. Vis. Sci.* **44,** 2588–2596.

Tripathi, R. C., Li, J., Borisuth, N. S., and Tripathi, B. J. (1993). Trabecular cells of the eye express messenger RNA for transforming growth factor-beta 1 and secrete this cytokine. *Invest. Ophthalmol. Vis. Sci.* **34,** 2562–2569.

Tripathi, R. C., Chan, W. F., Li, J., and Tripathi, B. J. (1994a). Trabecular cells express the TGF-beta 2 gene and secrete the cytokine. *Exp. Eye Res.* **58,** 523–528.

Tripathi, R. C., Li, J., Chan, W. F., and Tripathi, B. J. (1994b). Aqueous humor in glaucomatous eyes contains an increased level of TGF-beta 2. *Exp. Eye Res.* **59,** 723–727.

Troger, J., Neyer, S., Heufler, C., Huemer, H., Schmid, E., Griesser, U., Kralinger, M., Kremser, B., Baldissera, I., and Kieselbach, G. (2001). Substance P and vasoactive intestinal polypeptide in the streptozotocin-induced diabetic rat retina. *Invest. Ophthalmol. Vis. Sci.* **42,** 1045–1050.

Valpuesta, J. M., Martin-Benito, J., Gomez-Puertas, P., Carrascosa, J. L., and Willison, K. R. (2002). Structure and function of a protein folding machine: The eukaryotic cytosolic chaperonin CCT. *FEBS Lett.* **529,** 11–16.

Vittal, V., Rose, A., Gregory, K. E., Kelley, M. J., and Acott, T. S. (2005). Changes in gene expression by trabecular meshwork cells in response to mechanical stretching. *Invest. Ophthalmol. Vis. Sci.* **46,** 2857–2868.

Vittitow, J., and Borrás, T. (2002). Expression of optineurin, a glaucoma-linked gene, is influenced by elevated intraocular pressure. *Biochem. Biophys. Res. Commun.* **298,** 67–74.

Vittitow, J., and Borrás, T. (2004). Genes expressed in the human trabecular meshwork during pressure-induced homeostatic response. *J. Cell. Physiol.* **201,** 126–137.

Vittitow, J. L., Garg, R., Rowlette, L. L., Epstein, D. L., O'Brien, E. T., and Borrás, T. (2002). Gene transfer of dominant-negative RhoA increases outflow facility in perfused human anterior segment cultures. *Mol. Vis.* **8,** 32–44.

Wang, N., Chintala, S. K., Fini, M. E., and Schuman, J. S. (2001). Activation of a tissue-specific stress response in the aqueous outflow pathway of the eye defines the glaucoma disease phenotype. *Nat. Med.* **7,** 304–309.

Webb, J. G., Shearer, T. W., Yates, P. W., Mukhin, Y. V., and Crosson, C. E. (2003). Bradykinin enhancement of PGE2 signalling in bovine trabecular meshwork cells. *Exp. Eye Res.* **76,** 283–289.

Weinreb, R. N., Polansky, J. R., Kramer, S. G., and Baxter, J. D. (1985). Acute effects of dexamethasone on intraocular pressure in glaucoma. *Invest. Ophthalmol. Vis. Sci.* **26,** 170–175.

Werner, E. B. (1996). Normal-tension glaucoma. *In* "The Glaucomas" (R. Ritch, M. B. Shields, and T. Krupin, eds.), pp. 769–800. Mosby, St. Louis, MO.

Whitmore, A. V., Libby, R. T., and John, S. W. (2005). Glaucoma: Thinking in new ways—A role for autonomous axonal self-destruction and other compartmentalised processes? *Prog. Retin. Eye Res.* **24,** 639–662.

Wirtz, M. K., Samples, J. R., Xu, H., Severson, T., and Acott, T. S. (2002). Expression profile and genome location of cDNA clones from an infant human trabecular meshwork cell library. *Invest. Ophthalmol. Vis. Sci.* **43,** 3698–3704.

Won, K. A., Schumacher, R. J., Farr, G. W., Horwich, A. L., and Reed, S. I. (1998). Maturation of human cyclin E requires the function of eukaryotic chaperonin CCT. *Mol. Cell. Biol.* **18,** 7584–7589.

Xue, W., Comes, N., and Borrás, T. (2007). Presence of an established calcification marker in trabecular meshwork tissue of glaucoma donors. *Invest. Ophthalmol. Vis. Sci.* **48,** 3184–3194.

Yan, D. B., Coloma, F. M., Metheetrairut, A., Trope, G. E., Heathcote, J. G., and Ethier, C. R. (1994). Deformation of the lamina cribrosa by elevated intraocular pressure. *Br. J. Ophthalmol.* **78,** 643–648.

Yang, P., Agapova, O., Parker, A., Shannon, W., Pecen, P., Duncan, J., Salvador-Silva, M., and Hernandez, M. R. (2004). DNA microarray analysis of gene expression in human optic nerve head astrocytes in response to hydrostatic pressure. *Physiol. Genomics* **17,** 157–169.

Yorio, T., Krishnamoorthy, R., and Prasanna, G. (2002). Endothelin: Is it a contributor to glaucoma pathophysiology? *J. Glaucoma* **11,** 259–270.

Yoshimura, N., Kikuchi, T., Kuroiwa, S., and Gaun, S. (2003). Differential temporal and spatial expression of immediate early genes in retinal neurons after ischemia-reperfusion injury. *Invest. Ophthalmol. Vis. Sci.* **44,** 2211–2220.

You, L., and Kruse, F. E. (2002). Differential effect of activin A and BMP-7 on myofibroblast differentiation and the role of the Smad signaling pathway. *Invest. Ophthalmol. Vis. Sci.* **43,** 72–81.

Yun, A. J., Murphy, C. G., Polansky, J. R., Newsome, D. A., and Alvarado, J. A. (1989). Proteins secreted by human trabecular cells. Glucocorticoid and other effects. *Invest. Ophthalmol. Vis. Sci.* **30,** 2012–2022.

Zhang, X., Clark, A. F., and Yorio, T. (2006). Heat shock protein 90 is an essential molecular chaperone for nuclear transport of glucocorticoid receptor beta. *Invest. Ophthalmol. Vis. Sci.* **47,** 700–708.

Zhang, X., Ognibene, C. M., Clark, A. F., and Yorio, T. (2007). Dexamethasone inhibition of trabecular meshwork cell phagocytosis and its modulation by glucocorticoid receptor beta. *Exp. Eye Res.* **84,** 275–284.

Zhao, X., Pearson, K. E., Stephan, D. A., and Russell, P. (2003). Effects of prostaglandin analogues on human ciliary muscle and trabecular meshwork cells. *Invest. Ophthalmol. Vis. Sci.* **44,** 1945–1952.

Zhao, X., Ramsey, K. E., Stephan, D. A., and Russell, P. (2004). Gene and protein expression changes in human trabecular meshwork cells treated with transforming growth factor-beta. *Invest. Ophthalmol. Vis. Sci.* **45,** 4023–4034.

Zou, J., Guo, Y., Guettouche, T., Smith, D. F., and Voellmy, R. (1998). Repression of heat shock transcription factor HSF1 activation by HSP90 (HSP90 complex) that forms a stress-sensitive complex with HSF1. *Cell* **94,** 471–480.

CHAPTER 12

Molecular Approaches to Glaucoma: Intriguing Clues for Pathology

Ernst R. Tamm
Institute of Human Anatomy and Embryology, University of Regensburg, 93053 Regensburg, Germany

I. OVERVIEW

Despite intensive research efforts over the past decades, the molecular events that cause damage to retinal ganglion cells (RGC) and their axons in primary open angle glaucoma (POAG) have not been substantially clarified. There is evidence though that intraocular pressure (IOP) plays a critical role, as an IOP that is too high for the health of the optic nerve axons has been identified as the most critical risk factor for glaucomatous RGC damage in several prospective, randomized, multi-center clinical studies (Collaborative Normal-Tension Glaucoma Study Group, 1998a,b; The AGIS Investigators, 2000; Gordon *et al.*, 2002; Leske *et al.*, 2003; Higginbotham *et al.*, 2004). IOP is generated in the trabecular meshwork (TM) outflow pathways (Johnson and Erickson, 2000; Johnson, 2006), which show an abnormally high

Current Topics in Membranes, Volume 62

1063-5823/08 $35.00
DOI: 10.1016/S1063-5823(08)00412-2

resistance for aqueous humor outflow in POAG (Grant, 1963). As of today, there is uncertainty on the nature of the molecular changes that lead to an increase in intraocular pressure (IOP) in POAG. Over the recent years, several molecules have been identified that appear to be involved in the pathogenesis of POAG. Some of them such as transforming growth factor-β (TGF-β), thrombospondin-1 (TSP-1), connective tissue growth factor (CTGF), and bone morphogenetic protein-7 (BMP-7) are involved in the control mechanisms of extracellular matrix (ECM) turnover in the TM outflow pathways. Changes in ECM turnover in the extracellular spaces of the TM very likely play a role in POAG, since an increase in fibrillar ECM in the juxtacanalicular region (JCT) of the TM is the most characteristic pathological finding in patients with POAG (Rohen and Witmer, 1972; Lütjen-Drecoll *et al.*, 1986; Rohen *et al.*, 1993; Lütjen-Drecoll and Rohen, 2001). Other molecules such as myocilin, optineurin, and WD repeat domain 36 are the products of the *MYOC, OPTN*, and *WDR36* genes, which have been identified by linkage analysis and positional cloning as causative for some forms of POAG (Stone *et al.*, 1997; Rezaie *et al.*, 2002; Monemi *et al.*, 2005). This article will focus on the data that are currently available on the function of all of these proteins as related to POAG. It is the strong believe of the author that a more complete understanding of these proteins will substantially help to elucidate the molecular mechanisms that govern aqueous humor outflow through the TM and the pathogenetic mechanisms of POAG.

II. TRANSFORMING GROWTH FACTOR-B

A very characteristic structural change in the TM of eyes with POAG is an increase in ECM in the juxtacanalicular region (JCT), which is localized directly adjacent to the inner wall endothelium of Schlemm's canal (Lütjen-Drecoll and Rohen, 2001). In contrast to the inner parts of the TM, the JCT does not form lamellae, but is rather a typical loose connective tissue with resident cells that are embedded in a loosely arranged fibrillar ECM. JCT cells form long cellular processes that attach to one other, to JCT ECM fibrils or to the endothelial cells of Schlemm's canal. Together with the endothelium of Schlemm's canal and its basement membrane, the JCT forms the inner wall region (Johnson, 2006). It is generally agreed upon that the inner wall region is the site of trabecular outflow resistance in the normal eye and in that with POAG (Johnson and Erickson, 2000; Johnson, 2006; Tamm *et al.*, 2007), and it seems more than likely that the ECM increase in the JCT of eyes with POAG is causatively related to the processes that cause the pathological increase in outflow resistance. The predominant change in JCT ECM in POAG involves the accumulation of banded fibrillar elements that are embedded in different glycoproteins, and which have been termed "plaque material" (Rohen and Witmer, 1972). Plaque material derives from thickened sheaths of elastic fibers in the JCT, and increases in correlation with axonal damage in the optic nerve head (Rohen *et al.*, 1981; Lütjen-Drecoll *et al.*, 1986; Gottanka *et al.*, 1997). The molecular composition of plaque material is largely unclear. Eyes with ocular hypertension or early glaucoma do not show a substantial increase in this material, indicating that plaque material *per se* is rather a symptom than a cause of increased outflow resistance in POAG (Gottanka *et al.*, 1997). Nevertheless, the symptom is very characteristic and strongly indicates that changes in the amount and quality of the JCT ECM are causatively involved in or linked to the pathogenetic processes in POAG that finally lead to the increase of trabecular outflow resistance in POAG.

The available information on the nature of factors, which modulate ECM turnover in the normal TM and its increase in POAG, indicates that transforming growth factors (TGF)-β1 and -2 are very likely involved. TGF-β2 is found in the normal aqueous humor at relatively high concentrations (Granstein *et al.*, 1990; Jampel *et al.*, 1990; Cousins *et al.*, 1991) and several independent studies reported on a higher than normal concentration of TGF-β2 in the aqueous humor of patients with POAG (Tripathi *et al.*, 1994b; Inatani *et al.*, 2001; Picht *et al.*, 2001; Ochiai and Ochiai, 2002). TGF-β2 in the aqueous humor is very likely secreted from the non-pigmented and pigmented ciliary epithelium (Helbig *et al.*, 1991), and from cells of the lens (Allen *et al.*, 1998; Gordon-Thomson *et al.*, 1998). Moreover, cells of the TM have been shown to secrete TGF-β2 and its isoform TGF-β1 (Tripathi *et al.*, 1993b, 1994a), and to express receptors for both factors (Borisuth *et al.*, 1992; Tripathi *et al.*, 1993a). In a variety of disorders throughout the body, TGF-β signaling mediates fibrosis and a pathological increase in ECM deposition (Ihn, 2002; Schnaper *et al.*, 2003; Huggins and Sahn, 2004; Bataller and Brenner, 2005; Gressner and Weiskirchen, 2006; Liu, 2006; Willis and Borok, 2007). A substantial number of *in vitro* studies suggest a similar role of TGF-β2 for the increase in TM ECM deposition in POAG. Treatment of cultured human TM cells with TGF-β2 causes an increase in the synthesis of fibronectin and of a variety of other ECM molecules (Li *et al.*, 2000; Zhao *et al.*, 2004; Zhao and Russell, 2005; Fuchshofer *et al.*, 2007). In addition, TGF-β2 treatment of human TM cells *in vitro* induces a substantial and irreversible cross-linking of fibronectin by the action of tissue transglutaminase (Welge-Lüssen *et al.*, 2000), and a decrease in the activity of MMPs (Fuchshofer *et al.*, 2003). In anterior segment perfusion cultures, perfusion with TGF-β2 promotes a focal accumulation of fine fibrillar extracellular material in the TM (Gottanka *et al.*, 2004) and an increase in fibronectin synthesis (Bachmann *et al.*, 2006; Fleenor *et al.*,

2006), effects that are correlated with a reduction in outflow facility. Following treatment of TM cells with TGF-β1, the expression of myocilin is increased, an extracellular glycoprotein in the TM which will be discussed in detail in a separate paragraph of this article (Tamm *et al.*, 1999).

TGF-β signaling is also modulating the expression of matrix Gla protein, which is downregulated in TM cells following treatment with TGF-β1 (Vittitow and Borras, 2004). The gene encoding matrix Gla protein (*MGP*) has been found to be the fifth-highest expressed gene in an unnormalized cDNA library from fresh human trabecular meshwork (Tomarev *et al.*, 2003). It is also very highly expressed in libraries established from cultured TM cells (Gonzalez *et al.*, 2000; Wirtz *et al.*, 2002). Matrix Gla protein is a small, 79-amino acid ECM protein that contains posttranslationally modified γ-carboxylated glutamic acid residues resulting from Vitamin K-dependant carboyxlation of the protein in the endoplasmic reticulum (Price and Williamson, 1985). In general, an important function of matrix Gla protein is to inhibit ectopic calcification in certain soft tissues, probably by binding calcium ions through the γ-carboxylated glutamic acid residues, or by antagonizing signaling of bone morphogenetic proteins (BMPs). Matrix Gla protein in the TM apparently is able to perform both of these functions (Xue *et al.*, 2006). Knockout mice deficient in *Mgp* ($Mgp^{-/-}$) develop severe vascular calicification that occurs in all elastic and muscular arteries, and also ectopically in the cartilage (Luo *et al.*, 1997). In a recent study by Xue and colleagues, the calcification marker alkaline phosphate was found to have a higher activity in the TM of five patients with POAG, while the expression of matrix Gla protein was found to be reduced (Xue *et al.*, 2007). Treatment with TGF-β2 significantly induced the activity of alkaline phosphatase in cultured TM cells (Xue *et al.*, 2007). While the available histopathological data on the human TM in POAG very clearly indicate that a major calcification process comparable to that in atherosclerosis is absent in the TM (Rohen and Witmer, 1972; Tripathi, 1972; Fine *et al.*, 1981; Rohen *et al.*, 1981; Alvarado and Murphy, 1992), it is tempting to speculate that a more subtle mineralization of the TM ECM is involved in the structural changes of the TM in POAG, and that TGF-β signaling is causatively involved.

In addition to its action on ECM proteins, TGF-β signaling does also increase the expression of intracellular proteins, such as αB-crystallin (Welge-Lüssen *et al.*, 1999; Bachmann *et al.*, 2006). αB-Crystallin is a member of the small heat shock protein family, which acts as molecular chaperone in multiple cell types, and has refractive functions in the lens (Horwitz, 2003). In the TM, αB-crystallin is preferentially expressed in cells of the JCT (Tamm *et al.*, 1996a), and is found in higher amounts in TM cells of patients with POAG (Lütjen-Drecoll *et al.*, 1998). In cells of the retinal pigmented epithelium, an increase in αB-crystallin protects against apoptosis due to oxidative stress, a function that would also be highly relevant for the TM (Alge *et al.*, 2002).

Another intracellular protein that is induced upon treatment of TM cells with TGF-β1 is α-smooth muscle actin (Tamm *et al.*, 1996b), an actin isoform that is typically expressed in vascular smooth muscle cells and myofibroblasts, cells that are present in healing wounds and scars (Wang *et al.*, 2006). In the normal human eye, several cells within the TM express immunoreactivity for α-smooth muscle actin (de Kater *et al.*, 1992; Flügel *et al.*, 1992), while in the scleral spur region, the site of the posterior attachment of the TM to the sclera, virtually all cells stain positively for this actin isoform (Tamm *et al.*, 1992). There is experimental evidence that TM cells influence the hydraulic conductivity of the inner wall region and outflow resistance not only by modulating ECM turnover, but also by actively changing cell shape and altering the geometry of the outflow pathways (Tian *et al.*, 2000; Wiederholt *et al.*, 2000). An increase in TM cell tone is correlated with an increase in outflow resistance (Wiederholt *et al.*, 1996). α-Smooth muscle actin expression induced by TGF-β1 has been shown to substantially enhance cell traction force in myofibroblasts (Chen *et al.*, 2007), a scenario that might be also true for TM cells thus generating an additional mechanism, in addition to modulating ECM turnover, by which TGF-β signaling might increase outflow resistance in POAG. Interestingly, cyclic mechanical stretch activates the promoter of TGF-β1 and induces its expression in TM cells (Liton *et al.*, 2005) as does treatment with TGF-β1 and TGF-β2 (Li *et al.*, 1996), indicating that under certain conditions self-amplifying mechanisms might substantially augment the adverse actions of TGF-β1 on the TM outflow pathways.

III. THROMBOSPONDIN-1

In addition to its effects on ECM turnover, TGF-β1 and 2 are involved in multiple extremely important biological processes throughout the body, including proliferation, apoptosis, and modulation of the immune system. Naturally, the activity of TGF-βs *in vivo* needs to be subject of tight control mechanisms that have to be modified in patients with POAG, in order to explain a causative role of TGF-β signaling. In general, TGF-βs are secreted as latent complexes, which are unable to interact with cellular receptors (Gleizes *et al.*, 1997; Annes *et al.*, 2003). The same is true for the aqueous humor in the normal eye and in that with POAG, where most of TGF-β 2 is found in its latent, inactive form (Tripathi *et al.*, 1994b; Picht *et al.*, 2001).

Under *in vitro* conditions, active TGF-βs are generated by extremes of pH, heat, or chaotropic agents, mechanisms that are not likely to be of physiological relevance for TGF-β activation *in vivo*. The physiological mechanisms of TGF-β activation *in vivo* are not well understood, but proteolytic processing by plasmin, exposure to reactive oxygen species, and/or binding to integrins might be involved (Gleizes *et al.*, 1997; Khalil, 1999; Munger *et al.*, 1999). A very potent activator of latent TGF-β *in vivo* and *in vitro* is the matricellular protein thrombospondin-1 (TSP-1) (Schultz-Cherry *et al.*, 1994; Crawford *et al.*, 1998; Murphy-Ullrich and Poczatek, 2000), which belongs to a small family of secreted glycoproteins (Adams, 2001). Matricellular proteins such as TSP-1 are secreted proteins that influence cell function by modulating cell-matrix interactions (Bornstein, 2001). Indeed, TSP-1 might also be critical for TGF-β activation in the TM, as there is a considerable constitutive TSP-1 expression in the JCT of the human TM (Fig. 1) (Flügel-Koch *et al.*, 2004). Moreover, in the TM of patients with POAG, an increase in TSP-1 immunoreactivity and staining for TSP-1 in all regions of the TM has been observed in about one-third of patient eyes that were investigated (Fig. 2) (Flügel-Koch *et al.*, 2004). The expression of TSP-1 in TM cells is induced upon treatment with TGF-β1 (Flügel-Koch *et al.*, 2004) and TGF-β2 (Fig. 3), indicating again the presence of self-amplifying mechanisms for TGF-β signaling in the TM outflow pathways.

IV. CONNECTIVE TISSUE GROWTH FACTOR

Connective tisse growth factor (CTGF) which is a member of the CCN (CTGF, cysteine-rich angiogenic protein 61, and nephroblastoma overexpression gene) family of regulatory proteins, has been found to be upregulated in a substantial number of disorders that are associated with a pathological increase in ECM including scleroderma, renal and pulmonary fibrosis, inflammatory bowel disease, and atherosclerosis (Ito *et al.*, 1998; Sato *et al.*, 2000; di Mola *et al.*, 2004; Cicha *et al.*, 2005; Yamamoto *et al.*, 2005). It has been identified as a critical downstream mediator of the fibrogenic action of TGF-β2 (Ihn, 2002; Leask and Abraham, 2004) which itself induces the expression of CTGF (Yang *et al.*, 1998; Fuchshofer *et al.*, 2005). In the eye, the expression of CTGF has been found in TM (Tomarev *et al.*, 2003), iris sphincter and ciliary muscle cells (Liang *et al.*, 2003), retinal vascular endothelial cells (Wunderlich *et al.*, 2000c), epi- and subretinal membranes (Meyer *et al.*, 2002), plaques of human anterior subcapsular cataracts (Wunderlich *et al.*, 2000a), corneal scars (Wunderlich *et al.*, 2000b), tear fluid (van Setten *et al.*, 2003b), and pterygia (van Setten *et al.*, 2003a). CTGF has also been detected in the aqueous humor (van Setten

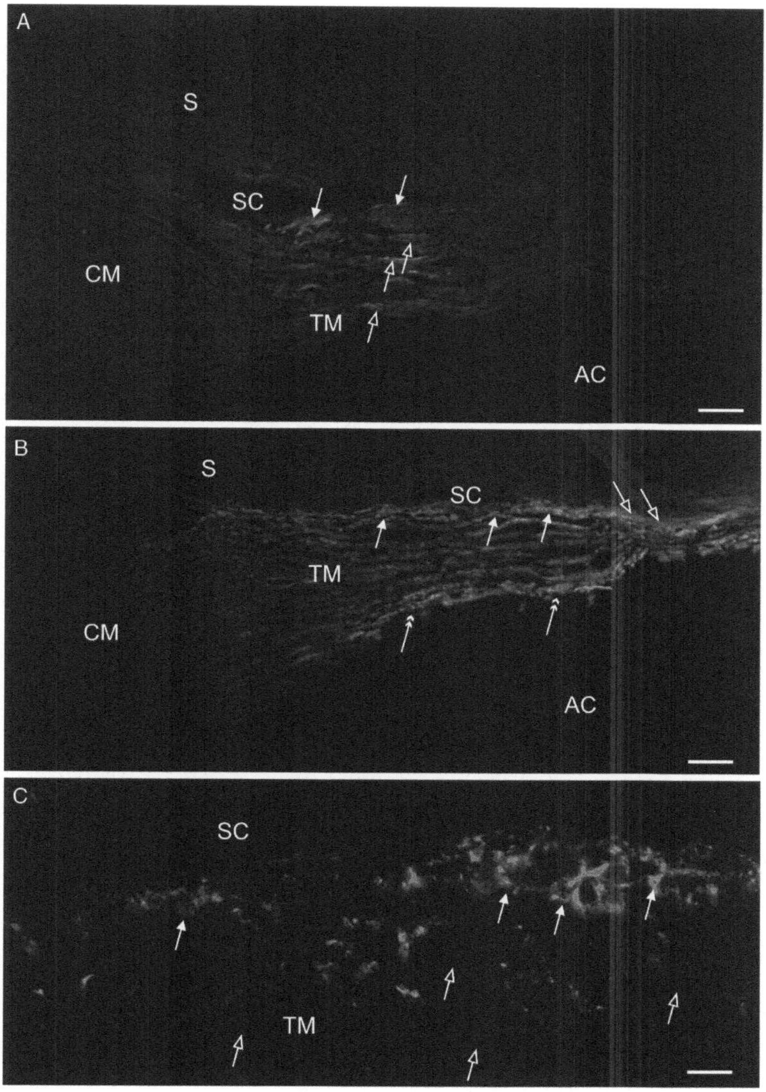

FIGURE 1 Immunohistochemistry for TSP-1 in the TM of normal human donors. (A) The TM of a 52-year old donor (case 281/98) shows immunoreactivity for TSP-1 in focal areas of the juxtacanalicular region (solid arrows). In addition, focal staining of uveal and corneoscleral TM cells is observed (open arrows). (B) In the TM of a 70-year old donor (case 15/99), there is continuous labeling for TSP-1 in the juxtacanalicular region (solid arrows), which extends to the area of Schwalbe's line (open arrows). In addition, there is intense labeling of the anterior uveal TM (double arrows). (C) Upon higher magnification, juxtacanalicular immunoreactivity for TSP-1 is predominately localized to extracellular areas surrounding juxtacanalicular TM cells (solid arrows). In contrast, the connective tissue core of the corneoscleral TM lamellae is largely negative for TSP-1 (open arrows). AC: Anterior chamber. SC: Schlemm's canal. CM: Ciliary muscle. S: Sclera. Magnification bars: 47 μm (A, B); 6.6 μm (C). From Flügel-Koch *et al.* (2004). (See Color Insert.)

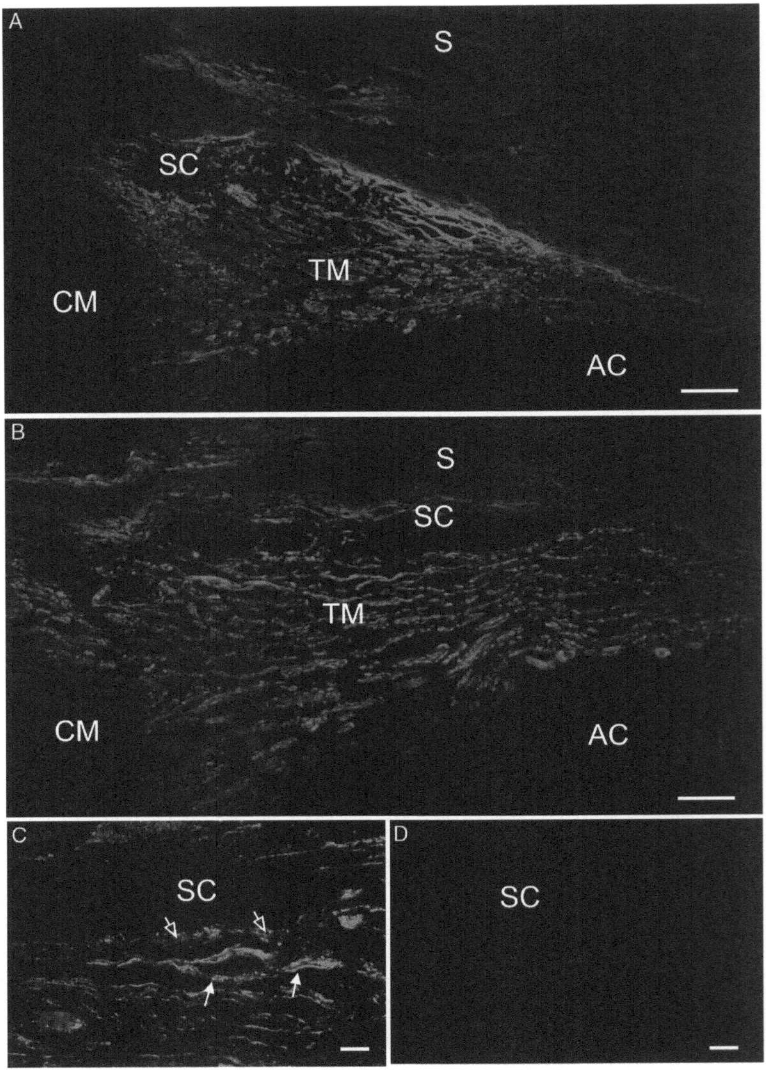

FIGURE 2 Immunohistochemistry for TSP-1 in the TM of a 67-year old patient (case 11/99) with POAG (A), and 71-year old patient (case 8/99) with POAG after long-term treatment with topical steroids (B). (A, B) In the TM of both patients, intense immunoreactivity for TSP-1 is observed throughout all regions of the TM. TSP-1 labeling is also observed in the scleral tissue that lines the outer wall of Schlemm's canal and the collector channels, which originate from it. (C) Upon higher magnification, the cells lining the corneoscleral and uveal trabecular lamellae are labeled (solid arrows). In addition, positive staining is observed in the connective tissue core of the lamellae (open arrows). (D) In a control section, no positive immunoreactivity for TSP-1 is observed. AC: Anterior chamber. SC: Schlemm's canal. CM: Ciliary muscle. S: Sclera. Magnification bars: 47 μm (A, B); 6.6 μm (C, D). From Flügel-Koch *et al.* (2004). (See Color Insert.)

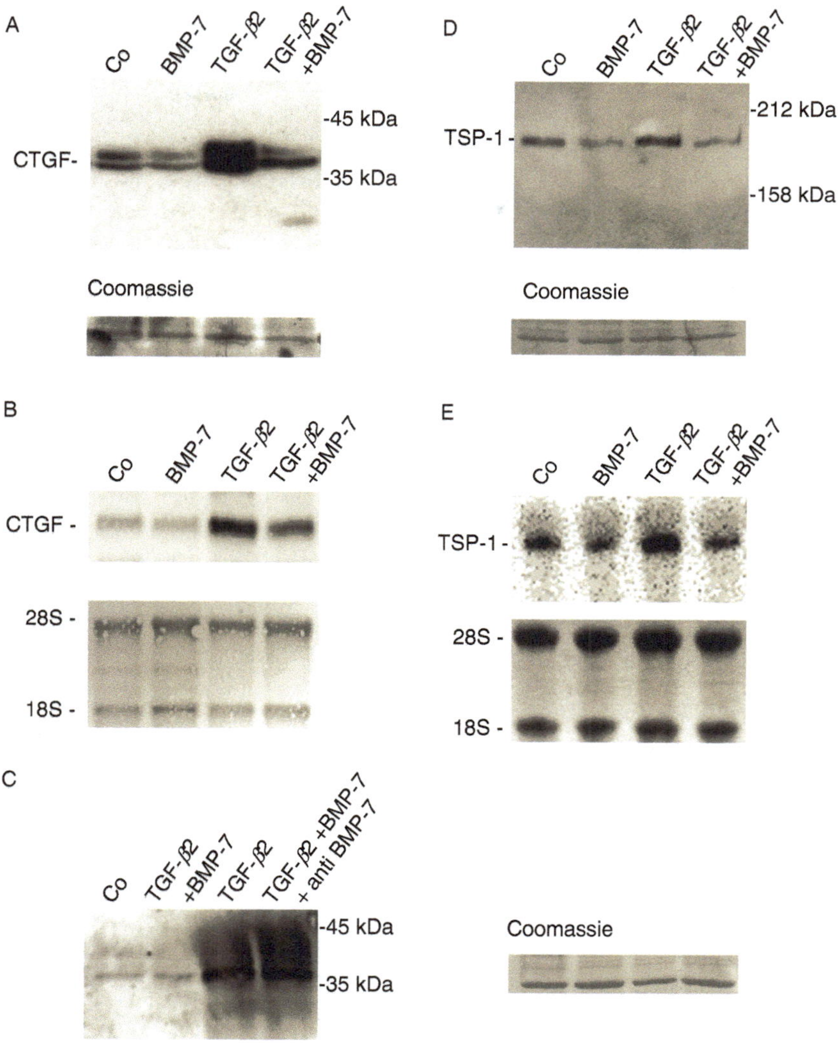

FIGURE 3 Western (A, C, D) and Northern blot (B, E) analysis of connective tissue growth factor (CTGF, A and C) and thrombospondin-1 (TSP-1, D), and their respective mRNAs (CTGF mRNA in B, and TSP-1 mRNA in E) in culture medium and RNA from cultured human trabecular meshwork cells following treatment with 300 pM bone morphogenetic protein-7 (BMP-7), 300 pM transforming growth factor-β2 (TGF-β2), a combination of both (TGF-β2 + BMP7), or TGF-β2 and BMP7 in the presence of neutralizing antibodies for BMP-7 (TGF-β2 + BMP-7 + anti BMP-7). Co: Control. For Western blots, membranes were stained with Coomassie blue to confirm equal loading of proteins. For Northern blot analyses, integrity of RNA and equal loading were controlled by staining ribosomal RNA with methylene blue. From Fuchshofer *et al.* (2007).

et al., 2002) and recent studies reported that the concentration of CTGF is increased in the aqueous humor of patients with pseudoexfoliation syndrome (Ho *et al.*, 2005). In the TM, the expression of CTGF is substantially upregulated following treatment with TGF-β2 (Fuchshofer *et al.*, 2007) (Fig. 3), strongly indicating that, as in other tissues, CTGF is directly involved in the increase of TM ECM deposition induced by TGF-β2. CTGF is also found in higher amounts in the TM following mechanical stretch (Chudgar *et al.*, 2006), an effect which is very likely induced by the action of TGF-β1 which is also upregulated in TM cells by stretch (Liton *et al.*, 2005). In vascular smooth muscle cells, the expression of CTGF has been shown to be induced following treatment with endothelin-1, an effect that is mediated via activation of the RhoA/Rho-kinase signaling pathway (Rodriguez-Vita *et al.*, 2005). A similar effect in the eye might be very relevant for the TM outflow pathways, as endothlin-1 which is secreted from tissues of the anterior eye (Eichhorn und Lütjen-Drecoll, 1993) into the aqueous humor (Lepple-Wienhues *et al.*, 1992) causes contraction of TM cells (Lepple-Wienhues *et al.*, 1991). Contraction of TM cells leads to an increase in TM outflow resistance (Wiederholt *et al.*, 2000), an effect that might be amplified by induction of CTGF synthesis and its subsequent action on ECM increase. Interestingly, endothelin-1 has been found to be elevated in the aqueous humor of patients with POAG (Noske *et al.*, 1997).

V. BONE MORPHOGENETIC PROTEIN-7

Recent studies indicate that the action of TGF-β2 and CTGF on ECM turnover in the TM is strongly antagonized by bone morphogenetic protein-7 (BMP-7) (Fuchshofer *et al.*, 2007) (Fig. 3), a growth factor of the bone morphogenetic protein family. BMP-7 is a 35-kDa homodimeric protein and member of the TGF-β superfamiliy of cysteine knot cytokines, which appears to counteract TGF-β action through direct antagonism involving the Smad signaling pathways (Chatziantoniou and Dussaule, 2005). In a corneal alkali injury model, BMP-7 has been shown to suppress TGF-β induced effects on corneal scarring (Saika *et al.*, 2005). In the kidney, BMP-7 counteracts an epithelial-to-mesenchymal transition that is induced by TGF-β1 and reverses chronic renal injury (Zeisberg *et al.*, 2003). In addition, BMP-7 antagonizes the TGF-β-dependant fibrogenesis in mesangial cells of the kidney glomerulus (Wang and Hirschberg, 2003, 2004). In embryonic life, BMP-7 plays a critical role during renal and eye development (Dudley *et al.*, 1995; Luo *et al.*, 1995). In the adult organism, the expression of BMP-7 is retained in the eye, in which its expression and that of its receptors has been shown in cornea, TM and optic nerve (You *et al.*, 1999; Wordinger *et al.*,

2002). We could recently show that BMP-7 strongly antagonizes *in vitro* the TGF-β2-induced expression of a broad panel of molecules, which would result in an accumulation of ECM in the TM *in situ* (Fuchshofer *et al.*, 2007). While treatment of human TM cells with TGF-β2 induced the expression of CTGF, TSP-1, fibronectin, collagen types IV and VI, and plasminogen activator inhibitor (PAI)-1, these effects were inhibited when TGF-β2 was added in combination with BMP-7 (Figs. 3 and 4). BMP-7 alone had no effects on the expression of all of these molecules. As BMP-7 is expressed in the adult TM *in situ*, it seems more than reasonable to assume that it similarly modulates and antagonizes the effects of TGF-β2 signaling on the TM *in vivo*. Wordinger and colleagues could show that BMP-4 apparently has very similar activities on the biology of TM cells as BMP-7 and is also counteracting the action of TGF-β (Wordinger *et al.*, 2007). Cultured TM cells treated with TGF-β2 significantly increased fibronectin levels, while BMP-4 blocked the induction of fibronectin. There are a number of molecules that regulate the signaling action of BMP family members, including that of BMP-7 and BMP-4, by binding to BMPs and preventing the ligand from interacting with the cellular receptor complex (Chen *et al.*, 2004). The expression of some of these BMP-associated molecules that act as BMP-antagonists like follistatin, gremlin and chordin has been observed in TM cells (Wordinger *et al.*, 2002). Applying gene chip analyses, Wordinger and coworkers profiled the expression of molecules involved in BMP-signaling and compared their expression between normal TM cells and those from donors with POAG (Wordinger *et al.*, 2007). As a result of this study, an upregulation of gremlin and its mRNA could be found in POAG samples. In subsequent experiments it was shown that gremlin blocked the negative effect of BMP-4 on the TGF-β2 induced upregulation of fibronectin. In anterior segment perfused organ cultures, gremlin added to the medium caused an elevation of outflow resistance strongly indicating that the BMP-signaling pathway is involved in the modulation of TM outflow resistance (Wordinger *et al.*, 2007). Needless to say, these experiments strongly indicate that the pharmacological activation of BMP-signaling in the TM might be a promising strategy to treat POAG.

VI. MYOCILIN

Myocilin is the gene product of the *MYOC* gene which is responsible for *GLC1A*-linked POAG. Originally, *MYOC* mutations were identified in families with autosomal-dominant juvenile POAG (Stone *et al.*, 1997), and subsequently also reported by numerous other researchers (Fingert *et al.*, 2002; Tamm, 2002). Patients harboring mutations in *MYOC* may suffer from

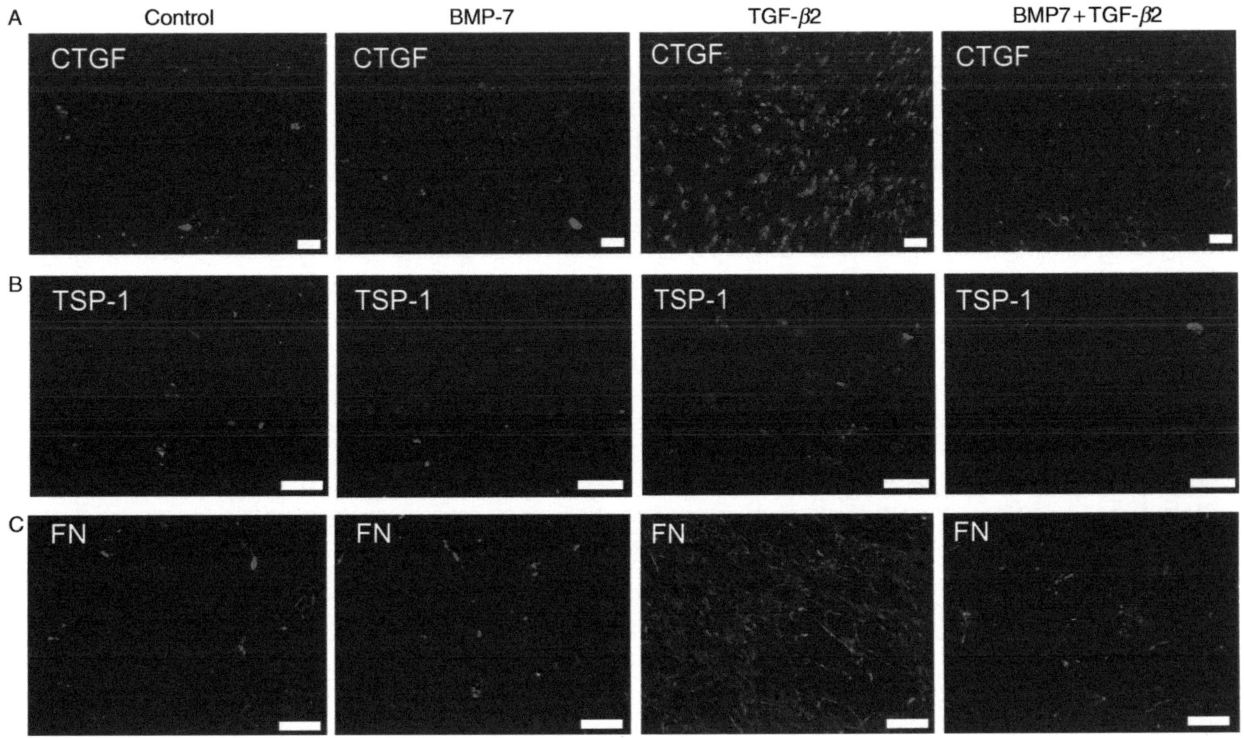

FIGURE 4 Immunoreactivity for connective tissue growth factor (CTGF, A), thrombospondin-1 (TSP-1, B), and fibronectin (FN, C) in cultured human trabecular meshwork cells following treatment with 300 pM bone morphogenetic protein-7 (BMP-7), 300 pM transforming growth factor-β 2 (TGF-β2), or a combination of both. Treatment with 300 pM BMP-7 caused no changes in immunoreactivity. In contrast, after treatment with 300 pM TGF-β2, the intensity of staining for CTGF, TSP-1 and FN was considerably enhanced, an effect that was markedly reduced following treatment with a combination of BMP-7 and TGF-β2. Inset in A shows immunoreactivity for CTGF in cellular vesicles (arrowhead). Magnification bars: 100 μm. From Fuchshofer *et al.* (2007). (See Color Insert.)

very high IOP, indicating that the TM outflow pathways are primarily affected (Wiggs *et al.*, 1995; Alward *et al.*, 1998). Overall, about 2–4% of POAG cases worldwide appear to be caused by mutations in *MYOC* (Fingert *et al.*, 1999). Myocilin is a 55–57 kDa glycoprotein with high expression in tissues of the anterior eye such as iris, ciliary body, cornea and sclera (Adam *et al.*, 1997; Karali *et al.*, 2000; Swiderski *et al.*, 2000). The trabecular meshwork shows the highest expression of myocilin in the eye (Adam *et al.*, 1997; Tamm *et al.*, 1999; Swiderski *et al.*, 2000), and according to data of an un-normalized cDNA library from fresh TM, myocilin is the third-highest expressed gene in the human TM (Tomarev *et al.*, 2003). Outside the eye, considerable amounts of myocilin and its mRNA are synthesized by Schwann cells of peripheral nerves (Ohlmann *et al.*, 2003), while substantially lower levels have been found in skeletal muscle, heart, optic nerve, brain and podocytes of the kidney (Ortego *et al.*, 1997; Fingert *et al.*, 1998; Nguyen *et al.*, 1998; Swiderski *et al.*, 1999; Noda *et al.*, 2000; Clark *et al.*, 2001; Ricard *et al.*, 2001; Goldwich *et al.*, 2005). No substantial differences with regard to expression pattern and localization of myocilin have been observed between humans and common laboratory animals such as mouse and rat (Abderrahim *et al.*, 1998; Takahashi *et al.*, 1998; Tomarev *et al.*, 1998; Taguchi *et al.*, 2000; Ahmed *et al.*, 2001; Torrado *et al.*, 2002; Knaupp *et al.*, 2004). With regard to the subcellular localization of myocilin, some earlier studies reported an intracellular location and the association of myocilin with microtubules (Kubota *et al.*, 1997; Mertts *et al.*, 1999), mitochondria (Wentz-Hunter *et al.*, 2002a), actin, vimentin, or the myosin regulatory light chain (Ueda *et al.*, 2000; Wentz-Hunter *et al.*, 2002b). Currently, the overall consensus of multiple laboratories appears to be that myocilin is an extracellular secreted protein which is found in the supernatant of cultured cells *in vitro* (Jacobson *et al.*, 2001; Sohn *et al.*, 2002; Goldwich *et al.*, 2003; Joe *et al.*, 2003; Shepard *et al.*, 2003; Gobeil *et al.*, 2004; Aroca-Aguilar *et al.*, 2005; Malyukova *et al.*, 2006; Vollrath and Liu, 2006), and in the aqueous humor of various species including that of humans *in vivo* (Rao *et al.*, 2000; Jacobson *et al.*, 2001; Russell *et al.*, 2001; Shepard *et al.*, 2003; Fautsch *et al.*, 2004; Gobeil *et al.*, 2004; Aroca-Aguilar *et al.*, 2005; Zillig *et al.*, 2005). Consistent with an extracellular role of myocilin are observations that show binding of myocilin to other extracellular proteins such as fibronectin (Filla *et al.*, 2002), optimedin (Torrado *et al.*, 2002), or hevin, a secretory protein of the BM-40/SPARC/osteonectin family (Li *et al.*, 2006).

Myocilin consists of two major domains: a coiled-coil domain containing a leucine zipper motif near the N-terminal, and an olfactomedin-like domain near the C-terminal (Tamm, 2002). The C-terminal olfactomedin domain is substantially conserved and defines the family of olfactomedin proteins that contains a number of secreted glycoproteins (Karavanich and Anholt,

1998a,b). The prototype olfactomedin has originally been identified as glycoprotein in the mucous layer covering the bullfrog olfactory neuroepithelium (Snyder *et al.*, 1991). In mammals, the homologous glycoprotein olfactomedin 1 (noelin, neuronal olfactomedin-related ER localized protein, pancortin) is highly expressed in the central nervous system (Danielson *et al.*, 1994). Other prominent members of the family are olfactomedin 3 (optimedin) (Torrado *et al.*, 2002), a secreted glycoprotein that is primarily expressed in the eye, and olfactomedin 4 (hGC-1) which is found in bone marrow, gastrointestinal tract and prostate (Zhang *et al.*, 2002). *In situ*, myocilin is found in a multimeric structure, which is caused by non-covalent interactions between the leucine zipper motifs (Fautsch and Johnson, 2001), and extensive disulfide bond formation using five cysteine residues (Fautsch *et al.*, 2004). Consequently, myocilin is found in the human aqueous humor in complexes of 120 and 180 kDa (Fautsch and Johnson, 2001), and in bovine and monkey aqueous humor in complexes of more than 200 kDa (Russell *et al.*, 2001). The presence of disulfide bonds in myocilin multimers also strongly argues for a role of myocilin as secretory protein, as disulfide bonds, which are usually formed in the endoplasmic reticulum, are frequently observed in secretory proteins, but not in proteins of the cytosol.

Some peculiarities as to the secretion of myocilin have been reported. Hardy and coworkers observed that in cultured trabecular meshwork cells myocilin associates with the extracellular membrane of lipid particles that have some biochemical characteristics of exosomes, and that the release of myocilin in the extracellular space occurs in association with exosome-like vesicles (Hardy *et al.*, 2005). The same group provided evidence that myocilin is associated with exosome-like material in the human aqueous humor (Perkumas *et al.*, 2007). It was suggested that this mode of secretion is specific for the trabecular meshwork, which appears to be unlikely as myocilin in the aqueous humor obviously does not derive from the chamber angle but rather from other sources such as iris and ciliary body. The binding of myocilin to the exosome membrane was reported to involve the coiled-coil domain, but not the olfactomedin domain (Stamer *et al.*, 2006). Some data from other laboratories give support to the observation of an association between myocilin and cell membranes. Ricard and coworkers found an association of canine myocilin with lipids of putatively cell membrane origin (Ricard *et al.*, 2006), while Joe and colleagues reported on data indicating an interaction between myocilin and flotillin-1, an integral membrane protein and constituent of lipid rafts (Joe *et al.*, 2005).

Another peculiarity of the secretory mechanism of myocilin refers to the fact that recombinant myocilin in transfected immortalized laboratory cell lines (e.g. COS1, HEK293) is often not only secreted as full-length protein, but also together with a C-terminal cleavage product that appears to result from intracellular endoproteolytic cleavage of myocilin (Goldwich *et al.*, 2003; Aroca-Aguilar *et al.*, 2005; Fautsch *et al.*, 2006). The site of cleavage differs between different cell lines and experimental settings and has been reported to be between amino acids 214/215 (Goldwich *et al.*, 2003; Fautsch *et al.*, 2006) or amino acids 226/227 (Aroca-Aguilar *et al.*, 2005). The differences regarding the cleavage site of myocilin can be explained by experiments involving site-directed mutagenesis, which show that the protease implicated in this processing has no strict amino acid sequence requirements at the cleavage site (Sanchez-Sanchez *et al.*, 2007). Recent studies indicate that calpain II is a myocilin processing protease that appears to be involved in intracellular myocilin cleavage (Sanchez-Sanchez *et al.*, 2007). C-terminal cleavage fragments should not associate with exosome-like membranes during their secretion, as they do not contain the coiled-coil domain of the full length protein. So far, the relevance (if any) of myocilin cleavage products for the living organism is unclear. Smaller fragments that stain with myocilin antibodies have been observed in aqueous humor samples (Russell *et al.*, 2001; Aroca-Aguilar *et al.*, 2005), but it is not clear, if these fragments result from intracellular endoproteolytic cleavage as in immortalized cell lines, or from extracellular proteolytic cleavage inside or outside the organism. Intracellular cleavage might be a process to regulate extracellular interactions of myocilin, assuming that the biological properties of complete myocilin differ from that of the C-terminal cleavage fragment. Indeed, data by Goldwich and collegues (Goldwich *et al.*, 2003) obtained in anterior segment perfused organ cultures provide functional evidence that the C-terminal fragment differs in its functional properties from full-length myocilin (see below).

Very limited data are available that indicate a specific function of myocilin. Peters and colleagues investigated the effects of recombinant myocilin on spreading and substrate adhesion of fibroblasts (Peters *et al.*, 2005). Fibroblast attached, but failed to spread on myocilin as substrate. In addition, spreading of fibroblasts on the Hep II domain of fibronectin as substrate was significantly inhibited in the presence of myocilin, as was focal adhesion formation and the incorporation of paxillin into focal adhesions. The data appear to indicate that myocilin could act as matricellular protein that modifies the number of contacts between cells and extracellular matrix in the trabecular meshwork. In general, matricellular proteins are secreted proteins that influence cell function by modulating cell–matrix interactions (Sage and Bornstein, 1991; Bornstein, 2001). Prominent matricellular proteins are the already discussed TSP-1 and SPARC. It is interesting to note that, similar to myocilin, both proteins inhibit spreading of fibroblasts under culture conditions (Murphy-Ullrich and Hook, 1989; Sage *et al.*, 1989), and are constitutively expressed (albeit at much lower amounts than myocilin) in the trabecular meshwork (Rhee *et al.*, 2003; Flügel-Koch *et al.*, 2004). The expression of counter-adhesive matricellular molecules in the trabecular meshwork might be important to facilitate a continuous remodeling of cell–matrix contacts in the inner wall region of the TM. Such a remodeling should facilitate the continuous formation of inter- and intracellular pores and/or giant vacuoles in the inner wall of Schlemm's canal endothelium, and constitute an important mechanism to modulate flow through the trabecular meshwork (Ethier, 2002; Johnson, 2006). Whatever will turn out to be the function of myocilin *in vivo*, it is obviously not a critical factor for the function of the outflow pathways (at least in the mouse eye) as knockout mice with a targeted deletion of myocilin do not develop an obvious ocular phenotype (Kim *et al.*, 2001).

There is experimental evidence to support the hypothesis that secreted myocilin plays a role in modulating the hydrodynamic outflow resistance in the TM, and that elevated amounts of myocilin in POAG could obstruct the trabecular outflow system and cause an increase in trabecular outflow resistance. The expression of myocilin is tremendously induced by treatment of cultured TM cells or TM in anterior segment perfused organ cultures with dexamethasone. The increase in myocilin expression following treatment with dexamethasone is observed in a time-dependant manner (Tamm *et al.*, 1999) (Fig. 5) and comparable to the time course that is observed during the development of steroid-induced ocular hypertension and glaucoma (Nguyen *et al.*, 1998). Moreover, an increase in immunostaining for myocilin has been reported in the TM of patients with POAG (Lütjen-Drecoll *et al.*, 1998). Furthermore, recombinant myocilin is very effective at blocking polycarbonate filters with a pore size similar to that of the TM (Goldwich *et al.*, 2003). In addition, myocilin in the aqueous humor is tightly bound to polycarbonate filters that become obstructed after perfusion with aqueous humor (Russell *et al.*, 2001). The strongest support so far for the hypothesis that elevated myocilin obstructs the outflow system comes from data provided by Fautsch and coworkers (Fautsch *et al.*, 2006). The authors purified human recombinant myocilin from an eukaryotic expression system and perfused human anterior eye segment organ cultures with 2 μg/ml of recombinant myocilin. When myocilin was perfused in porcine aqueous humor, a significant increase in outflow resistance was observed. Similar effects were observed when myocilin was preincubated with porcine aqueous humor. The maximum outflow resistance was obtained five to 17 hours after infusion and remained above baseline for more than three days. In contrast, only minimal effects were observed when myocilin was perfused with regular cell culture medium, indicating that myocilin needs to form a complex with proteins in the aqueous humor that enables it to bind specifically within the TM. So far, the protein(s) that interact(s) with myocilin in porcine aqueous humor have not been identified, but experimental data suggest that albumin,

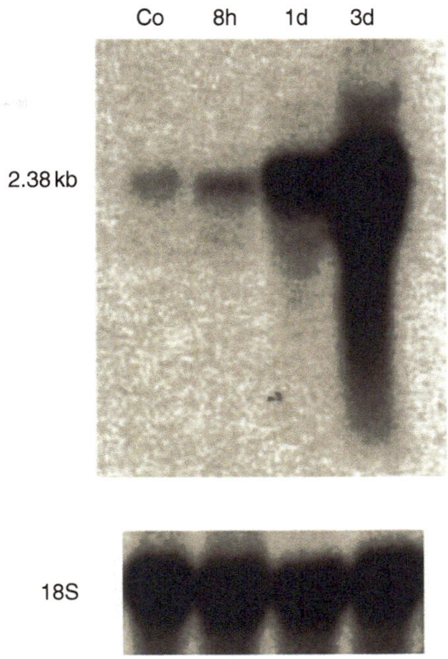

FIGURE 5 Northern blot analysis of myocilin mRNA in TM monolayer cell culture after treatment with dexamethasone (10^{-7} M) for 8 hours, 1 day, and 3 days. Twenty micrograms of total RNA were loaded per lane. The exposure time of the autoradiograph was 24 hours. Relative amounts and integrity of RNA that were loaded were controlled by reprobing the membrane with a cDNA probe specific for guinea pig 18S ribosomal RNA (lower panel). Relative densitometric intensities (normalized to 18S RNA) of the myocilin bands are as follows: lane 1, 1; lane 2, 3; lane 3, 38; and lane 4, 96. The size of a molecular marker is given in kilobases. Co: control. From Tamm *et al.* (1999).

a very abundant protein in aqueous humor, is not the binding partner that is necessary for myocilin-induced outflow resistance. In those eyes that showed an increase in outflow resistance, myocilin accumulated in high amounts in the JCT supporting the concept that outflow resistance can be modified by ECM compounds in the JCT. The results with eukaryotic myocilin differ markedly with those of a previous study using myocilin from a bacterial source (Fautsch *et al.*, 2000), and strongly emphasize that functional data on myocilin need to be obtained with eukaryotic myocilin. Its is interesting to note that for an influence of myocilin on outflow resistance, full length myocilin is required, whereas the C-terminal cleavage fragment previously mentioned has no effect. Goldwich *et al.* were able to express and purify substantial amounts of the C-terminal fragment (Goldwich *et al.*, 2003). When this fragment was added to the perfusate in anterior segment perfusion cultures, no effect on outflow facility was observed indicating substantial functional differences between full-length myocilin and the C-terminal fragment (Goldwich *et al.*, 2003).

While the data obtained with myocilin perfusion of anterior eye segment organ cultures are impressive, experimental evidence for a direct role of myocilin on outflow resistance in the living organism is lacking so far. Zillig and coworkers developed transgenic mice (*β*B1-crystallin-MYOC) that strongly express myocilin in their lenses under control of the lens specific *β*B1-crystallin promoter (Zillig *et al.*, 2005). The transgenic expression of myocilin from the lens resulted in an almost five-fold increase of secreted normal myocilin in the aqueous humor of the transgenic mice (Fig. 6). By immunohistochemistry, secreted transgenic myocilin could be observed on the surfaces of lens and corneal endothelium, and in high amounts in the chamber angle (Fig. 7). Nevertheless, the intraocular pressure of the transgenic animals did not differ from that of control mice. Comparable data were obtained in a different laboratory with another set of transgenic mice that had been genetically modified to overexpress myocilin, and that did also not show significant changes in intraocular pressure (Gould *et al.*, 2004). A possible explanation for the different results between transgenic animals and organ culture studies could be that myocilin is present at much lower concentrations in the living transgenic eye (0.2 μg/ml) (Zillig *et al.*, 2005) than in perfused organ cultures (2 μg/ml) (Fautsch *et al.*, 2006). Still, the relevance of these amounts for glaucoma in humans remains unclear, as it is not known, if eyes treated with dexamethasone and/or suffering from glaucoma can produce myocilin at those high amounts that were used for organ culture experiments. Another explanation for the differences between the living mouse eye and the organ-cultured human anterior segment could be species-related structural and/or biochemical differences between the trabecular meshwork of mice and men.

Data from multiple laboratories provide overwhelming evidence that recombinant myocilin harboring mutations, which would cause a severe glaucoma phenotype in humans, is not secreted from cultured cells (Caballero *et al.*, 2000; Jacobson *et al.*, 2001; Joe *et al.*, 2003; Gobeil *et al.*, 2004; Liu and Vollrath, 2004; Zillig *et al.*, 2005; Vollrath and Liu, 2006). The predominant subcellular localization of mutated myocilin *in vitro* has been shown to be the endoplasmic reticulum (Sohn *et al.*, 2002; Joe *et al.*, 2003; Liu and Vollrath, 2004; Malyukova *et al.*, 2006; Yam *et al.*, 2007). Comparable data were observed in transgenic mice (*β*B1-crystallin-Tyr437HisMYOC) with ectopic *in vivo* expression of mutated human Y437H myocilin in the lens under control of the lens-specific *β*B1-crystallin promoter (Zillig *et al.*, 2005). In patients, this mutation causes an aggressive form of juvenile-onset primary open angle glaucoma (Alward *et al.*, 1998). In contrast to wild-type

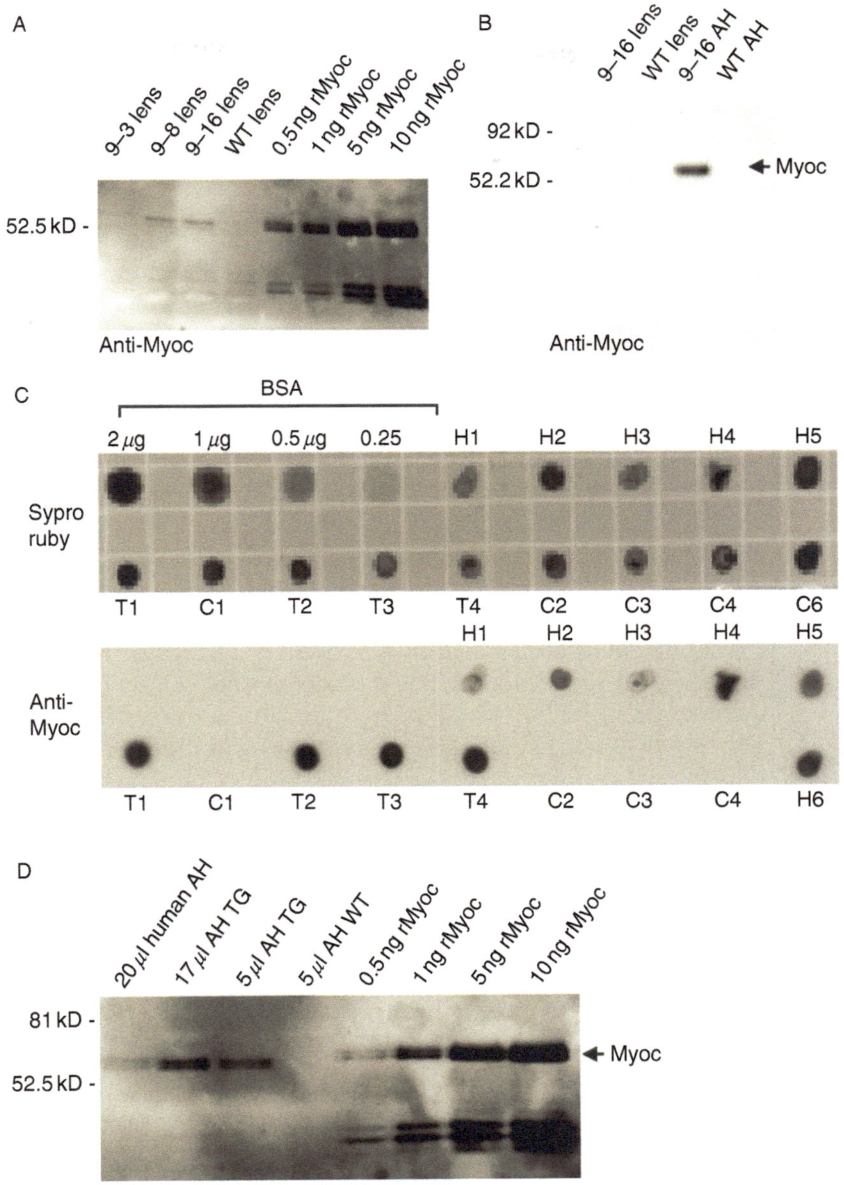

FIGURE 6 Secreted transgenic myocilin in βB1-crystallin-MYOC transgenic mice. (A) Western blot analysis for transgenic myocilin in lenses from lines 9–3, 9–8, 9–16 and wild-type (WT) littermates at P1. Different amounts of purified recombinant myocilin (rMyoc) were loaded for comparison. (B) Western blot for transgenic myocilin in lens and AH of 9–16 animals and wild-type littermates (WT) at P21. (C) Western dot blot for transgenic myocilin in AH

myocilin, mutated myocilin was not secreted into the aqueous humor of transgenic animals, but accumulated in the endoplasmic reticulum of lens fibers (Fig. 8) (Zillig *et al.*, 2005). The reason for the lack of secretion appears to be misfolding of mutated myocilin resulting in a highly aggregation-prone protein that forms large aggregates (Zhou and Vollrath, 1999; Liu and Vollrath, 2004). Interestingly, culturing cells at 30 °C, a condition known to facilitate protein folding, promotes secretion of mutant myocilin and normalizes cell morphology (Liu and Vollrath, 2004). Part of the mutated myocilin seems to aggregate with wild-type myocilin and to form heteromeric wild-type/mutant aggregates (Sohn *et al.*, 2002; Gobeil *et al.*, 2004; Vollrath and Liu, 2006), which are not secreted, an effect that results in a diminished secretion of extracellular wild-type myocilin (Caballero and Borrás, 2001; Jacobson *et al.*, 2001; Gobeil *et al.*, 2004). Overall, intracellular sequestration and temperature sensitive secretion is very characteristic and has been shown to be associated with the vast majority of glaucoma-causing mutations in myocilin (Gobeil *et al.*, 2006; Vollrath and Liu, 2006). Similar effects as in human mutated myocilin were observed in studies introducing glaucoma-causing mutations in mouse myocilin (Malyukova *et al.*, 2006).

It is not entirely clear, how the intracellular sequestration of mutated myocilin should cause glaucoma in patients. One possibility could be the formation of heteromeric wild-type/mutant aggregates resulting in lower levels of wild-type myocilin. The data from knockout mice with a targeted deletion of myocilin, which do not develop an obvious phenotype and glaucoma (Kim *et al.*, 2001), strongly argue against this possibility. Another possibility is that the intracellular sequestration of mutated myocilin initiates a cellular unfolded protein response, which results in cell stress and finally apoptotic cell death (Welihinda *et al.*, 1999; Hampton, 2000). Misfolded proteins are usually recognized by control systems in the rough endoplasmic reticulum, transported to the cytosol, and degraded by ubiquination and proteosomal degradation (Ellgaard and Helenius, 2003). Failure of this mechanism results in the accumulation of misfolded protein in the endoplasmic reticulum, in the congestion of the secretory pathway and dysfunction of the endoplasmic reticulum, and initiates the unfolded protein response that leads to cell death. Such a scenario is thought to cause cell death in several inherited neurodegenerative disorders that are – similar to glaucoma

samples from four transgenic animals (T1–T4), four wild-type littermates (C1–C4) and six human donors (H1–H6). Total protein was visualized by SYPRO-Ruby staining. Different amounts of bovine serum albumin (BSA) were loaded for comparison. (D) Western blot for transgenic myocilin in the AH from one human donor (20 μl), transgenic 9–16 animals (TG) and wild-type littermates (WT) at P21. 5 μl AH were obtained from one individual animal, and 17 μl were pooled from the eyes of four animals. Different amounts of purified recombinant myocilin (rMyoc) were loaded for comparison. From Zillig *et al.* (2005).

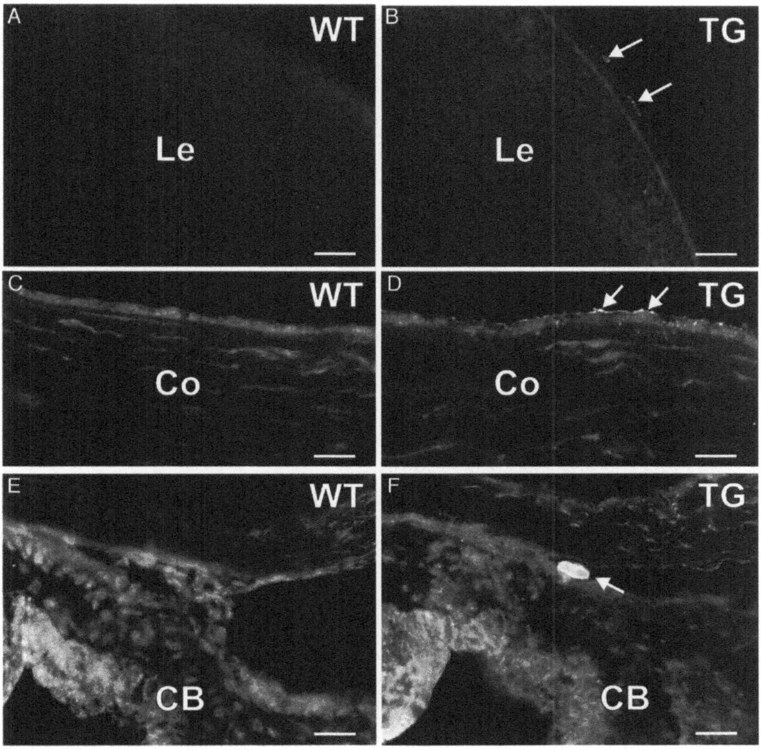

FIGURE 7 Immunostaining for myocilin in the eyes of βB1-crystallin-MYOC transgenic animals (TG; B, C, D) and wild-type littermates (WT; A, C, E). (A, B) Positive immunoreactivity for myocilin is detected in a thin layer (arrows) on the lens (Le) surface of transgenic animals (TG), but not in wild-type (WT) littermates. (C, D) Keratocytes and corneal endothelial cells are positively labeled in the cornea (Co) of both wild-type (WT) and transgenic (TG) animals. In addition, in transgenic animals, the inner surface of the cornea is covered with a thin layer of immunoreactive material (arrows). (E, F) Cells of the ciliary body (CB) and cells of the chamber angle are positively stained for myocilin in both transgenic (TG) and wild-type animals (WT). In addition, in transgenic animals, a homogenous mass with strong immunoreactivity for myocilin is observed in the chamber angle and covers the inner parts of the TM (arrow). Scale bars. 16 μm (A, B); 10 μm (C–F). From Zillig *et al.* (2005). (See Color Insert.)

associated with myocilin mutations – dominant, delayed disorders (Lambert *et al.*, 1998; Beuret *et al.*, 1999; Jana *et al.*, 2000; Johnston *et al.*, 2000). Indeed, the accumulation of mutated myocilin in the endoplasmic reticulum in cultured cells induces cytotoxic changes, cell death and apoptosis (Sohn *et al.*, 2002; Joe *et al.*, 2003; Liu and Vollrath, 2004; Yam *et al.*, 2007), but not necessarily a general block of the secretory pathway (Malyukova *et al.*, 2006). Comparable findings have been reported in a transgenic animal with

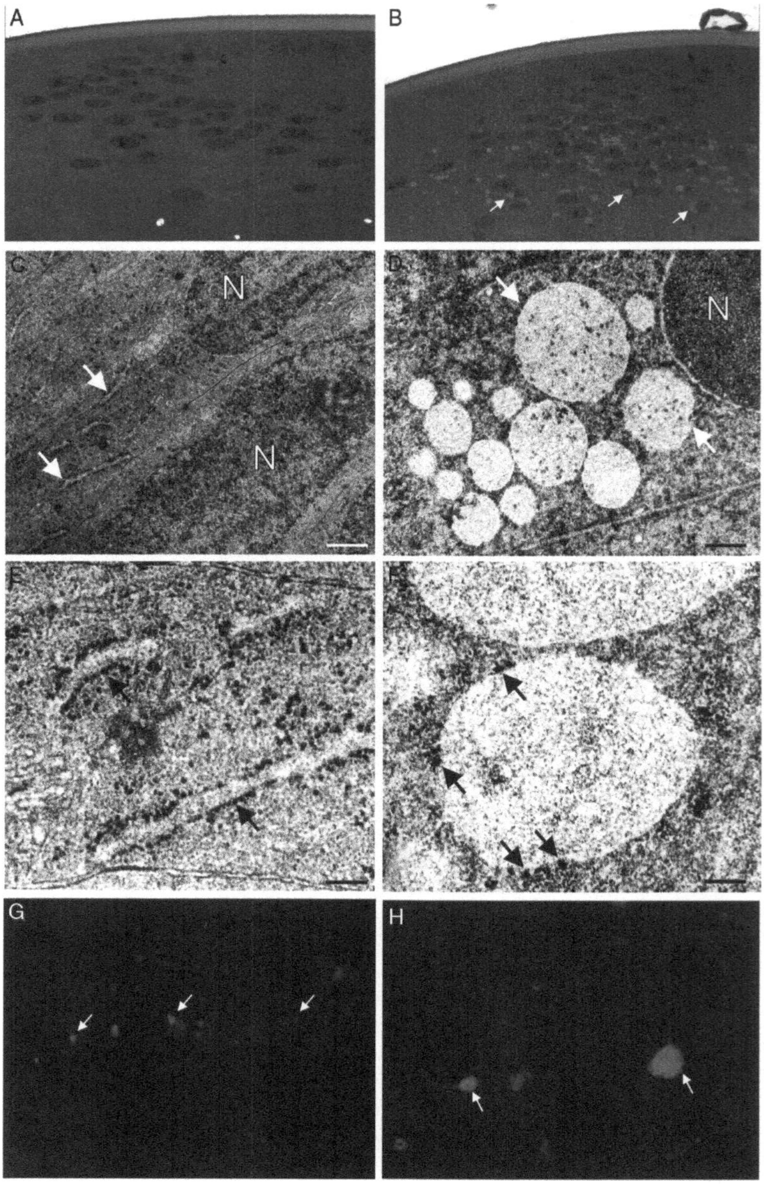

FIGURE 8 Light microscopy (A, B), electron microscopy (C–F) and immunostaining for myocilin (G, H) in lenses of βB1-crystallin-Tyr437HisMYOC transgenic mice (B, D, F, H) and wild-type littermates (A, C, E, G). (A, B) Multiple vesicles (arrows) are seen in lens fibers at the bow region of transgenic animals (B), but not in wild-type littermates (A). (C, D) By electron microscopy, the vesicles (arrow in D) are localized to the cytoplasm surrounding the nucleus (N)

ectopic expression of mutant myocilin in the lens (Zillig *et al.*, 2005), in which the accumulation of mutant proteins in the endoplasmic reticulum of lens fibers resulted in the formation of nuclear cataracts, loss of transparency, cell death and finally the rupture of the lens (Fig. 9). Recent important data by Shepard and coworkers indicate that in addition to an ER stress response, another factor might contribute to a dominant negative effect of mutated myocilin (Shepard *et al.*, 2007). The authors found evidence for binding of mutated myocilin to the peroxisomal targeting signal-1 receptor (PTS1R). Interestingly, mutations with a more severe early-onset POAG in patients showed a higher degree of association to PTS1R, while wild-type myocilin did not bind to PTS1R. A likely explanation for this observation appears to be that the mutations caused the exposure of a cryptic peroxisomal targeting sequence. In the same study, it was shown that adenovirus mediated gene transfer of mutated human myocilin to the mouse eye caused an increase in IOP. Unfortunately, the interpretation of this set of data is difficult, as besides showing data on IOP, Shepard and collegues did not provide any other structural or functional data on these mice.

Clearly, an unfolded protein response in patients with myocilin glaucoma could lead to cell death of trabecular meshwork cells resulting in structural changes of the outflow pathways. Such changes might cause a substantial increase in outflow resistance in the trabecular meshwork and finally lead to the very high levels of intraocular pressure, which are commonly observed in patients suffering from glaucoma because of mutations in the myocilin gene (Alward *et al.*, 1998). Still, data from mouse models do not entirely support this concept. Gould and colleagues developed mice carrying a mutant allele of the mouse myocilin gene (Gould *et al.*, 2006). The mutation Y423H was used which is analogous to the particular severe human mutation (Y437H) that causes an aggressive form of juvenile-onset primary open angle glaucoma (Alward *et al.*, 1998). As in cell culture studies, mutant myocilin was not secreted, but accumulated in cells of the chamber angle. Nevertheless, this accumulation did not lead to the initiation of an unfolded protein response, to an increase in intraocular pressure nor to glaucomatous changes. Taken

of lens fibers and are filled with electron dense granular material (D). In wild-type littermates (C), the same perinuclear area of lens fibers contains cisterns of rER, which appear to be of normal size (arrow in C). (E, F) Higher magnification of perinuclear vesicles in transgenic animals (F) shows that the vesicles are surrounded by a membrane that contains ribosomes (arrows in F), and confirms their origin from rER cisterns, which are of normal size in wild-type littermates (arrows in E). (G, H) Immunocytochemistry with antibodies specific for myocilin shows strong positive immunoreactivity in lens fiber vesicles (arrows) of transgenic animals, and confirms that the vesicles are caused by an accumulation of Tyr437His mutated myocilin in the rER. Scale bars: 16 μm (A, B); 690 nm (C, D); 166 nm (E, F); 6 μm (G); 4 μm (H). From Zillig *et al.* (2005). (See Color Insert.)

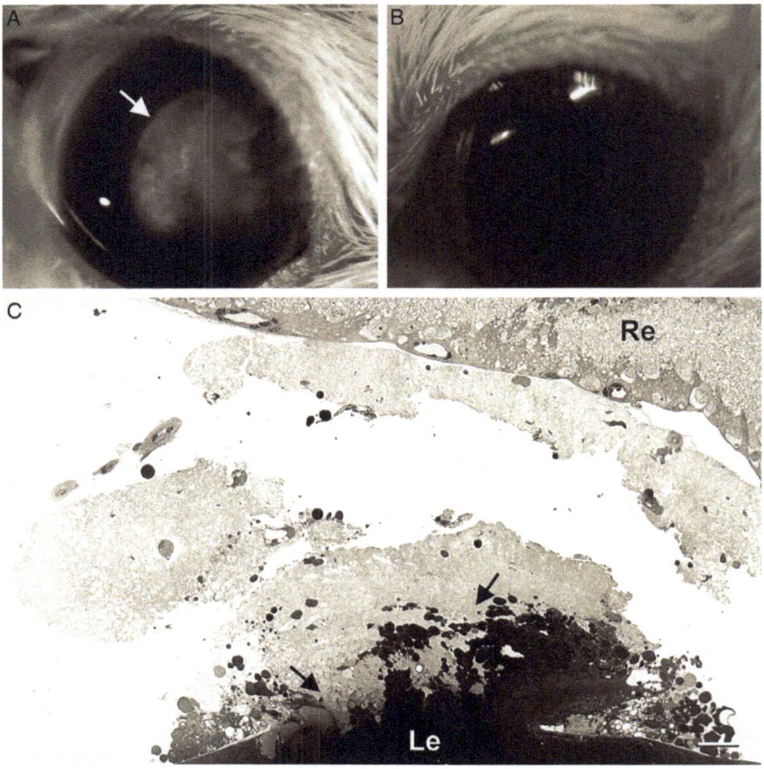

FIGURE 9 Lenses of βB1-crystallin-Tyr437HisMYOC transgenic mice develop cataracts. (A,B) Lenses of transgenic mice expressing Tyr437His mutated myocilin lose transparency because of nuclear catarcts (A), while lenses of wild-type littermates (B) are transparent. (C) Light microscopy of the posterior pole of a transgenic lens (Le) with cataract that is ruptured at its posterior pole (arrows). Re: Retina. Scale bars: 12 μm (C). Zillig *et al.*, (2005).

together, the data of Gould and coworkers indicate that apparent misfolding and non-secretion of mutant myocilin is not sufficient to cause glaucoma in the mouse eye. Surprisingly, data from another mouse model are at variance with those of Gould *et al.* Senatorov and coworkers developed transgenic mice with overexpression of mutated myocilin by introducing a bacterial artificial chromosome encoding the human Y437H mutation into the mouse genome (Senatorov *et al.*, 2006). Similar to findings in the other mouse models with this mutation (Zillig *et al.*, 2005; Gould *et al.*, 2006), mutated myocilin was not secreted (Senatorov *et al.*, 2006). Despite the absence of any significant pathological changes in the trabecular meshwork, the mice developed a moderate elevation of intraocular pressure, which was about 2 mmHg higher than in eyes of wild-type littermates. Correlated with the increase was a continuous loss of retinal ganglion cells similar to the situation in glaucoma. So far, it is completely unclear why the mouse models developed by Gould and coworkers and those generated by Senatorov and coworkers differ so substantially in their phenotypes, and why none of the models provides support for the unfolded protein response in the rough endoplasmic reticulum as causative for glaucoma associated with mutant myocilin. A possible mechanism as to why mutated human myocilin might cause POAG as opposed to mutated mouse myocilin could be binding to PTS1R, as mutation of mouse myocilin does not expose a cryptic peroxisomal targeting sequence (Shepard *et al.*, 2007). Still, so far, it is completely unclear how binding of mutated myocilin to PTS1R should lead to higher IOP and POAG.

A reasonable approach to shed some light on the role of mutant myocilin for the outflow pathways and the development of glaucoma could be a thorough histopathological analysis of the trabecular meshwork of affected patients. Still, despite the fact that the causative role of mutant myocilin for some forms of glaucoma is known for ten years, such data are missing until today.

VII. OPTINEURIN

Rezai and colleagues identified mutations in the *OPTN* gene as causative for autosomal-dominant inherited POAG at the *GLC1E* locus (Rezaie *et al.*, 2002). Most of the affected patients were reported to suffer from a rare familial form of normal tension glaucoma, a major subtype of glaucoma, in which IOP is constantly within the statistically normal range. Importantly, lowering intraocular pressure in patients with normal pressure glaucoma effectively slows down progress of the disease, indicating that IOP is also a major risk factor in patients with normal pressure glaucoma (Collaborative Normal-Tension Glaucoma Study Group, 1998a,b). Retinal ganglion cells of patients with normal tension glaucoma appear to be more vulnerable as those of normal subjects, as they undergo glaucomatous changes even though IOP is not elevated. Two of the three mutations that were originally reported were also found in other populations worldwide (Alward *et al.*, 2003; Aung *et al.*, 2003; Hauser *et al.*, 2006b; Ayala-Lugo *et al.*, 2007) and affected patients appear to have a more severe form of normal pressure glaucoma than those without mutation (Aung *et al.*, 2005). These mutations include the E50K alteration, in which the codon for glutamic acid is changed to that for lysine, and the c.691–692insAG change that results in a predicted premature stop codon in exon 6. Overall, the sequence alterations in the

OPTN gene are rare events and are found at a frequency of less than 1% in POAG populations (Alward *et al.*, 2003; Aung *et al.*, 2003; Ayala-Lugo *et al.*, 2007; Hauser *et al.*, 2006b). Several other sequence variations in the *OPTN* gene were also identified throughout the world in other cohorts of patients with POAG (Leung *et al.*, 2003; Baird *et al.*, 2004; Funayama *et al.*, 2004; Fuse *et al.*, 2004; Willoughby *et al.*, 2004; Weisschuh *et al.*, 2005).

As disease-causing mutations in the *OPTN* gene are preferentially found in patients with normal pressure glaucoma, the functional properties of the encoded protein optineurin are very likely not important for maintenance of intraocular pressure, but rather for the survival of retinal ganglion cells (RGC) and their axons. The localization of optineurin and its mRNA was analyzed in tissues of the human, monkey and mouse eye, and found in multiple sites such as cornea, ciliary body, trabecular meshwork, lens, retina and optic nerve (Rezaie *et al.*, 2002; Vittitow and Borras, 2002; Kamphuis and Schneemann, 2003; Rezaie and Sarfarazi, 2005; Rezaie *et al.*, 2005; Kroeber *et al.*, 2006) (Fig. 10). When relative amounts of optineurin and its mRNA are compared with each other in the different tissues of mouse eye, the expression of optineurin is found be highest in the retina (De Marco *et al.*, 2006; Kroeber *et al.*, 2006). Importantly, RGC are those neurons in the mouse retina (Fig. 10) that preferentially show immunoreactivity for optineurin (De Marco *et al.*, 2006; Kroeber *et al.*, 2006). Corroborating results were reported for the rat retina, in which, similar to the mouse eye, retinal ganglion cells are preferentially labeled for optineurin (Wang *et al.*, 2007). In the mouse eye, the expression of optineurin is observed as early as embryonic day 10.5 (E 10.5) (Rezaie and Sarfarazi, 2005; De Marco *et al.*, 2006). In rat RGC, the expression of optineurin is seen at E 17 and increases by two-fold until postnatal day 21 (P 21), when RGC are fully differentiated (Wang *et al.*, 2007). Outside the eye, optineurin is widely expressed and transcripts have been detected in RNA from heart, brain, lung, liver, skeletal muscle, pancreas, spleen, kidney, small intestine, placenta, and testis of various species including human, monkey, mouse and chicken (Li *et al.*, 1998; Stroissnigg *et al.*, 2002; Rezaie and Sarfarazi, 2005; Rezaie *et al.*, 2005). The expression of optineurin appears to be under control of tumor necrosis factor-α (TNF-α), as an increase in optineurin mRNA has been observed in HEK293 MCF-7 cells upon treatment with TNF-α (Li *et al.*, 1998), and in the trabecular meshwork of anterior-segment perfused organ cultures (Vittitow and Borras, 2002). Conflicting data have been reported for the action of dexamethasone on the expression of optineurin, as an increase in expression was observed in organ-cultured human trabecular meshwork (Vittitow and Borras, 2002), while a decrease was seen in porcine TM cells in monolayer cell culture (Obazawa *et al.*, 2004). An induction of TM optineurin expression by elevated pressure in anterior segment perfused organ cultures was described in one report

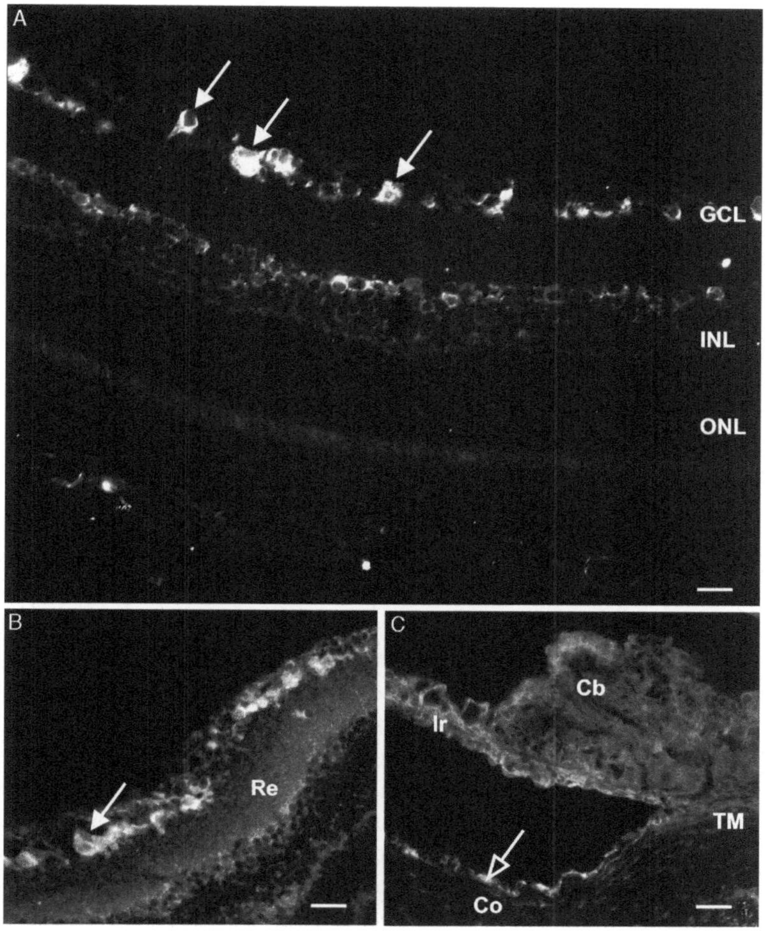

FIGURE 10 Immunohistochemistry for endogenous optineurin in the eyes of wild-type mice at P21. (A) In the posterior eye, an intense signal is present in retinal ganglion cells of the central retina (arrows). In addition, some neurons at the inner aspect of the inner nuclear layer (INL) are positively labeled. GCL. Ganglion cell layer; ONL. Outer nuclear layer. (B) Staining of retinal ganglion cells for optineurin (arrow) is also present in the peripheral retina (Re). (C) In the anterior eye, immunolabeling for optineurin is present in cells of the corneal endothelium (arrow), iris, ciliary body (Cb), and the trabecular meshwork (TM). Magnification bar: 40 mm. From Kroeber *et al.*, (2006). (See Color Insert.)

(Vittitow and Borras, 2002), but not found in another study (Kamphuis and Schneemann, 2003). Mechanical stretch or an increase in hydrostatic pressure had no significant influence on the expression of optineurin in cultured TM cells (Obazawa *et al.*, 2004).

Optineurin is a conserved 67-kD protein with multiple leucine zipper domains and a putative zinc finger domain at the COOH terminus. Several studies described a close subcellular association of optineurin with the Golgi complex (Li *et al.*, 1998; Rezaie *et al.*, 2002, 2005). Although it was originally reported to be secreted into the aqueous humor (Rezaie *et al.*, 2002), experimental studies overexpressing optineurin in the lens of transgenic mice *in vivo* (Kroeber *et al.*, 2006), or in cell cultures *in vitro* (Park *et al.*, 2006) did not observe secretion of optineurin into the aqueous humor or cell culture supernatant. There is considerable evidence for a critical functional role of optineurin in maintaining Golgi morphology and exocytosis, as optineurin binds myosin VI and appears to link it to the Golgi complex (Sahlender *et al.*, 2005). Myosin VI is a multifunctional motor protein that plays a major role in endocytic and secretory membrane traffic pathways. In addition, optineurin binds to huntingtin (Faber *et al.*, 1998) that interacts with HAP1 (huntingtin-associated protein), which binds directly to the dynactin subunit $p150^{Glued}$ at the dynactin complex (Engelender *et al.*, 1997). In addition, optineurin does also bind to Rab8, a protein that belongs to a large family of small GTPases which participate in and regulate intracellular membrane trafficking pathways (Hattula and Peranen, 2000). Based on these data, two major "linker" functions for optineurin have been suggested (Sahlender *et al.*, 2005) (Fig. 11): First, by directly interacting with myosin VI, optineurin is linked to the actin cytoskeleton and is involved in actin-based motor activity around the Golgi complex. By binding to huntingtin it is linked to HAP1 and the dynactin complex and via the minus end directed motor protein dynein to the microtubule network. According to this concept, optineurin could play a role in coordinating microtubule-based and actin-based motor activity around the Golgi complex. Second, optineurin might link myosin VI to Rab8, a regulatory protein that is involved in the exocytic pathway at the trans-Golgi-network and in membrane fusion at the plasma membrane. Importantly, an experimental knock-down of optineurin expression by means of small interfering-RNA disrupted the structure of the Golgi complex and reduced secretion and exocytosis (Sahlender *et al.*, 2005). A reduction in secretion, preferentially in RGC, could be the cause of the glaucoma phenotype in patients with the E50K mutation, as optineurin was observed in much lower levels around the Golgi complex of cultured fibroblasts obtained from affected patients (Rezaie *et al.*, 2002).

Another hypothesis on a functional role of optineurin has been based on data that show optineurin to interact with adenovirus E3–14.7K protein, an inhibitor of tumor necrosing factor-α (TNF-α)-induced cytolysis, and to interfere with the protective effect of E3–14.7K against TNF-α-mediated cell death (Li *et al.*, 1998). TNF-α induces apoptosis of cultured RGC *in vitro* (Tezel and Wax, 2000), and is upregulated in the optic nerve head

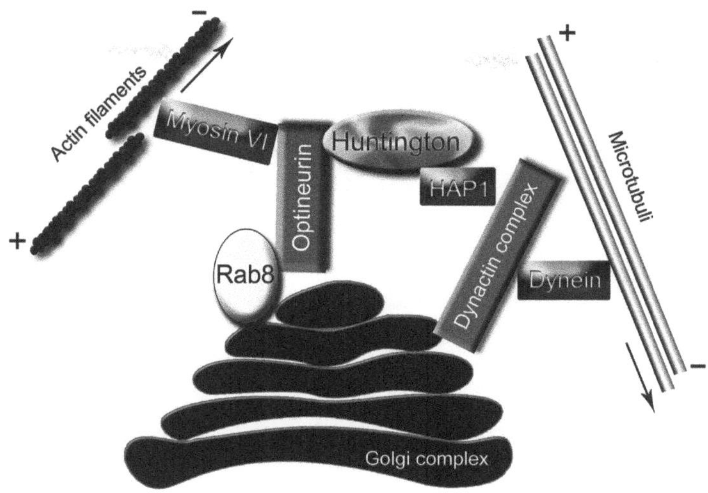

FIGURE 11 Motor protein complexes at the Golgi complex. Optineurin may play a central role in coordinating actin-based and microtubule-based motor function for maintaining Golgi morphology. The optineurin-binding partner huntingtin has been shown to interact with HAP1, which in turn was reported to form a complex with the dynactin subunit p150glued and modulate/ regulate the dynein–dynactin complex. Loss of minus end–directed dynein motor activity causes in fragmentation of the Golgi ribbon structure. On the other hand loss of myosin VI results in a reduction in the size of the Golgi complex, because now the dynein complex is still active and able to retract the Golgi complex toward the MTOC. However, if optineurin is depleted, both motor complexes are nonfunctional and the Golgi complex is fragmented. Modified from Sahlender *et al.* (2005). This model is based on studies using non-neuronal cells, but might also be applicable to neurons such as retinal ganglion cells.

(Yuan and Neufeld, 2000) and retina of glaucomatous eyes (Tezel *et al.*, 2001). Based on these observations, it has been suggested that wild-type optineurin plays a neuroprotective role in the eye, but when defective, contributes to the glaucomatous neuropathy (Rezaie *et al.*, 2002). An interaction between TNF-α and optineurin is supported by findings that show an increase in the expression of optineurin in trabecular meshwork cells following TNF-α treatment (Vittitow and Borras, 2002), and a possible interaction between polymorphisms in the OPTN and TNF-α genes that appear to increase the risk for glaucoma (Funayama *et al.*, 2004). To study the protective effects of optineurin on apoptosis in an animal model, βB1-crystallin-OPTN mice were generated with ectopic overexpression of optineurin in the lens under control of the strong lens-specific βB1-crystallin promoter (Kroeber *et al.*, 2006) (Fig. 12). Subsequently, βB1-crystallin-OPTN mice were crossed with transgenic βB1-crystallin-TGFβ1 mice that have been modified to overexpress activated TGF-β1 in the lens. βB1-crystallin-TGFβ1

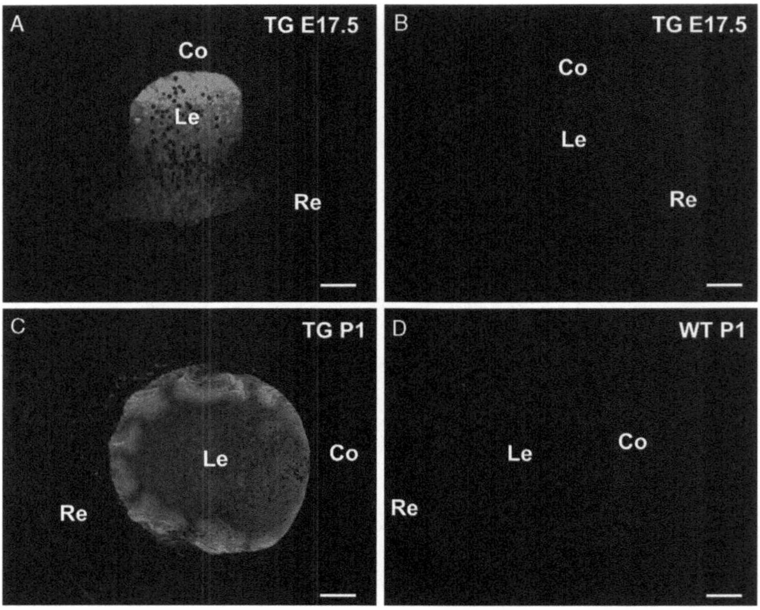

FIGURE 12 Immunohistochemistry for optineurin in the eyes of transgenic βB1-crystallin-OPTN (TG) and wild-type (WT) mice at P1 (A, B) and E 17 (C, D). An intense signal for transgenic optineurin is present in the cytoplasm of lens fibers in transgenic animals of both ages (A, C), but not in wild-type eyes at P1 (B). In addition, staining for endogenous optineurin is present in cells at the inner aspect of the retina both in transgenic and wild-type animals at P1 (arrows). Staining for optineurin is completely blocked by preabsorption with immunizing peptide (D). Le: Lens; Co: Cornea; Re: Retina. Magnification bars. 10 μm (A), 40 μm (B, C). From Kroeber *et al.* (2006). (See Color Insert.)

mice show substantial apoptotic cell death in lens fibers and epithelial cells starting at around E17.5 (Flügel-Koch *et al.*, 2002), and it was expected that this system could help to identify any anti-apoptotic roles of optineurin. Still, no differences regarding extent and time course of apoptotic cell death between the lenses of βB1-crystallin-TGFβ1 mice and those of double βB1-crystallin-OPTN /βB1-crystallin-TGFβ1 mice were observed (Fig. 13). It was concluded that optineurin does not modify TGF-β1 induced cell death in this system; although transgenic optineurin is present in considerable amounts in the lens fibers and the time course of optineurin and TGFβ1 expression is similar because of the usage of the same promoter. TNF-α induced apoptosis is mediated through the death receptor pathway involving binding to TNF receptor I (Gupta, 2002). In contrast, TGF-β1 can use multiple pathways in inducing apoptosis that are cell type and context-dependent. TGF-β1 can interact with TNF-α to induce apoptosis, or mediate

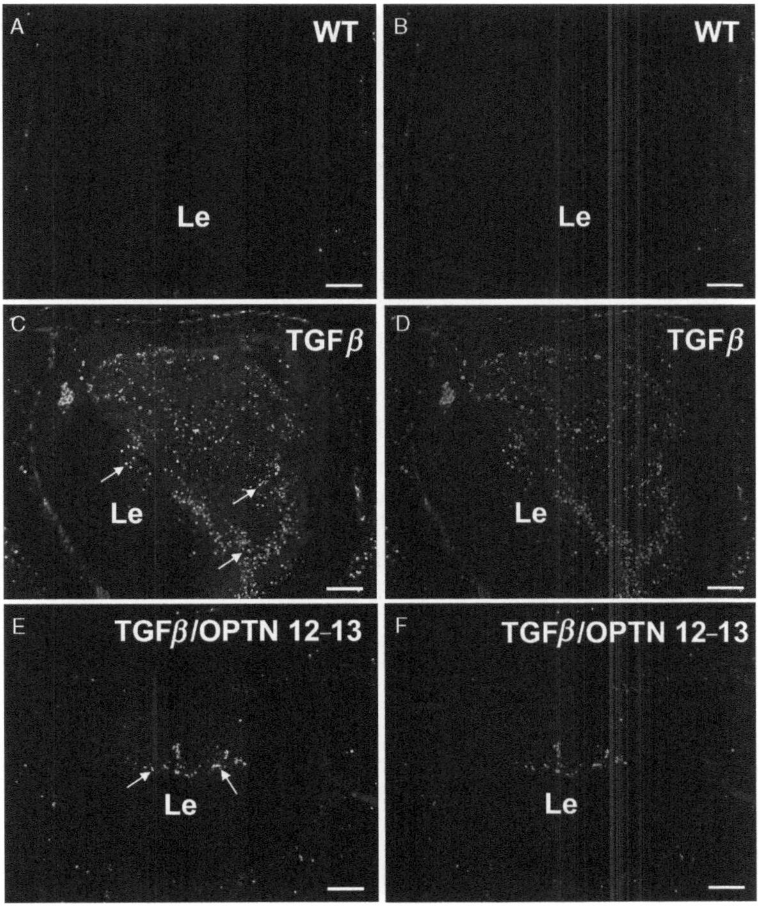

FIGURE 13 TUNEL-labeling for the detection of apoptosis in lens (Le) fibers at E17.5. A, C, E. TUNEL-labeling; B, D, F. nuclear counterstain. TUNEL-positive nuclei are not observed in lenses of wild-type animals, (WT, A, B), but are abundant (arrows) in lenses of both βB1-crystallin-TGFβ1 (TGFβ1, C, D), or double βB1-crystallin-OPTN/βB1-crystallin-TGFβ1 (TGFβ1/OPTN 12–13) mice (E, F). Magnification bar. 20 μm. From Kroeber *et al.* (2006). (See Color Insert.)

apoptosis through mitogen-activated protein kinase (MAPK) signaling or the mitochondrial apoptotic pathway (Sanchez-Capelo, 2005). Therefore, it cannot be excluded that the pathways of TGFβ1-induced apoptosis in the system applied by Kroeber and coworkers lacked the appropriate molecular elements, which are required for presumably anti-apoptotic effects of optineurin. Nevertheless, the available *in vivo* data do currently not support the presence of anti-apoptotic functions of optineurin. Comparable findings were recently observed in an *in vitro* study using an immortalized rat retinal ganglion cell line (RGC-5) to study the role of wild-type and mutated optineurin for apoptotic cell death of RGC (Chalasani *et al.*, 2007). In this study, TNF-α induced cell death was not inhibited by overexpression of wild-type optineurin nor its E50K mutant. In marked contrast, overexpression of wild-type optineurin increased apoptotic cell death, an effect that was even more pronounced when E50K mutated optineurin was overexpressed. Moreover, mutated E50K optineurin did induce apoptosis even in absence of TNF-α. Reactive oxygen species (ROS) were produced upon expression of E50K optineurin, and antioxidative treatment abolished ROS production and cell death. In another study on non-neuronal COS7 and NIH3T3 cells, overexpression of optineurin did actually protect cells from H_2O_2 induced cell death (De Marco *et al.*, 2006). In response to this apoptotic stimulus, optineurin changed its subcellular localization and translocated from the Golgi to the nucleus. E50K mutated optineurin lost the ability to translocate to the nucleus and, when overepressed, compromised the mitochondrial membrane potential. The functional role for the translocation of optineurin into the nucleus is not clear, but it might be related to transcriptional control, as binding of optineurin to transcription factor IIIA has been described (Moreland *et al.*, 2000).

In summary, the available data point to an important role of optineurin in several functional pathways. Currently, it is not clear, how the different pathways are affected in the living eye of patients with POAG caused by mutated optineurin, and which of the pathways is more important for the health of RGC.

VIII. WD REPEAT DOMAIN 36

WD repeat domain 36 is the product of the *WDR36* gene which was identified as causative for *GLC1G*-linked glaucoma by Monemi and coworkers (Monemi *et al.*, 2005). Currently, the importance of *WDR36* for the onset of POAG is not clear, as in the meantime several groups were unable to confirm mutations in *WDR36* as causative for POAG in their cohorts of patients (Hauser *et al.*, 2006a; Hewitt *et al.*, 2006; Fingert *et al.*, 2007; Weisschuh *et al.*, 2007). In one of these studies (Hauser *et al.*, 2006a), an association of *WDR36* sequence variants with a more severe disease in affected individuals was observed, suggesting that defects in the *WDR36* gene may contribute to POAG and that *WDR36* may be a glaucoma modifier gene.

WDR36 consists of 23 exons and encodes for a 931 amino acid protein that contains multiple G-beta WD40 repeats as characteristic motifs. WD40 repeats are minimally conserved regions of approximately 40 amino acids typically bracketed by gly-his and trp-asp (GH-WD), which may facilitate formation of heterotrimeric or multiprotein complexes. Family members are involved in a variety of cellular processes, such as proliferation, signal transduction, apoptosis, and gene regulation. Two mRNA transcripts for WDR36 were observed at 5.9 and 2.5 kb, which were preferentially expressed in heart, placenta, liver and smooth muscle (Monemi *et al.*, 2005). By RT-PCR, expression was found in basically all ocular tissues (Monemi *et al.*, 2005). *WDR36* was identified as one of the genes that are upregulated upon T-cell activation and was highly co-regulated with interleukin 2 (Mao *et al.*, 2004). A specific function is unknown, but may be related to the biogenesis of ribosomal RNA (rRNA), as its homologue in yeast, Utp21, is a protein of the small-subunit (SSU) processome which is a large ribonucleoprotein required for the biogenesis of 18S rRNA and with putative roles in both pre-rRNA processing and ribosome assembly (Bernstein *et al.*, 2004).

IX. CONCLUSION

As of today, several molecules have been identified that appear to be involved in the pathogenetic processes of POAG. Since an increase in fibrillar ECM in the JCT of the TM is the most characteristic pathological finding in patients with POAG, many of these molecules are involved in the control mechanisms of ECM turnover in the TM outflow pathways. Other molecules have been identified from genetic studies applying linkage analysis and positional cloning. Studies that lead to more complete understanding of these molecules should provide the key to clarify the molecular mechanisms that govern aqueous humor outflow through the TM and the pathogenetic mechanisms of POAG.

References

Abderrahim, H., Jaramillo-Babb, V. L., Zhou, Z., and Vollrath, D. (1998). Characterization of the murine TIGR/myocilin gene. *Mamm. Genome* **9**, 673–675.

Adam, M. F., Belmouden, A., Binisti, P., Brézin, A. P., Valtot, F., Béchetoille, A., Dascotte, J.-C., Copin, B., Gomez, L., Chaventre, A., Bach, J.-F., and Garchon, H.-J. (1997). Recurrent mutations in a single exon encoding the evolutionarily conserved olfactomedin-homology domain of TIGR in familial open-angle glaucoma. *Hum. Mol. Gen.* **12**, 2091–2097.

Adams, J. C. (2001). Thrombospondins: Multifunctional regulators of cell interactions. *Annu. Rev. Cell Dev. Biol.* **17**, 25–51.

Ahmed, F., Torrado, M., Johnson, E., Morrison, J., and Tomarev, S. I. (2001). Changes in mRNA levels of the Myoc/Tigr gene in the rat eye after experimental elevation of intraocular pressure or optic nerve transection. *Invest. Ophthalmol. Vis. Sci.* **42**, 3165–3172.

Alge, C. S., Priglinger, S. G., Neubauer, A. S., Kampik, A., Zillig, M., Bloemendal, H., and Welge-Lussen, U. (2002). Retinal pigment epithelium is protected against apoptosis by alphaB-crystallin. *Invest. Ophthalmol. Vis. Sci.* **43**, 3575–3582.

Allen, J. B., Davidson, M. G., Nasisse, M. P., Fleisher, L. N., and McGahan, M. C. (1998). The lens influences aqueous humor levels of transforming growth factor-beta 2. *Graefes Arch. Clin. Exp. Ophthalmol.* **236,** 305–311.

Alvarado, J. A., and Murphy, C. G. (1992). Outflow obstruction in pigmentary and primary open angle glaucoma. *Arch. Ophthalmol.* **110,** 1769–1778.

Alward, W. L., Fingert, J. H., Coote, M. A., Johnson, A. T., Lerner, S. F., Junqua, D., Durcan, F. J., McCartney, P. J., Mackey, D. A., Sheffield, V. C., and Stone, E. M. (1998). Clinical features associated with mutations in the chromosome 1 open-angle glaucoma gene. *N. Engl. J. Med.* **338,** 1022–1027.

Alward, W. L., Kwon, Y. H., Kawase, K., Craig, J. E., Hayreh, S. S., Johnson, A. T., Khanna, C. L., Yamamoto, T., Mackey, D. A., Roos, B. R., Affatigato, L. M., Sheffield, V. C., *et al.* (2003). Evaluation of optineurin sequence variations in 1,048 patients with open-angle glaucoma. *Am. J. Ophthalmol.* **136,** 904–910.

Annes, J. P., Munger, J. S., and Rifkin, D. B. (2003). Making sense of latent TGFβ activation. *J. Cell Sci.* **116,** 217–224.

Aroca-Aguilar, J. D., Sanchez-Sanchez, F., Ghosh, S., Coca-Prados, M., and Escribano, J. (2005). Myocilin mutations causing glaucoma inhibit the intracellular endoproteolytic cleavage of myocilin between amino acids Arg226 and Ile227. *J. Biol. Chem.* **280,** 21043–21051.

Aung, T., Ebenezer, N. D., Brice, G., Child, A. H., Prescott, Q., Lehmann, O. J., Hitchings, R. A., and Bhattacharya, S. S. (2003). Prevalence of optineurin sequence variants in adult primary open angle glaucoma: Implications for diagnostic testing. *J. Med. Genet.* **40,** e101.

Aung, T., Rezaie, T., Okada, K., Viswanathan, A. C., Child, A. H., Brice, G., Bhattacharya, S. S., Lehmann, O. J., Sarfarazi, M., and Hitchings, R. A. (2005). Clinical features and course of patients with glaucoma with the E50K mutation in the optineurin gene. *Invest. Ophthalmol. Vis. Sci.* **46,** 2816–2822.

Ayala-Lugo, R. M., Pawar, H., Reed, D. M., Lichter, P. R., Moroi, S. E., Page, M., Eadie, J., Azocar, V., Maul, E., Ntim-Amponsah, C., Bromley, W., Obeng-Nyarkoh, E., *et al.* (2007). Variation in optineurin (OPTN) allele frequencies between and within populations. *Mol. Vis.* **13,** 151–163.

Bachmann, B., Birke, M., Kook, D., Eichhorn, M., and Lütjen-Drecoll, E. (2006). Ultrastructural and biochemical evaluation of the porcine anterior chamber perfusion model. *Invest. Ophthalmol. Vis. Sci.* **47,** 2011–2020.

Baird, P. N., Richardson, A. J., Craig, J. E., Mackey, D. A., Rochtchina, E., and Mitchell, P. (2004). Analysis of optineurin (OPTN) gene mutations in subjects with and without glaucoma: The Blue Mountains Eye Study. *Clin. Experiment. Ophthalmol.* **32,** 518–522.

Bataller, R., and Brenner, D. A. (2005). Liver fibrosis. *J. Clin. Invest.* **115,** 209–218.

Bernstein, K. A., Gallagher, J. E., Mitchell, B. M., Granneman, S., and Baserga, S. J. (2004). The small-subunit processome is a ribosome assembly intermediate. *Eukaryot. Cell* **3,** 1619–1626.

Beuret, N., Rutishauser, J., Bider, M. D., and Spiess, M. (1999). Mechanism of endoplasmic reticulum retention of mutant vasopressin precursor caused by a signal peptide truncation associated with diabetes insipidus. *J. Biol. Chem.* **274,** 18965–18972.

Borisuth, N. S., Tripathi, B. J., and Tripathi, R. C. (1992). Identification and partial characterization of TGF-beta 1 receptors on trabecular cells. *Invest. Ophthalmol. Vis. Sci.* **33,** 596–603.

Bornstein, P. (2001). Thrombospondins as matricellular modulators of cell function. *J. Clin. Invest.* **107,** 929–934.

Caballero, M., and Borrás, T. (2001). Inefficient processing of an olfactomedin-deficient myocilin mutant: Potential physiological relevance to glaucoma. *Biochem. Biophys. Res. Commun.* **282,** 662–670.

Caballero, M., Rowlette, L. L., and Borrás, T. (2000). Altered secretion of a TIGR/MYOC mutant lacking the olfactomedin domain. *Biochim. Biophys. Acta* **1502,** 447–460.

Chalasani, M. L., Radha, V., Gupta, V., Agarwal, N., Balasubramanian, D., and Swarup, G. (2007). A glaucoma-associated mutant of optineurin selectively induces death of retinal ganglion cells which is inhibited by antioxidants. *Invest. Ophthalmol. Vis. Sci.* **48,** 1607–1614.

Chatziantoniou, C., and Dussaule, J. C. (2005). Insights into the mechanisms of renal fibrosis: Is it possible to achieve regression? *Am. J. Physiol. Renal. Physiol.* **289,** F227–F234.

Chen, D., Zhao, M., and Mundy, G. R. (2004). Bone morphogenetic proteins. *Growth Factors* **22,** 233–241.

Chen, J., Li, H., SundarRaj, N., and Wang, J. H. (2007). Alpha-smooth muscle actin expression enhances cell traction force. *Cell. Motil. Cytoskeleton* **64,** 248–257.

Chudgar, S. M., Deng, P., Maddala, R., Epstein, D. L., and Rao, P. V. (2006). Regulation of connective tissue growth factor expression in the aqueous humor outflow pathway. *Mol. Vis.* **12,** 1117–1126.

Cicha, I., Yilmaz, A., Klein, M., Raithel, D., Brigstock, D. R., Daniel, W. G., Goppelt-Struebe, M., and Garlichs, C. D. (2005). Connective tissue growth factor is overexpressed in complicated atherosclerotic plaques and induces mononuclear cell chemotaxis *in vitro*. *Arterioscler. Thromb. Vasc. Biol.* **25,** 1008–1013.

Clark, A. F., Kawase, K., English-Wright, S., Lane, D., Steely, H. T., Yamamoto, T., Kitazawa, Y., Kwon, Y. H., Fingert, J. H., Swiderski, R. E., Mullins, R. F., Hageman, G. S., *et al.* (2001). Expression of the glaucoma gene myocilin (MYOC) in the human optic nerve head. *FASEB J.* **15,** 1251–1253.

Collaborative Normal-Tension Glaucoma Study Group. (1998a). Comparison of glaucomatous progression between untreated patients with normal-tension glaucoma and patients with therapeutically reduced intraocular pressures. *Am. J. Ophthalmol.* **126,** 487–497.

Collaborative Normal-Tension Glaucoma Study Group. (1998b). The effectiveness of intraocular pressure reduction in the treatment of normal-tension glaucoma. Collaborative Normal-Tension Glaucoma Study Group. *Am. J. Ophthalmol.* **126,** 498–505.

Cousins, S. W., McCabe, M. M., Danielpour, D., and Streilein, J. W. (1991). Identification of transforming growth factor-beta as an immunosuppresive factor in aqueous humor. *Invest. Ophthalmol. Vis. Sci.* **32,** 2201–2211.

Crawford, S. E., Stellmach, V., Murphy-Ullrich, J. E., Ribeiro, S. M., Lawler, J., Hynes, R. O., Boivin, G. P., and Bouck, N. (1998). Thrombospondin-1 is a major activator of TGF-β1 *in vivo*. *Cell* **93,** 1159–1170.

Danielson, P. E., Forss-Petter, S., Battenberg, E. L., deLecea, L., Bloom, F. E., and Sutcliffe, J. G. (1994). Four structurally distinct neuron-specific olfactomedin-related glycoproteins produced by differential promoter utilization and alternative mRNA splicing from a single gene. *J. Neurosci. Res.* **38,** 468–478.

de Kater, A. W., Shahsafaei, A., and Epstein, D. L. (1992). Localization of smooth muscle and nonmuscle actin isoforms in the human aqueous outflow pathway. *Invest. Ophthalmol. Vis. Sci.* **33,** 424–429.

De Marco, N., Buono, M., Troise, F., and Diez-Roux, G. (2006). Optineurin increases cell survival and translocates to the nucleus in a Rab8-dependent manner upon an apoptotic stimulus. *J. Biol. Chem.* **281,** 16147–16156.

di Mola, F. F., Di Sebastiano, P., Gardini, A., Innocenti, P., Zimmermann, A., Buchler, M. W., and Friess, H. (2004). Differential expression of connective tissue growth factor in inflammatory bowel disease. *Digestion* **69,** 245–253.

Dudley, A. T., Lyons, K. M., and Robertson, E. J. (1995). A requirement for bone morphogenetic protein-7 during development of the mammalian kidney and eye. *Genes Dev.* **9,** 2795–2807.

Eichhorn, M., and Lütjen-Drecoll, E. (1993). Distribution of endothelin-like immunoreactivity in the human ciliary epithelium. *Curr.Eye Res.* **12,** 753–757.

Ellgaard, L., and Helenius, A. (2003). Quality control in the endoplasmic reticulum. *Nat. Rev. Mol. Cell Biol.* **4,** 181–191.

Engelender, S., Sharp, A. H., Colomer, V., Tokito, M. K., Lanahan, A., Worley, P., Holzbaur, E. L., and Ross, C. A. (1997). Huntingtin-associated protein 1 (HAP1) interacts with the p150Glued subunit of dynactin. *Hum. Mol. Genet.* **6,** 2205–2212.

Ethier, C. R. (2002). The inner wall of Schlemm's canal. *Exp. Eye Res.* **74,** 161–172.

Faber, P. W., Barnes, G. T., Srinidhi, J., Chen, J., Gusella, J. F., and MacDonald, M. E. (1998). Huntingtin interacts with a family of WW domain proteins. *Hum. Mol. Genet.* **7,** 1463–1474.

Fautsch, M. P., and Johnson, D. H. (2001). Characterization of myocilin–myocilin interactions. *Invest. Ophthalmol. Vis. Sci.* **42,** 2324–2331.

Fautsch, M. P., Bahler, C. K., Jewison, D. J., and Johnson, D. H. (2000). Recombinant TIGR/MYOC increases outflow resistance in the human anterior segment. *Invest. Ophthalmol. Vis. Sci.* **41,** 4163–4168.

Fautsch, M. P., Vrabel, A. M., Peterson, S. L., and Johnson, D. H. (2004). *In vitro* and *in vivo* characterization of disulfide bond use in myocilin complex formation. *Mol. Vis.* **10,** 417–425.

Fautsch, M. P., Bahler, C. K., Vrabel, A. M., Howell, K. G., Loewen, N., Teo, W. L., Poeschla, E. M., and Johnson, D. H. (2006). Perfusion of his-tagged eukaryotic myocilin increases outflow resistance in human anterior segments in the presence of aqueous humor. *Invest. Ophthalmol. Vis. Sci.* **47,** 213–221.

Filla, M. S., Liu, X., Nguyen, T. D., Polansky, J. R., Brandt, C. R., Kaufman, P. L., and Peters, D. M. (2002). *In vitro* localization of TIGR/MYOC in trabecular meshwork extracellular matrix and binding to fibronectin. *Invest. Ophthalmol. Vis. Sci.* **43,** 151–161.

Fine, B. S., Yanoff, M., and Stone, R. A. (1981). A clinicopathologic study of four cases of primary open-angle glaucoma compared to normal eyes. *Am. J. Ophthalmol.* **91,** 88–105.

Fingert, J. H., Ying, L., Swiderski, R., Nystuen, A. M., Arbour, N. C., Alward, W. L. M., Sheffiled, V. C., and Stone, E. M. (1998). Characterization and comparison of the human and mouse GLC1A glaucoma genes. *Genome Res.* **8,** 377–384.

Fingert, J. H., Heon, E., Liebmann, J. M., Yamamoto, T., Craig, J. E., Rait, J., Kawase, K., Hoh, S. T., Buys, Y. M., Dickinson, J., Hockey, R. R., Williams-Lyn, D., *et al.* (1999). Analysis of myocilin mutations in 1703 glaucoma patients from five different populations. *Hum. Mol. Genet.* **8,** 899–905.

Fingert, J. H., Stone, E. M., Sheffield, V. C., and Alward, W. L. (2002). Myocilin glaucoma. *Surv. Ophthalmol.* **47,** 547–561.

Fingert, J. H., Alward, W. L., Kwon, Y. H., Shankar, S. P., Andorf, J. L., Mackey, D. A., Sheffield, V. C., and Stone, E. M. (2007). No association between variations in the WDR36 gene and primary open-angle glaucoma. *Arch. Ophthalmol.* **125,** 434–436.

Fleenor, D. L., Shepard, A. R., Hellberg, P. E., Jacobson, N., Pang, I. H., and Clark, A. F. (2006). TGFbeta2-induced changes in human trabecular meshwork: Implications for intraocular pressure. *Invest. Ophthalmol. Vis. Sci.* **47,** 226–234.

Flügel, C., Tamm, E., Lütjen-Drecoll, E., and Stefani, F. H. (1992). Age-related loss of α-smooth muscle actin in normal and glaucomatous human trabecular meshwork of different age groups. *J. Glaucoma* **1,** 165–173.

Flügel-Koch, C., Ohlmann, A., Piatigorsky, J., and Tamm, E. R. (2002). Overexpression of TGF-β 1 alters early development of cornea and lens in transgenic mice. *Dev. Dyn.* **225,** 111–125.

Flügel-Koch, C., Ohlmann, A., Fuchshofer, R., Welge-Lüssen, U., and Tamm, E. R. (2004). Thrombospondin-1 in the trabecular meshwork: Localization in normal and glaucomatous eyes, and induction by TGF-beta1 and dexamethasone *in vitro*. *Exp. Eye Res.* **79,** 649–663.

Fuchshofer, R., Welge-Lüssen, U., and Lütjen-Drecoll, E. (2003). The effect of TGF-beta2 on human trabecular meshwork extracellular proteolytic system. *Exp. Eye Res.* **77,** 757–765.

Fuchshofer, R., Birke, M., Welge-Lüssen, U., Kook, D., and Lütjen-Drecoll, E. (2005). Transforming growth factor-beta 2 modulated extracellular matrix component expression in cultured human optic nerve head astrocytes. *Invest. Ophthalmol. Vis. Sci.* **46,** 568–578.

Fuchshofer, R., Yu, A. H., Welge-Lüssen, U., and Tamm, E. R. (2007). Bone morphogenetic protein-7 is an antagonist of transforming growth factor-beta2 in human trabecular meshwork cells. *Invest. Ophthalmol. Vis. Sci.* **48,** 715–726.

Funayama, T., Ishikawa, K., Ohtake, Y., Tanino, T., Kurosaka, D., Kimura, I., Suzuki, K., Ideta, H., Nakamoto, K., Yasuda, N., Fujimaki, T., Murakami, A., *et al.* (2004). Variants in optineurin gene and their association with tumor necrosis factor-alpha polymorphisms in Japanese patients with glaucoma. *Invest. Ophthalmol. Vis. Sci.* **45,** 4359–4367.

Fuse, N., Takahashi, K., Akiyama, H., Nakazawa, T., Seimiya, M., Kuwahara, S., and Tamai, M. (2004). Molecular genetic analysis of optineurin gene for primary open-angle and normal tension glaucoma in the Japanese population. *J. Glaucoma* **13,** 299–303.

Gleizes, P. E., Munger, J. S., Nunes, I., Harpel, J. G., Mazzieri, R., Noguera, I., and Rifkin, D. B. (1997). TGF-β latency: Biological significance and mechanisms of activation. *Stem Cells* **15,** 190–197.

Gobeil, S., Rodrigue, M. A., Moisan, S., Nguyen, T. D., Polansky, J. R., Morissette, J., and Raymond, V. (2004). Intracellular sequestration of hetero-oligomers formed by wild-type and glaucoma-causing myocilin mutants. *Invest. Ophthalmol. Vis. Sci.* **45,** 3560–3567.

Gobeil, S., Letartre, L., and Raymond, V. (2006). Functional analysis of the glaucoma-causing TIGR/myocilin protein: Integrity of amino-terminal coiled-coil regions and olfactomedin homology domain is essential for extracellular adhesion and secretion. *Exp. Eye Res.* **82,** 1017–1029.

Goldwich, A., Ethier, C. R., Chan, D. W., and Tamm, E. R. (2003). Perfusion with the olfactomedin domain of myocilin does not affect outflow facility. *Invest. Ophthalmol. Vis. Sci.* **44,** 1953–1961.

Goldwich, A., Baulmann, D. C., Ohlmann, A., Flügel-Koch, C., Schocklmann, H., and Tamm, E. R. (2005). Myocilin is expressed in the glomerulus of the kidney and induced in mesangioproliferative glomerulonephritis. *Kidney Int.* **67,** 140–151.

Gonzalez, P., Epstein, D. L., and Borras, T. (2000). Characterization of gene expression in human trabecular meshwork using single-pass sequencing of 1060 clones. *Invest. Ophthalmol. Vis. Sci.* **41,** 3678–3693.

Gordon, M. O., Beiser, J. A., Brandt, J. D., Heuer, D. K., Higginbotham, E. J., Johnson, C. A., Keltner, J. L., Miller, J. P., Parrish, R. K., 2nd, Wilson, M. R., and Kass, M. A. (2002). The Ocular Hypertension Treatment Study: Baseline factors that predict the onset of primary open-angle glaucoma. *Arch. Ophthalmol.* **120,** 714–720discussion 829–30.

Gordon-Thomson, C., de Iongh, R. U., Hales, A. M., Chamberlain, C. G., and McAvoy, J. W. (1998). Differential cataractogenic potency of TGF-beta1, -beta2, and -beta3 and their expression in the postnatal rat eye. *Invest. Ophthalmol. Vis. Sci.* **39,** 1399–1409.

Gottanka, J., Johnson, D. H., Martus, P., and Lütjen-Drecoll, E. (1997). Severity of optic nerve damage in eyes with POAG is correlated with changes in the trabecular meshwork. *J. Glaucoma* **6,** 123–132.

Gottanka, J., Chan, D., Eichhorn, M., Lütjen-Drecoll, E., and Ethier, C. R. (2004). Effects of TGF-β2 in perfused human eyes. *Invest. Ophthalmol. Vis. Sci.* **45,** 153–158.

Gould, D. B., Miceli-Libby, L., Savinova, O. V., Torrado, M., Tomarev, S. I., Smith, R. S., and John, S. W. (2004). Genetically increasing Myoc expression supports a necessary pathologic role of abnormal proteins in glaucoma. *Mol. Cell Biol.* **24,** 9019–9025.

Gould, D. B., Reedy, M., Wilson, L. A., Smith, R. S., Johnson, R. L., and John, S. W. (2006). Mutant myocilin nonsecretion *in vivo* is not sufficient to cause glaucoma. *Mol. Cell Biol.* **26,** 8427–8436.

Granstein, R. D., Staszewski, R., Knisley, T. L., Zeira, E., Nazareno, R., Latina, M., and Albert, D. M. (1990). Aqueous humor contains transforming growth factor-β and a small (<3500 daltons) inhibitor of thymocyte proliferation. *J. Immunol.* **144,** 3021–3027.

Grant, W. M. (1963). Experimental aqueous perfusion in enucleated human eyes. *Arch. Ophthalmol.* **69,** 783–801.

Gressner, A. M., and Weiskirchen, R. (2006). Modern pathogenetic concepts of liver fibrosis suggest stellate cells and TGF-beta as major players and therapeutic targets. *J. Cell. Mol. Med.* **10,** 76–99.

Gupta, S. (2002). A decision between life and death during TNF-alpha-induced signaling. *J. Clin. Immunol.* **22,** 185–194.

Hampton, R. Y. (2000). ER stress response: Getting the UPR hand on misfolded proteins. *Curr. Biol.* **10,** R518–R521.

Hardy, K. M., Hoffman, E. A., Gonzalez, P., McKay, B. S., and Stamer, W. D. (2005). Extracellular trafficking of myocilin in human trabecular meshwork cells. *J. Biol. Chem.* **280,** 28917–28926.

Hattula, K., and Peranen, J. (2000). FIP-2, a coiled-coil protein, links Huntingtin to Rab8 and modulates cellular morphogenesis. *Curr. Biol.* **10,** 1603–1606.

Hauser, M. A., Allingham, R. R., Linkroum, K., Wang, J., LaRocque-Abramson, K., Figueiredo, D., Santiago-Turla, C., del Bono, E. A., Haines, J. L., Pericak-Vance, M. A., and Wiggs, J. L. (2006a). Distribution of WDR36 DNA sequence variants in patients with primary open-angle glaucoma. *Invest. Ophthalmol. Vis. Sci.* **47,** 2542–2546.

Hauser, M. A., Sena, D. F., Flor, J., Walter, J., Auguste, J., Larocque-Abramson, K., Graham, F., Delbono, E., Haines, J. L., Pericak-Vance, M. A., Rand Allingham, R., and Wiggs, J. L. (2006b). Distribution of optineurin sequence variations in an ethnically diverse population of low-tension glaucoma patients from the United States. *J. Glaucoma* **15,** 358–363.

Helbig, H., Kittredge, K. L., Coca-Prados, M., Davis, J., Palestine, A. G., and Nussenblatt, R. B. (1991). Mammalian ciliary-body epithelial cells in culture produce transforming growth factor-beta. *Graefes Arch. Clin. Exp. Ophthalmol.* **229,** 84–87.

Hewitt, A. W., Dimasi, D. P., Mackey, D. A., and Craig, J. E. (2006). A glaucoma case-control study of the WDR36 gene D658G sequence variant. *Am. J. Ophthalmol.* **142,** 324–325.

Higginbotham, E. J., Gordon, M. O., Beiser, J. A., Drake, M. V., Bennett, G. R., Wilson, M. R., and Kass, M. A. (2004). The Ocular Hypertension Treatment Study: Topical medication delays or prevents primary open-angle glaucoma in African American individuals. *Arch. Ophthalmol.* **122,** 813–820.

Ho, S. L., Dogar, G. F., Wang, J., Crean, J., Wu, Q. D., Oliver, N., Weitz, S., Murray, A., Cleary, P. E., and O'Brien, C. (2005). Elevated aqueous humour tissue inhibitor of matrix metalloproteinase-1 and connective tissue growth factor in pseudoexfoliation syndrome. *Br. J. Ophthalmol.* **89,** 169–173.

Horwitz, J. (2003). Alpha-crystallin. *Exp. Eye Res.* **76,** 145–153.

Huggins, J. T., and Sahn, S. A. (2004). Causes and management of pleural fibrosis. *Respirology* **9,** 441–447.

Ihn, H. (2002). Pathogenesis of fibrosis: Role of TGF-β and CTGF. *Curr. Opin. Rheumatol.* **14,** 681–685.

Inatani, M., Tanihara, H., Katsuta, H., Honjo, M., Kido, N., and Honda, Y. (2001). Transforming growth factor-$\beta 2$ levels in aqueous humor of glaucomatous eyes. *Graefes Arch. Clin. Exp. Ophthalmol.* **239,** 109–113.

Ito, Y., Aten, J., Bende, R. J., Oemar, B. S., Rabelink, T. J., Weening, J. J., and Goldschmeding, R. (1998). Expression of connective tissue growth factor in human renal fibrosis. *Kidney Int.* **53,** 853–861.

Jacobson, N., Andrews, M., Shepard, A. R., Nishimura, D., Searby, C., Fingert, J. H., Hageman, G., Mullins, R., Davidson, B. L., Kwon, Y. H., Alward, W. L., Stone, E. M., *et al.* (2001). Non-secretion of mutant proteins of the glaucoma gene myocilin in cultured trabecular meshwork cells and in aqueous humor. *Hum. Mol. Genet.* **10,** 117–125.

Jampel, H. D., Roche, N., Stark, W. J., and Roberts, A. B. (1990). Transforming growth factor-β in human aqueous humor. *Curr. Eye Res.* **9,** 963–969.

Jana, N. R., Tanaka, M., Wang, G., and Nukina, N. (2000). Polyglutamine length-dependent interaction of Hsp40 and Hsp70 family chaperones with truncated N-terminal huntingtin: Their role in suppression of aggregation and cellular toxicity. *Hum. Mol. Genet.* **9,** 2009–2018.

Joe, M. K., Sohn, S., Hur, W., Moon, Y., Choi, Y. R., and Kee, C. (2003). Accumulation of mutant myocilins in ER leads to ER stress and potential cytotoxicity in human trabecular meshwork cells. *Biochem. Biophys. Res. Commun.* **312,** 592–600.

Joe, M. K., Sohn, S., Choi, Y. R., Park, H., and Kee, C. (2005). Identification of flotillin-1 as a protein interacting with myocilin: Implications for the pathogenesis of primary open-angle glaucoma. *Biochem. Biophys. Res. Commun.* **336,** 1201–1206.

Johnson, M. (2006). 'What controls aqueous humour outflow resistance?'. *Exp. Eye Res.* **82,** 545–557.

Johnson, M., and Erickson, K. (2000). Mechanisms and routes of aqueous humor drainage. *In* "Principles and Practice of Ophthalmology" (D. M. Albert and F. A. Jakobiec, eds.), pp. 2577–2595. Saunders, Philadelphia.

Johnston, J. A., Dalton, M. J., Gurney, M. E., and Kopito, R. R. (2000). Formation of high molecular weight complexes of mutant Cu, Zn-superoxide dismutase in a mouse model for familial amyotrophic lateral sclerosis. *Proc. Natl. Acad. Sci. USA* **97,** 12571–12576.

Kamphuis, W., and Schneemann, A. (2003). Optineurin gene expression level in human trabecular meshwork does not change in response to pressure elevation. *Ophthalmic Res.* **35,** 93–96.

Karali, A., Russell, P., Stefani, F. H., and Tamm, E. R. (2000). Localization of myocilin/trabecular meshwork inducible glucocorticoid response protein in the human eye. *Invest. Ophthalmol. Vis. Sci.* **41,** 729–740.

Karavanich, C., and Anholt, R. R. (1998a). Evolution of olfactomedin. Structural constraints and conservation of primary sequence motifs. *Ann. N.Y. Acad. Sci.* **855,** 294–300.

Karavanich, C. A., and Anholt, R. R. (1998b). Molecular evolution of olfactomedin. *Mol. Biol. Evol.* **15,** 718–726.

Khalil, N. (1999). TGF-β: From latent to active. *Microbes Infect.* **1,** 1255–1263.

Kim, B. S., Savinova, O. V., Reedy, M. V., Martin, J., Lun, Y., Gan, L., Smith, R. S., Tomarev, S. I., John, S. W., and Johnson, R. L. (2001). Targeted disruption of the myocilin gene (Myoc) suggests that human glaucoma-causing mutations are gain of function. *Mol. Cell Biol.* **21,** 7707–7713.

Knaupp, C., Flügel-Koch, C., Goldwich, A., Ohlmann, A., and Tamm, E. R. (2004). The expression of myocilin during murine eye development. *Graefes Arch. Clin. Exp. Ophthalmol.* **242,** 339–345.

Kroeber, M., Ohlmann, A., Russell, P., and Tamm, E. R. (2006). Transgenic studies on the role of optineurin in the mouse eye. *Exp. Eye Res.* **82,** 1075–1085.

Kubota, R., Noda, S., Wang, Y., Minoshima, S., Asakawa, S., Kudoh, J., Mashima, Y., Oguchi, Y., and Shimizu, N. (1997). A novel myosin-like protein (myocilin) expressed in the connecting cilium of the photoreceptor: Molecular cloning, tissue expression and chromosomal mapping. *Genomics* **41,** 360–369.

Lambert, M. P., Barlow, A. K., Chromy, B. A., Edwards, C., Freed, R., Liosatos, M., Morgan, T. E., Rozovsky, I., Trommer, B., Viola, K. L., Wals, P., Zhang, C., *et al.* (1998). Diffusible, nonfibrillar ligands derived from Abeta1–42 are potent central nervous system neurotoxins. *Proc. Natl. Acad. Sci. USA* **95,** 6448–6453.

Lepple-Wienhues, A., Becker, M., Stahl, F., Berweck, S., Hensen, J., Noske, W., Eichhorn, M., and Wiederholt, M. (1992). Endothelin-like immunoreactivity in the aqueous humour and in conditioned medium from cultured ciliary epithelial cells. *Curr. Eye Res.* **11,** 1041–1046.

Lepple-Wienhues, A., Stahl, F., Willner, U., Schafer, R., and Wiederholt, M. (1991). Endothelin-evoked contractions in bovine ciliary muscle and trabecular meshwork: interaction with calcium, nifedipine and nickel. *Curr. Eye Res.* **10,** 983–989.

Leask, A., and Abraham, D. J. (2004). TGF-beta signaling and the fibrotic response. *Faseb J.* **18,** 816–827.

Leske, M. C., Heijl, A., Hussein, M., Bengtsson, B., Hyman, L., and Komaroff, E. (2003). Factors for glaucoma progression and the effect of treatment: The early manifest glaucoma trial. *Arch. Ophthalmol.* **121,** 48–56.

Leung, Y. F., Fan, B. J., Lam, D. S., Lee, W. S., Tam, P. O., Chua, J. K., Tham, C. C., Lai, J. S., Fan, D. S., and Pang, C. P. (2003). Different optineurin mutation pattern in primary open-angle glaucoma. *Invest. Ophthalmol. Vis. Sci.* **44,** 3880–3884.

Li, J., Tripathi, B. J., Chalam, K. V., and Tripathi, R. C. (1996). Transforming growth factor-beta 1 and -beta 2 positively regulate TGF-beta 1 mRNA expression in trabecular cells. *Invest. Ophthalmol. Vis. Sci.* **37,** 2778–2782.

Li, Y., Kang, J., and Horwitz, M. S. (1998). Interaction of an adenovirus E3 14.7-kilodalton protein with a novel tumor necrosis factor alpha-inducible cellular protein containing leucine zipper domains. *Mol. Cell Biol.* **18,** 1601–1610.

Li, J., Tripathi, B. J., and Tripathi, R. C. (2000). Modulation of pre-mRNA splicing and protein production of fibronectin by TGF-β2 in porcine trabecular cells. *Invest. Ophthalmol. Vis. Sci.* **41,** 3437–3443.

Li, Y., Aroca-Aguilar, J. D., Ghosh, S., Sanchez-Sanchez, F., Escribano, J., and Coca-Prados, M. (2006). Interaction of myocilin with the C-terminal region of hevin. *Biochem. Biophys. Res. Commun.* **339,** 797–804.

Liang, Y., Li, C., Guzman, V. M., Evinger, A. J., 3rd, Protzman, C. E., Krauss, A. H., and Woodward, D. F. (2003). Comparison of prostaglandin F2alpha, bimatoprost (prostamide), and butaprost (EP2 agonist) on Cyr61 and connective tissue growth factor gene expression. *J. Biol. Chem.* **278,** 27267–27277.

Liton, P. B., Liu, X., Challa, P., Epstein, D. L., and Gonzalez, P. (2005). Induction of TGF-beta1 in the trabecular meshwork under cyclic mechanical stress. *J. Cell. Physiol.* **205,** 364–371.

Liu, Y. (2006). Renal fibrosis: New insights into the pathogenesis and therapeutics. *Kidney Int.* **69,** 213–217.

Liu, Y., and Vollrath, D. (2004). Reversal of mutant myocilin non-secretion and cell killing: Implications for glaucoma. *Hum. Mol. Genet.* **13,** 1193–1204.

Luo, G., Hofmann, C., Bronckers, A. L., Sohocki, M., Bradley, A., and Karsenty, G. (1995). BMP-7 is an inducer of nephrogenesis, and is also required for eye development and skeletal patterning. *Genes Dev.* **9,** 2808–2820.

Luo, G., Ducy, P., McKee, M. D., Pinero, G. J., Loyer, E., Behringer, R. R., and Karsenty, G. (1997). Spontaneous calcification of arteries and cartilage in mice lacking matrix GLA protein. *Nature* **386,** 78–81.

Lütjen-Drecoll, E., and Rohen, J. W. (2001). Functional morphology of the trabecular meshwork. *In* "Duane's Foundations of Clinical Ophthalmology" (W. Tasman and E. A. Jaeger, eds.), pp. 1–30. Lippincott, Philadelphia.

Lütjen-Drecoll, E., Shimizu, T., Rohrbach, M., and Rohen, J. W. (1986). Quantitative analysis of "plaque material" in the inner and outer wall of Schlemm's canal in normal and glaucomatous eyes. *Exp. Eye Res.* **42,** 443–455.

Lütjen-Drecoll, E., May, C. A., Polansky, J. R., Johnson, D. H., Bloemendal, H., and Nguyen, T. D. (1998). Localization of the stress proteins αB-crystallin and trabecular meshwork inducible glucocorticoid response protein in normal and glaucomatous trabecular meshwork. *Invest. Ophthalmol. Vis. Sci.* **39,** 517–525.

Malyukova, I., Lee, H. S., Fariss, R. N., and Tomarev, S. I. (2006). Mutated mouse and human myocilins have similar properties and do not block general secretory pathway. *Invest. Ophthalmol. Vis. Sci.* **47,** 206–212.

Mao, M., Biery, M. C., Kobayashi, S. V., Ward, T., Schimmack, G., Burchard, J., Schelter, J. M., Dai, H., He, Y. D., and Linsley, P. S. (2004). T lymphocyte activation gene identification by coregulated expression on DNA microarrays. *Genomics* **83,** 989–999.

Mertts, M., Garfield, S., Tanemoto, K., and Tomarev, S. I. (1999). Identification of the region in the N-terminal domain responsible for the cytoplasmic localization of Myoc/Tigr and its association with microtubules. *Lab. Invest.* **79,** 1237–1245.

Meyer, P., Wunderlich, K., Kain, H. L., Prunte, C., and Flammer, J. (2002). Human connective tissue growth factor mRNA expression of epiretinal and subretinal fibrovascular membranes: A report of three cases. *Ophthalmologica* **216,** 284–291.

Monemi, S., Spaeth, G., DaSilva, A., Popinchalk, S., Ilitchev, E., Liebmann, J., Ritch, R., Heon, E., Crick, R. P., Child, A., and Sarfarazi, M. (2005). Identification of a novel adult-onset primary open-angle glaucoma (POAG) gene on 5q22.1. *Hum. Mol. Genet.* **14,** 725–733.

Moreland, R. J., Dresser, M. E., Rodgers, J. S., Roe, B. A., Conaway, J. W., Conaway, R. C., and Hanas, J. S. (2000). Identification of a transcription factor IIIA-interacting protein. *Nucleic Acids Res.* **28,** 1986–1993.

Munger, J. S., Huang, X., Kawakatsu, H., Griffiths, M. J., Dalton, S. L., Wu, J., Pittet, J. F., Kaminski, N., Garat, C., Matthay, M. A., Rifkin, D. B., and Sheppard, D. (1999). The integrin αvβ6 binds and activates latent TGFβ1: A mechanism for regulating pulmonary inflammation and fibrosis. *Cell* **96,** 319–328.

Murphy-Ullrich, J. E., and Hook, M. (1989). Thrombospondin modulates focal adhesions in endothelial cells. *J. Cell Biol.* **109,** 1309–1319.

Murphy-Ullrich, J. E., and Poczatek, M. (2000). Activation of latent TGF-β by thrombospondin-1: Mechanisms and physiology. *Cytokine Growth Factor Rev.* **11,** 59–69.

Nguyen, T. D., Chen, P., Huang, W. D., Chen, H., Johnson, D., and Polansky, J. R. (1998). Gene structure and properties of TIGR, an olfactomedin-related glycoprotein cloned from glucocorticoid-induced trabecular meshwork cells. *J. Biol. Chem.* **273,** 6341–6350.

Noda, S., Mashima, Y., Obazawa, M., Kubota, R., Oguchi, Y., Kudoh, J., Minoshima, S., and Shimizu, N. (2000). Myocilin expression in the astrocytes of the optic nerve head. *Biochem. Biophys. Res. Commun.* **276,** 1129–1135.

Noske, W., Hensen, J., and Wiederholt, M. (1997). Endothelin-like immunoreactivity in aqueous humor of patients with primary open-angle glaucoma and cataract. *Graefes Arch. Clin. Exp. Ophthalmol.* **235,** 551–552.

Obazawa, M., Mashima, Y., Sanuki, N., Noda, S., Kudoh, J., Shimizu, N., Oguchi, Y., Tanaka, Y., and Iwata, T. (2004). Analysis of porcine optineurin and myocilin expression in trabecular meshwork cells and astrocytes from optic nerve head. *Invest. Ophthalmol. Vis. Sci.* **45,** 2652–2659.

Ochiai, Y., and Ochiai, H. (2002). Higher concentration of transforming growth factor-β in aqueous humor of glaucomatous eyes and diabetic eyes. *Jpn. J. Ophthalmol.* **46,** 249–253.

Ohlmann, A., Goldwich, A., Flügel-Koch, C., Fuchs, A. V., Schwager, K., and Tamm, E. R. (2003). Secreted glycoprotein myocilin is a component of the myelin sheath in peripheral nerves. *Glia* **43,** 128–140.

Ortego, J., Escribano, J., and Coca-Prados, M. (1997). Cloning and characterization of subtracted cDNAs from human ciliary body library encoding TIGR, a protein involved in juvenile open angle glaucoma with homology to myosin and olfactomedin. *FEBS Lett.* **413,** 349–353.

Park, B. C., Shen, X., Samaraweera, M., and Yue, B. Y. (2006). Studies of optineurin, a glaucoma gene: Golgi fragmentation and cell death from overexpression of wild-type and mutant optineurin in two ocular cell types. *Am. J. Pathol.* **169,** 1976–1989.

Perkumas, K. M., Hoffman, E. A., McKay, B. S., Allingham, R. R., and Stamer, W. D. (2007). Myocilin-associated exosomes in human ocular samples. *Exp. Eye Res.* **84,** 209–212.

Peters, D. M., Herbert, K., Biddick, B., and Peterson, J. A. (2005). Myocilin binding to Hep II domain of fibronectin inhibits cell spreading and incorporation of paxillin into focal adhesions. *Exp. Cell Res.* **303,** 218–228.

Picht, G., Welge-Luessen, U., Grehn, F., and Lütjen-Drecoll, E. (2001). Transforming growth factor-$\beta 2$ levels in the aqueous humor in different types of glaucoma and the relation to filtering bleb development. *Graefes Arch. Clin. Exp. Ophthalmol.* **239,** 199–207.

Price, P. A., and Williamson, M. K. (1985). Primary structure of bovine matrix Gla protein, a new vitamin K-dependent bone protein. *J. Biol. Chem.* **260,** 14971–14975.

Rao, P. V., Allingham, R. R., and Epstein, D. L. (2000). TIGR/myocilin in human aqueous humor. *Exp. Eye Res.* **71,** 637–641.

Rezaie, T., and Sarfarazi, M. (2005). Molecular cloning, genomic structure, and protein characterization of mouse optineurin. *Genomics* **85,** 131–138.

Rezaie, T., Child, A., Hitchings, R., Brice, G., Miller, L., Coca-Prados, M., Heon, E., Krupin, T., Ritch, R., Kreutzer, D., Crick, R. P., and Sarfarazi, M. (2002). Adult-onset primary open-angle glaucoma caused by mutations in optineurin. *Science* **295,** 1077–1079.

Rezaie, T., Waitzman, D. M., Seeman, J. L., Kaufman, P. L., and Sarfarazi, M. (2005). Molecular cloning and expression profiling of optineurin in the rhesus monkey. *Invest. Ophthalmol. Vis. Sci.* **46,** 2404–2410.

Rhee, D. J., Fariss, R. N., Brekken, R., Sage, E. H., and Russell, P. (2003). The matricellular protein SPARC is expressed in human trabecular meshwork. *Exp. Eye Res.* **77,** 601–607.

Ricard, C. S., Agapova, O. A., Salvador-Silva, M., Kaufman, P. L., and Hernandez, M. R. (2001). Expression of myocilin/TIGR in normal and glaucomatous primate optic nerves. *Exp. Eye Res.* **73,** 433–447.

Ricard, C. S., Mukherjee, A., Silver, F. L., and Wagenknecht, P. L. (2006). Canine myocilin is associated with lipid modified by palmitic acid. *Mol. Vis.* **12,** 1427–1436.

Rodriguez-Vita, J., Ruiz-Ortega, M., Ruperez, M., Esteban, V., Sanchez-Lopez, E., Plaza, J. J., and Egido, J. (2005). Endothelin-1, via ETA receptor and independently of transforming growth factor-beta, increases the connective tissue growth factor in vascular smooth muscle cells. *Circ. Res.* **97,** 125–134.

Rohen, J. W., and Witmer, R. (1972). Electron microscopic studies on the trabecular meshwork in glaucoma simplex. *Albrecht Von. Graefes Arch. Klin. Exp. Ophthalmol.* **183,** 251–266.

Rohen, J. W., Futa, R., and Lütjen-Drecoll, E. (1981). The fine structure of the cribriform meshwork in normal and glaucomatous eyes as seen in tangential sections. *Invest. Ophthalmol. Vis. Sci.* **21,** 574–585.

Rohen, J. W., Lütjen-Drecoll, E., Flügel, C., Meyer, M., and Grierson, I. (1993). Ultrastructure of the trabecular meshwork in untreated cases of primary open-angle glaucoma (POAG). *Exp. Eye Res.* **56,** 683–692.

Russell, P., Tamm, E. R., Grehn, F. J., Picht, G., and Johnson, M. (2001). The presence and properties of myocilin in the aqueous humor. *Invest. Ophthalmol. Vis. Sci.* **42,** 983–986.

Sage, E. H., and Bornstein, P. (1991). Extracellular proteins that modulate cell-matrix interactions. SPARC, tenascin, and thrombospondin. *J. Biol. Chem.* **266,** 14831–14834.

Sage, H., Vernon, R. B., Funk, S. E., Everitt, E. A., and Angello, J. (1989). SPARC, a secreted protein associated with cellular proliferation, inhibits cell spreading *in vitro* and exhibits Ca^{2+}-dependent binding to the extracellular matrix. *J. Cell Biol.* **109,** 341–356.

Sahlender, D. A., Roberts, R. C., Arden, S. D., Spudich, G., Taylor, M. J., Luzio, J. P., Kendrick-Jones, J., and Buss, F. (2005). Optineurin links myosin VI to the Golgi complex and is involved in Golgi organization and exocytosis. *J. Cell. Biol.* **169,** 285–295.

Saika, S., Ikeda, K., Yamanaka, O., Flanders, K. C., Nakajima, Y., Miyamoto, T., Ohnishi, Y., Kao, W. W., Muragaki, Y., and Ooshima, A. (2005). Therapeutic effects of adenoviral gene transfer of bone morphogenic protein-7 on a corneal alkali injury model in mice. *Lab. Invest.* **85,** 474–486.

Sanchez-Capelo, A. (2005). Dual role for TGF-beta1 in apoptosis. *Cytokine Growth Factor Rev.* **16,** 15–34.

Sanchez-Sanchez, F., Martinez-Redondo, F., Aroca-Aguilar, J. D., Coca-Prados, M., and Escribano, J. (2007). Characterization of the intracellular proteolytic cleavage of myocilin and identification of calpain II as a myocilin processing protease. *J. Biol. Chem.* **282**(38), 27810–27824.

Sato, S., Nagaoka, T., Hasegawa, M., Tamatani, T., Nakanishi, T., Takigawa, M., and Takehara, K. (2000). Serum levels of connective tissue growth factor are elevated in patients with systemic sclerosis: Association with extent of skin sclerosis and severity of pulmonary fibrosis. *J. Rheumatol.* **27,** 149–154.

Schnaper, H. W., Hayashida, T., Hubchak, S. C., and Poncelet, A. C. (2003). TGF-β signal transduction and mesangial cell fibrogenesis. *Am. J. Physiol. Renal Physiol.* **284,** F243–F252.

Schultz-Cherry, S., Lawler, J., and Murphy-Ullrich, J. E. (1994). The type 1 repeats of thrombospondin 1 activate latent transforming growth factor-beta. *J. Biol. Chem.* **269,** 26783–26788.

Senatorov, V., Malyukova, I., Fariss, R., Wawrousek, E. F., Swaminathan, S., Sharan, S. K., and Tomarev, S. (2006). Expression of mutated mouse myocilin induces open-angle glaucoma in transgenic mice. *J. Neurosci.* **26,** 11903–11914.

Shepard, A. R., Jacobson, N., Sui, R., Steely, H. T., Lotery, A. J., Stone, E. M., and Clark, A. F. (2003). Characterization of rabbit myocilin: Implications for human myocilin glycosylation and signal peptide usage. *BMC Genet.* **4,** 5.

Shepard, A. R., Jacobson, N., Millar, J. C., Pang, I. H., Steely, H. T., Searby, C. C., Sheffield, V. C., Stone, E. M., and Clark, A. F. (2007). Glaucoma-causing myocilin mutants require the Peroxisomal targeting signal-1 receptor (PTS1R) to elevate intraocular pressure. *Hum. Mol. Genet.* **16,** 609–617.

Snyder, D. A., Rivers, A. M., Yokoe, H., Menco, B. P., and Anholt, R. R. (1991). Olfactomedin: Purification, characterization, and localization of a novel olfactory glycoprotein. *Biochemistry* **30,** 9143–9153.

Sohn, S., Hur, W., Joe, M. K., Kim, J. H., Lee, Z. W., Ha, K. S., and Kee, C. (2002). Expression of wild-type and truncated myocilins in trabecular meshwork cells: Their subcellular localizations and cytotoxicities. *Invest. Ophthalmol. Vis. Sci.* **43,** 3680–3685.

Stamer, W. D., Perkumas, K. M., Hoffman, E. A., Roberts, B. C., Epstein, D. L., and McKay, B. S. (2006). Coiled-coil targeting of myocilin to intracellular membranes. *Exp. Eye Res.* **83,** 1386–1395.

Stone, E. M., Fingert, J. H., Alward, W. L. M., Nguyen, T. D., Polansky, J. R., Sunden, S. L. F., Nishimura, D., Clark, A. F., Nystuen, A., Nichols, B. E., Mackey, D. A., Ritch, R., *et al.* (1997). Identification of a gene that causes primary open angle glaucoma. *Science* **275,** 668–670.

Stroissnigg, H., Repitz, M., Miloloza, A., Linhartova, I., Beug, H., Wiche, G., and Propst, F. (2002). FIP-2, an IkappaB-kinase-gamma-related protein, is associated with the Golgi apparatus and translocates to the marginal band during chicken erythroblast differentiation. *Exp. Cell Res.* **278,** 133–145.

Swiderski, R. E., Ying, L., Cassell, M. D., Alward, W. L., Stone, E. M., and Sheffield, V. C. (1999). Expression pattern and *in situ* localization of the mouse homologue of the human MYOC (GLC1A) gene in adult brain. *Brain Res. Mol. Brain Res.* **68,** 64–72.

Swiderski, R. E., Ross, J. L., Fingert, J. H., Clark, A. F., Alward, W. L., Stone, E. M., and Sheffield, V. C. (2000). Localization of MYOC transcripts in human eye and optic nerve by *in situ* hybridization. *Invest. Ophthalmol. Vis. Sci.* **41,** 3420–3428.

Taguchi, M., Kanno, H., Kubota, R., Miwa, S., Shishiba, Y., and Ozawa, Y. (2000). Molecular cloning and expression profile of rat myocilin. *Mol. Genet. Metab.* **70,** 75–80.

Takahashi, H., Noda, S., Imamura, Y., Nagasawa, A., Kubota, R., Mashima, Y., Kudoh, J., Oguchi, Y., and Shimizu, N. (1998). Mouse myocilin (Myoc) gene expression in ocular tissues. *Biochem. Biophys. Res. Commun.* **248,** 104–109.

Tamm, E., Flügel, C., Stefani, F. H., and Rohen, J. W. (1992). Contractile cells in the human scleral spur. *Exp. Eye Res.* **54,** 531–543.

Tamm, E., Russell, P., Epstein, D. L., Johnson, D. H., and Piatigorsky, J. (1999). Modulation of Myocilin/TIGR expression in human trabecular meshwork. *Invest. Ophthalmol. Vis. Sci.* **40,** 2577–2582.

Tamm, E. R. (2002). Myocilin and glaucoma: Facts and ideas. *Prog. Retin. Eye Res.* **21,** 395–428.

Tamm, E. R., Russell, P., Johnson, D. H., and Piatigorsky, J. (1996a). Human and monkey trabecular meshwork accumulate alpha B-crystallin in response to heat shock and oxidative stress. *Invest. Ophthalmol. Vis. Sci.* **37,** 2402–2413.

Tamm, E. R., Siegner, A., Baur, A., and Lütjen-Drecoll, E. (1996b). Transforming growth factor-$\beta 1$ induces α-smooth muscle-actin expression in cultured human and monkey trabecular meshwork. *Exp. Eye Res.* **62,** 389–397.

Tamm, E. R., Toris, C. B., Crowston, J. G., Sit, A., Lim, S., Lambrou, G., and Alm, A. (2007). Basic science of intraocular pressure. *In* "Intraocular Pressure. Reports and Consensus Statements of the 4th Global AIGS Consensus Meeting on Intraocular Pressure" (R. N. Weinreb, J. D. Brandt, D. Garway-Heath, and F. Medeiros, eds.), pp. 1–14. Kugler Publications, Amsterdam.

Tezel, G., and Wax, M. B. (2000). Increased production of tumor necrosis factor-alpha by glial cells exposed to simulated ischemia or elevated hydrostatic pressure induces apoptosis in cocultured retinal ganglion cells. *J. Neurosci.* **20,** 8693–8700.

Tezel, G., Li, L. Y., Patil, R. V., and Wax, M. B. (2001). TNF-alpha and TNF-alpha receptor-1 in the retina of normal and glaucomatous eyes. *Invest. Ophthalmol. Vis. Sci.* **42,** 1787–1794.

The AGIS Investigators (2000). The advanced glaucoma intervention study (AGIS): 7. The relationship between control of intraocular pressure and visual field deterioration. *Am. J. Ophthalmol.* **130,** 429–440.

Tian, B., Geiger, B., Epstein, D. L., and Kaufman, P. L. (2000). Cytoskeletal involvement in the regulation of aqueous humor outflow. *Invest. Ophthalmol. Vis. Sci.* **41,** 619–623.

Tomarev, S. I., Tamm, E. R., and Chang, B. (1998). Characterization of the mouse *myoc/tigr* gene. *Biochem. Biophys. Res. Commun.* **245,** 887–893.

Tomarev, S. I., Wistow, G., Raymond, V., Dubois, S., and Malyukova, I. (2003). Gene expression profile of the human trabecular meshwork: NEIBank sequence tag analysis. *Invest. Ophthalmol. Vis. Sci.* **44,** 2588–2596.

Torrado, M., Trivedi, R., Zinovieva, R., Karavanova, I., and Tomarev, S. I. (2002). Optimedin: A novel olfactomedin-related protein that interacts with myocilin. *Hum. Mol. Genet.* **11,** 1291–1301.

Tripathi, R. C. (1972). Aqueous outflow pathway in normal and glaucomatous eyes. *Br. J. Ophthalmol.* **56,** 157–174.

Tripathi, R. C., Borisuth, N. S., Kolli, S. P., and Tripathi, B. J. (1993a). Trabecular cells express receptors that bind TGF-$\beta 1$ and TGF- $\beta 2$: A qualitative and quantitative characterization. *Invest. Ophthalmol. Vis. Sci.* **34,** 260–263.

Tripathi, R. C., Li, J. P., Borisuth, N. S. C., and Tripathi, B. J. (1993b). Trabecular cells of the eye express messenger RNA for transforming growth factor-$\beta 1$ and secrete this cytokine. *Invest. Ophthalmol. Vis. Sci.* **34,** 2562–2569.

Tripathi, R. C., Chan, W. F. A., Li, J., and Tripathi, B. J. (1994a). Trabecular cells express the TGF-$\beta 2$ gene and secrete this cytokine. *Exp. Eye Res.* **58,** 523–528.

Tripathi, R. C., Li, J., Chan, W. F. A., and Tripathi, B. J. (1994b). Aqueous humor in glaucomatous eyes contains an increased level of TGF-$\beta 2$. *Exp. Eye Res.* **58,** 723–727.

Ueda, J., Wentz-Hunter, K. K., Cheng, E. L., Fukuchi, T., Abe, H., and Yue, B. Y. (2000). Ultrastructural localization of myocilin in human trabecular meshwork cells and tissues. *J. Histochem. Cytochem.* **48,** 1321–1330.

van Setten, G. B., Blalock, T. D., Grotendorst, G., and Schultz, G. S. (2002). Detection of connective tissue growth factor in human aqueous humor. *Ophthalmic Res.* **34,** 306–308.

van Setten, G., Aspiotis, M., Blalock, T. D., Grotendorst, G., and Schultz, G. (2003a). Connective tissue growth factor in pterygium: Simultaneous presence with vascular endothelial growth factor – Possible contributing factor to conjunctival scarring. *Graefes Arch. Clin. Exp. Ophthalmol.* **241,** 135–139.

van Setten, G. B., Blalock, T. D., Grotendorst, G., and Schultz, G. S. (2003b). Detection of connective tissue growth factor (CTGF) in human tear fluid: Preliminary results. *Acta Ophthalmol. Scand.* **81,** 51–53.

Vittitow, J., and Borras, T. (2002). Expression of optineurin, a glaucoma-linked gene, is influenced by elevated intraocular pressure. *Biochem. Biophys. Res. Commun.* **298,** 67–74.

Vittitow, J., and Borras, T. (2004). Genes expressed in the human trabecular meshwork during pressure-induced homeostatic response. *J. Cell. Physiol.* **201,** 126–137.

Vollrath, D., and Liu, Y. (2006). Temperature sensitive secretion of mutant myocilins. *Exp. Eye Res.* **82,** 1030–1036.

Wang, S., and Hirschberg, R. (2003). BMP7 antagonizes TGF-beta-dependent fibrogenesis in mesangial cells. *Am. J. Physiol. Renal. Physiol.* **284,** F1006–F1013.

Wang, S., and Hirschberg, R. (2004). Bone morphogenetic protein-7 signals opposing transforming growth factor beta in mesangial cells. *J. Biol. Chem.* **279,** 23200–23206.

Wang, J., Zohar, R., and McCulloch, C. A. (2006). Multiple roles of alpha-smooth muscle actin in mechanotransduction. *Exp. Cell Res.* **312,** 205–214.

Wang, J. T., Kunzevitzky, N. J., Dugas, J. C., Cameron, M., Barres, B. A., and Goldberg, J. L. (2007). Disease gene candidates revealed by expression profiling of retinal ganglion cell development. *J. Neurosci.* **27,** 8593–8603.

Weisschuh, N., Neumann, D., Wolf, C., Wissinger, B., and Gramer, E. (2005). Prevalence of myocilin and optineurin sequence variants in German normal tension glaucoma patients. *Mol. Vis.* **11,** 284–287.

Weisschuh, N., Wolf, C., Wissinger, B., and Gramer, E. (2007). Variations in the WDR36 gene in German patients with normal tension glaucoma. *Mol. Vis.* **13,** 724–729.

Welge-Lüssen, U., May, C. A., Eichhorn, M., Bloemendal, H., and Lütjen-Drecoll, E. (1999). AlphaB-crystallin in the trabecular meshwork is inducible by transforming growth factor-beta. *Invest. Ophthalmol. Vis. Sci.* **40,** 2235–2241.

Welge-Lüssen, U., May, C. A., and Lütjen-Drecoll, E. (2000). Induction of tissue transglutaminase in the trabecular meshwork by TGF- β1 and TGF-β2. *Invest. Ophthalmol. Vis. Sci.* **41,** 2229–2238.

Welihinda, A. A., Tirasophon, W., and Kaufman, R. J. (1999). The cellular response to protein misfolding in the endoplasmic reticulum. *Gene Expr.* **7,** 293–300.

Wentz-Hunter, K., Ueda, J., Shimizu, N., and Yue, B. Y. (2002a). Myocilin is associated with mitochondria in human trabecular meshwork cells. *J. Cell Physiol.* **190,** 46–53.

Wentz-Hunter, K., Ueda, J., and Yue, B. Y. (2002b). Protein interactions with myocilin. *Invest. Ophthalmol. Vis. Sci.* **43,** 176–182.

Wiederholt, M., Bielka, S., Schweig, F., Lütjen-Drecoll, E., and Lepple-Wienhues, A. (1996). Regulation of outflow rate and resistance in the perfused anterior segment of the bovine eye. *Exp. Eye Res.* **61,** 223–234.

Wiederholt, M., Thieme, H., and Stumpff, F. (2000). The regulation of trabecular meshwork and ciliary muscle contractility. *Prog. Retin. Eye Res.* **19,** 271–295.

Wiggs, J. L., Del Bono, E. A., Schuman, J. S., Hutchinson, B. T., and Walton, D. S. (1995). Clinical features of five pedigrees genetically linked to the juvenile glaucoma locus on chromosome 1q21-q31. *Ophthalmology* **102,** 1782–1789.

Willis, B. C., and Borok, Z. (2007). TGF-{beta}-induced EMT: Mechanisms and implications for fibrotic lung disease. *Am. J. Physiol. Lung Cell. Mol. Physiol.* .

Willoughby, C. E., Chan, L. L., Herd, S., Billingsley, G., Noordeh, N., Levin, A. V., Buys, Y., Trope, G., Sarfarazi, M., and Heon, E. (2004). Defining the pathogenicity of optineurin in juvenile open-angle glaucoma. *Invest. Ophthalmol. Vis. Sci.* **45,** 3122–3130.

Wirtz, M. K., Samples, J. R., Xu, H., Severson, T., and Acott, T. S. (2002). Expression profile and genome location of cDNA clones from an infant human trabecular meshwork cell library. *Invest. Ophthalmol. Vis. Sci.* **43,** 3698–3704.

Wordinger, R. J., Agarwal, R., Talati, M., Fuller, J., Lambert, W., and Clark, A. F. (2002). Expression of bone morphogenetic proteins (BMP), BMP receptors, and BMP associated proteins in human trabecular meshwork and optic nerve head cells and tissues. *Mol. Vis.* **8,** 241–250.

Wordinger, R. J., Fleenor, D. L., Hellberg, P. E., Pang, I. H., Tovar, T. O., Zode, G. S., Fuller, J. A., and Clark, A. F. (2007). Effects of TGF-beta2, BMP-4, and gremlin in the trabecular meshwork: Implications for glaucoma. *Invest. Ophthalmol. Vis. Sci.* **48,** 1191–1200.

Wunderlich, K., Pech, M., Eberle, A. N., Mihatsch, M., Flammer, J., and Meyer, P. (2000a). Expression of connective tissue growth factor (CTGF) mRNA in plaques of human anterior subcapsular cataracts and membranes of posterior capsule opacification. *Curr. Eye Res.* **21,** 627–636.

Wunderlich, K., Senn, B. C., Reiser, P., Pech, M., Flammer, J., and Meyer, P. (2000b). Connective tissue growth factor in retrocorneal membranes and corneal scars. *Ophthalmologica* **214,** 341–346.

Wunderlich, K., Senn, B. C., Todesco, L., Flammer, J., and Meyer, P. (2000c). Regulation of connective tissue growth factor gene expression in retinal vascular endothelial cells by angiogenic growth factors. *Graefes Arch. Clin. Exp. Ophthalmol.* **238,** 910–915.

Xue, W., Wallin, R., Olmsted-Davis, E. A., and Borras, T. (2006). Matrix GLA protein function in human trabecular meshwork cells: Inhibition of BMP2-induced calcification process. *Invest. Ophthalmol. Vis. Sci.* **47,** 997–1007.

Xue, W., Comes, N., and Borras, T. (2007). Presence of an established calcification marker in trabecular meshwork tissue of glaucoma donors. *Invest. Ophthalmol. Vis. Sci.* **48,** 3184–3194.

Yam, G. H., Gaplovska-Kysela, K., Zuber, C., and Roth, J. (2007). Aggregated myocilin induces russell bodies and causes apoptosis: Implications for the pathogenesis of myocilin-caused primary open-angle glaucoma. *Am. J. Pathol.* **170,** 100–109.

Yamamoto, T., Sawada, Y., Katayama, I., and Nishioka, K. (2005). Nodular scleroderma: Increased expression of connective tissue growth factor. *Dermatology* **211,** 218–223.

Yang, D. H., Kim, H. S., Wilson, E. M., Rosenfeld, R. G., and Oh, Y. (1998). Identification of glycosylated 38-kDa connective tissue growth factor (IGFBP-related protein 2) and proteolytic fragments in human biological fluids, and up-regulation of IGFBP-rP2 expression by TGF-beta in Hs578T human breast cancer cells. *J. Clin. Endocrinol. Metab.* **83,** 2593–2596.

You, L., Kruse, F. E., Pohl, J., and Volcker, H. E. (1999). Bone morphogenetic proteins and growth and differentiation factors in the human cornea. *Invest. Ophthalmol. Vis. Sci.* **40,** 296–311.

Yuan, L., and Neufeld, A. H. (2000). Tumor necrosis factor-alpha: A potentially neurodestructive cytokine produced by glia in the human glaucomatous optic nerve head. *Glia* **32,** 42–50.

Zeisberg, M., Hanai, J., Sugimoto, H., Mammoto, T., Charytan, D., Strutz, F., and Kalluri, R. (2003). BMP-7 counteracts TGF-beta1-induced epithelial-to-mesenchymal transition and reverses chronic renal injury. *Nat. Med.* **9,** 964–968.

Zhang, J., Liu, W. L., Tang, D. C., Chen, L., Wang, M., Pack, S. D., Zhuang, Z., and Rodgers, G. P. (2002). Identification and characterization of a novel member of olfactomedin-related protein family, hGC-1, expressed during myeloid lineage development. *Gene* **283,** 83–93.

Zhao, X., and Russell, P. (2005). Versican splice variants in human trabecular meshwork and ciliary muscle. *Mol. Vis.* **11,** 603–608.

Zhao, X., Ramsey, K. E., Stephan, D. A., and Russell, P. (2004). Gene and protein expression changes in human trabecular meshwork cells treated with transforming growth factor-beta. *Invest. Ophthalmol. Vis. Sci.* **45,** 4023–4034.

Zhou, Z., and Vollrath, D. (1999). A cellular assay distinguishes normal and mutant TIGR/myocilin protein. *Hum. Mol. Genet.* **8,** 2221–2228.

Zillig, M., Wurm, A., Grehn, F. J., Russell, P., and Tamm, E. R. (2005). Overexpression and properties of wild-type and Tyr437His mutated myocilin in the eyes of transgenic mice. *Invest. Ophthalmol. Vis. Sci.* **46,** 223–234.

CHAPTER 13

Outflow Signaling Mechanisms and New Therapeutic Strategies for the Control of Intraocular Pressure

Iok-Hou Pang* and Abbot F. Clark†

*Glaucoma Research, Alcon Research, Ltd., Fort Worth, Texas, USA

†Department of Cell Biology and Genetics University North Texas Health Science Center, Fort Worth, Texas, USA

Current Topics in Membranes, Volume 62

1063-5823/08 $35.00
DOI: 10.1016/S1063-5823(08)00413-4

I. OVERVIEW

Tens of millions of the world's population are inflicted with glaucoma, a sight-threatening disease. One major risk factor for the development and progression of glaucoma is abnormally elevated intraocular pressure, which is a result of impeded outflow of aqueous humor. Understanding the regulation of aqueous outflow and its improvement have become urgent needs in the development of future treatments for this disease. This review discusses the potential pathological events involved in primary open angle glaucoma, while focusing on relevant molecular and cellular mechanisms. This review further describes the various on-going and novel therapeutic strategies being designed and evaluated for enhancement of aqueous outflow.

II. INTRODUCTION

Glaucoma is one of the leading causes of blindness in the world, estimated to affect 60 million individuals worldwide in 2010 and 80 million by 2020. Among them, 14% will develop bilateral blindness (Quigley and Broman, 2006). This disease is a complex, age-related, and inherited optic neuropathy with characteristic slow progressive loss of retinal ganglion cells and excavation of the optic disc. Among the risk factors, such as age, race, and family history, that are associated with glaucoma, elevated intraocular pressure (IOP) is the most pivotal. Although not all patients with elevated IOP (>21 mmHg) develop glaucoma, the occurrence of glaucoma increases significantly with increased IOP. Ocular hypertension in glaucoma is a result of a reduction in aqueous humor outflow facility, concomitant with biochemical and morphological changes in the trabecular meshwork (TM).

A. Ocular Hypertension: A Major Risk Factor of Glaucoma

Many prospective and randomized clinical trials have consistently demonstrated that lowering IOP is important in slowing the progression of glaucoma, as well as preventing and delaying its onset. The Advanced Glaucoma Intervention Study (AGIS) showed that, regardless of treatment strategies, patients with higher mean IOPs had a faster disease progression than those with lower IOPs. Most importantly, the subset of patients with IOPs below 18 mmHg at all visits had no or minimal progression during the 6-year follow-up period (AGIS-Investigators, 2000). Likewise, in the Collaborative Initial Glaucoma Treatment Study (CIGTS), in which newly diagnosed glaucoma patients were randomized to initial treatment with topical ocular medicines or glaucoma filtration surgery, there was little disease progression over the course of 5 years in individuals with the greatest reduction in IOP (Lichter *et al.*, 2001). The Early Manifest Glaucoma Treatment Trial (EMGTT) evaluated the effect of IOP-lowering therapy in patients with early glaucoma. The treated group had approximately half the risk for glaucoma progression relative to the untreated patients (Leske *et al.*, 2003). The clinical benefit of controlling IOP extends to patients with normal tension glaucoma, whose IOP is considered within the normal range (<21 mmHg). The Collaborative Normal Tension Glaucoma Study (CNTGS) demonstrated that lowering IOP in this patient population also reduced disease progression (CNTGS-Group, 1998). In addition to slowing down its development and progression, lowering IOP prevents or delays the onset of glaucoma as well. In the Ocular Hypertension Treatment Study (OHTS), ~50% of the enrolled ocular hypertensive patients received IOP-lowering treatment, while the other patients were untreated. After a 5-year follow-up period, the treated individuals were twofold less likely to develop glaucoma, indicating that lowering IOP prevented or delayed the onset of glaucoma (Kass *et al.*, 2002). The obvious conclusion from all these well-designed clinical studies is that high IOP is associated with an increased risk of glaucomatous damage and lowering it reduces such risk. There should also be no doubt that a robust IOP reduction is a necessary and effective treatment in most glaucoma patients.

B. Fluctuation of IOP

In addition to high IOP, IOP fluctuation has also been proposed as an independent risk factor for glaucoma. In the AGIS, long-term (months and years) fluctuation of IOP was shown to be a strong and independent predictor of visual field deterioration (Nouri-Mahdavi *et al.*, 2004). Many other prospective studies also found that ocular hypertension and IOP fluctuation are correlated with glaucoma progression (O'Brien *et al.*, 1991; Bergea *et al.*, 1999; Stewart *et al.*, 2000). Recently, Hong and colleagues further demonstrated that even in glaucoma patients with low IOP (<18 mmHg; controlled by surgery), long-term IOP fluctuation was a risk factor for the decline of visual field (Hong *et al.*, 2007). However, not all studies support this correlation. Analyses of data of the EMGTT and OHTS did not detect an independent link between IOP fluctuation and glaucoma progression (Bengtsson *et al.*, 2007; Gordon *et al.*, 2007). To reconcile these differences, Caprioli hypothesized that IOP fluctuation is the prominent risk factor in patients with lower IOP, but when the IOP is high (as those in the EMGTT and OHTS), the mean IOP became the predominant risk factor (Caprioli, 2007).

Currently, the mechanism of glaucomatous damage induced by IOP fluctuation is yet unclear. However, as compared to a constant state of stress, the large IOP fluctuation may cause ever-changing stress levels that the affected ocular tissues cannot compensate effectively and thus become damaged.

C. Regulation of IOP

Pressure in the eye is balanced by an equilibrium between aqueous humor formation and its outflow, as described by the Goldmann equation.

$$\text{Goldmann equation : IOP} = \left[\frac{(F - U)}{C}\right] + P_{\mathrm{e}}$$

where F is the aqueous humor formation rate, U is the uveoscleral outflow rate, C is the trabecular outflow facility, and P_{e} is the episcleral venous pressure.

An excessive production of aqueous humor and/or a reduction of its outflow could cause ocular hypertension. However, there are no clinically relevant differences in rates of aqueous humor production between glaucomatous and normal individuals. In glaucoma and ocular hypertensive patients, various studies indicated that the cause of IOP elevation is a reduction in aqueous outflow (Langham, 1979; Segawa, 1979; Rohen, 1983).

D. Aqueous Outflow Pathways

After being produced in the ciliary epithelium, aqueous humor travels from the posterior chamber, through the pupil, then enters the anterior chamber. Along the route, it helps to maintain the metabolic homeostasis of the neighboring ocular tissues. Aqueous humor leaves the eye through two major aqueous outflow pathways: the trabecular pathway and the uveoscleral pathway.

1. Trabecular Pathway

The trabecular pathway, which is also called the conventional outflow, involves the TM and the Schlemm's canal in the eye. The TM is located at the anterior chamber angle bordered by the cornea and iris. It is a meshwork formed by strands of collagenous sheets and beams populated with TM cells, with microscopically open spaces between the beams. The Schlemm's canal is a ring-like channel of irregular diameter. It has an endothelium-lined lumen surrounded by a thin discontinuous basement membrane. The aqueous humor percolates through the TM, the juxtacanalicular tissue (JCT), passes through the inner wall endothelium of Schlemm's canal into the canal lumen, and subsequently drains into collector channels and the episcleral veins. The principal source of resistance in this pathway is at the JCT and inner wall of Schlemm's canal. The inverse of this resistance is called outflow facility, or C in the Goldmann equation. The outflow facility, together with IOP, determines the flow rate of aqueous humor through the trabecular pathway. Outflow facility was shown to be reduced in primary open angle glaucoma (POAG) and ocular hypertensive patients (Larsson *et al.*, 1995; Toris *et al.*, 2002). Untreated ocular hypertensive patients also developed a progressive decrease in facility over a 10-year period (Linner, 1976).

2. Uveoscleral Pathway

In additional to the trabecular pathway, a fraction of the aqueous humor leaves the eye through the intercellular spaces of the iris root and ciliary muscle, and eventually empties into the scleral substance, the perivascular and perineural scleral spaces, and into the episcleral and orbital tissues. This outflow pathway is generally known as the uveoscleral or unconventional outflow. Its flow rate is designated as U in the Goldmann equation. At pressure levels greater than 7–10 mmHg, aqueous outflow through the uveoscleral pathway has a very low dependence of IOP, which is often not significantly different from zero. This relative pressure-independence is likely an integrated result of the complex resistance and capacitance characteristics of the multiple fluid compartments in the ocular tissues along the route. Recently, some feel that certain pharmacological agents, such as prostanoids, can alter uveoscleral outflow by increasing its pressure dependence (Weinreb, 2000). In ocular hypertensive patients, the calculated uveoscleral outflow rate was significantly lower than that in the age-matched controls (Toris *et al.*, 2002).

E. Measurement of Outflow Rates

Aqueous outflow in the living eye can be measured by several experimental methods. In animal studies, the most popular technique is the Bárány's two-level constant pressure perfusion technique (Bárány, 1964). This method requires cannulation of the anterior chamber. The inserted cannula is connected to a reservoir containing an artificial aqueous humor or other appropriate physiological solution. The reservoir is then elevated to generate raised IOP in the eye, which can be calculated by the density of the perfusion solution and the relative height of the reservoir, as well as confirmed by a second cannula connected to a pressure transducer. Driven by this pressure, the artificial aqueous humor in the reservoir slowly flows through the outflow pathways in the eye. The flow rate can be monitored by weighing the reservoir continuously. This procedure is then repeated with a different reservoir height (hence the term "two-level constant pressure"). The difference between the flow rates at the two pressure levels, divided by the pressure difference, is the outflow facility, which by definition is the pressure-sensitive fraction of aqueous outflow, and mainly contributed by the trabecular outflow.

In order to directly assess uveoscleral outflow, a labeled molecule, such as ^{125}I-albumin, is injected or continuously perfused into the anterior chamber. Animals are subsequently euthanized at different time points, ocular tissues of the uveoscleral pathway dissected, and levels of the labeled molecule in these tissues evaluated. Based on the initial concentration of the label in the anterior chamber and its rate of appearance in the ocular tissues, the uveoscleral outflow rate can be extrapolated (Bill, 1966). This method is cumbersome and labor-intensive. It also requires careful separation of the anatomical structures involved in the two outflow pathways. An alternative method to estimate uveoscleral outflow is by subtracting the pressure-dependent outflow rate from the aqueous formation rate, obtained from techniques such as the time-dependent dilution of a tracer molecule in the anterior chamber. This calculated and thus indirect approximation of uveoscleral outflow is typically less accurate or reproducible than data using direct determinations.

In human subjects, a noninvasive method, based on the theoretic work of Friedenwald (Friedenwald, 1937), has been used. In this technique, a tonograph monitors the decrease in IOP continuously while a known weight rests on the cornea. From the rate of IOP change and assuming that there are no alterations in the aqueous formation rate or episcleral venous pressure, the pressure-dependent aqueous outflow can be estimated. This method and its various modifications are not designed to measure uveoscleral outflow.

Based on these descriptions, it is quite clear that experimental determinations of aqueous outflow rate and facility are rather difficult, often with questionable accuracies. Most of these techniques cannot provide a direct assessment of uveoscleral outflow. These imperfections frequently are the sources of controversies in understanding of the biochemical, cellular, and physiological mechanisms in outflow regulation, as well as the exact effects of pharmacological agents on aqueous outflow.

F. Pathological Changes to Outflow Pathway in Glaucoma

The exact mechanism responsible for the decrease in aqueous outflow in POAG is still controversial. Nonetheless, an excessive amount of extracellular matrix (ECM) material accumulates in the TM of POAG eyes, which can cause a partial blockade of aqueous outflow (Segawa, 1979; Rohen, 1983; Lütjen-Drecoll *et al.*, 1986; Acott *et al.*, 1988). Several biochemical changes in ECM proteins and glycosaminoglycans (GAGs) occur in the glaucomatous TM. There are increased levels of the ECM protein fibronectin (Babizhayev and Brodskaya, 1989), decreased levels of hyaluronic acid and increased chondroitin sulfate and GAG degrading enzyme resistant material (Knepper *et al.*, 1996a), and increased levels of the ECM cross-linking enzyme tissue transglutaminase (Tovar-Vidales *et al.*, 2008). In addition, aqueous humor levels of the protease inhibitor plasminogen activator inhibitor-1 (PAI-1) are elevated in glaucoma patients (Dan *et al.*, 2005), which would also contribute to increased ECM deposition in the glaucomatous TM. Moreover, glucocorticoids, which have long been associated with POAG, also increase the synthesis and secretion of ECM molecules in cultured human and bovine TM cells and tissues (Johnson *et al.*, 1990; Steely *et al.*, 1992; Fujisawa, 1994; Dickerson *et al.*, 1998). Further support for the role of the ECM in glaucoma comes from the finding that transgenic mice that have mutations introduced into the Col1A1 gene to make this collagen more resistant to degradation accumulate collagen in the aqueous outflow pathways as well as develop elevated IOP and glaucomatous optic neuropathy (Aihara *et al.*, 2003; Mabuchi *et al.*, 2004). In addition, inhibition of matrix metalloproteinases (MMPs), enzymes that hydrolyze ECM, in the TM elevated IOP (Bradley *et al.*, 1998) and activation of MMPs decreased IOP (Bradley *et al.*, 1998; Pang *et al.*, 2003b) in perfusion cultured human eyes.

At the present time, the cause of ECM increase in the glaucomatous TM is not fully understood. Popular hypotheses include: (1) the TM cells in glaucoma patients are less active in their phagocytic activity, which leads to a reduced clearance of ECM (Bill, 1975); and (2) glaucoma patients have fewer TM cells (Alvarado *et al.*, 1984), which can result in a slower degradation of ECM. It is also possible that other functional abnormalities of the TM cells contribute to the accumulation of ECM. Interestingly, in addition to the TM, the ciliary muscle, especially the anterior tip and the surrounding elastic fibers, of glaucoma patients has a higher amount of ECM as well (Gabelt and Kaufman, 2005).

In addition to the enhanced accumulation of ECM, abnormal cytoskeletal changes in the TM may be involved in the pathogenesis of glaucoma as well. When cultured human TM cells were exposed to glucocorticoids, the F-actin microfilaments in cells progressively reorganize, forming geodesic dome-like structures, called cross-linked actin networks (CLANs) (Clark *et al.*, 1994). A similar development of CLANs was also reported in outflow tissues, such as the TM and Schlemm's canal endothelium, perfused with glucocorticoids (Clark *et al.*, 2005). More importantly, the CLANs are more abundant in

TM cells (Clark *et al.*, 1995) and TM tissues (Hoare *et al.*, 2008; Read *et al.*, 2007) derived from glaucoma donors. Furthermore, glaucoma eyes also appear to have more tangled F-actin fibers in the JCT and Schlemm's canal endothelial cells and more regions with punctuate actin distributions (Read *et al.*, 2007). Since cytoskeleton is vital to many cell functions, these anomalies in the outflow pathways of glaucoma patients likely contribute to the reduction in outflow facility.

Genetic studies in the past decade have identified specific mutations of the myocilin gene (*MYOC*) that are linked directly to juvenile- and adult-onset POAG (Stone *et al.*, 1997). Expression of myocilin in human TM cells can be enhanced by treatment with the glucocorticoid dexamethasone (Nguyen *et al.*, 1998; Clark *et al.*, 2001). Elevated IOP in *MYOC* glaucoma is a gain-of-function phenotype, since haploinsufficiency does not cause elevated IOP or glaucomatous optic neuropathy. Several recent studies indicate how *MYOC* mutations cause elevated IOP. Wild type (normal) myocilin is a secreted glycoprotein in the TM, and myocilin is found in the aqueous humor (Jacobson *et al.*, 2001). Glaucomatous mutations in *MYOC* cause myocilin to be retained within TM cells and prevent myocilin secretion (Jacobson *et al.*, 2001). This can lead to myocilin retention within the endoplasmic reticulum causing a stress response (Joe *et al.*, 2003). A second study has shown that mutant myocilin interacts with PTS1R due to mutation-induced misfolding and exposure of a cryptic peroxisomal targeting signal (PTS1) on the carboxy terminus of myocilin. The degree of this association between mutant myocilin and PTS1R correlates well with the clinical IOP phenotypes of *MYOC* glaucoma patients, with the more severe early-onset POAG mutations having a higher degree of association (Shepard *et al.*, 2007). More importantly, transduction of mouse eyes with mutant, but not wild type, human myocilin elevated IOP in mouse eyes, and IOP elevation was dependent on mutation-induced exposure of the normally cryptic PTS1 signal (Shepard *et al.*, 2007). This work provides the first true animal model of human glaucoma.

G. Current Glaucoma Therapies

Even though glaucoma manifests as an optic neuropathy and retinopathy, there are no clinically approved methods for direct neuroprotection or treatment of these aspects of the disease. Instead, all glaucoma therapies, both pharmacological and surgical, are presently directed at lowering IOP. As indicated by the abovementioned clinical trials, IOP reduction is effective in preventing and delaying the onset and progression of glaucoma.

Pharmacological agents have been successfully used to lower pressure in the eye for more than a century. Currently, glaucoma medications can be divided into six pharmacological classes (Table I). These drugs decrease IOP by either suppressing aqueous humor production or increasing aqueous outflow (Fig. 1).

H. Aqueous Production Suppressing Agents

Compounds that are known to reduce aqueous production include the β-adrenergic receptor antagonists (β-blockers), carbonic anhydrase inhibitors (CAIs), and α2-adrenergic receptor agonists. They are efficacious and safe, hence widely used in the treatment of glaucoma.

TABLE I

Current Glaucoma Therapies

Drug class	Compounds	Mechanisms of action
β-Blockers	Betaxolol Carteolol Levobunolol Metipranolol Timolol	• Block activation of β-adrenergic receptor • Decrease aqueous production
Carbonic anhydrase inhibitors	Acetazolamide Brinzolamide Dorzolamide	• Inhibit carbonic anhydrase • Decrease aqueous production
α2-Agonist	Apraclonidine Brimonidine	• Activate α2-adrenergic receptor • Decrease aqueous production • Increase trabecular/uveoscleral outflow
Cholinergics	Carbachol Echothiophate iodide Physostigmine Pilocarpine	• Activate muscarinic receptor • Increase trabecular outflow
Epinephrine and analogs	Dipivefrin Epinephrine	• Activate various adrenergic receptors • Decrease aqueous production • Increase outflow
Prostaglandin analogs	Bimatoprost Latanoprost Travoprost	• Activate FP prostaglandin receptor • Increase uveoscleral outflow

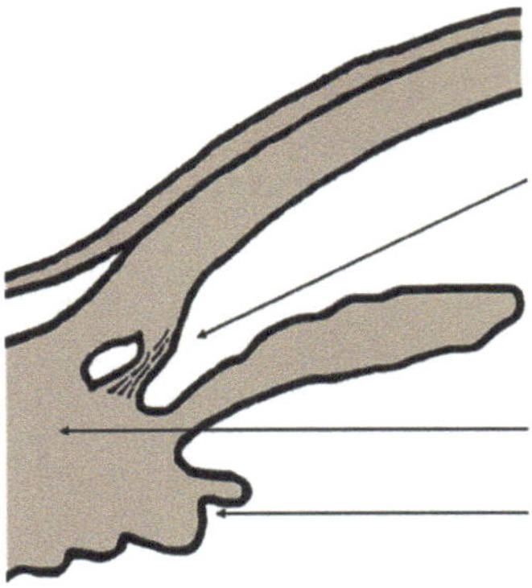

FIGURE 1 Mechanisms involved in pharmacological regulation of intraocular pressure.

1. β-Blockers

β-Blockers, such as betaxolol, carteolol, levobunolol, metipranolol, and timolol, are some of the most commonly used drugs in glaucoma therapy (Novack, 1987). They are competitive antagonists of β-adrenergic receptors. These agents block the binding of endogenous adrenergic neurotransmitters, i.e., norepinephrine and epinephrine, to the receptors in the ciliary processes and prevent their activation. The β-adrenergic receptors are coupled to adenylyl cyclase. Blockade of receptor activation prevents activation of the adenylyl cyclase, and thus a reduction in cyclic AMP levels in the ciliary epithelial cells, which subsequently suppresses the formation of aqueous humor. The cellular pathway(s) involved in the regulation of aqueous formation by cyclic AMP is still unclear. However, β-blockers are known to inhibit Na–K–Cl cotransport and the Na–K–ATPase in the ciliary epithelium. Moreover, this class of drugs can also reduce the blood–aqueous flux of ascorbate and inhibit plasma flow to the ciliary processes. All of these biological effects of β-blockers can contribute, at least partly, to their reduction in aqueous humor production.

2. Carbonic Anhydrase Inhibitors

For many years, oral administration of CAIs, such as acetazolamide, lowered IOP effectively. Unfortunately, their neurological, gastrointestinal, and metabolic untoward effects limit their acceptance by patients. The discovery and development of topically active CAIs, such as brinzolamide and dorzolamide, have minimized the systemic side effects and revitalized the use of this class of compounds in glaucoma treatment (Sugrue, 2000; Herkel and Pfeiffer, 2001). CAIs inhibit carbonic anhydrase, mainly isozyme II, in the ciliary epithelium and reduce the production of bicarbonate ion, which is a critical component for active ion transport in aqueous formation. A reduction in bicarbonate by CAIs diminishes sodium and fluid transport across the ciliary epithelium, and decreases aqueous humor production.

3. α_2-Agonists

α_2-Adrenergic agonists, e.g., apraclonidine and brimonidine, are effective IOP-lowering agents for both open and closed angle glaucomas (Robin, 1997; Adkins and Balfour, 1998). They selectively activate the α_2-adrenergic receptor of the ciliary epithelium. Activation of this receptor activates an inhibitory GTP-binding protein, which then inhibits the adenylyl cyclase. This leads to a reduction in intracellular cyclic AMP levels and eventually suppressed aqueous humor production. Studies also demonstrated that apraclonidine increases trabecular outflow and brimonidine stimulates uveoscleral outflow. The molecular and cellular mechanisms of these outflow effects are uncertain, but speculated to involve changes in contractility of the TM and ciliary muscle.

I. Aqueous Outflow-Increasing Agents

Drugs that increase aqueous outflow include the cholinergics, epinephrine analogs, and prostaglandin analogs (PGAs). They encompass the oldest and the most recent clinically approved compounds: physicians have been treating glaucoma with pilocarpine, a cholinergic agonist, for more than 100 years, whereas, the PGAs were approved for glaucoma treatment in recent years.

1. Cholinergics

The cholinergics are safe and effective IOP-lowering compounds (Hoyng and van Beek, 2000). They include muscarinic cholinergic agonists, such as pilocarpine and carbachol, and cholinesterase inhibitors, such as physostigmine and echothiophate iodide. These compounds activate the muscarinic cholinergic receptor, either directly as receptor agonists (e.g., pilocarpine and carbachol), or indirectly by reducing the enzymatic degradation of the endogenous agonist acetylcholine (e.g., physostigmine and echothiophate iodide). How receptor activation leads to reduction in IOP is still not clear. It has been hypothesized that activation of the muscarinic receptor causes contraction of certain ocular smooth muscles, notably the ciliary muscle and iris sphincter. Contraction of the longitudinal ciliary muscle pulls the scleral spur and TM posteriorly, enlarges the extracellular space in TM, and enhances trabecular outflow.

2. Epinephrine and Analogs

Epinephrine binds to and activates various adrenergic receptor subtypes in the eye. It and its prodrug dipivefrin lower IOP by both suppressing aqueous production and increasing aqueous outflow (Hoyng and van Beek, 2000). The multiple, complex cellular mechanisms involved in these pharmacological actions are yet to be fully delineated.

3. Prostaglandin Analogs

PGAs, such as latanoprost, travoprost, and bimatoprost, are very efficacious in lowering IOP, and therefore are popular compounds for the treatment of glaucoma (Hejkal and Camras, 1999; Linden, 2001). Latanoprost, travoprost, and bimatoprost are prodrugs. Their metabolic products activate the FP prostaglandin receptor with high affinity. In contrast, whether the fourth PGA, unoprostone, activates the FP receptor is still controversial (Griffin *et al.*, 1997; Bhattacherjee *et al.*, 2001). When compared with other PGAs, this compound is less efficacious with mean IOP reduction consistently less than latanoprost. Latanoprost and travoprost stimulate uveoscleral outflow without significantly affecting trabecular outflow or aqueous production. Bimatoprost slightly increases both the trabecular outflow and aqueous production, in addition to enhancing uveoscleral outflow.

Agonists of the FP receptor have been shown to cause changes in two biological functions in ocular structures related to aqueous outflow. First, FP receptor activation induces relaxation of the TM and ciliary muscle (Thieme *et al.*, 2006). This effect reduces tension and changes the topography of outflow pathways, which theoretically can improve uveoscleral outflow. Second, FP receptor agonists also upregulate the expression of MMPs, enzymes responsible for the hydrolysis of excessive ECM, in cultured human and monkey ciliary muscle cells. Activation of MMPs augments the rate of ECM degradation, which should open up extracellular space and decrease resistance to aqueous humor traveling through these spaces. After receiving PGA treatment for a year, monkey ciliary muscle had significant expansion in optically empty spaces between muscle bundles compared to untreated or vehicle-treated control animals (Richter *et al.*, 2003). These cellular and morphological changes likely play a role in the PGAs' effect on uveoscleral outflow.

J. Surgical Therapy

Surgical therapy for POAG is usually performed when medications have failed or are poorly tolerated and progressive glaucoma damage is still occurring. Results from the CIGTS show that both initial medical therapy and initial surgical therapy are valuable in lowering IOP and delaying progression of glaucomatous damage. There are many glaucoma surgical procedures, most of which are designed to improve aqueous outflow. They can be classified as incisional or laser surgical techniques. The incisional techniques, such as trabeculectomy and its variations, non-penetrating filtration procedures, and drainage tube implants, etc, aim to create new physical outflow pathways for aqueous humor. In contrast, the laser techniques, also known as laser trabeculoplasty, do not produce holes in the TM or burn into the lumen of Schlemm's canal. Instead, laser treatment releases cytokines, especially interleukin-1β and tumor necrosis factor-α, from TM cells. These cytokines modify various TM cell functions, including induction of MMP expression and degradation of ECM (Bradley *et al.*, 2000). Furthermore, laser-treated TM cells were observed to be more active in their phagocytic, migratory, and proliferative activities (Bylsma *et al.*, 1988; Alexander *et al.*, 1989). These cellular effects of laser treatment may be responsible, at least partly, for the increase of aqueous outflow and reduction in IOP. In addition, laser-induced scars may cause contraction of treated areas and consequently stretching of adjacent regions. This may produce enlarged extracellular spaces in the TM and improve aqueous outflow.

III. NEW APPROACHES FOR IOP LOWERING

A. Cytoskeleton-Disrupting Agents

Cells within the aqueous outflow pathway, such as the TM cells and the endothelial cells lining the Schlemm's canal, have an extensive cytoskeleton. The cytoskeleton is a complex network of protein filaments that extends throughout the cytoplasm. It can be classified into three principal types of protein filaments: actin microfilaments, microtubules, and intermediate filaments. Each type of cytoskeleton is formed by a different protein monomer and can be arranged into various structures according to its associated proteins. For example, certain associated proteins regulate the assembly of actin filaments and microtubules by controlling the rate and direction of polymerization. Other associated proteins connect filaments to one another or to other cell components, such as the plasma membrane, thus forming unique cytoskeletal architecture. Still other associated proteins interact with filaments to allow movements.

The ability of eukaryotic cells to preserve and perform their many coordinated cell functions depends on the cytoskeleton. It is responsible for the maintenance of cell shape, cell–cell junctions, cell–matrix interaction, adhesion, contraction, movement, as well as transport of intracellular organelles and molecules. As discussed earlier in this chapter, TM cells and Schlemm's canal endothelial cells in glaucoma eyes were shown to have abnormal cytoskeletal structures, which may be responsible, at least partially, for the reduction in aqueous outflow facility seen in POAG patients. Therefore, compounds that disrupt the cytoskeleton may modify these cell functions and local topography of the outflow pathway and consequently affect aqueous outflow (Tian *et al.*, 2000b). Indeed, drugs with this pharmacological action were shown effectively lower IOP in animal studies (Table II).

1. Cytochalasins

Cytochalasins block the polymerization and elongation of actin microfilaments by capping the barbed ends of the filaments. In perfused human eyes, cytochalasin D increased outflow facility with a duration of action of at least 14 h (Johnson, 1997). Perfusion of the anterior chamber of anesthetized monkeys with cytochalasin B doubled the aqueous outflow (Kaufman and Bárány, 1977; Robinson and Kaufman, 1991). Morphological evaluation of the treated eyes showed that these compounds caused TM distension and ruptures in the inner wall of Schlemm's canal, thereby enhancing outflow and washout of ECM (Svedbergh *et al.*, 1978).

2. Latrunculins

The latrunculins bind to monomeric G-actin and cause the disorganization of actin filaments. In human ocular tissues and cells, these compounds induced many cytoskeletal changes, such as reorganization of intermediate filaments in Schlemm's canal inner wall cells, disruption of actin microfilament integrity in TM cells, and substantial expansion of the space between the of Schlemm's canal inner wall and the trabecular collagen beams (Cai *et al.*, 1999; Cai *et al.*, 2000; Sabanay *et al.*, 2006). In addition, latrunculin B dose-dependently relaxed the ciliary muscle (Okka *et al.*, 2004a). All these actions can contribute to the enhanced outflow effect of latrunculins. In anesthetized monkeys and cultured porcine and human eyes, latrunculin A and/or B significantly improved aqueous humor outflow and decreased IOP for up to 24 h (Okka *et al.*, 2004b; Fan *et al.*, 2005; Ethier *et al.*, 2006).

3. Swinholide A

Swinholide A is a marine macrolide that severs actin filaments and sequesters actin dimers. Perfusion of this compound in the anterior chamber increased aqueous outflow facility in anesthetized monkeys (Tian *et al.*, 2001).

TABLE II

Cytoskeleton-Disrupting Agents

Compounds	Cellular mechanisms	Pharmacological effects
Cytochalasins	• Block actin filament elongation	• Cause TM distension and rupture of Schlemm's canal inner wall • Increase outflow in perfused human and monkey eyes
Latrunculins	• Bind to G-actin • Disrupt actin filaments	• Expand space between the Schlemm's canal inner wall and TM beams • Increase outflow in perfused monkey, pig, and human eyes
Swinholide A	• Severs actin filaments • Sequesters actin dimers	• Increases outflow in perfused monkey eyes
Ethacrynic acid and analogs	• Inhibit microtubule assembly • Reduce phosphorylation of focal adhesion kinase and paxillin • Disrupt TM cytoskeleton • Alter TM cell shape • Decrease TM focal adhesion	• Increase outflow facility in perfused bovine and human eyes • Lower IOP in rabbits, monkeys, and advanced glaucoma patients
H-7	• Inhibits protein kinases • Causes TM cytoskeleton reorganization • Relaxes TM cell	• Increases outflow in perfused human eyes
ROCK inhibitors	• Inhibit ROCK • Decrease actin stress fibers • Reduce myosin light-chain phosphorylation • Change TM cell shapes	• Widen the extracellular spaces in the TM • Increase outflow perfused porcine eyes • Lower IOP in the rabbit and monkey

4. Ethacrynic Acid

Ethacrynic acid inhibits microtubule assembly and reduces phosphorylation of focal adhesion kinase and paxillin, both focal adhesion proteins. Focal adhesion kinase and paxillin are important components in the integrin-mediated cell adhesion signaling pathways. In cultured TM cells, ethacrynic acid and analogs disrupted cytoskeleton, altered cell shape irreversibly, and decreased of focal adhesion (Shimazaki *et al.*, 2004b; Rao *et al.*, 2005c). Furthermore, ethacrynic acid inhibited the Na–K–Cl cotransport mechanism on TM cell membrane, which affected cell volume and permeability of the TM (O'Donnell *et al.*, 1995). Intracameral administration of ethacrynic acid lowered IOP in rabbits, monkeys, and advanced glaucoma patients (Melamed *et al.*, 1992). It increased outflow facility in *ex vivo* bovine and human eyes. Unfortunately, this compound does not penetrate the cornea well; its efficacy after topical ocular administration was minimal. And most importantly, its long-term use caused significant local untoward effects in animals. Recently, new and more efficacious derivatives of ethacrynic acid were synthesized and shown to lower IOP in cats and monkeys after intracameral injection (Shimazaki *et al.*, 2004a).

5. Protein Kinase Inhibitors

Even though many protein kinase inhibitors lower IOP in animal studies, their mechanism of action has not been conclusively demonstrated. Histological assessment suggests that they affect cytoskeleton of the TM or Schlemm's canal endothelial cells and increase outflow rate of aqueous humor. Initial work employed kinase inhibitors that have broad spectrum activity, inhibiting many kinases. Recently, a family of rho-associated coiled coil-forming kinase (ROCK) inhibitors was found to effectively lower IOP (Honjo *et al.*, 2001b).

6. Broad Spectrum Kinase Inhibitors

H-7, a broad spectrum protein kinase inhibitor effective in inhibiting the activities of many kinases, including protein kinase A, protein kinase C, protein kinase G, and ROCK, was shown to increase aqueous outflow in perfused human anterior segments (Bahler *et al.*, 2004) and in anesthetized monkey eyes (Tian *et al.*, 2004). In the TM of perfused eyes, cytoskeleton reorganization and cell relaxation were observed. The Schlemm's canal inner wall also exhibited protrusion and partial loss of endothelial cells (Bahler *et al.*, 2004; Sabanay *et al.*, 2004; Hu *et al.*, 2006). Other broad spectrum kinase inhibitors, HA1077, ML-7, ML-9, and chelerythrine, stimulated outflow facility in various animal models (Tian *et al.*, 2000a; Honjo *et al.*, 2001a, 2002).

7. ROCK Inhibitors

Recently, ROCK inhibitors, such as Y-27632, Y-39983, and H-1152, have been found to lower IOP efficaciously. Y-27632 and H-1152 increased outflow facility of aqueous humor in enucleated porcine eyes (Rao *et al.*, 2001; Rao *et al.*, 2005a). Y-27632 and Y-39983 also lowered IOP in the rabbit and monkey (Honjo *et al.*, 2001b; Waki *et al.*, 2001; Tian and Kaufman, 2005; Tokushige *et al.*, 2007).

ROCKs are kinases that can be activated by a cell signaling molecule Rho. Once activated, ROCKs modify functions of various proteins by phosphorylation. These target proteins, such as adducin, ezrin–radixin–moesin proteins, intermediate filament proteins, LIM kinases, myosin light chain phosphatase, sodium hydrogen exchanger NHE1, etc, play significant roles in cell shape, contractility, and focal adhesion. Consequently, compounds that inhibit ROCK activity are expected to affect these cell functions. In cultured human TM and Schlemm's canal cells, ROCK inhibitors were reported to decrease actin stress fibers, reduce myosin light-chain phosphorylation, and change cell shape, leading to a widening of the extracellular spaces in the TM, especially of the JCT (Rao *et al.*, 2005b; Rosenthal *et al.*, 2005; Koga *et al.*, 2006). These cellular changes likely contribute to the ocular hypotensive effect of ROCK inhibitors. It is important to note that ROCKs are present in many other tissues, notably vascular cells, and ROCK inhibitors may affect these other tissues which may produce side effects. In fact, conjunctival hyperemia and sporadic punctate subconjunctival hemorrhage have been observed in animals receiving topical administration of ROCK inhibitors (Tokushige *et al.*, 2007).

B. Activators of ECM Hydrolysis

As described earlier in this chapter, an excessive accumulation of ECM in the TM may be responsible for ocular hypertension seen in glaucoma patients. Reduction of the excessive ECM by stimulating its degradation should improve aqueous outflow and consequently lower IOP. ECM turnover in the TM is regulated by a family of zinc-containing extracellular neutral proteinases, called matrix metalloproteinases (MMPs) (Alexander *et al.*, 1991; Acott, 1992; Samples *et al.*, 1993). These enzymes are involved in normal development, reproduction, wound healing, and tissue remodeling, as well as in disease conditions, such as angiogenesis, tumor metastasis, arthritis, Sorsby's fundus dystrophy, and age-related macular degeneration (Clark, 1998). MMPs, synthesized as proenzymes, require proteolytic cleavage for activation. Their enzymatic activities are inhibited by endogeneous peptides known as tissue inhibitors of metalloproteinases (TIMPs). In the TM, several MMPs, e.g., MMP-1, MMP-2, MMP-3, and MMP-9, as well as TIMPs, such as TIMP-1 and TIMP-2, were detected (Alexander *et al.*, 1991; Samples *et al.*, 1993; Parshley *et al.*, 1996; Alexander *et al.*, 1998; Pang *et al.*, 2003b). MMPs are also present in human aqueous humor (Ando *et al.*, 1993).

The involvement of MMPs in the regulation of aqueous outflow has been demonstrated in numerous studies. *Ex vivo* perfusion of the human eye with purified MMP-3 alone or together with MMP-2 and MMP-9 increased outflow facility considerably (Bradley *et al.*, 1998). Correspondingly, inhibitors of MMPs, such as the TIMPs, minocylcine, or L-tryptophan hydroxamate, reduced aqueous outflow (Bradley *et al.*, 1998). Interleukin-1α, a cytokine known to increase MMP expression in the TM (Alexander *et al.*, 1991; Samples *et al.*, 1993; Pang *et al.*, 2003a), also enhanced outflow facility in human and rat eyes (Kee and Seo, 1997; Bradley *et al.*, 1998). Furthermore, an increase in the ocular expression of MMPs was proposed to mediate the ocular hypotensive effects of laser trabeculoplasty and PGAs (Parshley *et al.*, 1995, 1996; Lindsey *et al.*, 1996, 1997).

In addition to laser treatment and prostaglandins, there are other means to increase MMP activity in ocular tissues (Table III). Recently, it was discovered that stimulation of the activator protein-1 (AP-1) pathway in cultured human TM cell upregulated MMP-3 expression (Fleenor *et al.*, 2003). Subsequently, tert-butylhydroquinone, a small molecule AP-1 stimulator, was found to improve aqueous outflow in glaucoma and non-glaucoma donor eyes (Pang *et al.*, 2003a). This outflow effect correlated with an increase in MMP-3 levels in the TM cells. These data suggest that small molecules that can increase MMP expression are potentially valuable approaches to IOP regulation. It is interesting to note that the AP-1 pathway is by no means the only cell signaling pathway involved in MMP production. JNK and p38 MAP kinases were also shown to play important roles in the modulation of MMP expression (Kelley *et al.*, 2007b). Compounds that stimulate these molecular mechanisms may also prove useful.

A subset of ECM molecules, the glycosaminoglycans (GAGs), can be hydrolyzed by GAG-degrading enzymes. GAGs comprise hyaluronic acid, chondroitin sulfate, dermatan sulfate, keratan sulfate, heparan sulfate, and others. The GAGs likely contribute to IOP elevation in glaucoma. For example, the GAG profile in human glaucoma TM is significantly different

TABLE III

Activators of Extracellular Matrix Hydrolysis

Compounds	Cellular mechanisms	Pharmacological effects
Tert-butylhydro-quinone	• Activates AP-1 • Increases MMP3 expression in TM cell	• Increases outflow in perfused human eyes
AL-3037A	• Stimulates GAG hydrolysis	• Increases outflow in perfused bovine and human eyes • Lowers IOP in rabbits

from that of normal donors (Knepper *et al.*, 1996a,b). Similarly, experimental ocular hypertension caused changes in GAG profiles in the TM of several animal species (Knepper *et al.*, 1978, 1985a). Hence, Francois hypothesized that an increase in GAG in the TM is a causative factor of IOP elevation in POAG patients (Francois, 1975). This hypothesis is further supported by GAG-degrading enzymes, such as hyaluronidases and chondroitinases, that produced a decrease in IOP when perfused into bovine (Bárány and Scotchbrook, 1954; Pedler, 1956; Sawaguchi *et al.*, 1993), rabbit (Knepper, *et al.*,1985b), guinea pig (Melton and DeVille, 1960), dog (Van Buskirk and Brett, 1978), or monkey (Peterson and Jocson, 1974; Sawaguchi *et al.*, 1992) eyes. However, Hubbard and colleagues were unable to detect any outflow effect of intracameral injection of chrondroitinase or hyaluronidase in the monkey (Hubbard *et al.*, 1997).

Degradation of GAGs can also be catalyzed by metal ions in the presence of ascorbate. For example, a small molecule with a chelated ferric ion, sodium ferri ethylenediaminetetraacetate (AL-3037A), together with ascorbate, accelerated GAG depolymerization (Pang *et al.*, 2001). In perfused bovine, normal human, or glaucomatous human eyes, AL-3037A plus ascorbate produced marked increases in aqueous outflow (Pang *et al.*, 2001). In normal or dexamethasone-induced hypertensive rabbits, topical ocular administration of AL-3037A was effective in lowering IOP (Pang *et al.*, 2001). Addition of ascorbate is not necessary in rabbit studies, because the aqueous humor already contains approximately 1.1 mM of ascorbate. These results suggest that stimulation of GAG degradation may represent another new and practical method to treat glaucoma.

C. Adenosine Receptor Agonists and Antagonists

Three adenosine receptor subtypes, the A_1, A_{2A}, and A_3 receptors, have been shown to participate in the regulation of aqueous production and outflow. Compounds that activate or antagonize the activation of these receptors affect IOP in various species (Table IV). For example, A_1 receptor agonists, such as *N*-6-cyclohexyladenosine (CHA) and (*R*)-phenylisopropyladenosine (R-PIA), lower IOP in the mouse (Avila *et al.*, 2001), rabbit (Crosson, 1992, 1995, 2001; Crosson and Gray, 1994), and monkey (Tian *et al.*, 1997). In some studies, the ocular hypotensive effects of these compounds were preceded by a transient increase in IOP (Crosson and Gray, 1994; Tian *et al.*, 1997; Crosson, 2001). This was likely a result of non-specific activation of the A_{2A} receptor, because A_{2A} receptor antagonists abolished the initial ocular hypertensive effect without affecting the longer-lasting hypotensive effect of CHA and R-PIA (Tian *et al.*, 1997; Crosson, 2001).

TABLE IV
Adenosine Receptor Agonists and Antagonists

Drug Class	Cellular mechanisms	Pharmacological effects
A_1 Agonist	• Activates MMP	• Increases outflow in bovine and monkey eyes
	• Affects Schlemm's canal and TM cell function	• Lowers IOP in mouse, rabbit, and monkey
A_{2A} Agonist	• Affects Schlemm's canal and TM cell function	• Increases aqueous humor formation • Increases aqueous outflow • Increases or decreases IOP
A_3 Agonist	• Activates chloride channels on non-pigmented ciliary epithelial cells	• Increases aqueous humor formation • Increases IOP in mouse
A_3 Antagonist	• Blocks A3 receptor-mediated chloride channel activation on non-pigmented ciliary epithelial cells	• Decreases aqueous humor formation • Decreases IOP in mouse and monkey

Agonists of A_1 receptor lower IOP by increasing aqueous outflow, as demonstrated in perfused bovine (Crosson *et al.*, 2005) and monkey eyes (Tian *et al.*, 1997). The outflow effect in bovine eyes was reduced by a non-selective MMP inhibitor GM-6001, suggesting that MMP may play a role in the outflow effect of A_1 receptor agonists (Crosson *et al.*, 2005). In addition, A_1 receptor agonists also increase whole cell currents in cultured human Schlemm's canal inner-wall cells (Karl *et al.*, 2005) and increase calcium influx and decrease cell volume in cultured human TM cells (Fleischhauer *et al.*, 2003). Changes in functions of these cells may also contribute to the decrease in outflow resistance induced by the compounds.

Several laboratories demonstrated that A_{2A} receptor agonists, such as 2-*p*-(2-carboxyethyl)-phenethylamino-5′-benzyl)-adenosine (CGS-21680) and 2-(1-hexyn-1-yl)-adenosine (2-H-Ado), increase IOP in the mouse (Avila *et al.*, 2001), rabbit (Crosson, 1995; Crosson and Gray, 1996; Konno *et al.*, 2005b), and cat (Crosson and Gray, 1996), accompanied by an increase in aqueous production (Crosson and Gray, 1996; Konno *et al.*, 2005b). However, Konno and colleagues showed that CGS-21680 and other A_{2A} receptor agonists enhanced aqueous outflow and consequently lowered IOP in rabbits (Konno *et al.*, 2004, 2005a). The reason for this discrepancy in results is currently not understood. A_{2A} agonists have opposite effect compared to the

A_1 agonists in whole cell currents in Schlemm's canal inner-wall cells (Karl *et al.*, 2005), but increase calcium influx and decrease cell volume, similar to A_1 agonists, in cultured human TM cells (Fleischhauer *et al.*, 2003).

Agonists of the A_3 receptor, such as N^6-(3-iodobenzyl)-adenosine-5′-*N*-methyl-uronamide (IB-MECA), have been shown to increase IOP in the mouse (Avila *et al.*, 2001, 2002; Civan, 2003; Civan and Macknight, 2004; Yang *et al.*, 2005; Do and Civan, 2006). The cellular mechanism of action involves activation of chloride channels on the plasma membrane of non-pigmented ciliary epithelial cells, which leads to stimulation of aqueous humor formation (Mitchell *et al.*, 1999; Civan, 2003; Civan and Macknight, 2004; Do and Civan 2004). Apparently, the A_3 receptor-mediated aqueous humor formation is active in normal eyes, because antagonists of the A_3 receptor lower IOP in the mouse (Avila *et al.*, 2002; Do and Civan, 2006; Yang *et al.*, 2005) and monkey (Okamura *et al.*, 2004). Similarly, A3 receptor-knockout mice had a significantly lower IOP than their wild-type controls (Avila *et al.*, 2002).

D. Serotonergic Agonists

Serotonergic receptor agonists and antagonists have long been studied for their IOP effects. However, because of the multiple serotonergic receptor subtypes and the lack of specificity of many of the agents tested, pharmacological actions of many of these compounds on IOP and aqueous hydrodynamics are unclear and controversial. Recently, a 5-HT_2 agonist, R-(−)-1-(4-iodo-2,5-dimethoxyphenyl)-2-aminopropane (R-DOI), was shown to lower IOP in laser-induced ocular hypertensive and normotensive monkeys (May *et al.*, 2003a; Gabelt *et al.*, 2005). The effect of this compound was largely mediated by an increase in uveoscleral outflow (Gabelt *et al.*, 2005). These findings, together with the discoveries of 5-HT_2 receptors in the human ciliary body (Chidlow *et al.*, 2004) and cultured human TM cells (Sharif *et al.*, 2006), sparked a renewed interest in this pharmacological class of compounds. A series of selective 5-HT_2 agonists, such as S-(+)-1-(2-aminopropyl)-8,9-dihydropyrano[3,2-e]indole, the 1R,2R isomer of 1-(4-bromo-2,5-dimethoxyphenyl)-2-aminopropan-1-ol, 1-((S)-2-aminopropyl)-1H-indazol-6-ol, and tetrahydrobenzodifuran analogs, were synthesized and demonstrated to have high binding affinities for the 5-HT_{2A}, 5-HT_{2B}, and 5-HT_{2C} receptor subtypes, but not other receptors (May *et al.*, 2003a, 2006; Glennon *et al.*, 2004; Feng *et al.*, 2007). They were highly efficacious in lowering IOP in lasered monkey eyes (May *et al.*, 2003b, 2006; Glennon *et al.*, 2004; Feng *et al.*, 2007). Although their mechanism of action is still unknown, they are speculated to improve uveoscleral outflow analogous to R-DOI.

These new potential therapeutic compounds, cytoskeleton-disrupting agents, kinase inhibitors, ECM hydrolysis stimulators, adenosine analogs, and serotonin agonists, are exciting innovations in the arsenal for glaucoma treatment. Their pharmacological actions differ from the existing therapies, hence may provide additional or complementary advantages to the existing therapies. For example, these new drug classes may lower IOP in glaucoma patients who are refractory to the currently available medications. Alternatively, they may induce further reduction in IOP when used as an adjunctive agent. Furthermore, some of these compounds, such as the ECM hydrolysis stimulators, are designed to correct pathological changes (e.g., excessive accumulation of ECM) in the TM. They should be able to modify the underlying disease process instead of just managing the symptoms. Nevertheless, these new compounds may have their limitations as well. Since most of them have not been tested in humans, it is uncertain what untoward effects they may produce. Their pharmacological actions are unlikely specific to only the tissues involved in IOP regulation. This is especially true for the cytoskeleton-disrupting agents, kinase inhibitors, and ECM hydrolysis stimulators. These cellular targets are ubiquitous in most tissues. Chronic exposure to these compounds may generate unacceptable ocular or systemic toxicity. The adenosine analogs and serotonergic agonists, depending on the distribution of their respective receptors, may be more specific in their effects. However, they are also known to affect cardiovascular and neurological functions. Their future clinical value will be determined by the balance between beneficial and possible side effects.

It is important to note that, in addition to the pharmacological agents described here, others, such as compounds related to the angiotensin-renin system, cyclic AMP- and cyclic GMP-stimulating agents, have also been explored extensively and shown to regulate IOP. Nonetheless, because there is no or only minimal new development reported in recent years, they are not discussed in this chapter. Interested readers are encouraged to consult earlier publications.

The above-mentioned new therapeutic approaches have demonstrated their proof-of-concept and usefulness with specific compounds in animal models. Most recently, other novel, potential approaches for glaucoma treatment have been unveiled. They are still at the "therapeutic target" developmental stage. In other words, at the present time, no or only limited pharmacological agents are recognized to selectively modify these targets and proven to affect IOP in animals. These targets are identified as probably involved in the pathogenesis of glaucoma. Functional modification of them may correct the underlying disease etiology and/or pathology. Future studies on these new targets (described below), including growth factors, cytokines, and other cell signaling molecules, are expected to lead to revolutionary, innovative treatment principles.

E. Growth Factors

The TM makes and secretes a wide variety of growth factors and cytokines (Wordinger *et al.*, 1998; Wordinger and Clark, 2008). The TM is often a target of these growth factors and cytokines, which regulate a number of cellular functions and activities. Many growth factors and cytokines are involved in controlling ECM metabolism (both synthesis and degradation) as well as regulating TM cell proliferation and phagocytosis. Altered growth factor signaling has important implications in the pathogenesis of glaucomatous damage to the aqueous outflow system (Table V).

1. TGFβ

A key central mediator that regulates TM cell function is transforming growth factor beta (TGFβ). TM cells make and secrete both TGFβ1 and TGFβ2 (Tripathi *et al.*, 1993a,b), as well as express functional TGFβ receptors (Wordinger *et al.*, 1998). TGFβ also appears to play important roles in glaucoma pathogenesis. A number of studies have shown that aqueous humor levels of TGFβ2 are elevated in POAG patients (Lütjen-Drecoll, 2005), while aqueous humor levels of TGFβ1 are elevated in patients with pseudoexfoliation syndrome (Schlotzer Schrehardt *et al.*, 2001). Mechanical stress (e.g., cellular stretch or elevated IOP) can induce TGFβ expression in the TM (Liton *et al.*, 2005). Overall, TGFβ modifies TM cell ECM metabolism promoting ECM deposition, making it an interesting candidate for mediating glaucomatous damage to the TM.

The effects of TGFβ on the TM are manifold. Treating TM cells with TGFβ1 or TGFβ2 alters the expression of hundreds of genes (Zhao *et al.*, 2004) (Shepard AR *et al.* personal communication). TGFβ2 increases the expression of TM cell fibronectin, PAI-1, collagen types I, III, IV, thrombospondin-1 (TSP-1) (Fuchshofer *et al.*, 2007; Wordinger *et al.*, 2007), tissue transglutaminase (Welge-Lussen *et al.*, 2000; Tovar-Vidales *et al.*, 2007), hyaluronan synthase (Usui *et al.*, 2003), and proteoglycans such as versican (Zhao and Russell, 2005). TSP-1 activates latent TGFβ *in vivo*, and its induction by TGFβ further amplifies TGFβ effects. The increased synthesis of ECM components, their cross-linking by transglutaminase, and decreased degradation via elevated PAI-1 would lead to increased ECM deposition, which is a key feature in glaucomatous TM tissues (Rohen, 1983). In addition, TGFβ can change the composition of the TM ECM by differentially

TABLE V

Newly Identified Signaling Pathways

Pathways	Expression	Cellular mechanisms	Pharmacological effects
TGFβ	• Expressed by TM cells • Increased level in glaucoma aqueous humor	• Alters TM gene expression • Alters ECM metabolism • Alters TM cytoskeleton • Increases TM phagocytosis	• Decreases outflow in perfused human eyes • Over-expression increases mouse IOP
CTGF	• Present in aqueous humor • Increased expression in TM by TGFβ2	• Regulates ECM metabolism • Affects TM gene expression	• Over-expression increases mouse IOP
BMP/Gremlin	• BMPs, receptors, and antagonists are present in TM • TM expression of gremlin is increased in glaucoma	• Affect ECM metabolism	• BMP4 deficient mouse has elevated IOP • Gremlin increases IOP in perfused human eyes
Wnt/sFRP1	• Wnts, receptors, and antagonists are present in TM • TM expression of sFRP1 is increased in glaucoma	• Regulate cell differentiation • Alter TM gene expression	• sFRP1 decreases outflow in perfused human eyes • Over-expression of sFRP1 increases mouse IOP
ELAM-1	• Increased expression in glaucoma TM		
CD44	• Increased sCD44 in glaucoma aqueous humor	• sCD44 is toxic to TM cells	
Cochlin	• Increased expression in glaucoma TM	• Increases TM aggregation	
SAA2	• Increased expression in glaucoma TM and aqueous humor	• Affects TM gene expression	• Decreases outflow in perfused human eyes

altering the synthesis and turnover of specific ECM molecules and by inducing alternative splice variants of ECM molecules such as fibronectin (Li *et al.*, 2000) and versican (Zhao and Russell, 2005). In addition to effects on the ECM, TGFβ inhibits TM cell proliferation (Wordinger *et al.*, 1998), increases the expression of smooth muscle α-actin (Tamm *et al.*, 1996), and increases TM cell phagocytosis (Cao *et al.*, 2003). The TGFβ-induced alteration of the TM ECM, cytoskeleton, cell proliferation, and rate of phagocytosis all could account for the decreased TM cellularity seen in glaucoma (Alvarado *et al.*, 1984). TGFβ receptor inhibitors block the TGFβ2-induction of PAI-1 and fibronectin in TM cells (Fleenor *et al.*, 2006), but it is still not known whether this signaling in the TM is via the canonical Smad pathway or a non-Smad pathway(s).

In addition to this circumstantial evidence, there is even more compelling evidence for the involvement of TGFβ in glaucomatous damage to the outflow pathway. Perfusion culture of human and porcine anterior segments with TGFβ2 increased outflow resistance and elevated IOP (Gottanka *et al.*, 2004; Bachmann *et al.*, 2006; Fleenor *et al.*, 2006). The TGFβ2 increased outflow resistance was accompanied by the induction of fibronectin and PAI-1 (Bachmann *et al.*, 2006; Fleenor *et al.*, 2006) and the accumulation of fine fibrillar material within the JCT (Gottanka *et al.*, 2004). Very recent studies have shown that transduction of rat and mouse eyes by intraocular injection with an adenoviral expression vector encoding a bioactivated form of TGFβ2 significantly elevated IOP over the course of weeks (Clark *et al.*, 2006). It is hoped that this will be a valuable animal model that mimics many features of human glaucoma.

2. CTGF

Connective Tissue Growth Factor (CTGF) is one of the genes induced by TGFβ treatment of TM cells (Shepard *et al.*, 2003; Fuchshofer *et al.*, 2007). CTGF is also present in the aqueous humor (van Setten *et al.*, 2002), and aqueous humor levels of CTGF are elevated in patients with pseudoexfoliation glaucoma (Ho *et al.*, 2005). Although CTGF plays important roles in development, it is generally expressed in adult tissues only during pathological states of fibrogenesis. CTGF can regulate ECM metabolism, and its effects are often synergistic to TGFβ. In fact, some of the TGFβ effects on the TM may be mediated by CTGF. Over-expression of CTGF in the anterior segment of rodent eyes elevates IOP (Shepard, personal communication), further supporting a role for CTGF in the regulation of aqueous humor outflow.

3. BMP

Bone morphogenic proteins (BMPs) were originally identified as osteogenic growth factors, but they are expressed in a variety of tissues where they regulate embryogenesis and other cell functions (Wordinger and Clark, 2007). Four BMPs, all three BMP receptors, and several BMP antagonists are expressed in adult TM cells and tissues (Wordinger *et al.*, 2002). Although *Bmp4* null mice are not viable, mice with a heterozygous *Bmp4* deficiency develop elevated IOP (Chang *et al.*, 2001). Several recent studies support an important functional role for BMPs in the TM. BMPs can block the effects of TGFβ on ECM metabolism in cultured TM cells. Fuchshofer and colleagues showed that BMP7 inhibited TGFβ2 induction of CTGF, fibronectin, TSP-1, collagens, and PAI-1 (Fuchshofer *et al.*, 2007). Wordinger *et al.* showed that BMP4 was also able to block TGFβ2 induction of fibronectin and PAI-1 (Wordinger *et al.*, 2007). Further evidence for the involvement of BMP signaling in the TM is the finding of increased expression of the BMP antagonist Gremlin in glaucomatous TM cells (Wordinger *et al.*, 2007). The addition of Gremlin to the medium of perfusion cultured human anterior segments significantly increased IOP, demonstrating that perturbation of BMP signaling affects the outflow pathway.

4. Wnt

Another new signaling pathway that regulates IOP has been recently discovered. The Wnt signaling pathway plays important roles in embryogenesis and morphogenesis, including development of the eye. Wnt is a secreted extracellular protein that binds to frizzled membrane receptors to signal via three different pathways. The canonical pathway involves β-catenin and Tcf transcription factors to regulate gene expression. Adult TM cells and tissues express all the components required for Wnt signaling (Wnts, FZDs, coreceptor LRP5, β-catenin, Tcf, and several Wnt antagonists) (Clark *et al.*, 2007; Wang *et al.*, 2008). Increased expression of the Wnt signaling antagonist secreted frizzled-related protein-1 (sFRP1) was found in studies comparing gene expression between normal and glaucomatous TM (Clark *et al.*, 2007; Wang *et al.*, 2008). To determine whether the TM has a functional Wnt signaling pathway and whether Wnt regulates IOP, sFRP1 was added to the perfusion medium of *ex vivo* perfusion cultured human anterior segments. Perfusion of human anterior segments with sFRP1 decreased the aqueous outflow facility and also decreased β-catenin protein levels in the TM. Suppression of Wnt signaling would cause decreased β-catenin levels due to enhanced β-catenin phosphorylation by glycogen synthase kinase-3β (GSK3β) and subsequent degradation of β-catenin by the proteosome. Increased expression of sFRP1 by transduction of mouse eyes with an Ad5. sFRP1 expression vector caused elevated IOP, and the degree of IOP

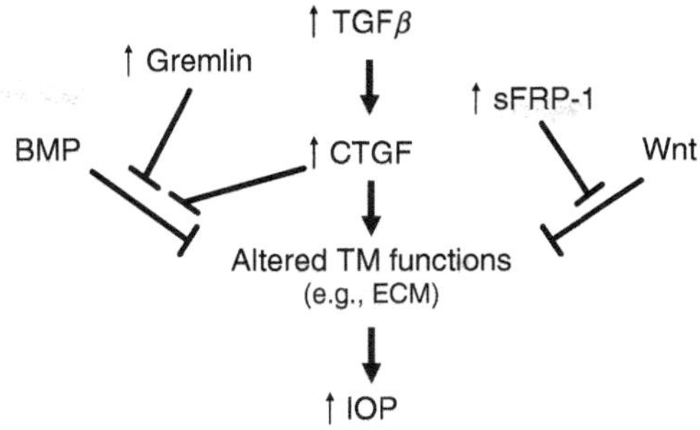

FIGURE 2 Potential interactions between TGFβ, CTGF, Wnt, and BMP pathways in the regulation of intraocular pressure.

elevation correlated with aqueous humor levels of sFRP1. Topical ocular administration of a GSK3β inhibitor reversed this sFRP-1-mediated ocular hypertension (Wang *et al.*, 2008). These results clearly demonstrate that the TM contains a functional Wnt signaling pathway that regulates IOP, which appears to be altered in glaucoma.

Growth factors in most cases do not work independently, and there often is a complex interplay between the growth factor signaling pathways. This appears to be the case for TGFβ, CTGF, BMP, and Wnt signaling, at least during development. During chondorogenesis and osteogenesis, BMP2 can induce β-catenin mediated signaling via Wnt ligands (Chen *et al.*, 2007), and CTGF is an important target of Wnt and BMP signaling in the differentiation of mesenchymal stem cells (Luo *et al.*, 2004). Wnt and BMP pathways also appear to interact in cancer cell differentiation and tumor suppression (Nishanian *et al.*, 2004). Interactions between the TGFβ and BMP pathways already have been shown in the adult TM (Fuchshofer *et al.*, 2007; Wordinger *et al.*, 2007), and studies are underway to determine whether there are similar interactions involving the Wnt and BMP/TGFβ signaling pathways in the TM (Fig. 2).

F. Cytokines and Other New Pathways

1. IL-1

Interleukin-1 (IL-1) is one of the cytokines induced in the anterior segment by laser trabeculoplasty (Bradley *et al.*, 2000; Alvarado *et al.*, 2005), and IL-1 plays a key role in regulating matrix metalloproteinase expression in the TM via several different signaling pathways, including AP-1 (Fleenor *et al.*, 2003), JNK (Hosseini *et al.*, 2006), and p38MAPK (Kelley *et al.*, 2007a).

Commensurate with this MMP activation, IL-1 also increases trabecular outflow *ex vivo* in perfusion cultured anterior segments (Bradley *et al.*, 1998) and *in vivo* when injected into rat eyes (Kee and Seo, 1997). IL-1 also induces ELAM-1 (E-Selectin), and ELAM-1 mRNA and protein expression is increased in glaucomatous TM cells and tissues (Wang *et al.*, 2001). Liton and colleagues independently confirmed that ELAM-1 gene expression was elevated in glaucomatous TM tissues (Liton *et al.*, 2006). Interestingly, a constitutively active IL-1 signaling pathway is present in glaucomatous TM cells, which is no longer regulated by IL-1 autocrine signaling (Zhang *et al.*, 2006). IL-1/NF-κB signaling also protects cultured TM cells from oxidation-induced apoptosis (Wang *et al.*, 2001).

2. CD44

Levels of the glycosaminoglycan hyaluronan are lower in glaucomatous TM tissues (Knepper *et al.*, 1996a), which has led to additional investigation of the hyaluronan receptor CD44. Immunohistochemical analysis of normal and glaucomatous eyes showed decreased levels of membrane associated CD44H in the glaucomatous TM (Knepper *et al.*, 1998). This was accompanied by increased aqueous humor levels of soluble CD44 (sCD44) in POAG patients compared with non-glaucomatous controls (Knepper *et al.*, 2005; Nolan *et al.*, 2007). Aqueous humor sCD44 concentrations in POAG patients were significantly correlated with the degree of visual field loss in POAG patients (Nolan *et al.*, 2007). There are multiple isoforms of sCD44 in aqueous humor due to varying degrees of phosphorylation, and there are greater levels of hypophosphorylated sCD44 in glaucomatous aqueous humor (Knepper *et al.*, 2005). sCD44 is toxic to cultured TM cells and retinal ganglion cells but not several other cell types (Choi *et al.*, 2005), with hypophosphorylated sCD44 being more toxic than standard sCD44 (Knepper *et al.*, 2005). Recently, Shepard and colleagues used viral vectors to over express CD44 in the anterior segments of mouse eyes, causing a significant increase in IOP that correlated with aqueous humor levels of sCD44 (Shepard *et al.*, 2008). This intriguing new pathway warrants additional study to determine whether sCD44 is directly involved in the generation of glaucomatous ocular hypertension and if so, to discover the molecular mechanisms involved.

3. Cochlin

Proteomics analysis of normal versus glaucomatous TM tissues led to the discovery of increased cochlin levels in the glaucoma samples. Cochlin is an ECM protein highly expressed in the inner ear, but is also expressed in the eye. PAGE/MS analysis of TM proteins extracted from normal donor eyes and from trabeculectomy specimens from glaucoma patients showed elevated cochlin in the glaucoma specimens, and this finding was confirmed by Western immunoblot and immunohistochemical analyses (Bhattacharya *et al.*, 2005a). The addition of purified cochlin to cultured TM cells caused TM cell aggregation. In addition, levels of cochlin protein also were increased in the TM of glaucomatous DBA/2J mice compared to non-glaucomatous mouse strains (Bhattacharya *et al.*, 2005a,b). However, no differences in normal versus glaucomatous TM tissue cochlin gene expression were found in a study comparing gene expression in TM tissues (Liton *et al.*, 2006). Further studies are needed to determine whether cochlin plays a causal pathogenic role in the development of ocular hypertension and glaucoma.

4. SAA

Serum amyloid A2 (SAA2) gene and protein expression are increased in glaucomatous TM cells and tissues. SAA2 is an acute phase response apolipoprotein made by the liver, and serum levels of SAA2 can increase >1000 fold during acute trauma or infection. However, chronically elevated levels of SAA2 can lead to amyloid deposition and amyloidosis. SAA2 is also expressed in the eye, and Wang and colleagues have recently shown increased expression of SAA2 mRNA in glaucomatous TM cells and tissues (Wang *et al.*, 2008). SAA protein levels are also significantly elevated in TM tissues and in the aqueous humor of glaucomatous patients compared to controls. The addition of recombinant SAA to the medium of perfusion cultured human eyes elevated IOP (Wang *et al.*, 2008), and transduction of mouse eyes with an SAA viral expression vector also elevated IOP (Wang, unpublished observation). In theory, SAA could be causing increased outflow resistance via amyloid deposition, but there was no evidence of amyloid deposition in the outflow pathway of either human glaucoma eyes or the SAA-induced ocular hypertensive mouse eyes (Wang, unpublished observation). *In vivo* or *in vitro* treatment of TM cells with SAA altered TM gene expression (Wang *et al.*, 2008), which may be responsible for the glaucomatous changes to the TM.

IV. FUTURE THERAPEUTIC OPPORTUNITIES

Almost all current glaucoma therapy lowers IOP either by suppressing aqueous humor formation or by increasing uveoscleral or trabecular outflow. However, none of these therapies address the underlying cause of the increased outflow resistance that occurs in glaucomatous eyes. Our inability to alter glaucomatous disease progression in the glaucomatous outflow pathway may be one important reason that glaucoma patients become "resistant" to their medical therapies over time because outflow damage continues to progress.

There is an opportunity for a paradigm shift in the treatment of glaucoma. A more thorough understanding of glaucomatous pathogenic damage to the aqueous outflow pathways should lead to the discovery and development of disease modifying therapeutic agents, which halt or perhaps even reverse the disease process and restore normal aqueous outflow. There would be a number of therapeutic advantages to this approach. Glaucoma patients have a greater degree of diurnal IOP fluctuation (Caprioli, 2007), in large degree because of their compromised outflow facility. Therapeutic normalization of the outflow facility would eliminate this glaucomatous flux in pressure. In addition, noncompliance (failure to adhere to therapy) common among glaucoma patients is a major issue in the effectiveness of current hypotensive therapy (Olthoff *et al.*, 2005; Schwartz, 2005). Long term compliance may be less of an issue if the disease process is reversed. Once the outflow facility is normalized, it may take months to years for the disease process to cause sufficient damage to again elevate IOP.

There also are significant opportunities to improve glaucoma drug delivery. Almost all glaucoma medications are administered as topical ocular drops once to three times a day. However, overall compliance (a patient's adherence to therapy) is a major issue and significantly impacts therapeutic success. In addition, compliance decreases with the administration of multiple medications, and a good percentage of glaucoma patients are not adequately controlled by a single medication. Physician administered sustained delivery of glaucoma medications would remove compliance from being a major issue in the successful treatment of patients. Although there have been some efforts in alternative strategies for the delivery of glaucoma medications, a significant amount of additional work will be required.

It is now technically possible to deliver potentially therapeutic genes into the anterior segment using viral expression vectors that selectively transduce TM cells. Several different viral vectors with tropism for the TM in rats, monkeys, and humans, have been identified including Ad5 (Borras *et al.*, 1999, 2001), HSV (Liu *et al.*, 1999), scAAV2 (Borras *et al.*, 2006), and FIV (Loewen *et al.*, 2001). Some anterior segment inflammation and limited duration transgene expression occur with both adenovirus and herpes simplex virus expression vectors (Hoffman *et al.*, 1997; Kaufman *et al.*, 1999). In contrast, there is less inflammation and longer term transgene expression with AAV and FIV vectors. Viral delivery of several transgenes, including stromelysin in rat eyes (Kee *et al.*, 2001), a dominant negative form of Rho kinase in cultured human anterior segments (Rao *et al.*, 2005a), exoenzyme C3 transferase in organ cultured monkey eyes (Liu *et al.*, 2005), and caldesmon in cultured human and monkey anterior segments (Gabelt *et al.*, 2006), have increased aqueous outflow, providing proof-of-principle for this potential new therapeutic approach. However, a number of basic questions remain with this therapeutic approach, including selection of the appropriate transgene to express, titration of the amount of gene expression to provide the desired efficacy, and long-term safety.

References

Acott, T. S. (1992). Trabecular extracellular matrix regulation. *In* "Pharmacology of Glaucoma" (S. M. Drance, E. M. Van Buskirk, and A. H. Neufeld, eds.), pp. 125–127. Williams & Wilkins, Baltimore.

Acott, T. S., Kingsley, P. D., Samples, J. R., and Van Buskirk, E. M. (1988). Human trabecular meshwork organ culture: Morphology and glycosaminoglycan synthesis. *Invest. Ophthalmol. Vis. Sci.* **29,** 90–100.

Adkins, J. C., and Balfour, J. A. (1998). Brimonidine. A review of its pharmacological properties and clinical potential in the management of open-angle glaucoma and ocular hypertension. *Drugs Aging* **12,** 225–241.

AGIS-Investigators. (2000). The advanced glaucoma intervention study (AGIS): 7. The relationship between control of intraocular pressure and visual field deterioration. *Am. J. Ophthalmol.* **130,** 429–440.

Aihara, M., Lindsey, J. D., and Weinreb, R. N. (2003). Ocular hypertension in mice with a targeted type I collagen mutation. *Invest. Ophthalmol. Vis. Sci.* **44,** 1581–1585.

Alexander, R. A., Grierson, I., and Church, W. H. (1989). The effect of argon laser trabeculoplasty upon the normal human trabecular meshwork. *Graefes Arch. Clin. Exp. Ophthalmol.* **227,** 72–77.

Alexander, J. P., Samples, J. R., Van Buskirk, E. M., and Acott, T. S. (1991). Expression of matrix metalloproteinases and inhibitor by human trabecular meshwork. *Invest. Ophthalmol. Vis. Sci.* **32,** 172–180.

Alexander, J. P., Samples, J. R., and Acott, T. S. (1998). Growth factor and cytokine modulation of trabecular meshwork matrix metalloproteinase and TIMP expression. *Curr. Eye Res.* **17,** 276–285.

Alvarado, J., Murphy, C., and Juster, R. (1984). Trabecular meshwork cellularity in primary open-angle glaucoma and nonglaucomatous normals. *Ophthalmology* **91,** 564–579.

Alvarado, J. A., Alvarado, R. G., Yeh, R. F., Franse Carman, L., Marcellino, G. R., and Brownstein, M. J. (2005). A new insight into the cellular regulation of aqueous outflow: How trabecular meshwork endothelial cells drive a mechanism that regulates the permeability of Schlemm's canal endothelial cells. *Br. J. Ophthalmol.* **89,** 1500–1505.

Ando, H., Twining, S. S., Yue, B. Y., Zhou, X., Fini, M. E., Kaiya, T., Higginbotham, E. J., and Sugar, J. (1993). MMPs and proteinase inhibitors in the human aqueous humor. *Invest. Ophthalmol. Vis. Sci.* **34,** 3541–3548.

Avila, M. Y., Stone, R. A., and Civan, M. M. (2001). A1-, A2A- and A3-subtype adenosine receptors modulate intraocular pressure in the mouse. *Br. J. Pharmacol.* **134,** 241–245.

Avila, M. Y., Stone, R. A., and Civan, M. M. (2002). Knockout of A3 adenosine receptors reduces mouse intraocular pressure. *Iovs* **43,** 3021–3026.

Babizhayev, M. A., and Brodskaya, M. W. (1989). Fibronectin detection in drainage outflow system of human eyes in ageing and progression of open-angle glaucoma. *Mech. Ageing Dev.* **47,** 145–157.

Bachmann, B., Birke, M., Kook, D., Eichhorn, M., and Lütjen-Drecoll, E. (2006). Ultrastructural and biochemical evaluation of the porcine anterior chamber perfusion model. *Invest. Ophthalmol. Vis. Sci.* **47,** 2011–2020.

Bahler, C. K., Hann, C. R., Fautsch, M. P., and Johnson, D. H. (2004). Pharmacologic disruption of Schlemm's canal cells and outflow facility in anterior segments of human eyes. *Invest. Ophthalmol. Vis. Sci.* **45,** 2246–2254.

Bárány, E. H. (1964). Simultaneous measurements of changing intraocular pressure and outflow facility in the vervet monkey by constant pressure infusion. *Invest. Ophthalmol. Vis. Sci.* **2,** 135–143.

Bárány, E. H., and Scotchbrook, S. (1954). Influence of testicular hyaluronidase on the resistance to flow through the angle of the anterior chamber. *Acta Physiol. Scand.* **30,** 240–248.

Bengtsson, B., Leske, M. C., Hyman, L., Heijl, A., and Early Manifest Glaucoma Trial Group. (2007). Fluctuation of intraocular pressure and glaucoma progression in the early manifest glaucoma trial. *Ophthalmology* **114,** 205–209.

Bergea, B., Bodin, L., and Svedbergh, B. (1999). Impact of intraocular pressure regulation on visual fields in open-angle glaucoma. *Ophthalmology* **106,** 997–1005.

Bhattacharya, S. K., Annangudi, S. P., Salomon, R. G., Kuchtey, R. W., Peachey, N. S., and Crabb, J. W. (2005a). Cochlin deposits in the trabecular meshwork of the glaucomatous DBA/2J mouse. *Exp. Eye Res.* **80,** 741–744.

Bhattacharya, S. K., Rockwood, E. J., Smith, S. D., Bonilha, V. L., Crabb, J. S., Kuchtey, R. W., Robertson, N. G., Peachey, N. S., Morton, C. C., and Crabb, J. W. (2005b). Proteomics reveal Cochlin deposits associated with glaucomatous trabecular meshwork. *J. Biol. Chem.* **280,** 6080–6084.

Bhattacherjee, P., Paterson, C. A., and Percicot, C. (2001). Studies on receptor binding and signal transduction pathways of unoprostone isopropyl. *J. Ocul. Pharmacol. Ther.* **17,** 433–441.

Bill, A. (1966). Conventional and uveoscleral drainage of aqueous humor in the cynomolgus monkey (Macaca irus) at normal and high intraocular pressures. *Exp. Eye Res.* **5,** 45–54.

Bill, A. (1975). Editorial: The drainage of aqueous humor. *Invest. Ophthalmol.* **14,** 1–3.

Borras, T., Rowlette, L. L., Erzurum, S. C., and Epstein, D. L. (1999). Adenoviral reporter gene transfer to the human trabecular meshwork does not alter aqueous humor outflow. Relevance for potential gene therapy of glaucoma. *Gene Ther.* **6,** 515–524.

Borras, T., Gabelt, B. T., Klintworth, G. K., Peterson, J. C., and Kaufman, P. L. (2001). Non-invasive observation of repeated adenoviral GFP gene delivery to the anterior segment of the monkey eye *in vivo*. *J. Gene Med.* **3,** 437–449.

Borras, T., Xue, W., Choi, V. W., Bartlett, J. S., Li, G., Samulski, R. J., and Chisolm, S. S. (2006). Mechanisms of AAV transduction in glaucoma-associated human trabecular meshwork cells. *J. Gene Med.* **8,** 589–602.

Bradley, J. M., Vranka, J., Colvis, C. M., Conger, D. M., Alexander, J. P., Fisk, A. S., Samples, J. R., and Acott, T. S. (1998). Effect of matrix metalloproteinases activity on outflow in perfused human organ culture. *Invest. Ophthalmol. Vis. Sci.* **39,** 2649–2658.

Bradley, J. M., Anderssohn, A. M., Colvis, C. M., Parshley, D. E., Zhu, X. H., Ruddat, M. S., Samples, J. R., and Acott, T. S. (2000). Mediation of laser trabeculoplasty-induced matrix metalloproteinase expression by IL-1beta and TNFalpha. *Invest. Ophthalmol. Vis. Sci.* **41,** 422–430.

Bylsma, S. S., Samples, J. R., Acott, T. S., and Van Buskirk, E. M. (1988). Trabecular cell division after argon laser trabeculoplasty. *Arch. Ophthalmol.* **106,** 544–547.

Cai, S., Liu, X., Glasser, A., Croft, M. A., Polansky, J. R., Fauss, D. J., and Kaufman, P. L. (1999). Latrunculin (Lat)-A effects on human trabecular meshwork (HTM) cells. *Invest. Opthalmol. Vis. Sci.* **40,** S505.

Cai, S., Liu, X., Glasser, A., Volberg, T., Filla, M., Geiger, B., Polansky, J. R., and Kaufman, P. L. (2000). Effect of latrunculin-A on morphology and actin-associated adhesions of cultured human trabecular meshwork cells. *Mol. Vis.* **6,** 132–143.

Cao, Y., Wei, H., Pfaffl, M. W., and Da, B. (2003). Effect of transforming growth factor beta 2 on phagocytosis in cultured bovine trabecular meshwork cells. *Ophthalmologe* **100,** 535–538.

Caprioli, J. (2007). Intraocular pressure fluctuation: An independent risk factor for glaucoma? *Arch. Ophthalmol.* **125,** 1124–1125.

Chang, B., Smith, R. S., Peters, M., Savinova, O. V., Hawes, N. L., Zabaleta, A., Nusinowitz, S., Martin, J. E., Davisson, M. L., Cepko, C. L., Hogan, B. L., and John, S. W. (2001). Haploinsufficient Bmp4 ocular phenotypes include anterior segment dysgenesis with elevated intraocular pressure. *BMC Genet.* **2,** 18.

Chen, Y., Whetstone, H. C., Youn, A., Nadesan, P., Chow, E. C., Lin, A. C., and Alman, B. A. (2007). Beta-catenin signaling pathway is crucial for bone morphogenetic protein 2 to induce new bone formation. *J. Biol. Chem.* **282,** 526–533.

Chidlow, G., Hiscott, P. S., and Osborne, N. N. (2004). Expression of serotonin receptor mRNAs in human ciliary body: A polymerase chain reaction study. *Graefes Arch. Clin. Exp. Ophthalmol.* **242,** 259–264.

Choi, J., Miller, A. M., Nolan, M. J., Yue, B. Y., Thotz, S. T., Clark, A. F., Agarwal, N., and Knepper, P. A. (2005). Soluble CD44 is cytotoxic to trabecular meshwork and retinal ganglion cells *in vitro*. *Invest. Ophthalmol. Vis. Sci.* **46,** 214–222.

Civan, M. M. (2003). The fall and rise of active chloride transport: Implications for regulation of intraocular pressure. *J. Exp. Zool.* **300a,** 5–13.

Civan, M. M., and Macknight, A. D. (2004). The ins and outs of aqueous humour secretion. *Exp. Eye Res.* **78,** 625–631.

Clark, A. F. (1998). New discoveries on the roles of matrix metalloproteinases in ocular cell biology and pathology. *Invest. Ophthalmol. Vis. Sci.* **39,** 2514–2516.

Clark, A. F., Wilson, K., McCartney, M. D., Miggans, S. T., Kunkle, M., and Howe, W. (1994). Glucocorticoid-induced formation of cross-linked actin networks in cultured human trabecular meshwork cells. *Invest. Ophthalmol. Vis. Sci.* **35,** 281–294.

Clark, A. F., Miggans, S. T., Wilson, K., Browder, S., and McCartney, M. D. (1995). Cytoskeletal changes in cultured human glaucoma trabecular meshwork cells. *J. Glaucoma* **4,** 183–188.

Clark, A. F., Steely, H. T., Dickerson, J. E., Jr., English Wright, S., Stropki, K., McCartney, M. D., Jacobson, N., Shepard, A. R., Clark, J. I., Matsushima, H., Peskind, E. R., Leverenz, J. B., *et al.* (2001). Glucocorticoid induction of the glaucoma gene MYOC in human and monkey trabecular meshwork cells and tissues. *Invest. Ophthalmol. Vis. Sci.* **42,** 1769–1780.

Clark, A. F., Brotchie, D., Read, A. T., Hellberg, P., English Wright, S., Pang, I.-H., Ethier, C. R., and Grierson, I. (2005). Dexamethasone alters F-actin architecture and promotes cross-linked actin network formation in human trabecular meshwork tissue. *Cell Motil. Cytoskeleton* **60,** 83–95.

Clark, A. F., Millar, J. C., Pang, I.-H., Jacobson, N., and Shepard, A. (2006). Adenoviral gene transfer of active human transforming growth factor-β2 induces elevated intraocular pressure in rats. *ARVO Abstract* 4771.

Clark, A. F., McNatt, L. G., Hellberg, P. E., Millar, J. C., Rubin, J. S., Pang, I.-H., and Wang, W.-H. (2007). A functional Wnt pathway in the trabecular meshwork that regulates intraocular pressure. *ARVO Abstract* 1142.

CNTGS Group. (1998). Comparison of glaucomatous progression between untreated patients with normal-tension glaucoma and patients with therapeutically reduced intraocular pressures. *Am. J. Ophthalmol.* **126,** 487–497.

Crosson, C. E. (1992). Ocular hypotensive activity of the adenosine agonist (R)-phenylisopropyladenosine in rabbits. *Curr. Eye Res.* **11,** 453–458.

Crosson, C. E. (1995). Adenosine receptor activation modulates intraocular pressure in rabbits. *J. Pharmacol. Exp. Ther.* **273,** 320–326.

Crosson, C. E. (2001). Intraocular pressure responses to the adenosine agonist cyclohexyladenosine: Evidence for a dual mechanism of action. *Invest. Opthalmol. Vis. Sci.* **42,** 1837–1840.

Crosson, C. E., and Gray, T. (1994). Modulation of intraocular pressure by adenosine agonists. *J. Ocul. Pharmacol.* **10,** 379–383.

Crosson, C. E., and Gray, T. (1996). Characterization of ocular hypertension induced by adenosine agonists. *Invest. Ophthalmol. Vis. Sci.* **37,** 1833–1839.

Crosson, C. E., Sloan, C. F., and Yates, P. W. (2005). Modulation of conventional outflow facility by the adenosine A1 agonist N6-cyclohexyladenosine. *Invest. Ophthalmol. Vis. Sci.* **46,** 3795–3799.

Dan, J., Belyea, D., Gertner, G., Leshem, I., Lusky, M., and Miskin, R. (2005). Plasminogen activator inhibitor-1 in the aqueous humor of patients with and without glaucoma. *Arch. Ophthalmol.* **123,** 220–224.

Dickerson, J. E., Jr., Steely, H. T., Jr., English Wright, S. L., and Clark, A. F. (1998). The effect of dexamethasone on integrin and laminin expression in cultured human trabecular meshwork cells. *Exp. Eye Res.* **66,** 731–738.

Do, C. W., and Civan, M. M. (2004). Basis of chloride transport in ciliary epithelium. *J. Membr. Biol.* **200,** 1–13.

Do, C. W., and Civan, M. M. (2006). Swelling-activated chloride channels in aqueous humour formation: On the one side and the other. *Acta Physiol.* **187,** 345–352.

Ethier, C. R., Read, A. T., and Chan, D. W. (2006). Effects of latrunculin-B on outflow facility and trabecular meshwork structure in human eyes. *Invest. Ophthalmol. Vis. Sci.* **47,** 1991–1998.

Fan, H., Rao, S. K., Zhou, Y. S., and Lam, D. S. (2005). Effect of latrunculin B on intraocular pressure in the monkey eye. *Arch. Ophthalmol.* **123,** 1456–1457.

Feng, Z., Mohapatra, S., Klimko, P. G., Hellberg, M. R., May, J. A., Kelly, C., Williams, G., McLaughlin, M. A., and Sharif, N. A. (2007). Novel benzodifuran analogs as potent 5-HT2A receptor agonists with ocular hypotensive activity. *Bioorg. Med. Chem. Lett.* **17,** 2998–3002.

Fleenor, D. L., Pang, I.-H., and Clark, A. F. (2003). Involvement of AP-1 in interleukin-1alpha-stimulated MMP-3 expression in human trabecular meshwork cells. *Invest. Ophthalmol. Vis. Sci.* **44,** 3494–3501.

Fleenor, D. L., Shepard, A. R., Hellberg, P. E., Jacobson, N., Pang, I.-H., and Clark, A. F. (2006). TGF beta 2-induced changes in human trabecular meshwork: Implications for intraocular pressure. *Invest. Ophthalmol. Vis. Sci.* **47,** 226–234.

Fleischhauer, J. C., Mitchell, C. H., Stamer, W. D., Karl, M. O., Peterson Yantorno, K., and Civan, M. M. (2003). Common actions of adenosine receptor agonists in modulating human trabecular meshwork cell transport. *J. Membr. Biol.* **193,** 121–136.

Francois, J. (1975). The importance of the mucopolysaccharides in intraocular pressure regulation. *Invest. Ophthalmol.* **14,** 173–176.

Friedenwald, J. S. (1937). Contribution to the theory and practice of tonometry. *Am. J. Ophthalmol.* **20,** 985–1024.

Fuchshofer, R., Yu, A. H., Welge-Lussen, U., and Tamm, E. R. (2007). Bone morphogenetic protein-7 is an antagonist of transforming growth factor-beta2 in human trabecular meshwork cells. *Invest. Ophthalmol. Vis. Sci.* **48,** 715–726.

Fujisawa, N. (1994). Dexamethasone effects of protein synthesis on organ-cultured human trabecular meshwork-autoradiographical, biochemical, and immunohistochemical study. *Nippon Ganka Gakkai Zasshi* **98,** 31–37.

Gabelt, B. T., and Kaufman, P. L. (2005). Changes in aqueous humor dynamics with age and glaucoma. *Prog. Retin Eye Res.* **24,** 612–637.

Gabelt, B. T., Okka, M., Dean, T. R., and Kaufman, P. L. (2005). Aqueous humor dynamics in monkeys after topical R-DOI. *Invest. Ophthalmol. Vis. Sci.* **46,** 4691–4696.

Gabelt, B. T., Hu, Y., Vittitow, J. L., Rasmussen, C. R., Grosheva, I., Bershadsky, A. D., Geiger, B., Borras, T., and Kaufman, P. L. (2006). Caldesmon transgene expression disrupts focal adhesions in HTM cells and increases outflow facility in organ-cultured human and monkey anterior segments. *Exp. Eye Res.* **82,** 935–944.

Glennon, R. A., Bondarev, M. L., Khorana, N., Young, R., May, J. A., Hellberg, M. R., McLaughlin, M. A., and Sharif, N. A. (2004). Beta-oxygenated analogues of the 5-HT2A serotonin receptor agonist 1-(4-bromo-2,5-dimethoxyphenyl)-2-aminopropane. *J. Med. Chem.* **47,** 6034–6041.

Gordon, M. O., Ocular Hypertension Treatment Study Group and European Prevention Treatment Study Group (2007). Validated prediction model for the development of primary open-angle glaucoma in individuals with ocular hypertension. *Ophthalmology* **114,** 10–19.

Gottanka, J., Chan, D., Eichhorn, M., Lütjen-Drecoll, E., and Ethier, C. R. (2004). Effects of TGF-beta2 in perfused human eyes. *Invest. Ophthalmol. Vis. Sci.* **45,** 153–158.

Griffin, B. W., Williams, G. W., Crider, J. Y., and Sharif, N. A. (1997). FP prostaglandin receptors mediating inositol phosphates generation and calcium mobilization in Swiss 3T3 cells: A pharmacological study. *J. Pharmacol. Exp. Ther.* **281,** 845–854.

Hejkal, T. W., and Camras, C. B. (1999). Prostaglandin analogs in the treatment of glaucoma. *Semin. Ophthalmol.* **14,** 114–123.

Herkel, U., and Pfeiffer, N. (2001). Update on topical carbonic anhydrase inhibitors. *Curr. Opin. Ophthalmol.* **12,** 88–93.

Ho, S. L., Dogar, G. F., Wang, J., Crean, J., Wu, Q. D., Oliver, N., Weitz, S., Murray, A., Cleary, P. E., and O'Brien, C. (2005). Elevated aqueous humour tissue inhibitor of matrix metalloproteinase-1 and connective tissue growth factor in pseudoexfoliation syndrome. *Br. J. Ophthalmol.* **89,** 169–173.

Hoare, M.-J., Grierson, I., Brotchie, D., Pollock, N., Cracknell, K., and Clark, A. F. (2008). Cross-linked actin networks (CLANs) in unmanipulated trabecular meshwork tissues of the human eye. Submitted.

Hoffman, L. M., Maguire, A. M., and Bennett, J. (1997). Cell-mediated immune response and stability of intraocular transgene expression after adenovirus-mediated delivery. *Invest. Ophthalmol. Vis. Sci.* **38,** 2224–2233.

Hong, S., Seong, G. J., and Hong, Y. J. (2007). Long-term intraocular pressure fluctuation and progressive visual field deterioration in patients with glaucoma and low intraocular pressures after a triple procedure. *Arch. Ophthalmol.* **125,** 1010–1013.

Honjo, M., Inatani, M., Kido, N., Sawamura, T., Yue, B. Y., Honda, Y., and Tanihara, H. (2001a). Effects of protein kinase inhibitor, HA1077, on intraocular pressure and outflow facility in rabbit eyes. *Arch. Ophthalmol.* **119,** 1171–1178.

Honjo, M., Tanihara, H., Inatani, M., Kido, N., Sawamura, T., Yue, B. Y., Narumiya, S., and Honda, Y. (2001b). Effects of rho-associated protein kinase inhibitor Y-27632 on intraocular pressure and outflow facility. *Invest. Ophthalmol. Vis. Sci.* **42,** 137–144.

Honjo, M., Inatani, M., Kido, N., Sawamura, T., Yue, B. Y., Honda, Y., and Tanihara, H. (2002). A myosin light chain kinase inhibitor, ML-9, lowers the intraocular pressure in rabbit eyes. *Exp. Eye Res.* **75,** 135–142.

Hosseini, M., Rose, A. Y., Song, K., Bohan, C., Alexander, J. P., Kelley, M. J., and Acott, T. S. (2006). IL-1 and TNF induction of matrix metalloproteinase-3 by c-Jun N-terminal kinase in trabecular meshwork. *Invest. Ophthalmol. Vis. Sci.* **47,** 1469–1476.

Hoyng, P. F., and van Beek, L. M. (2000). Pharmacological therapy for glaucoma: A review. *Drugs* **59,** 411–434.

Hu, Y., Gabelt, B. T., and Kaufman, P. L. (2006). Monkey organ-cultured anterior segments: Technique and response to H-7. *Exp. Eye Res.* **82,** 1100–1108.

Hubbard, W. C., Johnson, M., Gong, H., Gabelt, B. T., Peterson, J. A., Sawhney, R., Freddo, T., and Kaufman, P. L. (1997). Intraocular pressure and outflow facility are unchanged following acute and chronic intracameral chondroitinase ABC and hyaluronidase in monkeys. *Exp. Eye Res.* **65,** 177–190.

Jacobson, N., Andrews, M., Shepard, A. R., Nishimura, D., Searby, C., Fingert, J. H., Hageman, G., Mullins, R., Davidson, B. L., Kwon, Y. H., Alward, W. L., Stone, E. M., *et al.* (2001). Non-secretion of mutant proteins of the glaucoma gene myocilin in cultured trabecular meshwork cells and in aqueous humor. *Hum. Mol. Genet.* **10,** 117–125.

Joe, M. K., Sohn, S., Hur, W., Moon, Y., Choi, Y. R., and Kee, C. (2003). Accumulation of mutant myocilins in ER leads to ER stress and potential cytotoxicity in human trabecular meshwork cells. *Biochem. Biophys. Res. Commun.* **312,** 592–600.

Johnson, D. H. (1997). The effect of cytochalasin D on outflow facility and the trabecular meshwork of the human eye in perfusion organ culture. *Invest. Ophthalmol. Vis. Sci.* **38,** 2790–2799.

Johnson, D. H., Bradley, J. M., and Acott, T. S. (1990). The effect of dexamethasone on glycosaminoglycans of human trabecular meshwork in perfusion organ culture. *Invest. Ophthalmol. Vis. Sci.* **31,** 2568–2571.

Karl, M. O., Fleischhauer, J. C., Stamer, W. D., Peterson Yantorno, K., Mitchell, C. H., Stone, R. A., and Civan, M. M. (2005). Differential P1-purinergic modulation of human Schlemm's canal inner-wall cells. *Am. J. Physiol. Cell Physiol.* **288,** C784–C794.

Kass, M. A., Heuer, D. K., Higginbotham, E. J., Johnson, C. A., Keltner, J. L., Miller, J. P., Parrish, R. K., Wilson, M. R., and Gordon, M. O. (2002). The ocular hypertension treatment study—a randomized trial determines that topcial ocular hypotensive medication delays or prevents the onset of primary open angle glaucoma. *Arch. Ophthalmol.* **120,** 701–713.

Kaufman, P. L., and Bárány, E. H. (1977). Cytochalasin B reversibly increases outflow facility in the eye of the cynomolgus monkey. *Invest. Ophthalmol. Vis. Sci.* **16,** 47–53.

Kaufman, P. L., Jia, W. W., Tan, J., Chen, Z., Gabelt, B. T., Booth, V., Tufaro, F., and Cynader, M. (1999). A perspective of gene therapy in the glaucomas. *Surv. Ophthalmol.* **43**(Suppl 1), S91–S97.

Kee, C., and Seo, K. (1997). The effect of interleukin-1alpha on outflow facility in rat eyes. *J. Glaucoma* **6,** 246–249.

Kee, C., Sohn, S., and Hwang, J. M. (2001). Stromelysin gene transfer into cultured human trabecular cells and rat trabecular meshwork *in vivo*. *Invest. Ophthalmol. Vis. Sci.* **42,** 2856–2860.

Kelley, M. J., Rose, A., Song, K., Lystrup, B., Samples, J. W., and Acott, T. S. (2007a). p38 MAP kinase pathway and stromelysin regulation in trabecular meshwork cells. *Invest. Ophthalmol. Vis. Sci.* **48,** 3126–3137.

Kelley, M. J., Rose, A. Y., Song, K., Chen, Y., Bradley, J. M., Rookhuizen, D., and Acott, T. S. (2007b). Synergism of TNF and IL-1 in the induction of matrix metalloproteinase-3 in trabecular meshwork. *Invest. Ophthalmol. Vis. Sci.* **48,** 2634–2643.

Knepper, P. A., Breen, M., Weinstein, H. G., and Blacik, J. L. (1978). Intraocular pressure and glycosaminoglycan distribution in the rabbit eye: Effect of age and dexamethasone. *Exp. Eye Res.* **27,** 567–575.

Knepper, P. A., Collins, J. A., and Frederick, R. (1985a). Effects of dexamethasone, progesterone, and testosterone on IOP and GAGs in the rabbit eye. *Invest. Ophthalmol. Vis. Sci.* **26,** 1093–1100.

Knepper, P. A., Goossens, W., Hvizd, M., and Palmberg, P. F. (1996a). Glycosaminoglycans of the human trabecular meshwork in primary open-angle glaucoma. *Invest. Ophthalmol. Vis. Sci.* **37,** 1360–1367.

Knepper, P. A., Goossens, W., and Palmberg, P. F. (1996b). Glycosaminoglycan stratification of the juxtacanalicular tissue in normal and primary open-angle glaucoma. *Invest. Ophthalmol. Vis. Sci.* **37,** 2414–2425.

Knepper, P. A., Goossens, W., and Mayanil, C. S. (1998). CD44H localization in primary open-angle glaucoma. *Invest. Ophthalmol. Vis. Sci.* **39,** 673–680.

Knepper, P. A., and McLone, D. G. (1985b). Glycosaminoglycans and outflow pathways of the eye and brain. *Pediatr. Neurosci.* **12,** 240–251.

Knepper, P. A., Miller, A. M., Choi, J., Wertz, R. D., Nolan, M. J., Goossens, W., Whitmer, S., Yue, B. Y., Ritch, R., Liebmann, J. M., Allingham, R. R., and Samples, J. R. (2005). Hypophosphorylation of aqueous humor sCD44 and primary open-angle glaucoma. *Invest. Ophthalmol. Vis. Sci.* **46,** 2829–2837.

Koga, T., Koga, T., Awai, M., Tsutsui, J., Yue, B. Y., and Tanihara, H. (2006). Rho-associated protein kinase inhibitor, Y-27632, induces alterations in adhesion, contraction and motility in cultured human trabecular meshwork cells. *Exp. Eye Res.* **82,** 362–370.

Konno, T., Ohnuma, S. Y., Uemoto, K., Uchibori, T., Nagai, A., Kogi, K., Endo, K., Hosokawa, T., and Nakahata, N. (2004). Effects of 2-alkynyladenosine derivatives on intraocular pressure in rabbits. *Eur. J. Pharmacol.* **486,** 307–316.

Konno, T., Murakami, A., Uchibori, T., Nagai, A., Kogi, K., and Nakahata, N. (2005a). Involvement of adenosine A(2a) receptor in intraocular pressure decrease induced by 2-(1-octyn-1-yl)adenosine or 2-(6-cyano-1-hexyn-1-yl) adenosine. *J. Pharmacol. Sci.* **97,** 501–509.

Konno, T., Uchibori, T., Nagai, A., Kogi, K., and Nakahata, N. (2005b). 2-(1-Hexyn-1-yl) adenosine-induced intraocular hypertension is mediated via K+ channel opening through adenosine A(2A) receptor in rabbits. *Eur. J. Pharmacol.* **518,** 203–211.

Langham, M. E. (1979). The physiology and pathology of the intraocular pressure. *In* "Glaucoma: Contemporary International Concepts" (J. G. Bellows, ed.), pp. 24–48. Masson Publishing, New York.

Larsson, L. I., Rettig, E. S., and Brubaker, R. F. (1995). Aqueous flow in open-angle glaucoma. *Arch. Ophthalmol.* **113,** 283–286.

Leske, M. C., Heijl, A., Hussein, M., Bengtsson, B., Hyman, L., and Komaroff, E. (2003). Factors for glaucoma progression and the effect of treatment: The early manifest glaucoma trial. *Arch. Ophthalmol.* **121,** 48–56.

Li, J., Tripathi, B. J., and Tripathi, R. C. (2000). Modulation of pre-mRNA splicing and protein production of fibronectin by TGF-beta2 in porcine trabecular cells. *Invest. Ophthalmol. Vis. Sci.* **41,** 3437–3443.

Lichter, P. R., Musch, D. C., Gillespie, B. W., Guire, K. E., Janz, N. K., Wren, P. A., and Mills, R. P. (2001). Interim clinical outcomes in the Collaborative Initial Glaucoma Treatment Study comparing initial treatment randomized to medications or surgery. *Ophthalmology* **108,** 1943–1953.

Linden, C. (2001). Therapeutic potential of prostaglandin analogues in glaucoma. *Expert Opin. Investig. Drugs* **10,** 679–694.

Lindsey, J. D., Kashiwagi, K., Boyle, D., Kashiwagi, F., Firestein, G. S., and Weinreb, R. N. (1996). Prostaglandins increase proMMP-1 and proMMP-3 secretion by human ciliary smooth muscle cells. *Curr. Eye Res.* **15,** 869–875.

Lindsey, J. D., Kashiwagi, K., Kashiwagi, F., and Weinreb, R. N. (1997). Prostaglandin action on ciliary smooth muscle extracellular matrix metabolism: Implications for uveoscleral outflow. *Surv. Ophthalmol.* **41**(Suppl 2), S53–S59.

Linner, E. (1976). Ocular hypertension. I. The clinical course during ten years without therapy. Aqueous humour dynamics. *Acta Ophthalmol.* **54,** 707–720.

Liton, P. B., Liu, X., Challa, P., Epstein, D. L., and Gonzalez, P. (2005). Induction of TGF-beta1 in the trabecular meshwork under cyclic mechanical stress. *J. Cell Physiol.* **205,** 364–371.

Liton, P. B., Luna, C., Challa, P., Epstein, D. L., and Gonzalez, P. (2006). Genome-wide expression profile of human trabecular meshwork cultured cells, nonglaucomatous and primary open angle glaucoma tissue. *Mol. Vis.* **12,** 774–790.

Liu, X., Brandt, C. R., Gabelt, B. T., Bryar, P. J., Smith, M. E., and Kaufman, P. L. (1999). Herpes simplex virus mediated gene transfer to primate ocular tissues. *Exp. Eye Res.* **69,** 385–395.

Liu, X., Hu, Y., Filla, M. S., Gabelt, B. T., Peters, D. M., Brandt, C. R., and Kaufman, P. L. (2005). The effect of C3 transgene expression on actin and cellular adhesions in cultured human trabecular meshwork cells and on outflow facility in organ cultured monkey eyes. *Mol. Vis.* **11,** 1112–1121.

Loewen, N., Fautsch, M. P., Peretz, M., Bahler, C. K., Cameron, J. D., Johnson, D. H., and Poeschla, E. M. (2001). Genetic modification of human trabecular meshwork with lentiviral vectors. *Hum. Gene Ther.* **12,** 2109–2119.

Luo, Q., Kang, Q., Si, W., Jiang, W., Park, J. K., Peng, Y., Li, X., Luu, H. H., Luo, J., Montag, A. G., Haydon, R. C., and He, T. C. (2004). Connective tissue growth factor (CTGF) is regulated by Wnt and bone morphogenetic proteins signaling in osteoblast differentiation of mesenchymal stem cells. *J. Biol. Chem.* **279,** 55958–55968.

Lütjen-Drecoll, E. (2005). Morphological changes in glaucomatous eyes and the role of TGFbeta2 for the pathogenesis of the disease. *Exp. Eye Res.* **81,** 1–4.

Lütjen-Drecoll, E., Shimizu, T., Rohrbach, M., and Rohen, J. W. (1986). Quantitative analysis of 'plaque material' between ciliary muscle tips in normal- and glaucomatous eyes. *Exp. Eye Res.* **42,** 457–465.

Mabuchi, F., Lindsey, J. D., Aihara, M., Mackey, M. R., and Weinreb, R. N. (2004). Optic nerve damage in mice with a targeted type I collagen mutation. *Invest. Ophthalmol. Vis. Sci.* **45,** 1841–1845.

May, J. A., Chen, H. H., Rusinko, A., Lynch, V. M., Sharif, N. A., and McLaughlin, M. A. (2003a). A novel and selective 5-HT2 receptor agonist with ocular hypotensive activity: (S)-(+)-1-(2-aminopropyl)-8,9-dihydropyrano[3,2-e]indole. *J. Med. Chem.* **46,** 4188–4195.

May, J. A., McLaughlin, M. A., Sharif, N. A., Hellberg, M. R., and Dean, T. R. (2003b). Evaluation of the ocular hypotensive response of serotonin 5-HT1A and 5-HT2 receptor ligands in conscious ocular hypertensive cynomolgus monkeys. *J. Pharmacol. Exp. Ther.* **306,** 301–309.

May, J. A., Dantanarayana, A. P., Zinke, P. W., McLaughlin, M. A., and Sharif, N. A. (2006). 1-((S)-2-aminopropyl)-1H-indazol-6-ol: A potent peripherally acting 5-HT2 receptor agonist with ocular hypotensive activity. *J. Med. Chem.* **49,** 318–328.

Melamed, S., Kotas Neumann, R., Barak, A., and Epstein, D. L. (1992). The effect of intracamerally injected ethacrynic acid on intraocular pressure in patients with glaucoma. *Am. J. Ophthalmol.* **113,** 508–512.

Melton, C. E., and DeVille, W. B. (1960). Perfusion studies on eyes of four species. *Am. J. Ophthalmol.* **50,** 302–308.

Mitchell, C. H., Peterson Yantorno, K., Carre, D. A., McGlinn, A. M., Coca Prados, M., Stone, R. A., and Civan, M. M. (1999). A3 adenosine receptors regulate Cl- channels of nonpigmented ciliary epithelial cells. *Am. J. Physiol.* **276,** C659–C666.

Nguyen, T. D., Chen, P., Huang, W. D., Chen, H., Johnson, D., and Polansky, J. R. (1998). Gene structure and properties of TIGR, an olfactomedin-related glycoprotein cloned from glucocorticoid-induced trabecular meshwork cells. *J. Biol. Chem.* **273,** 6341–6350.

Nishanian, T. G., Kim, J. S., Foxworth, A., and Waldman, T. (2004). Suppression of tumorigenesis and activation of Wnt signaling by bone morphogenetic protein 4 in human cancer cells. *Cancer Biol. Ther.* **3,** 667–675.

Nolan, M. J., Giovingo, M. C., Miller, A. M., Wertz, R. D., Ritch, R., Liebmann, J. M., Allingham, R. R., Herndon, L. W., Wax, M. B., Smolyak, R., Hasan, F., Barnett, E. M., *et al.* (2007). Aqueous humor sCD44 concentration and visual field loss in primary open-angle glaucoma. *J. Glaucoma* **16,** 419–429.

Nouri-Mahdavi, K., Hoffman, D., Coleman, A. L., Liu, G., Li, G., Gaasterland, D., and Caprioli, J. (2004). Predictive factors for glaucomatous visual field progression in the advanced glaucoma intervention study. *Ophthalmology* **111,** 1627–1635.

Novack, G. D. (1987). Ophthalmic beta-blockers since timolol. *Surv. Ophthalmol.* **31,** 307–327.

O'Brien, C., Schwartz, B., Takamoto, T., and Wu, D. C. (1991). Intraocular pressure and the rate of visual field loss in chronic open-angle glaucoma. *Am. J. Ophthalmol.* **111,** 491–500.

O'Donnell, M. E., Brandt, J. D., and Curry, F. R. (1995). Na-K-Cl cotransport regulates intracellular volume and monolayer permeability of trabecular meshwork cells. *Am. J. Physiol.* **268,** C1067–C1074.

Okamura, T., Kurogi, Y., Hashimoto, K., Sato, S., Nishikawa, H., Kiryu, K., and Nagao, Y. (2004). Structure-activity relationships of adenosine A3 receptor ligands: New potential therapy for the treatment of glaucoma. *Bioorg. Med. Chem. Lett.* **14,** 3775–3779.

Okka, M., Tian, B., and Kaufman, P. L. (2004a). Effect of low-dose latrunculin B on anterior segment physiologic features in the monkey eye. *Arch. Ophthalmol.* **122,** 1482–1488.

Okka, M., Tian, B., and Kaufman, P. L. (2004b). Effects of latrunculin B on outflow facility, intraocular pressure, corneal thickness, and miotic and accommodative responses to pilocarpine in monkeys. *Trans. Am. Ophthalmol. Soc.* **102,** 251–257.

Olthoff, C. M., Schouten, J. S., van de Borne, B. W., and Webers, C. A. (2005). Noncompliance with ocular hypotensive treatment in patients with glaucoma or ocular hypertension an evidence-based review. *Ophthalmology* **112,** 953–961.

Pang, I. H., Moll, H., McLaughlin, M. A., Knepper, P. A., De Santis, L., Epstein, D. L., and Clark, A. F. (2001). Ocular hypotensive and aqueous outflow-enhancing effects of AL-3037A (sodium ferri ethylenediaminetetraacetate). *Exp. Eye Res.* **73,** 815–825.

Pang, I. H., Fleenor, D. L., Hellberg, P. E., Stropki, K., McCartney, M. D., and Clark, A. F. (2003a). Aqueous outflow-enhancing effect of tert-butylhydroquinone: Involvement of AP-1 activation and MMP-3 expression. *Invest. Ophthalmol. Vis. Sci.* **44,** 3502–3510.

Pang, I. H., Hellberg, P. E., Fleenor, D. L., Jacobson, N., and Clark, A. F. (2003b). Expression of matrix metalloproteinases and their inhibitors in human trabecular meshwork cells. *Invest. Ophthalmol. Vis. Sci.* **44,** 3485–3493.

Parshley, D. E., Bradley, J. M., Samples, J. R., Van Buskirk, E. M., and Acott, T. S. (1995). Early changes in matrix metalloproteinases and inhibitors after *in vitro* laser treatment to the trabecular meshwork. *Curr. Eye Res.* **14,** 537–544.

Parshley, D. E., Bradley, J. M., Fisk, A., Hadaegh, A., Samples, J. R., Van Buskirk, E. M., and Acott, T. S. (1996). Laser trabeculoplasty induces stromelysin expression by trabecular juxtacanalicular cells. *Invest. Ophthalmol. Vis. Sci.* **37,** 795–804.

Pedler, C. (1956). The relationship of hyaluronidase to aqueous outflow resistance. *Trans. Ophthalmol. Soc. UK* **76,** 51–63.

Peterson, W. S., and Jocson, V. L. (1974). Hyaluronidase effects on aqueous outflow resistance. Quantitative and localizing studies in the rhesus monkey eye. *Am. J. Ophthalmol.* **77,** 573–577.

Quigley, H. A., and Broman, A. T. (2006). The number of people with glaucoma worldwide in 2010 and 2020. *Br. J. Ophthalmol.* **90,** 262–267.

Rao, P. V., Deng, P. F., Kumar, J., and Epstein, D. L. (2001). Modulation of aqueous humor outflow facility by the Rho kinase-specific inhibitor Y-27632. *Invest. Ophthalmol. Vis. Sci.* **42,** 1029–1037.

Rao, P. V., Deng, P., Maddala, R., Epstein, D. L., Li, C. Y., and Shimokawa, H. (2005a). Expression of dominant negative Rho-binding domain of Rho-kinase in organ cultured human eye anterior segments increases aqueous humor outflow. *Mol. Vis.* **11,** 288–297.

Rao, P. V., Deng, P., Sasaki, Y., and Epstein, D. L. (2005b). Regulation of myosin light chain phosphorylation in the trabecular meshwork: Role in aqueous humour outflow facility. *Exp. Eye Res.* **80,** 197–206.

Rao, P. V., Shimazaki, A., Ichikawa, M., Alvarado, J. A., and Epstein, D. L. (2005c). Effects of novel ethacrynic acid derivatives on human trabecular meshwork cell shape, actin cytoskeletal organization, and transcellular fluid flow. *Biol. Pharm. Bull.* **28,** 2189–2196.

Read, A. T., Chan, D. W., and Ethier, C. R. (2007). Actin structure in the outflow tract of normal and glaucomatous eyes. *Exp. Eye Res.* **84,** 214–226.

Richter, M., Krauss, A. H., Woodward, D. F., and Lütjen-Drecoll, E. (2003). Morphological changes in the anterior eye segment after long-term treatment with different receptor selective prostaglandin agonists and a prostamide. *Invest. Ophthalmol. Vis. Sci.* **44,** 4419–4426.

Robin, A. L. (1997). The role of alpha-agonists in glaucoma therapy. *Curr. Opin. Ophthalmol.* **8,** 42–49.

Robinson, J. C., and Kaufman, P. L. (1991). Cytochalasin B potentiates epinephrine's outflow facility-increasing effect. *Invest. Ophthalmol. Vis. Sci.* **32,** 1614–1618.

Rohen, J. W. (1983). Why is intraocular pressure elevated in chronic simple glaucoma? Anatomical considerations. *Ophthalmology* **90,** 758–765.

Rosenthal, R., Choritz, L., Schlott, S., Bechrakis, N. E., Jaroszewski, J., Wiederholt, M., and Thieme, H. (2005). Effects of ML-7 and Y-27632 on carbachol- and endothelin-1-induced contraction of bovine trabecular meshwork. *Exp. Eye Res.* **80,** 837–845.

Sabanay, I., Tian, B., Gabelt, B. T., Geiger, B., and Kaufman, P. L. (2004). Functional and structural reversibility of H-7 effects on the conventional aqueous outflow pathway in monkeys. *Exp. Eye Res.* **78,** 137–150.

Sabanay, I., Tian, B., Gabelt, B. T., Geiger, B., and Kaufman, P. L. (2006). Latrunculin B effects on trabecular meshwork and corneal endothelial morphology in monkeys. *Exp. Eye Res.* **82,** 236–246.

Samples, J. R., Alexander, J. P., and Acott, T. S. (1993). Regulation of the levels of human trabecular matrix metalloproteinases and inhibitor by interleukin-1 and dexamethasone. *Invest. Ophthalmol. Vis. Sci.* **34,** 3386–3395.

Sawaguchi, S., Yue, B., Yeh, P., and Tso, M. O. M. (1992). Effects of intracameral injection of chondroitinase ABC *in vivo*. *Arch. Ophthalmol.* **110,** 110–117.

Sawaguchi, S., Lam, T. T., Yue, B. Y., and Tso, M. O. M. (1993). Effects of glycosaminoglycan-degrading enzymes on bovine trabecular meshwork in organ culture. *J. Glaucoma* **2,** 80–86.

Schlotzer Schrehardt, U., Zenkel, M., Kuchle, M., Sakai, L. Y., and Naumann, G. O. (2001). Role of transforming growth factor-beta1 and its latent form binding protein in pseudoexfoliation syndrome. *Exp. Eye Res.* **73,** 765–780.

Schwartz, G. F. (2005). Compliance and persistency in glaucoma follow-up treatment. *Curr. Opin. Ophthalmol.* **16,** 114–121.

Segawa, K. (1979). Electron microscopic changes in the trabecular tissue in primary open angle glaucoma. *In* "Glaucoma: Contemporary International Concepts" (J. G. Bellows, ed.), pp. 17–23. Masson Publishing, New York.

Sharif, N. A., Kelly, C. R., and McLaughlin, M. (2006). Human trabecular meshwork cells express functional serotonin-2A (5HT2A) receptors: Role in IOP reduction. *Invest. Ophthalmol. Vis. Sci.* **47,** 4001–4010.

Shepard, A. R., Fleenor, D., Jacobson, N., Pang, I.-H., and Clark, A. F. (2003). Connective tissue growth factor expression In glaucomatous human TM cells and aqueous humor and upregulation by TGF-beta2. *ARVO Abstract* 3161.

Shepard, A. R., Jacobson, N., Millar, J. C., Pang, I.-H., Steely, H. T., Searby, C. C., Sheffield, V. C., Stone, E. M., and Clark, A. F. (2007). Glaucoma-causing myocilin mutants require he Peroxisomal Targeting Signal-1 Receptor (PTS1R) to elevate intraocular pressure. *Hum. Mol. Genet.* **16,** 609–617.

Shepard, A. R., Nolan, M. J., Millar, J. C., Pang, I.-H., Luan, T., Jacobson, N., Wax, M. B., Clark, A. F., and Knepper, P. A. (2008). CD44 overexpression causes ocular hypertension in mouse. *ARVO Abstract* **2008,** 2880.

Shimazaki, A., Ichikawa, M., Rao, P. V., Kirihara, T., Konomi, K., Epstein, D. L., and Hara, H. (2004a). Effects of the new ethacrynic acid derivative SA9000 on intraocular pressure in cats and monkeys. *Biol. Pharm. Bull.* **27,** 1019–1024.

Shimazaki, A., Suhara, H., Ichikawa, M., Matsugi, T., Konomi, K., Takagi, Y., Hara, H., Rao, P. V., and Epstein, D. L. (2004b). New ethacrynic acid derivatives as potent cytoskeletal modulators in trabecular meshwork cells. *Biol. Pharm. Bull.* **27,** 846–850.

Steely, H. T., Browder, S. L., Julian, M. B., Miggans, S. T., Wilson, K. L., and Clark, A. F. (1992). The effects of dexamethasone on fibronectin expression in cultured human trabecular meshwork cells. *Invest. Ophthalmol. Vis. Sci.* **33,** 2242–2250.

Stewart, W. C., Kolker, A. E., Sharpe, E. D., Day, D. G., Holmes, K. T., Leech, J. N., Johnson, M., and Cantrell, J. B. (2000). Factors associated with long-term progression or stability in primary open-angle glaucoma. *Am. J. Ophthalmol.* **130,** 274–279.

Stone, E. M., Fingert, J. H., Alward, W. L. M., Nguyen, T. D., Polansky, J. R., Sunden, S. L. F., Nishimura, D., Clark, A. F., Nystuen, A., Nichols, B. E., Mackey, D. A., Ritch, R., *et al.* (1997). Identification of a gene that causes primary open angle glaucoma. *Science* **275,** 668–670.

Sugrue, M. F. (2000). Pharmacological and ocular hypotensive properties of topical carbonic anhydrase inhibitors. *Prog. Retin Eye Res.* **19,** 87–112.

Svedbergh, B., Lütjen-Drecoll, E., Ober, M., and Kaufman, P. L. (1978). Cytochalasin B-induced structural changes in the anterior ocular segment of the cynomolgus monkey. *Invest. Ophthalmol. Vis. Sci.* **17,** 718–734.

Tamm, E. R., Siegner, A., Baur, A., and Lütjen-Drecoll, E. (1996). Transforming growth factor-beta 1 induces alpha-smooth muscle-actin expression in cultured human and monkey trabecular meshwork. *Exp. Eye Res.* **62,** 389–397.

Thieme, H., Schimmat, C., Munzer, G., Boxberger, M., Fromm, M., Pfeiffer, N., and Rosenthal, R. (2006). Endothelin antagonism: Effects of FP receptor agonists prostaglandin F2alpha and fluprostenol on trabecular meshwork contractility. *Invest. Ophthalmol. Vis. Sci.* **47,** 938–945.

Tian, B., and Kaufman, P. L. (2005). Effects of the Rho kinase inhibitor Y-27632 and the phosphatase inhibitor calyculin A on outflow facility in monkeys. *Exp. Eye Res.* **80,** 215–225.

Tian, B., Gabelt, B. T., Crosson, C. E., and Kaufman, P. L. (1997). Effects of adenosine agonists on intraocular pressure and aqueous humor dynamics in cynomolgus monkeys. *Exp. Eye Res.* **64,** 979–989.

Tian, B., Brumback, L. C., and Kaufman, P. L. (2000a). ML-7, chelerythrine and phorbol ester increase outflow facility in the monkey Eye. *Exp. Eye Res.* **71,** 551–566.

Tian, B., Geiger, B., Epstein, D. L., and Kaufman, P. L. (2000b). Cytoskeletal involvement in the regulation of aqueous humor outflow. *Invest. Ophthalmol. Vis. Sci.* **41,** 619–623.

Tian, B., Kiland, J. A., and Kaufman, P. L. (2001). Effects of the marine macrolides swinholide A and jasplakinolide on outflow facility in monkeys. *Invest. Ophthalmol. Vis. Sci.* **42,** 3187–3192.

Tian, B., Wang, R. F., Podos, S. M., and Kaufman, P. L. (2004). Effects of topical H-7 on outflow facility, intraocular pressure, and corneal thickness in monkeys. *Arch. Ophthalmol.* **122,** 1171–1177.

Tokushige, H., Inatani, M., Nemoto, S., Sakaki, H., Katayama, K., Uehata, M., and Tanihara, H. (2007). Effects of topical administration of y-39983, a selective rho-associated protein kinase inhibitor, on ocular tissues in rabbits and monkeys. *Invest. Ophthalmol. Vis. Sci.* **48,** 3216–3222.

Toris, C. B., Koepsell, S. A., Yablonski, M. E., and Camras, C. B. (2002). Aqueous humor dynamics in ocular hypertensive patients. *J. Glaucoma* **11,** 253–258.

Tovar-Vidales, T., Roque, R., Clark, A. F., and Wordinger, R. J. (2008). Tissue transglutaminase expression and activity in normal and glaucomatous human trabecular meshwork cells and tissues. *Invest Ophthalmol. Vis. Sci.* **49,** 622–628.

Tripathi, R. C., Borisuth, N. S., Kolli, S. P., and Tripathi, B. J. (1993a). Trabecular cells express receptors that bind TGF-beta 1 and TGF-beta 2: A qualitative and quantitative characterization. *Invest. Ophthalmol. Vis. Sci.* **34,** 260–263.

Tripathi, R. C., Li, J., Borisuth, N. S., and Tripathi, B. J. (1993b). Trabecular cells of the eye express messenger RNA for transforming growth factor-beta 1 and secrete this cytokine. *Invest. Ophthalmol. Vis. Sci.* **34,** 2562–2569.

Usui, T., Nakajima, F., Ideta, R., Kaji, Y., Suzuki, Y., Araie, M., Miyauchi, S., Heldin, P., and Yamashita, H. (2003). Hyaluronan synthase in trabecular meshwork cells. *Br. J. Ophthalmol.* **87,** 357–360.

Van Buskirk, E. M., and Brett, J. (1978). The canine eye: *In vitro* studies of the intraocular pressure and facility of aqueous outflow. *Invest. Ophthalmol. Vis. Sci.* **17,** 373–377.

van Setten, G. B., Blalock, T. D., Grotendorst, G., and Schultz, G. S. (2002). Detection of connective tissue growth factor in human aqueous humor. *Ophthalmic. Res.* **34,** 306–308.

Waki, M., Yoshida, Y., Oka, T., and Azuma, M. (2001). Reduction of intraocular pressure by topical administration of an inhibitor of the Rho-associated protein kinase. *Curr. Eye Res.* **22,** 470–474.

Wang, N., Chintala, S. K., Fini, M. E., and Schuman, J. S. (2001). Activation of a tissue-specific stress response in the aqueous outflow pathway of the eye defines the glaucoma disease phenotype. *Nat. Med.* **7,** 304–309.

Wang, W.-H., McNatt, L. G., Pang, I.-H., Hellberg, P. E., Fingert, J. H., McCartney, M. D., and Clark, A. F. (2008). Increased expression of serum amyloid A in glaucoma and its effect on intraocular pressure. *Invest Ophthalmol. Vis. Sci.* **49,** 1916–1923.

Wang, W.-H., McNatt, L. G., Hellberg, P. E., Pang, I.-H., Millar, J. C., Steely, H. T., Rubin, J. S., Fingert, J. H., Sheffield, V. C., Stone, E. M., and Clark, A. F. (2008). Increased expression of the wnt antagonist/elevates intraocular pressure sFRP1 in glaucoma. *J. clin Invest* **118,** 1056–1064.

Weinreb, R. N. (2000). Uveoscleral outflow: The other outflow pathway. *J. Glaucoma* **9,** 343–345.

Welge-Lussen, U., May, C. A., and Lütjen-Drecoll, E. (2000). Induction of tissue transglutaminase in the trabecular meshwork by TGF-beta1 and TGF-beta2. *Invest. Ophthalmol. Vis. Sci.* **41,** 2229–2238.

Wordinger, R. J., and Clark, A. F. (2007). Bone morphogenetic proteins and their receptors in the eye. *Exp. Biol. Med.* **232,** 979–992.

Wordinger, R. J., and Clark, A. F. (2008). Growth factors and neurotrophic factors as targets. *In* "Ocular Therapeutics: Eye on New Discoveries" (T. Yorio, A. F. Clark, and M. B. Wax, eds.). Elsevier Ltd., Bostom. pp 87–116.

Wordinger, R. J., Clark, A. F., Agarwal, R., Lambert, W., McNatt, L., Wilson, S. E., Qu, Z., and Fung, B. K. (1998). Cultured human trabecular meshwork cells express functional growth factor receptors. *Invest. Ophthalmol. Vis. Sci.* **39,** 1575–1589.

Wordinger, R. J., Agarwal, R., Talati, M., Fuller, J., Lambert, W., and Clark, A. F. (2002). Expression of bone morphogenetic proteins (BMP), BMP receptors, and BMP associated proteins in human trabecular meshwork and optic nerve head cells and tissues. *Mol. Vis.* **8,** 241–250.

Wordinger, R. J., Fleenor, D. L., Pang, I.-H., Tovar, T., Zode, G., Fuller, J., and Clark, A. F. (2007). Gremlin inhibits Bone Morphogenetic Protein-4 (BMP-4) antagonism of TGFβ2 function in cultured human trabecular cells and in the perfused anterior eye organ culture model. *Invest. Ophthalmol. Vis. Sci.* **48,** 1191–1200.

Yang, H., Avila, M. Y., Peterson Yantorno, K., Coca Prados, M., Stone, R. A., Jacobson, K. A., and Civan, M. M. (2005). The cross-species A3 adenosine-receptor antagonist MRS 1292 inhibits adenosine-triggered human nonpigmented ciliary epithelial cell fluid release and reduces mouse intraocular pressure. *Curr. Eye Res.* **30,** 747–754.

Zhang, X., Schroeder, A., Callahan, E. M., Coyle, B. M., Wang, N., Erickson, K. A., Schuman, J. S., and Fini, M. E. (2006). Constitutive signalling pathway activity in trabecular meshwork cells from glaucomatous eyes. *Exp. Eye Res.* **82,** 968–973.

Zhao, X., and Russell, P. (2005). Versican splice variants in human trabecular meshwork and ciliary muscle. *Mol. Vis.* **11,** 603–608.

Zhao, X., Ramsey, K. E., Stephan, D. A., and Russell, P. (2004). Gene and protein expression changes in human trabecular meshwork cells treated with transforming growth factor-beta. *Invest. Ophthalmol. Vis. Sci.* **45,** 4023–4034.

Index

O

P

U

V

W

Z

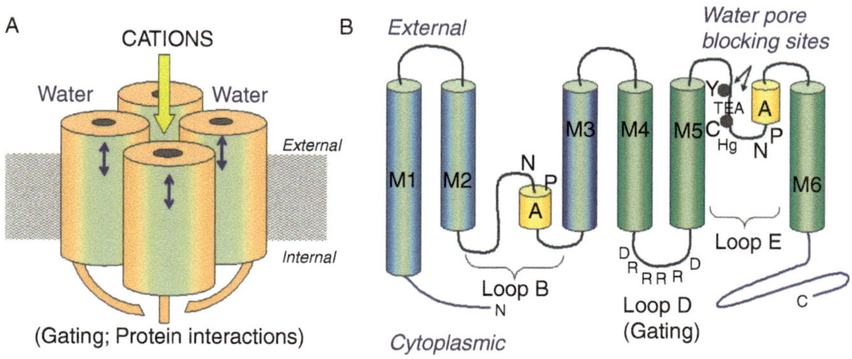

Please refer to Fig. 2.1 in text for figure legend.

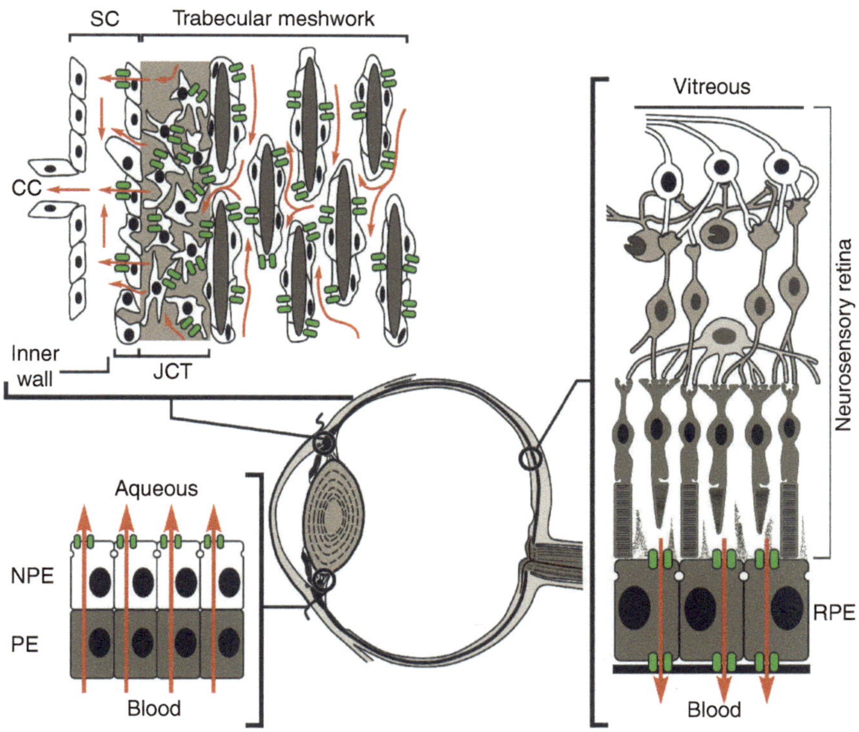

Please refer to Fig. 2.2 in text for figure legend.

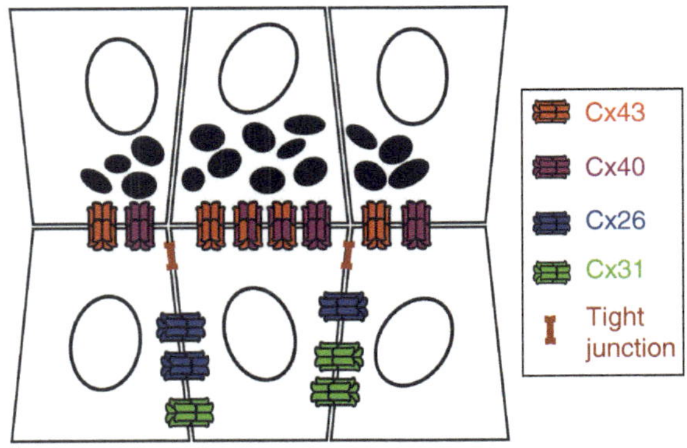

Please refer to Fig. 3.1 in text for figure legend.

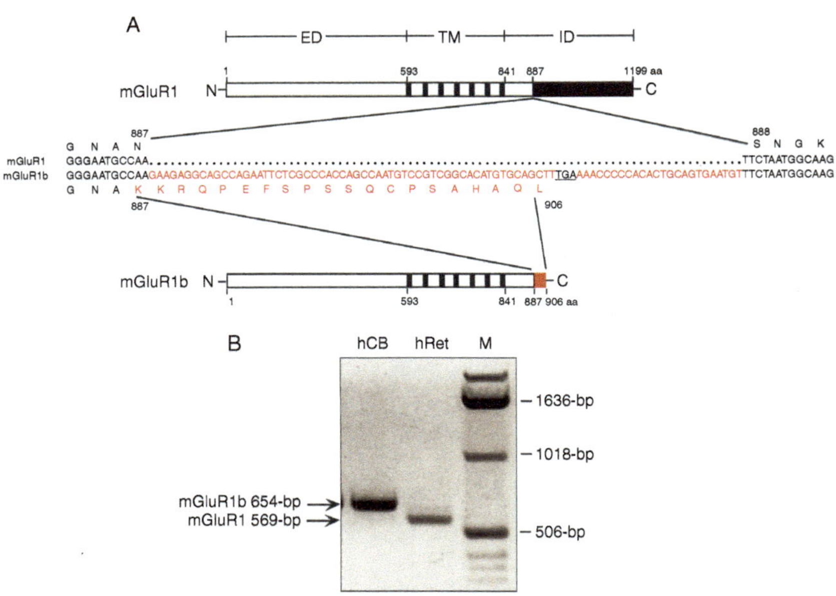

Please refer to Fig. 5.5 in text for figure legend.

Protein name	Glaucoma Tissue[1]	Normal Tissue[1]	Glaucoma transformed cells[2]	TGFβ-treat primary cells[3]	mRNA changed by glaucom conditions*	Basic Function
Adhesion / ECM						
Angiopoiet-like factor 7 / CDT6	●	---	---	---	●	Antiangiogenesis / ECM deposition
Cochlin	●	---	---	---	---	ECM architecture (deafness)
Syndecan 1	●	●	---	---	●	Transmembrane proteoglycan / adhesion
Thrombospondin1,3&4 precur	●	●	---	---	●	Cell-cell, cell-matrix interactions
Calcification						
Mimecan (Osteoglycin)	●	●	---	---	●	Induces bone formation with TFGβ
Cytoskeletal structure						
α–Actin	●	●	●	●	---	Microfilament / cell contraction
α–Tubulin	●	●	●	●	●	Microtubule
Transgelin (smooth muscle22)	---	●	---	●	●	Actin cross-linking
Tropomyosin	●	●	---	●	●	Actin binding protein /cell contraction
Vimentin	●	●	●	●	---	Intermediate filament
Metabolism						
Enolase	●	●	●	●	---	Metabolic enzyme
GAPDH	●	●	●	---	●	Cytosolic energy production
Pyruvate kinase	●	●	●	●	---	Metabolic enzyme
Protein modification						
Calreticulin	---	---	●	●	---	Calcium homeostasis / chaperone
Chaperonin containing TCP1	---	---	●	---	●	Protein folding
Protein disulfide isomerase	---	●	●	●	●	Protein folding
Signaling / cytokines						
Calgranulin A (S100 protein)	●	●	---	---	●	Calcium-dependent signaling
Stress / Defense						
αB-crystallin	---	●	---	---	●	Stress response
Heat-shock protein 70	---	●	●	---	●	Stress protection
Myocilin	●	---	---	---	●	Stress protein
Superoxide dismutase	---	●	---	●	●	Destroys superoxide radicals
Other						
Apolipoprotein D	---	●	---	---	●	Carrier protein / Lipid transport
Cornifin	●	---	---	---	●	Membrane cross-linking

From data on

1: Bhattacharya *et al*. JBC 7, 6080 (2005)

2: Steely *et al*. Mol Vis 12, 372 (2006)

3: Zhao *et al*. IOVS 45, 4023 (2004)

*: see multiple refs throughout the text (mechanical, DEX and TGFβ);

●: present or altered; ---: not present or not altered

Please refer to Fig. 11.5 in text for figure legend.

Function	1 hour IOP[a]	6 hours IOP[b]	2-4 days IOP[c]	7 days IOP[d]	12-48 hr Stretch[e]	Chromosome
Adhesion / ECM						
Angiopoietin-like factor7/CDT6	-	-	-	25.0	-	1p36
ICAM 4	-	-	-1.9	-	-	19p13-cen
Chitinase 3	NC	-	1.8	2.0	-	1q31-q32
Matrix remodeling asso 5 (aldican)	-	-	-	4.3	-	Xp22.33
MMP2	-	-	1.7	NC	NC	16q13-q21
MMP3	-	4	-	altered	-	11q22.3
Perlecan	NC	-	22.0	NC	-	1p36--p34
Syndecan 1	-	-	-	1.5	-	2p24.1
Syndecan 2	NC	-	-	-	1.5	8q22-q23
Tenascin C	NC	-	-	-2.1	2.0	9q32-q34
Calcification						
Biglycan	-	-	-	1.6	3.1	Xq28
Connect tiss growth factor (CTGF)	NC	-	-	1.3	1.7	6q23
Fibromodulin	NC	-	-	2.1	1.7	1q32
Matrix Gla	-	-	7.0	6.0	2.5	12p13-p12
Mimecan (Osteoglycin)	-	-	-	2.6	0.5	9q22
Osteoblast specific factor 2	NC	-	2.2	NC	0.5	13q13
Osteomodulin	NC	-	-1.9	NC	-	9q22.31
Osteonectin (SPARC)	NC	-	-	1.7	2.0	5q31-q32
Cytoskeleton / contraction						
α–Tubulin	NC	-	4.0	-	2.9	12q12-q14
Spectrin (erythrocyte)	-	-	-2.4	NC	-	1q21
Spectrin (non erythrocyte)	2.0	-	-	NC	-	9q33-q34
Transgelin (smooth muscle22)	0.7	-	4.4	-1.8	1.9	1q21-q25
Tropomodulin	NC	-	3.1	1.3	-	15q21-q21
Tropomyosin	0.7	-	-	altered	-	15q22
Protein modification						
Chaperonin containing TCP1	NC	-	1.9	altered	0.5	12q15
Protein disulfide isomerase	NC	-	-1.5	1.5	-	16p13
Proteasome 26S	-	-	2.4	NC	1.7	1q21
Signaling / cytokines						
Activin A receptor	up	-	-	NC	-	2q23-q24
Calgranulin A (S100 protein)	2.0	-	-1.2	NC	-	1q21
Chemokine (C-X-C motif) 11	4.3	-	-	4.3	-	4q21
Chemokine (C-X-C motif)* (2,3,12)		-	-	down	down	
Insul growth fact 1 (somatomedinC)	-	-	-	4.4	-	12q22-q23
Insulin growth fact bind prot*(2,3,4,7)	changed	-	2.6	-4.5	2.5	
Interleukin 6 (IL6)		4	1.8	-	-	7p21
Regulatory G-protein sign 11	NC	-	2.4	NC	-	16p13
Secretogranin II	NC	3.0	-	5.2	-	2q35-q36
Substance P precursor	NC	2.5	-	3.3	-	7q21-q22
Vasoact Intest Peptide (VIP)	NC	-	2.0	-1.5	-	6q25
VEGF	1.4	-	-	-1.7	-	6p12

Fig. 11.6 (Continued)

Function	1 hour IOP[a]	6 hours IOP[b]	2-4 days IOP[c]	7 days IOP[d]	12-48 hr Stretch[e]	Chromosome
Stress / Defense						
αB-crystallin	-	4.5	1.4	-	1.6	11q22-q23
Cytochrome P450* (26A,26C,1B)	1.9	-	2.0	-	2.6	
Heat-shock protein 70k 2	-	-	-	-4.0	-	14q24
Metallothioneins*	NC	4.0	4.0	-1.3	73.0	
Myocilin	NC	-	5.1	10.3	up	1q23-q24
Superoxide dismutase 1	NC	3.0	-	NC	-	21q22
Other						
Apolipoprotein D	NC	-	13.0	2.3	-	3q26-qter
Baculo IAP rep-cont 5 (survivin)	NC	-	2.0	NC	-	17q25
Cornifin (small proline-rich B)	6.1	-	-	-3.0	-	1q21-q22
Thrombomodulin	NC	-	-	1.8	-	20p11

Criteria: selected genes include those reported as most altered in at least one condition

a: Borrás (unpublished)

b: Gonzalez *et al*. (2000b)

c: Vittitow and Borrás (2004) and unpublished laboratory data

d: Borrás (unpublished)

e: Vittal *et al.* (2005)

*: family members pooled; -: not known

NC: no change

Please refer to Fig. 11.6 in text for figure legend.

Function	7 days[a] MicroMax 2400 (cDNA)	10 days[b] AffyU95Av2 12,627 (oligo)	10 days[c] MicroMax 2400 (cDNA)	21 days[d] AffyU133A 22,215 (oligo)	Gene Symbol	Chromosome
Adhesion / ECM						
Angiopoietin-like factor7 / CDT6	-	28.0	-	187.0	ANGPTL/CDT6	1p36
α1-Antichymotrypsin (serpin)	-	165.0	-	48.0	SERPINA3	14q32
Cadherin 2	-	-4.4	-	-	CDH2	18q11
Cadherin 4	-	-	2.4	-	CDH4	20q13
Chitinase 3	-	39.0	-	-6.1	CHI3L1	1q31-q32
Connective tissue growth factor	-	-	-	1.3	CTGF	6q23
Decorin	3.5	-	-	-1.5	DCN	12q21
Elastin	-	-66.0	-	-	ELN	7q11
Fibulin 1	2.8	7.6	-	-1.8	FBLN1	22q13
Fibronectin leucine rich transm prot 2	0.4	-	-	-2.5	FLRT2	14q24-q32
Procollagen C-proteinase enhancer	-	44.0	-	-	PCOLCE	7q22
Pigment epithelium-derived factor	24.0	46.0	-	-	PEDF	17p13
Serum amyloid A1	-	-	-	218.7	SAA1	11p15.1
Thrombospondin-2	-	-	-	-2.8	THBS2	6q27
Versican	-	-27.0	-	-	VCAN	5q14
Cytoskeleton / contraction						
γ2-actin, enteric smooth muscle	-	30.0	-	9.4	ACTG2	2p13.1
Transgelin (smooth muscle22)	-	-	1.5	3.5	TAGLN	1q21-q25
Tropomyosin 2	2.1	-	-	2.1	TPM2	9p13
Protein biosynthesis/ modification						
Aldo-keto reductase 1C1	-	126.0	-	-	AKR1C1	10p15-p14
Aldo-keto reductase 1C3	0.6	43.0	-	1.5	AKR1C3	10p15-p14
Aldo-keto reductase 1B10	3.5	175.0	-	1.3	AKR1B10	7q33
Arginigosuccinate synthetase 1	-	28.0	-	-	ASS1	9q34
Asparagine synthetase	0.3	-	down	-	ASNS	7q21
Cellular retinoic acid binding prot 1	-	-2.5	-	-	CRABP1	15q24
Cellular retinoic acid binding prot 2	0.3	-	-	-3.1	CRABP2	1q21
Cystatin C	2.8	-	-	-	CST3	20p11
24-dehydrocholesterol reductase	2.1	-	-	2.5	DHCR24	1p33-p31
Lactate dehydrogenase A	-	-	1.4	1.2	LDHA	11p15
Monoamine oxidase	2.4	-	-	13.1	MAOA	Xp11.3
Tryptophanyl-tRNA synthetase	0.5	-	0.3	-	WARS	14q32
Signaling / cytokines						
Insul-like growth fact1(somatomedinC)	-	-	-	-13.1	IGF1	12q22-q23
Insul-like growth fact binding protein 2	4.9	57.0	1.7	2.1	IGFBP2	2q33-q34
Insul-like growth fact binding protein 4	2.5	3.7	-	-1.3	IGFBP4	17q12-q21
Latent TGFβ binding protein 2	3.1	-	-	2.2	LTBP2	14q24
Gremlin	-	-39.0	-	-	GREM1	15q13-q15
Prostaglandin D2 synthase	-	38.0	-	-	PTGDS	9q34-q34
Secretogranin II	0.4	13.9	-1.6	-3.7	SCG2	2q35-q36

Fig. 11.7 (Continued)

Function	7 days[a] MicroMax 2400 (cDNA)	10 days[b] AffyU95Av2 12,627 (oligo)	10 days[c] MicroMax 2400 (cDNA)	21 days[e] AffyU133A 22,215 (oligo)	Gene Symbol	Chromosome
Stress / Defense						
Heat shock 70kDa protein 5	-	-	0.5	-1.3	HSPA5	9q33-q34
Metallothionein 1L	-	-	2.2	-	MT1L	16q13
Metallothionein 1X	-	-	-	3.7	MT1X	16q13
Myocilin	24.0	50.0	4.3	16.7	MYOC	1q23-q24
Prostacyclin synthase	-	-35.0	-	-	PTGIS	20q13
Other						
Apolipoprotein D	-	22.0	-	5.3	APOD	3q26-qter
Citrate transport protein	-	23.0	-	-	CTP	22q11
cAMP response elem bind prot	-	-	0.6	-	ATF4	22q13
Growth arrest&DNA damage-indprot	-	-	0.8	-	DDIT3/CHOP10	12q13
Growth arrest-specific 1	-	23.0	2.4	-	GAS1	9q21-q22
Ferritin light chain	3.6	-	-	-	FTL	19q13
Mortality factor 4 like 2	2.0	-	-	1.5	MORF4L2	Xq22
Potasium voltage-gated chanel	0.5	-	-	2.4	KCNB1	20q13
BCL2/adenovirus E1B19k int prot 2	-	-2.2	-	-	BNIP2	15q22
BCL2/adenovirus E1B19k int prot 3	2.2	-	-	2.3	BNIP3	10q26
Thrombomodulin	-	-	-	3.0	THBD	20p11

Criteria: selected genes are those present in the most altered list of one report report
or altered in more than one report

a: Ishibashi *et al.* (2002)

b: Lo *et al*. (2003); HTM specific (HTM-DEX vs ONH-astrocytes-DEX)

c: Leung *et al*. (2003)

d: Rozsa *et al*. (2006)

-: not reported

Please refer to Fig. 11.7 in text for figure legend.

Gene name	Mech strain[a]	DEX[b]	TGBβ2[c]	Glaucoma Proteome[d]	Glaucoma mRNA[e]	Basic Function
Altered under four conditions						
Angiopoietin-like factor7/CDT6	●	●●	●	●	-	Antiangiogenesis/ECM deposit
Protein disulfide isomerase	●●	●	●	●	-	Protein folding
Superoxide dismutase	●	●	●	●	-	Destroys superoxide radicals
Tropomyosin*	●●	●●	●	●	-	Stabilizes actin filaments
Altered under three conditions						
Aldo-keto reductases*	●	●●●	●	-	-	Antioxidants
Calgranulin A (S100 protein)	●●	●	-	●	-	Calcium binding/cytoskelt regul
Calreticulin	-	●	●	●	-	Calcium homeostasis/chaperone
Chaperonin containing TCP1	●●●	●	-	●	-	Protein folding
Chitinase 3	●●●	●●	●	-	-	Cartilage ECM
Connective tissue growth fact	●●	●	●	-	-	Differentiation of chondrocytes
Cytochrome P450*	●●●	●	-	-	●	Detoxification
Heat-shock proteins 70 kDa*	●	●●	-	●	-	Chaperone
Insul growth fact* (1,2)	●	●	-	-	●	Signaling hormone
Insulin-like growth fact bind proteins*	●●●●	●●●●	●	-	-	Bind to IGF, increase its half life
Matrix Gla	●●●	-	●	-	●	Inhibitor of calcification
Mimecan (Osteoglycin)	●●	-	●	●	-	Induces bone formation w/TGFβ
Myocilin	●●●	●●●●	-	●	-	Stress protein
Transgelin (smooth muscle22)	●●●●	●●	●	-	-	Actin crosslinking
Thrombomodulin	●	●	●	-	-	Binds thrombin/inhibits clotting
Thrombospondin*	●	●	-	●	-	Cell-cell, cell-matrix interactions
Versican	●●	●	●	-	-	Cell-cell and cell-ECM signaling
Altered under two conditions						
Apolipoprotein D	●●	●●	-	-	-	Carrier protein/lipid transport
αB-crystallin	●●●	-	●	-	-	Chaperone
α-1-antichymotrypsin (serpin)	●	●●	-	-	-	Serine protease inhibitor
Cadherin* (2,4,10,19,24)	-	●●	-	-	●	Calcium depend cell-cell adhes
Chemokine (C-X-C motif)*	●●●	-	-	-	●	Signaling cytokine/immunity
Cochlin	●	-	-	●	-	ECM architecture (deafness)
Cornifin (small proline-rich B)	●●	-	-	●	-	Crosslink to membrane
Fibulin 1	-	●●●	-	●	-	Fibrillar ECM/calcium binding
Interleukin 6, IL6	●●	●	-	-	-	Signaling cytokine/immunity
Metallothioneins*	●●●●	●●	-	-	-	Metal binding/detoxification
Osteoblast specific factor 2	●●	-	●	-	-	Cell attach, spread, adhesion
Pigment epithelium-derived factor	●	●●	-	-	-	Inhibitor of angiogenensis
Secretogranin II	●●	●●●●	-	-	-	Neuropeptide precur/secretion
Selectin-E (ELAM-1)	-	-	-	●	●	Cell adhesion
Altered under one condition						
Biglycan	●●	-	-	-	-	Osteoblast differenciation
MMP3	●	-	-	-	-	ECM remodeling
Serum amyloid A1	-	●	-	-	-	ECM deposits/inflamm
Tropomodulin	●●	-	-	-	-	Actin depolimerization blocker
Substance P precursor	●●	-	-	-	-	Signaling neurotransmitter

●: Expression altered in one independent study; -: not reported or not altered ; functional categories color coded

☐ ECM; ☐ protein synth/modification; ☐ stress/defense; ☐ cytoskeleton; ☐ signaling/cytokines; ☐ calcification

a: From data on Figure 5.

b: From data on Figure 6

c: Zhao *et al*. (2004)

d: From data on Figure 3

e: Diskin *et al*. (2006); Liton *et al*. (2006)

*: family members pooled

Please refer to Fig. 11.8 in text for figure legend.

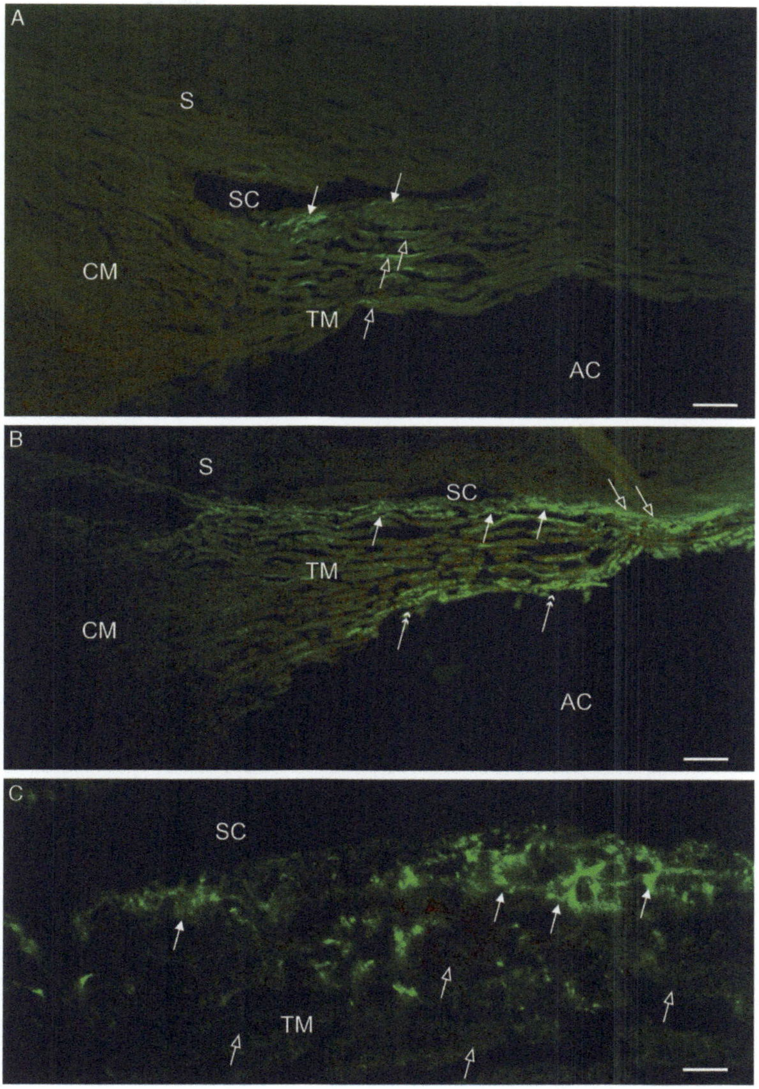

Please refer to Fig. 12.1 in text for figure legend.

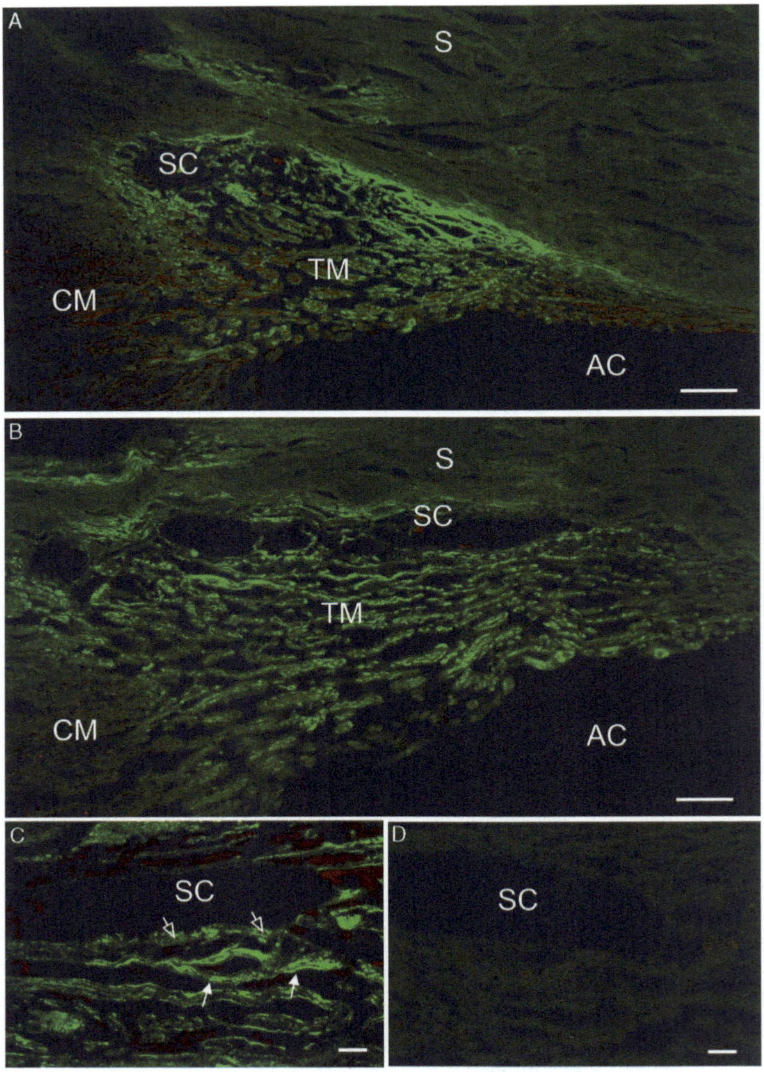

Please refer to Fig. 12.2 in text for figure legend.

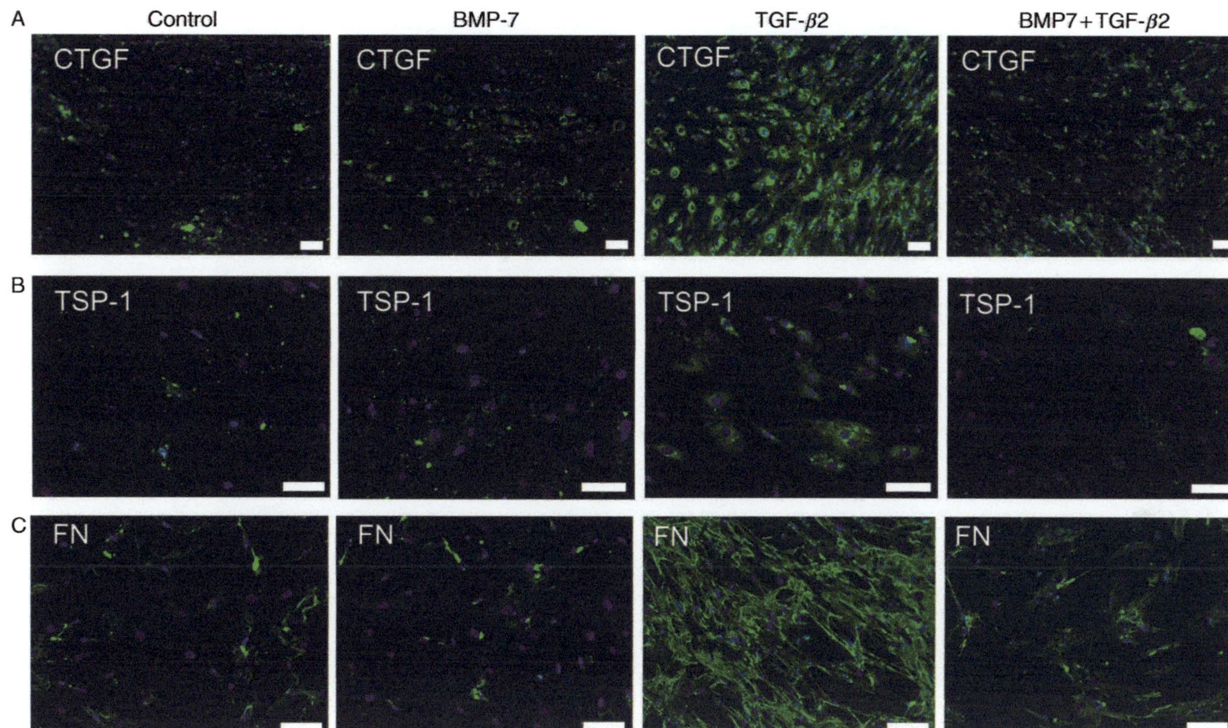

Please refer to Fig. 12.4 in text for figure legend.

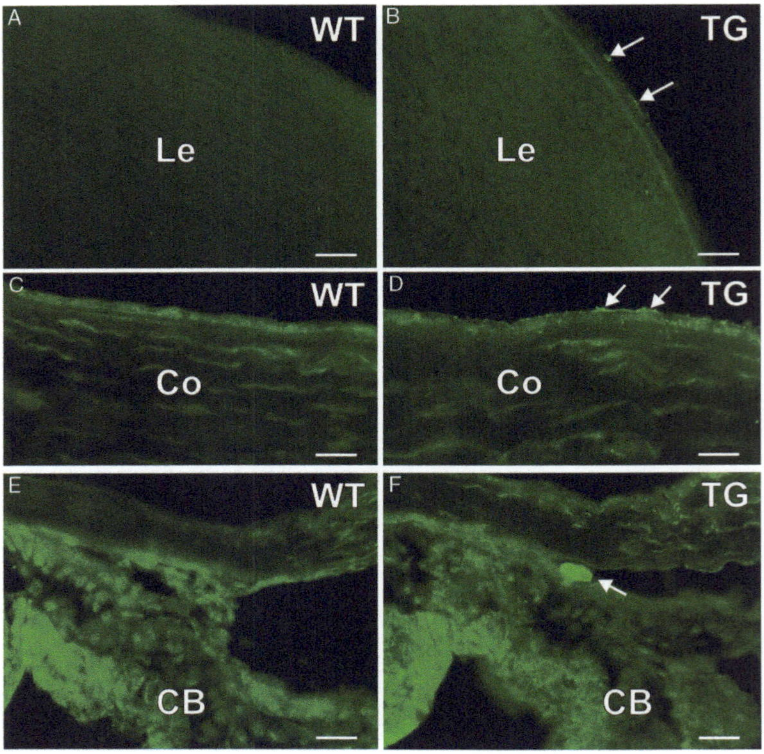

Please refer to Fig. 12.7 in text for figure legend.

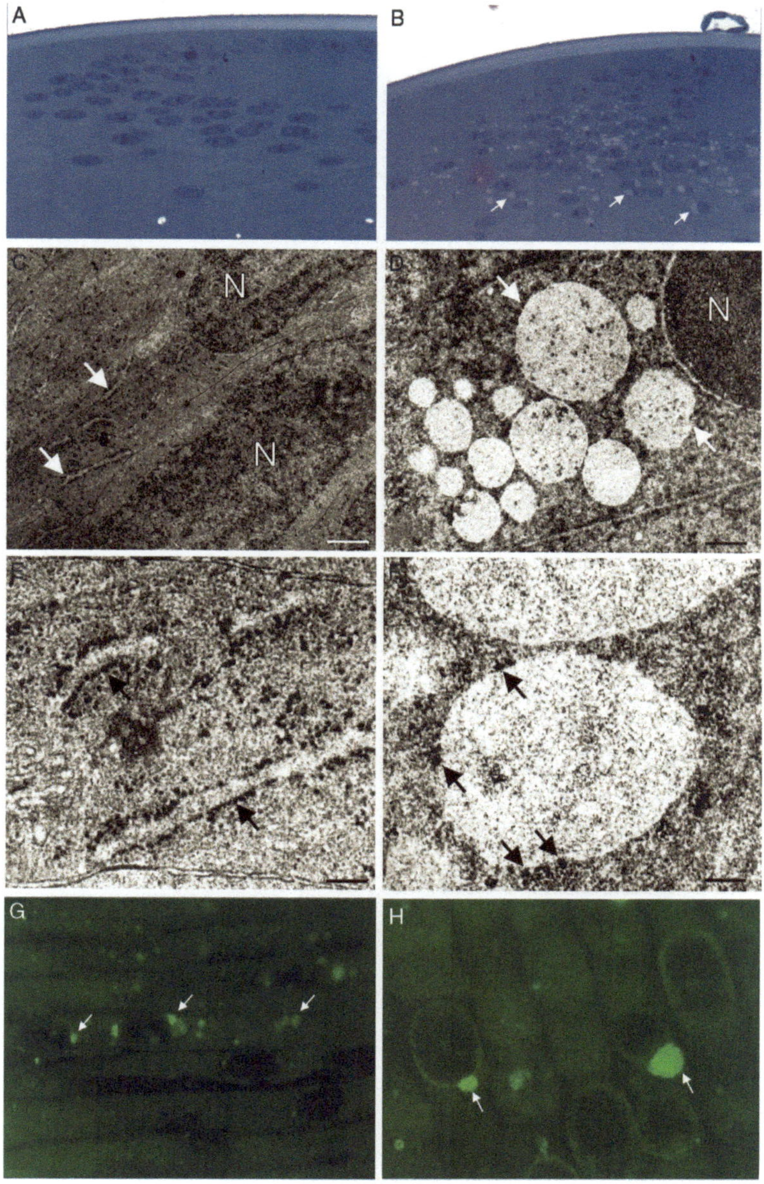

Please refer to Fig. 12.8 in text for figure legend.

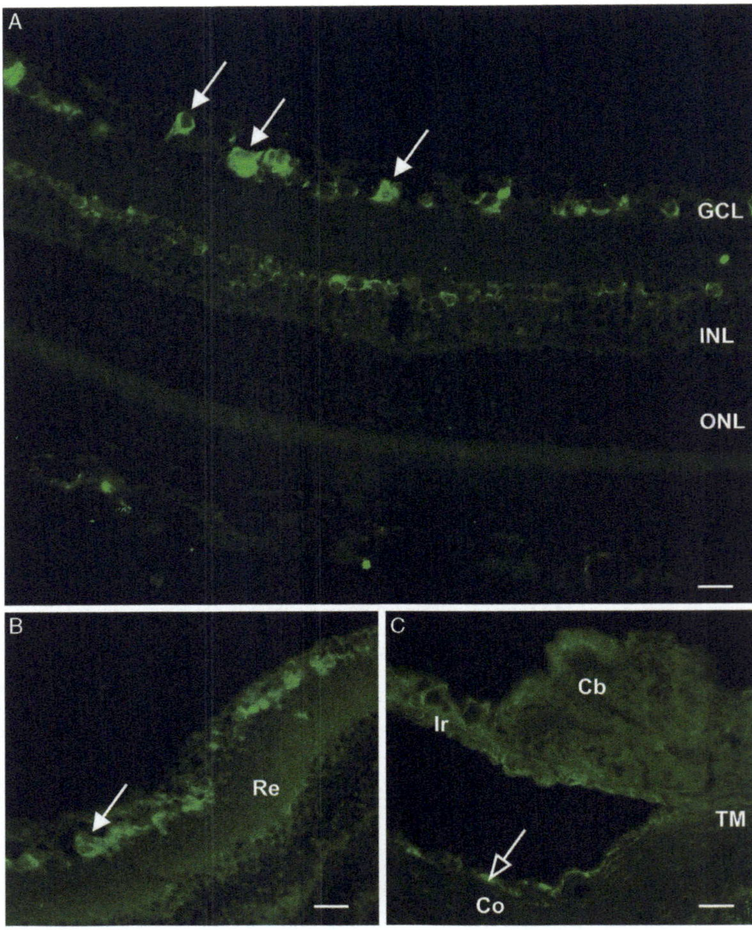

Please refer to Fig. 12.10 in text for figure legend.

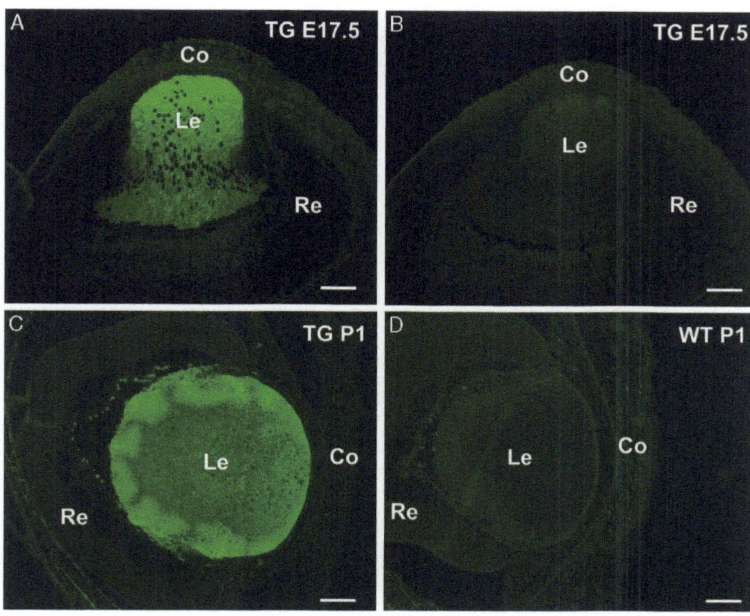

Please refer to Fig. 12.12 in text for figure legend.

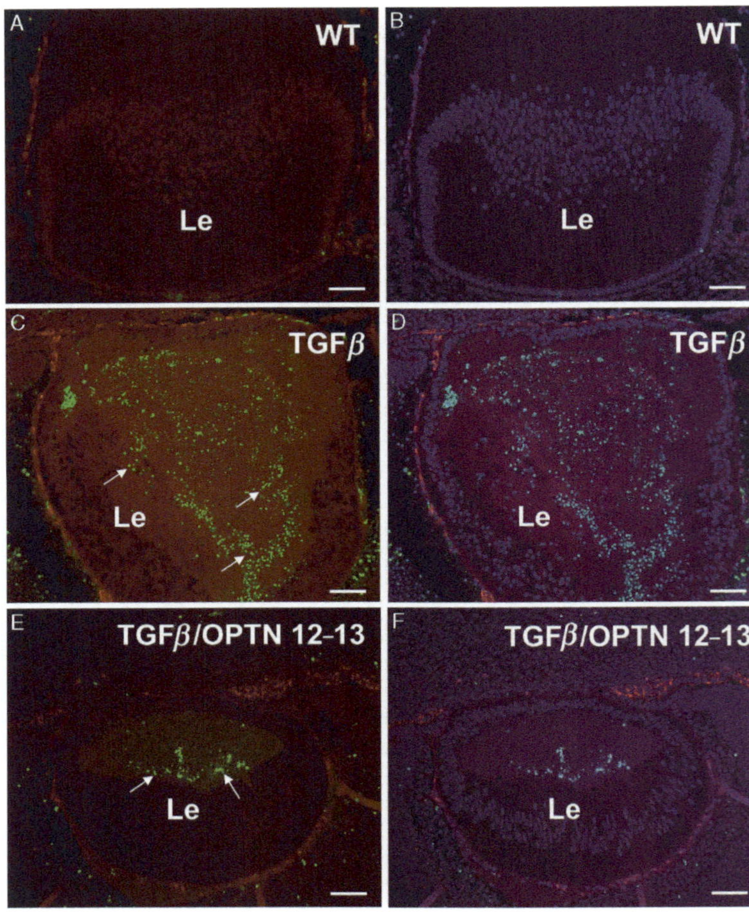

Please refer to Fig. 12.13 in text for figure legend.